Nanocellulose-Reinforced Thermoplastic Starch Composites

Also of interest

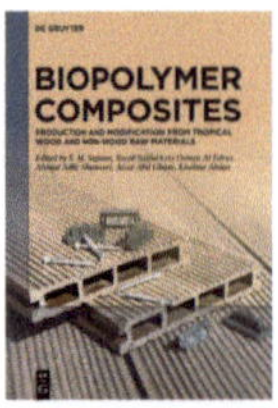

Biopolymer Composites
Production and Modification from Tropical Wood and Non-Wood Raw Materials
Sapuan, Al Edrus, Shamsuri, Abd Ghani, Abdan (Eds.), 2023
ISBN 978-3-11-076919-7, e-ISBN 978-3-11-076922-7

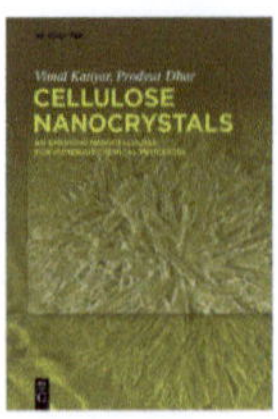

Cellulose Nanocrystals
An Emerging Nanocellulose for Numerous Chemical Processes
Katiyar, Dhar, 2020
ISBN 978-3-11-064452-4, e-ISBN 978-3-11-064801-0

Cellulose Composites
Processing and Characterization
Rakesh, Davim (Eds.), 2023
ISBN 978-3-11-076869-5, e-ISBN 978-3-11-076878-7

Polymeric Composites with Rice Hulls
An Introduction
Defonseka, 2019
ISBN 978-3-11-063968-1, 978-3-11-064320-6

Physical Sciences Reviews.
e-ISSN 2365-659X

Nanocellulose-Reinforced Thermoplastic Starch Composites

Sustainable Materials for Packaging

Edited by
R.A. Ilyas, S.M. Sapuan and Mohd Nor Faiz Norrahim

DE GRUYTER

Editors
Dr. R.A. Ilyas
Universiti Teknologi Malaysia
Faculty of Chemical and Energy Engineering
81310 UTM Johor Bahru
Malaysia

Prof. Ir. Dr. S.M. Sapuan
Universiti Putra Malaysia
Faculty of Engineering
43400 UPM Serdang Selangor
Malaysia

Dr. Mohd Nor Faiz Norrahim
Universiti Pertahanan Nasional Malaysia
Research Centre for Chemical Defence
Kem Perdana Sungai Besi
57000 Kuala Lumpur
Malaysia

ISBN 978-3-11-077356-9
e-ISBN (PDF) 978-3-11-077360-6
e-ISBN (EPUB) 978-3-11-077387-3

Library of Congress Control Number: 2023941014

Bibliographic information published by the Deutsche Nationalbibliothek
The Deutsche Nationalbibliothek lists this publication in the Deutsche Nationalbibliografie; detailed bibliographic data are available on the internet at http://dnb.dnb.de.

Cover image: AndreyGorulko/iStock/gettyimages
Typesetting: TNQ Technologies Pvt. Ltd.
Printing and binding: CPI books GmbH, Leck

www.degruyter.com

About the editors

R.A. Ilyas is a senior lecturer at the Faculty of Chemical and Energy Engineering, Universiti Teknologi Malaysia, Malaysia. He is also a Fellow of International Association of Advanced Materials (IAAM), Sweden, Fellow of International Society for Development and Sustainability (ISDS), Japan, a member of Royal Society of Chemistry, UK and Institute of Chemical Engineers (IChemE), UK. He received his Diploma in Forestry at Universiti Putra Malaysia, Bintulu Campus (UPMKB), Sarawak, Malaysia from Mei 2009 to April 2012. In 2012, he was awarded the Public Service Department (JPA) scholarship to pursue his bachelor's degree (BSc) in Chemical Engineering at Universiti Putra Malaysia (UPM). Upon completing his BSc. programme in 2016, he was again awarded the Graduate Research Fellowship (GRF) by the Universiti Putra Malaysia (UPM) to undertake a Ph.D. degree in the field of Biocomposite Technology & Design at the Institute of Tropical Forestry and Forest Products (INTROP) UPM. R.A. Ilyas was the recipient of the MVP Doctor of Philosophy Gold Medal Award UPM 2019, for Best Ph.D. Thesis and Top Student Award, INTROP, UPM. He was awarded with Outstanding Reviewer by Carbohydrate Polymers, Elsevier United Kingdom, Top Cited Article 2020-2021 Journal Polymer Composite, Wiley, 2022, and Best Paper Award at various International Conferences. R.A. Ilyas also was listed and awarded Among World's Top 2% Scientist (Subject-Wise) Citation Impact during the Single Calendar Year 2019, 2020, and 2021 by Stanford University, US, PERINTIS Publication Award 2021 and 2022 by Persatuan Saintis Muslim Malaysia, Emerging Scholar Award by Automotive and Autonomous Systems 2021, Belgium, Young Scientists Network - Academy of Sciences Malaysia (YSN-ASM) 2021, UTM Young Research Award 2021, UTM Publication Award 2021, and UTM Highly Cited Researcher Award 2021. In 2021, he won Gold Award and Special Award (Kreso Glavac (The Republic of Croatia) at the Malaysia Technology Expo (MTE2022), Gold Award dan Special Award at International Borneo Innovation, Exhibition & Competition 2022 (IBIEC2022), and a Gold Award at New Academia Learning Innovation (NALI2022). His main research interests are (1) Polymer Engineering (Biodegradable Polymers, Biopolymers, Polymer composites, Polymer-gels) and (2) Material Engineering (Natural fiber reinforced polymer composites, Biocomposites, Cellulose materials, Nanocomposites). To date he has authored or co-authored more than 431 publications (published/accepted): 188 Journals Indexed in JCR/ Scopus, 3 non-index Journal, 17 books, 104 book chapters, 78 conference proceedings/seminars, 4 research bulletins, 10 conference papers (abstract published in the book of abstract), 17 Guest Editor of Journal special issues and 10 Editor/ Co-Editor of Conference/Seminar Proceedings on green materials related subjects.

https://doi.org/10.1515/9783110773606-201

S.M. Sapuan is professor of composite materials at Universiti Putra Malaysia. He earned his B.Eng. degree in Mechanical Engineering from University of Newcastle, Australia in 1990, MSc from Loughborough University, UK in 1994 and Ph.D. from De Montfort University, UK in 1998. His research interests include natural fibre composites, materials selection, and concurrent engineering. To date he has authored or co-authored more than 1300 publications international journals (629 papers), books (17), edited books (13), chapters in books (91) and conference proceedings/seminars (597 papers). S.M. Sapuan was the recipient of Rotary Research Gold Medal Award 2012, The Alumni Medal for Professional Excellence Finalist, 2012 Alumni Awards, University of Newcastle, NSW, Australia, Khwarizmi International Award (KIA). In 2013 he was awarded with 5 Star Role Model Supervisor award by UPM. He has been awarded "Outstanding Reviewer" by Elsevier for his contribution in reviewing journal papers. He received Best Technical Paper Award in UNIMAS STEM International Engineering Conference in Kuching, Sarawak, Malaysia. S.M. Sapuan was recognized as the first Malaysian to be conferred Fellowship by the US-based Society of Automotive Engineers International (FSAE) in 2015. He was the 2015/2016 recipient of SEARCA Regional Professorial Chair. In 2016 ranking of UPM researchers based on the number of citations and h-index by SCOPUS, he is ranked the 6th from 100 researchers. In 2017, he was awarded with IOP Outstanding Reviewer Award by Institute of Physics, UK, National Book Award, The Best Journal Paper Award, UPM, Outstanding Technical Paper Award, Society of Automotive Engineers International, Malaysia, and Outstanding Researcher Award, UPM. He also received in 2017 Citation of Excellence Award from Emerald, UK, SAE Malaysia the Best Journal Paper Award, IEEE/TMU Endeavour Research Promotion Award, Best Paper Award by Chinese Defence Ordnance and Malaysia's Research Star Award (MRSA), from Elsevier.

M.N.F. Norrahim has research interests in nanotechnology, composites, materials science, polymers, chemical and biological defense, and biotechnology. He has conducted various studies on the benefits of nanocellulose and its applications. During his doctoral studies, he improved the nanocellulose production from oil palm biomass using mechanical treatment. Now, his research is focused on using nanocellulose for applications such as composites, adsorbents, and military. His expertise in this field has been proven by the high-impact publications he has produced related to nanocellulose research. To date, he has authored or co-authored more than 100 articles, including 40 book chapters in renowned journals on nanotechnology, materials science, chemistry, and biotechnology-related subjects. He also published 1 book in Elsevier which entitled "Industrial Application of Nanocellulose and its Nanocomposites". He has received several local and international innovation awards. His current H-index is Google Scholar is 34 and Scopus 29. Besides that, he has presented his research findings at several local and international conferences. He also received several innovation awards.

Preface

Researchers estimate that more than 8.3 billion tons of plastic have been produced since the early 1950s. Only 9% of all plastic waste ever produced has been recycled. About 12% has been incinerated, while the rest—79%—has accumulated in landfills, dumps, or the natural environment. A staggering 8 million tons of plastic end up in the world's oceans every year. How does it get there? A lot of it comes from the world's rivers, which serve as direct conduits of trash from the world's cities to the marine environment. Thus, ecological and health concerns are driving research efforts for developing biodegradable films. There are few alternatives that could reduce the environmental impact; and one of them is by replacing synthetic petroleum-based plastic with natural starch-based film. Thermoplastic starch has remarkable advantages, including abundance and ease availability, low cost, sustainability, and biodegradability, and capable of being modified or blended with other polymers. Nevertheless, low water resistance as well as low mechanical strength restrict its application in food packaging. On the other side, nanocellulose isolated from lignocellulosic fibers has attracted huge attention in the field of engineering and science due to its high aspect ratio, mechanical strength, thermal and crystallinity, unique morphology of hydroxyl group along with abundancy, biodegradability, and renewability. Nanocellulose as a reinforcer had proved to be a good option for fabricating bio nanocomposites for food packaging. This book will give a critical overview on the potential application of nanocellulose in food packaging and discuss new challenges and opportunities for starch biocomposites reinforced with nanocellulose.

- Comprehensively covers development and characterization of various starch biopolymer reinforced with nanocellulose that have huge potentially to replace conventional petroleum-based packaging.
- Includes commentary from leading industrial and academic experts in the field who present cutting-edge research on starch-biobased materials for packaging industry.
- Includes new potential starch-based materials to be used for food packaging applications for example Araucaria araucana, banana, barley, cassava, corn starch, horse chestnut, oat, pea, potato, sago, rice, sugar palm and wheat.
- Easily to be understand by readers from various background.
- Book covers economic issues of starch biobased packaging including socio-economic impact of biobased packaging bags life cycle cost analysis of biobased packaging products and the market for bio-based packaging: how consumers perceptions and preferences regarding bio-based packaging.

https://doi.org/10.1515/9783110773606-202

- Book covers economic issues and environmental issues of starch biobased packaging, the future of sustainable packaging, renewable source for packaging material, environmental advantages and challenges of biobased packaging materials, and life cycle assessment of biobased packaging products.

The Editors,
August 2023

Contents

List of contributing authors

W. Abotbina
Mechanical and Industrial Engineering Department
Faculty of Engineering
Al-Asmarya Islamic University
Zliten
Libya

Hairul Abral
Department of Mechanical Engineering
Universitas Negeri Padang
Padang 25173
Indonesia
And
Laboratory of Nanoscience and Technology
Department of Mechanical Engineering
Andalas University
Padang 25163
Indonesia

Nadia Adrus
Faculty of Chemical & Energy Engineering
Universiti Teknologi Malaysia
Skudai 81310
Malaysia

Mohd Shahrulnizam Ahmad
Faculty of Chemical & Energy Engineering
Universiti Teknologi Malaysia
Skudai 81310
Malaysia

Faris M. AL-Oqla
Department of Mechanical Engineering
Faculty of Engineering
The Hashemite University
P.O. Box 330127
Zarqa, 13133
Jordan

Fathilah Ali
Department of Chemical Engineering and Sustainability
Kulliyyah of Engineering, International Islamic University of Malaysia
P.O. Box 10
50728 Kuala Lumpur
Malaysia

Roshafima Rasit Ali
Department of Chemical and Environmental Engineering
Malaysia-Japan International Institute of Technology (MJIIT)
Universiti Teknologi Malaysia
Jalan Sultan Yahya Petra
54100 Kuala Lumpur
Malaysia

Nur Amalina Amirullah
Faculty of Science
University of Malaya
Institute of Biological Sciences
50603 Kuala Lumpur
Malaysia

Mochamad Asrofi
Department of Mechanical Engineering
Faculty of Engineering
Universitas Jember
Kalimantan 37
Jember 68121
Indonesia

Muhammad Rizal Muhammad Asyraf
Engineering Design Research Group (EDRG)
School of Mechanical Engineering
Faculty of Engineering
Universiti Teknologi Malaysia
81310 Johor Bahru, Johor
Malaysia
And
Centre for Advanced Composite Materials (CACM)
Universiti Teknologi Malaysia
81310 UTM Johor Bahru, Johor
Malaysia

M. S. N. Atikah
Department of Chemical and Environmental Engineering
Universiti Putra Malaysia
43400 UPM Serdang, Selangor
Malaysia

https://doi.org/10.1515/9783110773606-203

A. Atiqah
Institute of Microengineering and Nanoelectronics
Universiti Kebangsaan Malaysia
Bangi
43600, Selangor
Malaysia

Mohammad Saifulddin Mohd Azami
Faculty of Applied Sciences
Universiti Teknologi MARA (UiTM)
02600 Arau, Perlis
Malaysia
E-mail: saifulddin@uitm.edu.my

Ashraf Azmi
School of Chemical Engineering
College of Engineering
Universiti Teknologi MARA
40450, Shah
Alam, Selangor
Malaysia

Mohamad Irsyad Mohamad Azwardi
Department of Mechanical and Manufacturing Engineering
Advanced Engineering Materials and Composites (AEMC) Research Centre
Universiti Putra Malaysia (UPM)
Serdang, Selangor 43400
Malaysia

Nurul Huda Baharuddin
Faculty of Chemical and Energy Engineering
Universiti Teknologi Malaysia (UTM)
Skudai 81310
Malaysia
E-mail: nurul.huda@utm.my

Sneh Punia Bangar
Department of Food, Nutrition and Packaging Sciences
Clemson University
Clemson
USA

Claudia Barile
Department of Mechanics
Mathematics and Management
Polytechnic University of Bari
Via Orabona n.4
Bari 70125
Italy

Eduardo Valério de Barros Vilas Boas
Food Science Department
Federal University of Lavras
37200-900, Lavras
Minas Gerais
Brazil
E-mail: evbvboas@ufla.br

Julia Buchner
Laboratoire Innovation Matériau Bois Habitat (LIMBHA)
Ecole Supérieure du Bois
7 Rue Christian Pauc
44306 Nantes
France
E-mail: julia.buchner@esb-campus.fr

Quoc-Bao Bui
Sustainable Developments in Civil Engineering Research Group
Faculty of Civil Engineering
Ton Duc Thang University
Ho Chi Minh City
Vietnam
E-mail: buiquocbao@tdtu.edu.vn

Febio Dalanta
Department of Chemical Engineering
Faculty of Engineering
Diponegoro University
Semarang, 50275
Indonesia

Thi My Hanh Diep
Vietnam National University
Ho Chi Minh City, 700000
Vietnam
And
Center Research for Natural Resources Conservation
University of Science
Ho Chi Minh City
Vietnam
E-mail: mhg286@gmail.com

Muhammad Aqil Mohd Farizal
Faculty of Chemical & Energy Engineering
Universiti Teknologi Malaysia
Skudai 81310
Malaysia

Siti Khairunisah Ghazali
Faculty of Chemical & Energy Engineering
Universiti Teknologi Malaysia
Skudai 81310
Malaysia

A.A.N. Gunny
Centre of Excellence for Biomass Utilization
Universiti Malaysia Perlis
02600 Arau, Perlis
Malaysia
And
Faculty of Chemical Engineering Technology
Universiti Malaysia Perlis
Kompleks Pusat Pengajian Jejawi 3
Kawasan Perindustrian Jejawi
02600 Arau, Perlis
Malaysia

M. Y. S. Hamid
Department of Chemical Engineering
Faculty of Chemical and Energy Engineering
Universiti Teknologi Malaysia
81310 UTM Johor Bahru
Johor
Malaysia

Nur Izzah Nabilah Haris
Institute of Sustainable and Renewable Energy
Universiti Malaysia Sarawak
Sarawak
Malaysia

Moklis Muhammad Harussani
Energy Science and Engineering
Department of Transdisciplinary Science and Engineering
School of Environment and Society
Tokyo Institute of Technology
Meguro 152-8552
Tokyo
Japan

Rosnani Hasham
Faculty of Chemical & Energy Engineering
Universiti Teknologi Malaysia
Skudai 81310
Malaysia

C. S. Hassan
Faculty of Engineering
Technology and Built Environment
UCSI University
Kuala Lumpur
Malaysia

S. A. Hassan
Centre for Advanced Composite Materials
Universiti Teknologi Malaysia
81310 UTM Johor Bahru
Johor
Malaysia

Husain Siti Nor Hawanis
Department of Chemical Engineering
Faculty of Chemical and Energy Engineering
Universiti Teknologi Malaysia (UTM)
Skudai 81310, Johor
Malaysia

K. Z. Hazrati
German-Malaysian Institute
Taman Universiti
Jalan Ilmiah
Kajang 43000
Malaysia

Maria Lucila Hernández-Macedo
Industrial Biotechnology Graduation
Tiradentes University
Av. Murilo Dantas 300
49032-490, Aracaju, SE
Brazil
And
Molecular Biology Laboratory
Research and Technology Institute - ITP
Av. Murilo Dantas 300
Aracaju, SE
Brazil
E-mail: lucyherma@gmail.com
https://orcid.org/0000-0003-1050-9807

DongQuy Hoang
Faculty of Materials Science and Technology
University of Science
Ho Chi Minh City, 700000
Vietnam
And
Vietnam National University
Ho Chi Minh City, 700000
Vietnam
E-mail: htdquy@hcmus.edu.vn

Hazwani Husin
Research and Development Department
Nextgreen Pulp & Paper Sdn Bhd
Taman Tun Dr Ismail
60000 Kuala Lumpur
Malaysia
E-mail: hazwani.husin@nextgreenglobal.com

M.R.M. Huzaifah
Department of Crop Science
Faculty of Agricultural and Forestry Sciences
Universiti Putra Malaysia Bintulu Campus (UPMKB)
Bintulu, Sarawak
Malaysia

I. Idris
School of Chemical Engineering
College of Engineering
Universiti Teknologi MARA (UiTM)
40450 Shah
Alam, Selangor
Malaysia

Rushdan Ahmad Ilyas
Faculty of Chemical and Energy Engineering
Universiti Teknologi Malaysia (UTM)
Skudai 81310, Johor
Malaysia
And
Centre for Advanced Composite Materials (CACM)
Universiti Teknologi Malaysia (UTM)
Skudai 81310, Johor
Malaysia
And
Centre of Excellence for Biomass Utilisation
Universiti Malaysia Perlis
02600, Arau, Perlis
Malaysia
And
Institute of Tropical Forestry and Forest Products
Universiti Putra Malaysia
Serdang 43400, Selangor
Malaysia
E-mail: ahmadilyas@utm.my
https://orcid.org/0000-0001-6622-2632

Aleif Hakimi Ismail
Department of Mechanical and Manufacturing Engineering
Advanced Engineering Materials and Composites (AEMC) Research Centre
Universiti Putra Malaysia (UPM)
Serdang, Selangor 43400
Malaysia

Jamarosliza Jamaluddin
Faculty of Chemical & Energy Engineering
Universiti Teknologi Malaysia
Skudai 81310
Malaysia
E-mail: jamarosliza@utm.my
https://orcid.org/0000-0002-1147-7360

Nurjahirah Janudin
Research Centre for Chemical Defence
Universiti Pertahanan Nasional Malaysia
Kem Perdana Sungai Besi
57000 Kuala Lumpur
Malaysia

Mohd Azwan Jenol
Department of Bioprocess Technology
Faculty of Biotechnology and Biomolecular Sciences
Universiti Putra Malaysia
UPM, 43400 Serdang, Selangor
Malaysia

Dzun Noraini Jimat
Department of Biotechnology Engineering
Kulliyyah of Engineering
International Islamic University of Malaysia
53100, Selangor
Malaysia
E-mail: jnoraini@iium.edu.my

Ridhwan Jumaidin
Fakulti Teknologi Kejuruteraan Mekanikal dan Pembuatan
Universiti Teknikal Malaysia
Melaka, Hang Tuah Jaya
Durian Tunggal 76100
Malaysia

Suriani Mat Jusoh
Faculty of Ocean Engineering Technology and Informatics
Universiti Malaysia Terengganu
21030 Kuala Nerus, Terengganu
Malaysia

A. Khalina
Institute of Tropical Forestry and Forest Products
Universiti Putra Malaysia
Serdang 43400, Selangor
Malaysia

P. S. Khoo
Centre for Advanced Composite Materials
Universiti Teknologi Malaysia
81310 UTM Johor Bahru
Johor
Malaysia

Rafael Carvalho do Lago
Food Science Department
Federal University of Lavras
37200-900, Lavras
Minas Gerais
Brazil

Jorge Alberto López
Multidisciplinary Research Laboratory
Health Sciences Center
Federal University of Rio Grande do Norte
R. Gen. G. Cordeiro de Farias
s/n
59012-570, Natal RN
Brazil

Melbi Mahardika
Research Center for Biomass and Bioproducts
National Research and Innovation Agency (BRIN)
Cibinong 16911
Indonesia

Rohah A. Majid
Faculty of Chemical & Energy Engineering
Universiti Teknologi Malaysia
Skudai 81310
Malaysia

Mazween Mohamad Mazlan
Department of Process and Food Engineering,
Faculty of Engineering
Universiti Putra
Mukah
Malaysia

Dominic C. D. Midhun
Department of Chemistry
Sacred Heart College
Thevara
Pin 682013
Kochi, Kerala
India

Isabella Cícera Dias Miranda
Industrial Biotechnology Graduation
Universidade Tiradentes
Av. Murilo Dantas 300
49032-490, Aracaju, SE
Brazil
And
Molecular Biology Laboratory
Research and Technology Institute - ITP
Av. Murilo Dantas 300
Aracaju, SE
Brazil

Zurina Mohamad
Faculty of Chemical & Energy Engineering
Universiti Teknologi Malaysia
Skudai 81310
Malaysia

Dayangku Intan Munthoub
Faculty of Chemical & Energy Engineering
Universiti Teknologi Malaysia
Skudai 81310
Malaysia

Anne Aleesa Nazree
Faculty of Chemical & Energy Engineering
Universiti Teknologi Malaysia
Skudai 81310
Malaysia

Asmawi Nazrin
Advanced Engineering Materials and Composites Research Center (AEMC)
Universiti Putra Malaysia
Faculty of Engineering
43400 Seri Kembangan, Selangor
Malaysia
And
Laboratory of Biocomposite Technology
Institute of Tropical Forestry and Forest Products (INTROP)
Universiti Putra Malaysia
43400 UPM Serdang, Selangor
Malaysia
And
Institute of Power Engineering
Universiti Tenaga Nasional
Jalan IKRAM-UNITE
Kajang 43000, Selangor
Malaysia

Norzita Ngadi
Faculty of Chemical and Energy Engineering
Universiti Teknologi Malaysia (UTM)
Skudai 81310
Malaysia
E-mail: norzita@cheme.utm.my

Dang Mao Nguyen
Laboratoire Innovation Matériau Bois Habitat (LIMBHA)
Ecole Supérieure du Bois
7 Rue Christian Pauc
44306 Nantes
France
E-mail: mao.nguyen@esb-campus.fr.
https://orcid.org/0000-0002-0304-9227

Yusilawati Ahmad Nor
Department of Biotechnology Engineering
Kulliyyah of Engineering
International Islamic University of Malaysia
53100, Selangor
Malaysia

Abu Hassan Nordin
Department of Chemical Engineering
Faculty of Chemical and Energy Engineering
Universiti Teknologi Malaysia
81310 UTM Johor Bahru, Johor
Malaysia
And
Faculty of Applied Sciences
Universiti Teknologi MARA (UiTM)
02600 Arau, Perlis
Malaysia
E-mail: abuhassannordin@gmail.com
https://orcid.org/0000-0001-7546-3833

Muhammad Luqman Nordin
Department of Clinical Studies
Faculty of Veterinary Medicine
Universiti Malaysia Kelantan
Pengkalan Chepa
16100 Kota Bharu, Kelantan
Malaysia
E-mail: luqman.n@umk.edu.my

A. S. Norfarhana
Department of Chemical Engineering
Faculty of Chemical and Energy Engineering
Universiti Teknologi Malaysia
81310 UTM Johor Bahru, Johor
Malaysia

Mohd Nor Faiz Norrrahim
Green Polymer, Coatings and Packaging Cluster
School of Industrial Technology
Universiti Sains Malaysia
11800, Pulau Pinang
Malaysia
And
Research Center for Chemical Defence
Universiti Pertahanan Nasional Malaysia
UPNM Kem Perdana Sungai Besi
57000, Kuala Lumpur
Malaysia
E-mail: faiz@upnm.edu.my;
faiznorrrahim@gmail.com
https://orcid.org/0000-0003-2101-5642

Y. Nukman
Department of Mechanical Engineering
Faculty of Engineering
University of Malaya
50603 Kuala Lumpur
Malaysia

Norizan Mohd Nurazzi
Bioresource Technology Division
School of Industrial Technology
Universiti Sains Malaysia
11800 Pulau Pinang
Malaysia

Ana Lázara Matos de Oliveira
Food Science Department
Federal University of Lavras
37200-900, Lavras
Minas Gerais
Brazil

Abdoulhdi A. Borhana Omran
Department of Mechanical and Mechatronic Engineering
Faculty of Engineering
Sohar University
Sohar, P C-311
Oman

S. H. Othman
Department of Process and Food Engineering
Universiti Putra Malaysia
UPM
43400 Serdang, Selangor
Malaysia

Noor Illi Mohamad Puad
Department of Biotechnology Engineering
Kulliyyah of Engineering
International Islamic University of Malaysia
53100, Selangor
Malaysia

S. A. Rafiqah
Institute of Tropical Forest and Forest Products (INTROP)
Universiti Putra Malaysia
43400 UPM Serdang
Selangor
Malaysia

L. Rajeshkumar
Centre for Materials and Manufacturing Technology
KPR Institute of Engineering and Technology
Coimbatore 641407
Tamilnadu
India

Mohd Saiful Asmal Rani
School of Materials and Mineral Resources Engineering
Universiti Sains Malaysia
14300 Nibong Tebal
Pulau Pinang
Malaysia

R. Ridwan
Department of Environmental Engineering
Universitas Andalas
Padang
West Sumatera, 25163
Indonesia

Muhammad Huzaifah Mohd Roslim
Department of Crop Science
Universiti Putra Malaysia
Serdang 43400
Malaysia

Mohd Hafif Samsudin
Department of Bioprocess Technology
Faculty of Biotechnology and Biomolecular Sciences
Universiti Putra Malaysia
43400 UPM Serdang
Selangor
Malaysia

M. L. Sanyang
School of Agriculture and Environmental Science
University of The Gambia
P.O. Box 3530
MDI Road
Kanifing
The Gambia
And
Directorate of Research and Consultancy
University of The Gambia
P.O. Box 3530
MDI Road
Kanifing
The Gambia

Salit Mohd Sapuan
Advanced Engineering Materials and Composites Research Center (AEMC)
Universiti Putra Malaysia
Faculty of Engineering
43400 Seri Kembangan, Selangor
Malaysia
And
Laboratory of Biocomposite Technology
Institute of Tropical Forestry and Forest Products (INTROP)
Universiti Putra Malaysia
43400 UPM Serdang, Selangor
Malaysia
E-mail: sapuan@upm.edu.my

Nasmi Herlina Sari
Mechanical Engineering Department
Faculty of Engineering
University of Mataram
Mataram, West Nusa Tenggara
Indonesia

Nur Sharmila Sharip
Research and Development Department
Nextgreen Pulp & Paper Sdn Bhd
Taman Tun Dr Ismail
60000 Kuala Lumpur
Malaysia
E-mail: sharmila.sharip@nextgreenglobal.com
https://orcid.org/0000-0003-3322-9592

Vasi Uddin Siddiqui
Department of Mechanical and Manufacturing Engineering
Advanced Engineering Materials and Composites (AEMC) Research Centre
Universiti Putra Malaysia (UPM)
Serdang, Selangor 43400
Malaysia

S. Silviana
Department of Chemical Engineering
Faculty of Engineering
Diponegoro University
Semarang, 50275
Indonesia
E-mail: silviana@che.undip.ac.id

Noorashikin Soh Zularifin Soh
Department of Mechanical and Manufacturing Engineering
Advanced Engineering Materials and Composites (AEMC) Research Centre
Universiti Putra Malaysia (UPM)
Serdang, Selangor 43400
Malaysia

R. M. O. Syafiq
Department of Engineering Design and Manufacture
Faculty of Engineering
University of Malaya
50603 Kuala Lumpur
Malaysia
And
Laboratory of Biocomposite Technology
Institute of Tropical Forestry and Forest Products (INTROP)
Universiti Putra Malaysia
UPM
43400 Serdang, Selangor
Malaysia

Edi Syafri
Departement of Agricultural Technology
Politeknik Pertanian Negeri Payakumbuh
Payakumbuh
West Sumatera, 26271
Indonesia

Zatil Izzah Ahmad Tarmizi
Department of Chemical and Environmental Engineering
Malaysia-Japan International Institute of Technology (MJIIT)
Universiti Teknologi Malaysia
Jalan Sultan Yahya Petra
54100 Kuala Lumpur
Malaysia

Intan Syafinaz Mohamed Amin Tawakkal
Department of Process and Food Engineering
Universiti Putra Malaysia
43400 UPM Serdang, Selangor
Malaysia

Nurfatimah Mohd Thani
Department of Food Sciences
Faculty of Science and Technology
Universiti Kebangsaan Malaysia
UKM Bangi
Selangor, Malaysia
And
Innovation Center for Confectionery Technology
Faculty of Science and Technology
Universiti Kebangsaan Malaysia
UKM Bangi, Selangor
Malaysia
E-mail: nurfatimah@ukm.edu.my

S. M. K. Thiagamani
Department of Mechanical Engineering
Kalasalingam Academy of Research and Education
Srivilliputhur
India

Thien Huu Tran
VNUK - Institute for Research and Executive Education
The University of Da Nang
Da Nang
Vietnam
E-mail: thien.tran180316@vnuk.edu.vn

M. N. A. Uda
Institute of Nano Electronic Engineering
Universiti Malaysia Perlis
Kangar 01000
Malaysia

Farhana Aziz Ujang
Research and Development Department
Nextgreen Pulp & Paper Sdn Bhd
Taman Tun Dr. Ismail
60000 Kuala Lumpur
Malaysia

Mohd Hafizz Wondi
Faculty of Plantation and Agrotechnology
Universiti Teknologi MARA
Mukah, Sarawak
Malaysia

Tengku Arisyah Tengku Yasim-Anuar
Research and Development Department
Nextgreen Pulp & Paper Sdn Bhd
Taman Tun Dr Ismail
60000 Kuala Lumpur
Malaysia
E-mail: tengkuarisyah@nextgreenglobal.com

Mohamed Yusoff Mohd Zuhri
Advanced Engineering Materials and Composites Research Centre (AEMC)
Department of Mechanical and Manufacturing Engineering
Universiti Putra Malaysia
43400 UPM Serdang, Selangor
Malaysia
And
Laboratory of Biocomposite Technology
Institute of Tropical Forestry and Forest Products (INTROP)
Universiti Putra Malaysia
43400 UPM Serdang, Selangor
Malaysia
https://orcid.org/0000-0002-1069-7345

Asmawi Nazrin, Salit Mohd Sapuan, R. A. Ilyas*, H. S. N. Hawanis, A. Khalina, Ridhwan Jumaidin, M. R. M. Asyraf, N. Mohd Nurazzi, M. N. F. Norrrahim, L. Rajeshkumar and M. S. N. Atikah

1 Introduction to bio-based packaging materials

Abstract: Bio-based materials must be studied to replace polymers from petrochemical sources in packaging applications. However, using polymers from petrochemical sources has caused consumer and environmental concerns. Therefore, synthetic and non-synthetic materials that can be used for packaging applications, scale-up methods, industrial uses, sustainability assessments, and end-of-life alternatives will all be included in this study. Synthetic polymers, e.g., polylactic acid (PLA), polyethylene furanoate (PEF), polybutylene succinate (PBS), and non-synthetic polymers, including waxes, lipids,

***Corresponding author: R. A. Ilyas**, Faculty of Chemical and Energy Engineering, Universiti Teknologi Malaysia (UTM), Skudai 81310, Johor, Malaysia; Centre for Advanced Composite Materials (CACM), Universiti Teknologi Malaysia (UTM), Skudai 81310, Johor, Malaysia; Centre of Excellence for Biomass Utilisation, Universiti Malaysia Perlis, 02600, Arau, Perlis, Malaysia; and Institute of Tropical Forestry and Forest Products, Universiti Putra Malaysia, Serdang 43400, Selangor, Malaysia, E-mail: ahmadilyas@utm.my. https://orcid.org/0000-0001-6622-2632

Asmawi Nazrin, Advanced Engineering Materials and Composites Research Center (AEMC), Universiti Putra Malaysia, Faculty of Engineering, 43400 Seri Kembangan, Selangor, Malaysia; and Institute of Power Engineering, Universiti Tenaga Nasional, Jalan IKRAM-UNITEN, Kajang 43000, Selangor, Malaysia

Salit Mohd Sapuan, Advanced Engineering Materials and Composites Research Center (AEMC), Universiti Putra Malaysia, Faculty of Engineering, 43400 Seri Kembangan, Selangor, Malaysia

H. S. N. Hawanis, Department of Chemical Engineering, Faculty of Chemical and Energy Engineering, Universiti Teknologi Malaysia (UTM), Skudai 81310, Johor, Malaysia

A. Khalina, Institute of Tropical Forestry and Forest Products, Universiti Putra Malaysia, Serdang 43400, Selangor, Malaysia

Ridhwan Jumaidin, Fakulti Teknologi Kejuruteraan Mekanikal dan Pembuatan, Universiti Teknikal Malaysia Melaka, Hang Tuah Jaya, Durian Tunggal 76100, Malaysia

M. R. M. Asyraf, Centre for Advanced Composite Materials (CACM), Universiti Teknologi Malaysia (UTM), Skudai 81310, Johor, Malaysia; and Faculty of Mechanical Engineering, Universiti Teknologi Malaysia (UTM), Skudai 81310, Johor, Malaysia

N. Mohd Nurazzi, Bioresource Technology Division, School of Industrial Technology, Universiti Sains Malaysia, 11800, Pulau Pinang, Malaysia

M. N. F. Norrrahim, Green Polymer, Coatings and Packaging Cluster, School of Industrial Technology, Universiti Sains Malaysia, 11800, Pulau Pinang, Malaysia; and Research Center for Chemical Defence, Universiti Pertahanan Nasional Malaysia, UPNM Kem Perdana Sungai Besi, 57000, Kuala Lumpur, Malaysia

L. Rajeshkumar, Centre for Materials and Manufacturing Technology, KPR Institute of Engineering and Technology, Coimbatore 641407, Tamilnadu, India

M. S. N. Atikah, Department of Chemical and Environmental Engineering, Universiti Putra Malaysia, 43400 UPM Serdang, Selangor, Malaysia

As per De Gruyter's policy this article has previously been published in the journal Physical Sciences Reviews. Please cite as: A. Nazrin, S. M. Sapuan, R. A. Ilyas, H. S. N. Hawanis, A. Khalina, R. Jumaidin, M. R. M. Asyraf, N. Mohd Nurazzi, M. N. F. Norrrahim, L. Rajeshkumar and M. S. N. Atikah "Introduction to bio-based packaging materials" *Physical Sciences Reviews* [Online] 2023. DOI: 10.1515/psr-2022-0006 | https://doi.org/10.1515/9783110773606-001

proteins, starch, cellulose, and polyhydrodialkanoate (PHAs), are some of the bio-based compounds that will be covered in this work. Besides that, more attention is paid to surface modification techniques and coatings, multilayers, biocomposites, and additives used to modify material characteristics, particularly gas and moisture barriers and biodegradability. In sum, this research offers a comprehensive analysis of bio-based packaging materials, including processing, and an assessment of sustainability and available alternatives.

Keywords: bio-based packaging; cellulose; nanocellulose; starch.

1.1 Introduction

On a global scale, plastic usage in all sorts of applications is rapidly growing. The advancement of technology has developed easier and cheaper ways to manufacture plastic materials and products. The key function of packaging material is to preserve the food's nutritional value after the processing is completed and to maintain its condition during transportation and storage in the destined location. Nevertheless, the current packaging technology must regulate with economic standards and environmental awareness to be commercialized widely. Besides the product protection aspect, crucial subjects such as energy and material costs, pollution policy, and waste disposal management also must be considered. The ultimate packaging must protect the product with minimal cost that fulfills the industry requirements and consumer demand without intruding on an environmental issue. This paper offers the latest information on developing biodegradable biopolymers derived from natural sources, e.g., polylactic acid (PLA), chitosan, polyhydroxyalkanoates (PHAs), cellulose, and starch, specifically for their application in packaging, as shown in Figure 1.1. Furthermore, this review explores the possibility of utilising agricultural, brewery, and industrial wastes as raw materials

Figure 1.1: Cellulose, starch, chitosan, polylactic acid, and polyhydroxyalkanoates in sustainable biobased packaging applications.

for biopolymers production and adopting the integrated biorefinery concept. This approach also aims to reduce the overall production cost of biopolymers while maintaining environmental sustainability.

Referring to Reichert et al. [1] novel materials can be used in mono-or multi-layer structures while yet preserving biodegradability. New bio-based packaging materials should be made of bio-based materials, allowed for use in food contact, or biocompatible and biodegradable for medicinal uses. Ideally, bio-based sustainable packaging materials should come from renewable resources or byproducts of the manufacturing of agricultural or food products, as these sources are attracting a lot of attention in business and academia and do not compete with the production of food on a large scale. According to the category of biofuels, we may distinguish between first-, second-, and third-generation feedstocks depending on the sources utilised to produce packaging material or textiles. Generally, edible biomass, e.g., sugarcane, whey, or maize, is considered a first-generation feedstock. The second-generation feedstock consists mostly of lignocellulosic-based feedstocks, primarily non-edible, and comprises waste from the forestry, livestock, agricultural industries, and municipal wastes. Meanwhile, algal biomass is a third-generation feedstock. Most often, increases in the production of feedstock are related to decreases in price and competition with food, but a rise in the complexity of using the source. Hence, Figure 1.2 graphically explains an optimal cycle for bio-based packaging materials.

Hence, this article also highlights the potential of natural and synthetic biopolymers in contributing to sustainable development and presents future research opportunities in this area. Besides, this comprehensive work can serve as a valuable resource for scientists researching sustainable food packaging and corporations interested in scaling up biopolymer production for industrial use.

1.2 Natural biopolymer

Biomass is the primary source of natural biopolymers, widely applied in multidisciplinary applications. Limited and diminishing reserves of fossil fuel prompt biomass

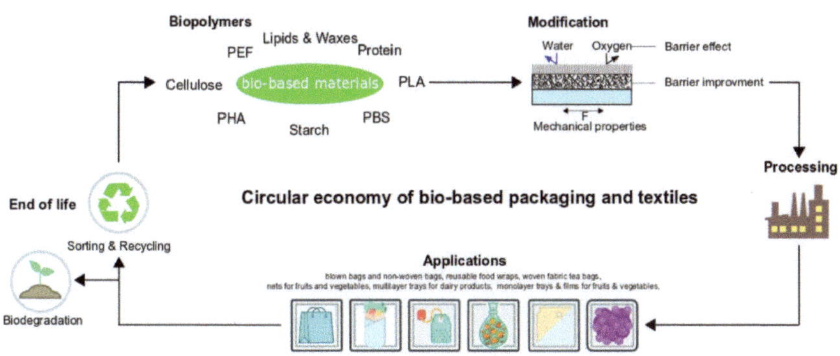

Figure 1.2: Schematic representation of an ideal cycle of bio-based materials for packaging application utilising PLA, PEF, PBS, PHA. Reproduced from [1].

feedstock as an abundant, renewable, and sustainable alternative resource. This eco-friendly material holds the potential to produce safer and cleaner packaging products reducing or in expectation of eradicating non-degradable petrochemical plastic [2].

1.2.1 Polysaccharides

Polysaccharides are regarded as an abundant source of biomaterial possessing unique physical, chemical, and biological properties. They comprise repeating units of monosaccharide copolymers linked through glycosidic bonds to form high molecular weight polymers. Polysaccharides are either found as a structural material in the construction of cell walls of plants (cellulose) and crustaceans (chitosan) or an energy storage system of starch (amylose and amylopectin). The most utilised polysaccharides in biopolymer production are starch, cellulose, and chitosan. Since polysaccharides are animal-and plant-based materials, escalating interest has led to them becoming the following polymer feedstocks. Figure 1.3 presents the most relevant polysaccharides from various sources and their applications for fabricating hydrogel-based bioinks for 3D bioprinting.

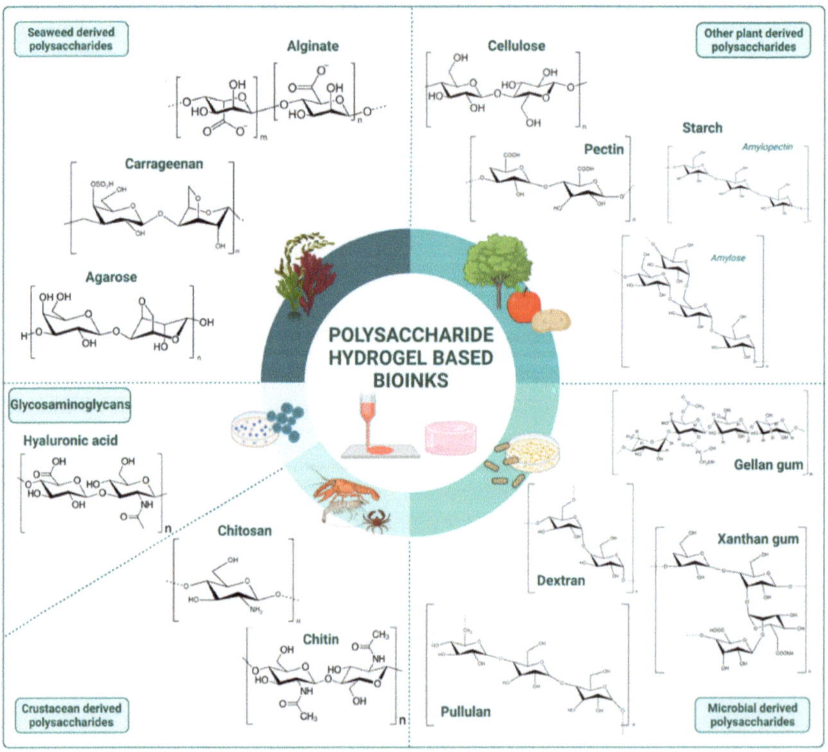

Figure 1.3: Most relevant polysaccharides from multiple sources and their applications for the fabrication of hydrogel-based bioinks for 3D bioprinting. Reproduced from ref. [3].

Starches are the major source of carbohydrate consumed by humans and animals as energy supply in their daily dietary [4, 5]. The sources are diverse from tubers, rhizomes, roots, stems, grains, and cereals [6]. The two primary components of starch, which are amylose and amylopectin act as an energy storage in plant. The different constitution of amylose and amylopectin in starches determine its molecular structure that is affecting the functional properties to develop biopolymer film. In Table 1.1, it can be observed that numerous modifications had been employed with sole purpose of improving the starch-based biopolymer's functional properties for packaging application. Due to low mechanical properties of biopolymer, filler are being included as reinforcement to enhance the mechanical properties of the biopolymers [16–22].

Table 1.1: Various starch-based biopolymers with different modifications.

Starch	Other compounds	Fabrication method	Results	Reference
Sugar palm	Glycerol/sorbitol/sugar palm nanocrystalline	Solution casting	TS (11.47 MPa), E (24.42 %), T_{max} (313.87 °C)	[7]
Chilean pine seed	i. Glycerol/husk powder ii. Glycerol/PLA/PHA iii. Glycerol/PLA/PHA/pehuen starch	Melt blending/injection moulding	i. TS (9 MPa), E (40 %), T_{max} (312 °C) ii. TS (10 MPa), E (10 %), T_{max} (316 °C) iii. TS (12 MPa), E (12 %), T_{max} (328 °C)	[8]
Arrowroot	Glycerol	Solution casting	TS (15.22 MPa), E (6.21%), T_{max} (313.02 °C)	[9, 10]
Dioscorea hispida	Sorbitol	Solution casting	TS (9.29 MPa), E (25.44%), T_{max} (327.2 °C)	[11]
Corn	i. Fructose ii. Glycerol iii. Fructose/glycerol	Solution casting	i. TS (17.15 MPa), E (65.59 %), WA (187.87 %) ii. TS (2.24 MPa), E (33.3 %), WA (98.82 %) iii. TS (10.66 MPa), E (57.67 %), WA (106.23 %)	[4]
Cassava	Cassava nanofibril	Solution casting	TS (14.2 MPa), E (20 %), WVP (2.5 g cm/(cm^2 s Pa))	[12]
Sweet potato	i. Glycerol/montmorillonite ii. Glycerol/montmorillonite/thyme essential oil	Solution casting	i. TS (4.53 MPa), E (44.91 %), WVP (0.022 gm/m^2/day) ii. TS (1.54 MPa), E (71.4 %), WVP (0.027 gm/m^2/day)	[13]
Brazilian pine seed	Glycerol/husk powder	Solution casting	TS (2.4 MPa), E (4.2 %), WVP (3.8 g mm/m^2 h kPa)	[14]
Pea	Glycerol/rice bran	Solution casting	TS (16 MPa), E (3.1 %), WVP (6.35 g mm/m^2 h kPa)	[15]

*TS, tensile strength; *E*, elongation at break, T_{max}, maximum degradation temperature; WA, water absorption; WVP, water vapour permeability.

Cellulose is mainly synthesized from the plant but is also produced by some bacteria. It is a sturdy fibrous polysaccharide responsible for maintaining plant cell wall structure. In plant structure, bundles of microfibrils make up cellulose chains. The capability of such biomaterial has high strength in aiding the stability of the plant structure and is potentially utilised in producing superior biopolymers. As most cellulose-based materials come from wood pulps and a cotton linter, the abundance of by-products from forestry, agricultural, and domestic wastes can also be viable cellulose extraction resources. In a work by Xia et al. [23], the cellulose-based film was successfully developed using old waste corrugated milk cartons (WCC). Although the film has a lower tensile strength (23.16 MPa) compared to traditional cellulose film (89 MPa), its brown and translucent appearance possesses better UV resistance due to phenolic lignin as a natural anti-UV radiation substance. Figure 1.4 elucidates the transparency of the two films from traditional cellulose and old waste corrugated milk cartons.

Furthermore, the water contact angle measurement recorded a higher value of 64.3° for WCC film, indicating more excellent wettability or hydrophobicity compared to 45.5° for traditional cellulose film. In quite a similar case, Xia et al. [24] obtained a highly transparent cellulose-based film from waste newspapers. They yielded a tensile strength of 80 MPa rivalling even the pure cellulose film. Meanwhile, Ai et al. [25] fabricated banana cellulose-based (BC) film with a prior delignification process and reported a more transparent film and higher tensile strength value of 32.8 MPa. The effectiveness of the BC film in preserving mango was verified when the pericarp recorded the lowest color index value (0.5), followed by unpackaged (5.25) and polyethylene film packaged (8.5). Furthermore, the high permeability of BC film allows ethylene gas to be released, which is responsible for accelerating mango ripening.

Chitosan is a form of amino polysaccharide processed when chitin from crustaceans and insects goes through alkali deacetylation. Extensive research has been conducted on chitosan as an antimicrobial agent in the agricultural, cosmetic, food and pharmaceutical industries. The nature of chitosan exhibits antibacterial and antifungal properties

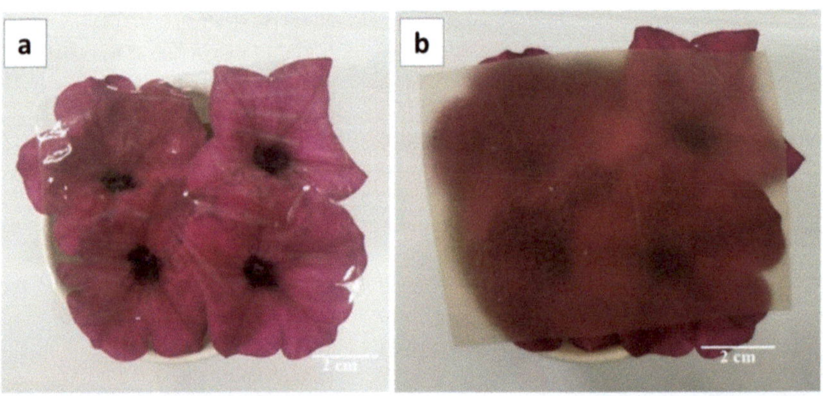

Figure 1.4: (a) Traditional cellulose and (b) old waste corrugated milk cartons (WCC) films. Reproduced from ref. [23].

depending on the degree of acetylation, molecular weight and pH value. Like most biomaterials, chitosan is safe, non-toxic, biocompatible, biodegradable, and sustainable, making it a good supply for biopolymer production. Furthermore, the total biomass from extracted shells of crabs, prawns, lobsters, and krills can be exploited rather than discarded. Ashrafi et al. [26] proved that chitosan/kombucha tea (KT) composite film possesses antimicrobial properties against *Escherichia coli* and *Staphylococcus aureus*.

Moreover, incorporating KT improved WVP (256.7×10^{-7} to 132.1×10^{-7} g mm/cm^2 h KPa), antioxidant activity (6.19–59.2 %, DPpH scavenging activity) and UV radiation resistance. When tested on packaging minced meat, the active film effectively preserved and extended the storage life in thwarting the microbial growth (5.36–2.11 log cfu/gr) on the 4th day of storage. Llanos and Tadini [27] reported that the addition of montmorillonite (MMT) and bamboo nanofibers (BNF) into chitosan films enhanced the elongation at break and reduced the WVP. Meanwhile, tensile strength was weakened. The minimal loading of both nanofillers could not bear the stress transfer during tension force but was sufficient enough to establish a tortuous path hindering the diffusion of water molecules. In that regard, flexibility was improved due to the plasticiser reducing the intermolecular forces.

1.2.2 Protein

Protein-based films derived from both animals (collagen and gelatin) and plants (soy, wheat, and whey) possess the capability of retarding biological invasion towards the food product. They provide protection against microbial infection owing to their higher barrier against gases, vapours and oils. In addition, they are infused with antioxidant and antimicrobial agents, extending the film's shelf-life and preserving the packaged product. The structure build of proteins comprises a linear sequence of amino acids and polar hydrogen bonds establishing a tortuous path preventing the diffusion of small molecules from passing through the film. Collagen films have better gas barrier capability than globular proteins like soy protein and wheat gluten, which have greater free volume due to their less linear structure and different amino acid side groups. Sabbah et al. [28] magnificently refined black cumin seed protein-based edible films to produce more homogeneous films possessing antimicrobial activity, improved barrier and mechanical properties desired for food packaging applications. By incorporating transglutaminase as a cross-linking enzyme and adjusting the pH value from 6 to 8, the elongation at break significantly increased from 16.4 % to 183.5 %.

1.3 Synthetic biopolymer

Compared to naturally occurring biopolymers forming intra- or extra-cellularly in living organisms such as starch, cellulose and chitosan, the biopolymer can also be synthesised chemically using biomasses.

1.3.1 Polylactic acid

Polylactic acid (PLA) is one of the most studied biopolymers as a candidate to substitute current petroleum-based polymers. It even rivals the commercial petroleum-based polymer such as polypropylene (PP), polyethylene (PE), polycarbonate (PC), polystyrene (PS), and polyethylene terephthalate (PET) in terms of mechanical strength, thermal stability and gas barrier properties. PLA's outstanding functional properties and good processability have been adapted in various applications, from as simple as mulching films and food packaging to complex as drug delivery capsules and tissue implantation devices. PLA can be synthesised through the fermentation of biomasses such as sugarcane bagasse, rice hulls or corn stover, which makes it a sustainable and renewable material. Moliner et al. [29] found that different water mediums significantly affected the degradation of PLA/sisal biocomposites. In 30 days, the alkaline medium recorded a maximum water absorption rate, while the acidic medium indicated a higher diffusivity. Regardless of sisal loading and pH value, the PLA backbone was unaffected by the hydrolysis signifying a lack of degradation activity. Figure 1.5 displays the physical appearance of biocomposite before and after submersion in various pH levels of mediums. Nazrin et al. [30] promoted the hydrophilicity of PLA by blending with TPS, resulting in more excellent solubility and WVP. In this case, a harmonised composition is required to establish adequate compatibility, which is acquired at 40 % compared to 30 % of TPS leading to lower WVP (1.10–1.02 × 10^{-11} g m/m^2 s Pa).

Meanwhile, Zhang and Zhen [31] compounded 0.5 % of modified fulvic acid (MFA) into PLA, increasing impact strength by 97.2 % and crystallinity from 4.9 % to 36.9 %. MFA acts as a heterogenous crystal nucleus providing a minimal surface-free energy barrier

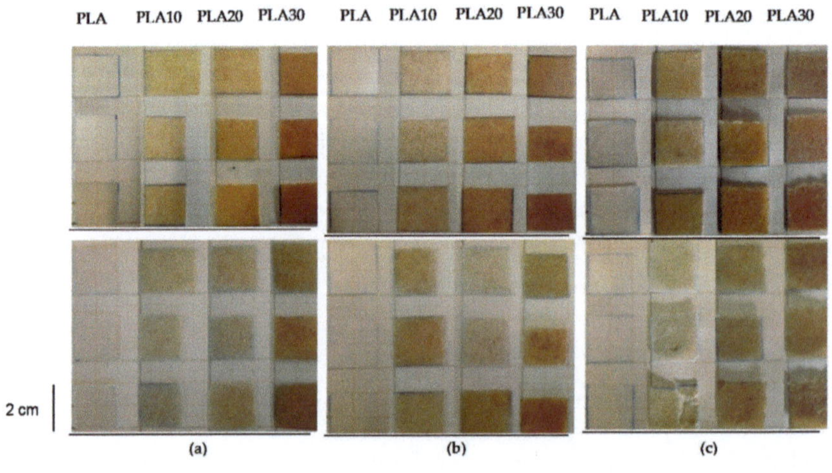

Figure 1.5: The physical appearance of PLA biocomposite before (above) and after soaked (below) in mediums at (a) pH = 2, (b) pH = 6, and (c) pH = 8. Reproduced from ref. [29].

for PLA nucleation. Relatively, enzymatic degradation by proteinase *K* was at a lower value for PLA/MFA (16.4 %) than pristine PLA (18 %). Chotiprayon et al. [32] studied the effect of various coir fibre (CF) loading towards PLA/TPS blend composites for bio-based and biodegradable products. Adding CF promotes the dispersion of PLA and TPS phases and functions as nucleating agents for the PLA in the blend. The crystallinity was enhanced from 7.24 % to 12.60 %, but higher CF loading decreased the crystallinity to 11.93 %. Perić et al. [33] reported a similar enhancement effect of nucleation on PLA/nanofibrillated cellulose (NFC) but with a significant improvement of crystallinity from 10.5 % to 18.4 %. The high aspect ratio of NFC also promotes impact strength and elongation at break.

1.3.2 Polybutylene succinate

Besides PLA, polybutylene succinate (PBS) is one of the biopolymers that has gained interest as a feedstock in developing food packaging material. Unlike PLA, PBS is commercially produced with the help of petrochemical materials, but recently, a new method of synthesising bio-succinic acid by bacteria called *Anaerobiospirillum succiniciproducens* had been discovered [1]. PBS has decent mechanical and thermal properties on par with PP's. A notable feature of PBS is that it can biodegrade well even if discarded carelessly in open waters, moist soil and compost soil.

Poly(butylene succinate) (PBS) is a type of biodegradable thermoplastic polymer that falls under the polyester family [34]. Its chemical structure comprises repeating units of butylene succinate ($C_8H_{12}O_4$). Also known as poly(tetramethylene succinate), this polymer has garnered significant attention due to its thermo-mechanical and physical properties equivalent to polypropylene [35]. PBS is a promising material among aliphatic biodegradable polyesters due to its good chemical resistance, thermoplastic behavior, biodegradability, and melt processability [36].

The biodegradation rate of PLA, PHO, PHB, TPS, PBS, PCL, and their blends, are shown in Figure 1.6. The figure provides a quantitative overview of the findings from biodegradation tests conducted in various environments such as marine, freshwater, anaerobic aquatic digestion, and soil environment. The figure includes a red dotted line indicating the minimum target of 90 % to be achieved to consider the standardised test a success and a green dotted line indicating 100 % biodegradability of the tested material under the specified conditions. Despite being the most commonly used materials in the industry, both PLA and PBS exhibit severely limited biodegradability in soil, aquatic, and marine environments, which significantly curtails their potential industrial applications [37].

1.3.3 Polyhydroxyalkanoates

Polyhydroxyalkanoates (PHA) is a unique division of biopolymers produced through biological activities of microorganisms, such as the fermentation of sugars and lipids.

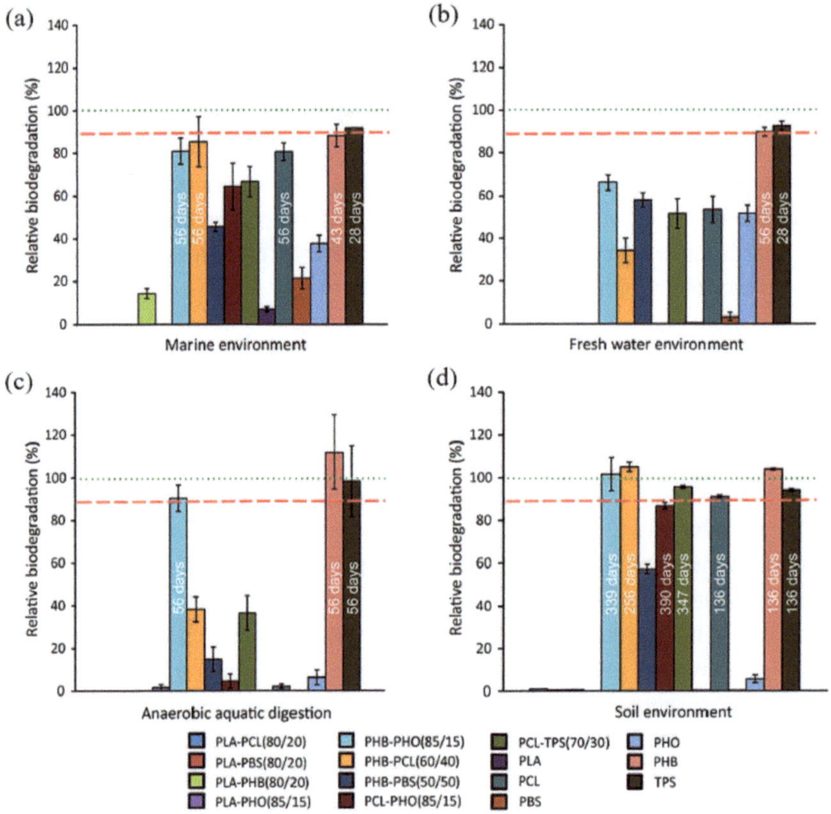

Figure 1.6: Biodegradation of PLA, PHB, PCL, PBS, PHO, PHB, TPS, and some of their blends in marine, water environments, and soil. Reproduced from ref. [37].

Various microorganisms are used to produce PHA: *Azotobacter, Aeromonas, Clostridium, Cupriavidus, Ralstonia, Methylobacterium, Syntrophomonas,* and *Pseudomonas,* but the common one used for a commercial purpose is *E. coli* [38]. The process of yielding PHA occurred when those bacteria accumulated archaea and cell walls as energy storage under a restricted growing environment. Therefore, the intracellular carbon source and energy storage are biodegradable in most environments. Moreover, PHA possesses good mechanical properties and thermal processability, which can be fabricated through industrial-scale production via compression and injection molding. In addition, PHA is biocompatible. Thus it was adapted widely in biomedical applications, including antimicrobial-releasing sutures, drug encapsulation and tissue scaffolds. However, several drawbacks, such as expensive production cost, highly flammable, and low thermal stability drag its future growth to be utilised in various applications compared with conventional polymers. The end products of PHA can be categorised by their chemical structure length consisting of short chain (3–5 carbon atoms), medium chain (6–14 carbon atoms), and long chain (15+ carbon atoms). The commercially available subset of PHA are poly (3-hydroxybutyrate) (PHB) and poly (3-hydroxybutyrate-co-3-hydroxyvalerate (PHBV)), which are short-chain structures.

In recent studies, PHB has been put forward as one of the candidates to be used in packaging applications, replacing PP and PE. To compensate for its low flame retardancy, renewable raw materials such as double-layered hydroxide, kenaf fiber, melamine phosphate-modified lignin, and halloysite nanotubes were incorporated [39]. This was the common practice employed in the polymer matrix to induce charring, consequently protecting the substrate against heat and oxygen aggression.

1.4 Recycle materials

Recycling bio-based materials entails efficient processing methods that result in same or new products from the components recovered as a result of sorting, which, as previously mentioned, is helpful for extending the material's lifespan. Through life cycle assessment studies, Cosate de Andrade et al. showed that recycling PLA has a lower environmental effect than composting [40]. In their review of PLA recycling, Badia et al. [41] examined the degradation of PLA, which can happen when recovered polymers are reprocessed via extrusion. Water-induced hydrolysis and heat degradation were taken into account. Due to chain scission during injection, PLA reprocessing may also lead to a rise in crystallisation throughout cooling with the number of injection cycles. PLA's hydrolysis is only susceptible to drying. According to Najafi et al. [42] who demonstrated the efficiency of three different chain extenders and clay at 2 % in an effort to regulate the melt viscosity, using chain extenders has been reported to assist manage the melt viscosity of PLA and its blends.

In order to mechanically recycle biopolyesters, a clean starting material is necessary. Additives incorporated into the biopolyester layers or matrix may cause contamination. The compatibility between the PLA matrix and the additive is often attained in the first scenario. As an example, employing copolymers as the compatibiliser in PLA/cellulose composites requires a compatibilisation technique when a separate layer is present [43] or cellulose esters in wood/PLA composites [44]. Radical agents or transesterification catalysts have been investigated as potential reactive processing techniques for enhancing compatibility in blends of various polyesters. The PLA4FOOD and BIO4MAP projects both investigated the recycling of mono-material PLA through the crushing of trays and sheets, washing of the scraps, drying, and extrusion to create pellets. The flowability of the melt and the mechanical properties of the materials produced are sustained during multiple processing cycles, according to the mechanical and rheological results. This is to ensure the possibility of using recycled PLA in blends with virgin PLA without the loss of material properties. There are numerous ways to separate layers in polymer-based multilayers, including delamination and selective dissolution-reprecipitation techniques, both of which have been in-depth discussed elsewhere [45].

As an alternative to mechanical recycling because of its low depolymerisation temperature and potential for impurities, PLA is a promising option for ecologically sound chemical recycling since it is easily hydrolysed with water to generate LA, which is then refined and polymerised to manufacture prime PLA. In addition, Piemonte et al. showed that the synthesis of LA via chemical depolymerisation of PLA is preferable from an energy standpoint to the production of LA from glucose fermentation using a life cycle assessment technique [46].

1.5 Conclusions

Both consumer demand and new regulations will drive the trend towards sustainable packaging. Biobased materials hold immense potential for use in packaging. Biobased materials, such as PLA, PBS, waxes, lipids, proteins, starch, cellulose, and PHAs, are well-positioned to capitalise on this market growth. While it is evident that high-end and specialised products such as organic goods offer excellent opportunities for their application, it is imperative to conduct more research in various aspects of biopackaging, including legislation, processing technology, and compatibility studies between food and packaging. However, there are still obstacles to overcome before bio-based packaging can fully replace petroleum-derived packaging. The availability of commercial-grade bioplastics is expected to increase, leading to broader acceptance of these alternative materials in sustainable packaging.

Acknowledgements: The authors would like to express gratitude for the financial support received from the Universiti Teknologi Malaysia for the project "The impact of Malaysian bamboos' chemical and fiber characteristics on their pulp and paper properties", grant number PY/2022/02318—Q.J130000.3851.21H99. The research has been carried out under the programme, Research Excellence Consortium (JPT (BPKI) 1000/016/018/25 (57)), provided by the Ministry of Higher Education Malaysia (MOHE).

References

1. Reichert CL, Bugnicourt E, Coltelli MB, Cinelli P, Lazzeri A, Canesi I, et al. Bio-based packaging: materials, modifications, industrial applications and sustainability. Polymers 2020;12:1–35.
2. Rajeshkumar L. 19 – biodegradable polymer blends and composites from renewable resources. In: Rangappa MS, Parameswaranpillai J, Siengchin S, Ramesh MBTBP, editors. Blends and composites, woodhead publishing series in composites science and engineering. Amsterdam, Netherlands: Woodhead Publishing; 2022:527–49 pp.
3. Teixeira MC, Lameirinhas NS, Carvalho JPF, Silvestre AJD, Vilela C, Freire CSR. A guide to polysaccharide-based hydrogel bioinks for 3D bioprinting applications. Int J Mol Sci 2022;23:6564.

4. Abotbina W, Sapuan SM, Sultan MTH, Alkbir MFM, Ilyas RA. Development and characterization of cornstarch-based bioplastics packaging film using a combination of different plasticizers. Polymers 2021; 13:3487.
5. Tarique J, Sapuan SM, Khalina A, Sherwani SFK, Yusuf J, Ilyas RA. Recent developments in sustainable arrowroot (Maranta arundinacea Linn) starch biopolymers, fibres, biopolymer composites and their potential industrial applications: a review. J Mater Res Technol 2021;13:1191–219.
6. Diyana ZN, Jumaidin R, Selamat MZ, Ghazali I, Julmohammad N, Huda N, et al. Physical properties of thermoplastic starch derived from natural resources and its blends: a review. Polymers 2021;13:1–20.
7. Ilyas RA, Sapuan SM, Ishak MR, Zainudin ES. Development and characterization of sugar palm nanocrystalline cellulose reinforced sugar palm starch bionanocomposites. Carbohydr Polym 2018;202: 186–202.
8. Castaño J, Rodríguez-Llamazares S, Carrasco C, Bouza R. Physical, chemical and mechanical properties of pehuen cellulosic husk and its pehuen-starch based composites. Carbohydr Polym 2012;90:1550–6.
9. Tarique J, Sapuan SM, Khalina A, Ilyas RA, Zainudin ES. Thermal, flammability, and antimicrobial properties of arrowroot (Maranta arundinacea) fiber reinforced arrowroot starch biopolymer composites for food packaging applications. Int J Biol Macromol 2022;213:1–10.
10. Tarique J, Zainudin ES, Sapuan SM, Ilyas RA, Khalina A. Physical, mechanical, and Morphological performances of arrowroot (Maranta arundinacea) fiber reinforced arrowroot starch Biopolymer composites. Polymers 2022;14:388.
11. Hazrati KZ, Sapuan SM, Zuhri MYM, Jumaidin R. Preparation and characterization of starch-based biocomposite films reinforced by Dioscorea hispida fibers. J Mater Res Technol 2021;15:1342–55.
12. Huang L, Zhao H, Yi T, Qi M, Xu H, Mo Q, et al. Reparation and properties of cassava residue cellulose nanofibril/cassava starch composite films. Nanomaterials 2020;10:1–19.
13. Issa AT, Schimmel KA, Worku M, Shahbazi A, Ibrahim SA, Tahergorabi R. Sweet potato starch-based nanocomposites: development, characterization, and biodegradability. Starch/Staerke 2018;70:1700273.
14. Daudt RM, Sinrod AJG, Avena-Bustillos RJ, Külkamp-Guerreiro IC, Marczak LDF, McHugh TH. Development of edible films based on Brazilian pine seed (Araucaria angustifolia) flour reinforced with husk powder. Food Hydrocolloids 2017;71:60–7.
15. Cano A, Jiménez A, Cháfer M, Gónzalez C, Chiralt A. Effect of amylose:amylopectin ratio and rice bran addition on starch films properties. Carbohydr Polym 2014;111:543–55.
16. Asyraf MRM, Ishak MR, Syamsir A, Nurazzi NM, Sabaruddin FA, Shazleen SS, et al. Mechanical properties of oil palm fibre-reinforced polymer composites: a review. J Mater Res Technol 2022;17:33–65.
17. Azman MA, Asyraf MRM, Khalina A, Petrů M, Ruzaidi CM, Sapuan SM, et al. Natural fiber reinforced composite material for product design: a short review. Polymers 2021;13:1917.
18. Norfarhana AS, Ilyas RA, Ngadi N. A review of nanocellulose adsorptive membrane as multifunctional wastewater treatment. Carbohydr Polym 2022;291:119563.
19. Norizan MN, Abdan K, Ilyas RA, Zin M, Muthukumar C, Rafiqah S, et al. Effect of fiber orientation and fiber loading on the mechanical and thermal properties of sugar palm yarn fiber reinforced unsaturated polyester resin composites. Polimery 2019;64:34–43.
20. Nurazzi NM, Sabaruddin FA, Harussani MM, Kamarudin SH, Rayung M, Asyraf MRM, et al. Mechanical performance and applications of CNTs reinforced polymer composites—a review. Nanomaterials 2021;11: 2186.
21. Sabaruddin FA, Paridah MT, Sapuan SM, Ilyas RA, Lee SH, Abdan K, et al. The effects of unbleached and bleached nanocellulose on the thermal and flammability of polypropylene-reinforced kenaf core hybrid polymer bionanocomposites. Polymers 2020;13:116.
22. Sharma S, Sudhakara P, Omran AAB, Singh J, Ilyas RA. Recent trends and developments in conducting polymer nanocomposites for multifunctional applications. Polymers 2021;13:2898.
23. Xia G, Zhou Q, Xu Z, Zhang J, Ji X, Zhang J, et al. Cellulose-based films with ultraviolet shielding performance prepared directly from waste corrugated pulp. Polymers 2021;13:1–14.

24. Xia G, Wan J, Zhang J, Zhang X, Xu L, Wu J, et al. Cellulose-based films prepared directly from waste newspapers via an ionic liquid. Carbohydr Polym 2016;151:223–9.
25. Ai B, Zheng L, Li W, Zheng X, Yang Y, Xiao D, et al. Biodegradable cellulose film prepared from banana pseudo-stem using an ionic liquid for mango preservation. Front Plant Sci 2021;12:1–10.
26. Ashrafi A, Jokar M, Nafchi AM. Preparation and characterization of biocomposite film based on chitosan and kombucha tea as active food packaging. Int J Biol Macromol 2018;108:444–54.
27. Llanos JHR, Tadini CC. Preparation and characterization of bio-nanocomposite films based on cassava starch or chitosan, reinforced with montmorillonite or bamboo nanofibers. Int J Biol Macromol 2018;107: 371–82.
28. Sabbah M, Altamimi M, Di Pierro P, Schiraldi C, Cammarota M, Porta R. Black edible films from protein-containing defatted cake of Nigella sativa seeds. Int J Mol Sci 2020;21:1–13.
29. Moliner C, Finocchio E, Arato E, Ramis G, Lagazzo A. Influence of the degradation medium on water uptake, morphology, and chemical structure of poly (lactic acid)-sisal bio-composites. Basel, Switzerland: MDPI; 2020, 13:1–19 pp.
30. Nazrin A, Sapuan SM, Zuhri MYM, Tawakkal ISMA, Ilyas RA. Water barrier and mechanical properties of sugar palm crystalline nanocellulose reinforced thermoplastic sugar palm starch (TPS)/poly(lactic acid) (PLA) blend bionanocomposites. Nanotechnol Rev 2021;10:431–42.
31. Zhang H, Zhen W. Performance, rheological behavior and enzymatic degradation of poly(lactic acid)/ modified fulvic acid composites. Int J Biol Macromol 2019;139:181–90.
32. Chotiprayon P, Chaisawad B, Yoksan R. Thermoplastic cassava starch/poly(lactic acid) blend reinforced with coir fibres. Int J Biol Macromol 2020. https://doi.org/10.1016/j.ijbiomac.2020.04.121.
33. Perić M, Putz R, Paulik C. Influence of nanofibrillated cellulose on the mechanical and thermal properties of poly(lactic acid). Eur Polym J 2019;114:426–33.
34. Luyt AS, Malik SS. Can biodegradable plastics solve plastic solid waste accumulation? In: Plastics to energy. Amsterdam, Netherlands: Elsevier; 2019:403–23 pp.
35. Liu GC, Zhang WQ, Wang XL, Wang YZ. Synthesis and performances of poly(butylene-succinate) with enhanced viscosity and crystallization rate via introducing a small amount of diacetylene groups. Chin Chem Lett 2017;28:354–7.
36. Rudnik E. Compostable polymer properties and packaging applications. In: Plastic films in food packaging. Amsterdam, Netherlands: Elsevier; 2013:217–48 pp.
37. Narancic T, Verstichel S, Chaganti SR, Morales-Gamez L, Kenny ST, De Wilde B, et al. Biodegradable plastic blends create new possibilities for end-of-life management of plastics but they are not a panacea for plastic pollution. Environ Sci Technol 2018;52:10441–52.
38. Meereboer KW, Misra M, Mohanty AK. Review of recent advances in the biodegradability of polyhydroxyalkanoate (PHA) bioplastics and their composites. Green Chem 2020;22:5519–58.
39. Vahabi H, Michely L, Moradkhani G, Akbari V, Cochez M, Vagner C, et al. Thermal stability and flammability behavior of poly (3-hydroxybutyrate)(PHB) based composites. Materials 2019;12:1–14.
40. Cosate de Andrade MF, Souza PMS, Cavalett O, Morales AR. Life cycle assessment of poly(lactic acid) (PLA): comparison between chemical recycling, mechanical recycling and composting. J Polym Environ 2016;24: 372–84.
41. Badia JD, Ribes-Greus A. Mechanical recycling of polylactide, upgrading trends and combination of valorization techniques. Eur Polym J 2016;84:22–39.
42. Najafi N, Heuzey MC, Carreau PJ, Wood-Adams PM. Control of thermal degradation of polylactide (PLA)-clay nanocomposites using chain extenders. Polym Degrad Stabil 2012;97:554–65.
43. Phuong VT, Gigante V, Aliotta L, Coltelli MB, Cinelli P, Lazzeri A. Reactively extruded ecocomposites based on poly(lactic acid)/bisphenol A polycarbonate blends reinforced with regenerated cellulose microfibers. Compos Sci Technol 2017;139:127–37.
44. Takatani M, Ikeda K, Sakamoto K, Okamoto T. Cellulose esters as compatibilizers in wood/poly(lactic acid) composite. J Wood Sci 2008;54:54–61.

45. Kaiser K, Schmid M, Schlummer M. Recycling of polymer-based multilayer packaging: a review. Recycling 2018;3:1–26.
46. Piemonte V, Sabatini S, Gironi F. Chemical recycling of pla: a great opportunity towards the sustainable development? J Polym Environ 2013;21:640–7.

Mohd Shahrulnizam Ahmad, Roshafima Rasit Ali, Zurina Mohamad, Zatil Izzah Ahmad Tarmizi, Siti Khairunisah Ghazali, Dayangku Intan Munthoub, Rohah A. Majid, Fathilah Ali, Rosnani Hasham, Anne Aleesa Nazree, Nadia Adrus, Muhammad Aqil Mohd Farizal and Jamarosliza Jamaluddin*

2 Fabrication of starch-based packaging materials

Abstract: This chapter aims to provide the reader with some information about the possibility of starch as a suitable substitute for synthetic polymers in biodegradable food packaging. This is due to the starch has good characteristics which are great biodegradability, low cost and also easy to gain from natural resources. However, some of technical challenges are also introduced before starch-based polymers can be used in more applications. These technical challenges involved preparation methods and incorporation of additives and these are being summarized in this topic. Hence, the enhancement of starch can be done in order to prepare innovative starch-based biodegradable materials.

2.1 Introduction and classification of starch biopolymer

Packaging is essential and it has been used since the first human civilization began in making use of tools. Plastics are a common example to portray packaging application as polymeric materials that have been a common man-made innovation since many centuries ago, which can be proved by the cellulose nitrate-based plastic prepared by A. Parker in 1838. The development of plastics experienced a rapid growth over the years and its application has been widely extensive since 1930. Today, the production of synthetic plastics for packaging

***Corresponding author: Jamarosliza Jamaluddin,** Faculty of Chemical & Energy Engineering, Universiti Teknologi Malaysia, Skudai 81310, Malaysia, E-mail: jamarosliza@utm.my. https://orcid.org/0000-0002-1147-7360

Mohd Shahrulnizam Ahmad, Zurina Mohamad, Siti Khairunisah Ghazali, Dayangku Intan Munthoub, Rohah A. Majid, Rosnani Hasham, Anne Aleesa Nazree, Nadia Adrus and Muhammad Aqil Mohd Farizal, Faculty of Chemical & Energy Engineering, Universiti Teknologi Malaysia, Skudai 81310, Malaysia

Roshafima Rasit Ali and Zatil Izzah Ahmad Tarmizi, Department of Chemical and Environmental Engineering, Malaysia-Japan International Institute of Technology (MJIIT), Universiti Teknologi Malaysia, Jalan Sultan Yahya Petra, 54100 Kuala Lumpur, Malaysia

Fathilah Ali, Department of Chemical Engineering and Sustainability, Kulliyyah of Engineering, International Islamic University of Malaysia, P.O. Box 10, 50728 Kuala Lumpur, Malaysia

As per De Gruyter's policy this article has previously been published in the journal Physical Sciences Reviews. Please cite as: M. S. Ahmad, R. R. Ali, Z. Mohamad, Z. I. A. Tarmizi, S. K. Ghazali, D. I. Munthoub, R. A. Majid, F. Ali, R. Hasham, A. A. Nazree, N. Adrus, M. A. M. Farizal and J. Jamaluddin "Fabrication of starch-based packaging materials" *Physical Sciences Reviews* [Online] 2023.
DOI: 10.1515/psr-2022-0010 | https://doi.org/10.1515/9783110773606-002

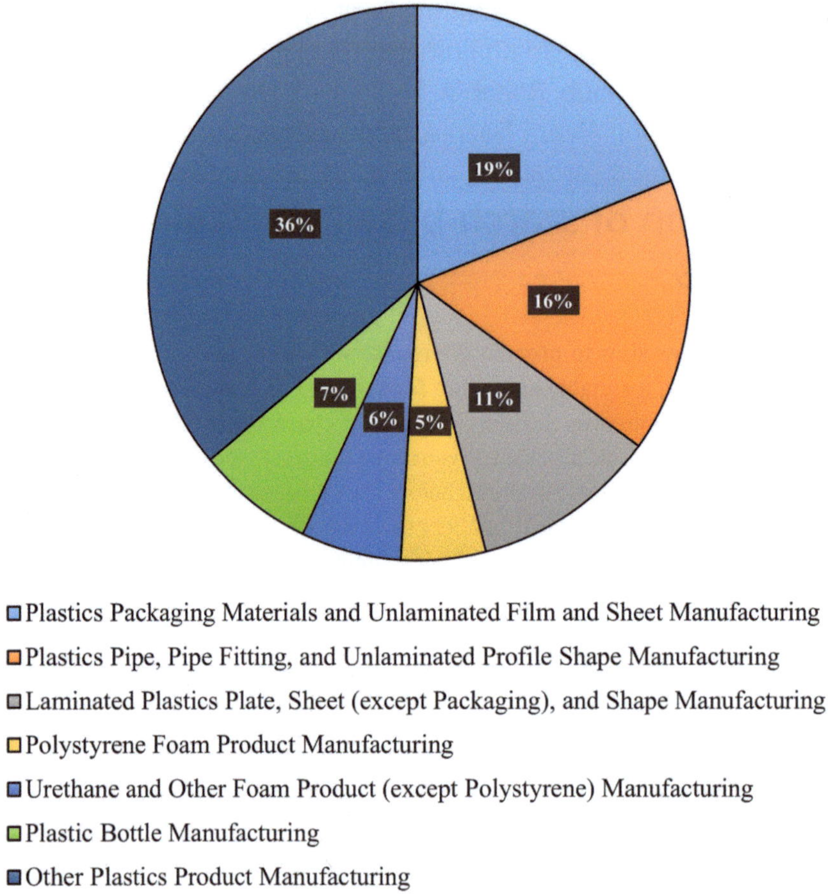

Figure 2.1: Markets for plastic product manufacturing worldwide pie chart in 2016 [1].

applications has reached more than 200 million metric tons throughout the world by hundreds of companies. Interestingly, after the World War II, plastic materials were introduced to replace paper as packaging applications. Nowadays, most of the production of plastics is more useful for packaging where it is reported that by 2016, plastic packaging materials and unlaminated film and sheet manufacturing held the biggest segment in the total global market for plastic products which is about 36% as shown in Figure 2.1 [1].

Polyethylene (PE) was the first thermoplastic to be produced and commercially available due to its low-cost production, ease of fabrication and manufacture, and its lightweight characteristic. PE becomes the leading packaging plastic because of its greater force behind the expansion to form a resealable package and a transparent material. Another common example of packaging thermoplastic is low-density polyethylene (LDPE), which is a branched PE that has low percentage of crystallinity in the range of 40–60%, with a density of about 0.940–0.970 g/cm^3. LDPE has excellent flexibility and easily adapts to blown film, cast film, extrusion coating, injection molding, and blow molding processes. Eventually, these conventional plastics give more harm than good for long-term effects, especially to nature. Although they are cheap, they are non-biodegradable which brings suffocations to the landfills.

Bio-based packaging materials are naturally made from an organic source which possess better properties than conventional ones and give a positive impact on the environment. The big difference between synthetic polymers and natural ones is the presence of oxygen and nitrogen in the natural polymers that allow the polymer to degrade naturally [2]. One of the sources to fabricate bio-based packaging materials is from starch, which approximately 50% of them are used to make bioplastics [3] because starch is widely abundant and a renewable biomass. From the preparation to its disposal method, starch-based bioplastics serve a holistic potential in maintaining our environment clean and healthy which may reduce carbon footprint, save energy as well as save cost consumption. These desired outcomes attract researchers and industries to focus more on the production of starch-based packaging materials and enhance their properties by introducing additives into the products such as plasticizers, stabilizers, cross-linkers, antimicrobial agents, and antioxidants.

Starch is a renewable natural polymer and has been getting much more attention since 1970s other than cellulose, soya and chitosan. Starch is a carbohydrate stored in various parts of plant, such as cereal grains, roots, tubers, stem-piths, leaves, seeds, fruit, and pollen, and is one of nature's energy reserves [4]. Due to its low cost, availability, and high manufacturing capabilities from renewable resources, starch is the most appealing material among these renewable natural polymers [5–8]. However, starch has its own limitations in aspects of low water resistance and high brittleness [6]. Furthermore, carbon dioxide (CO_2), water, and sunlight are the example of renewable sources of starch in plants.

Starch is a mixture of two glucans; amylose and amylopectin. Most starches contain amylose (water soluble) and amylopectin (water insoluble) about 10–20% and 80–90%, respectively [6, 9, 10]. Thus, this higher content of amylopectin gives a good hydrophilicity characteristic in starch and also can be used to improve the degradation rate of biopolymer. Corn and other cereals that have waxy or glutinous starch, which has little or no amylase, but a sugary mutant corn and some legumes have more amylose than amylopectin. Figure 2.2 illustrates the chain structure of amylose and amylopectin, where amylose has the glucose unit of 1,4-α-D that linked together in straight chains and amylopectin, in which has the glucose chain unit of 1,6-α-D that is highly branched [10].

2.2 Starch based packaging materials

2.2.1 Introduction to packaging materials

Plastic manufacturing and use have expanded dramatically over the world, exacerbating the trash disposal challenge. The increased concern about the environmental impact of

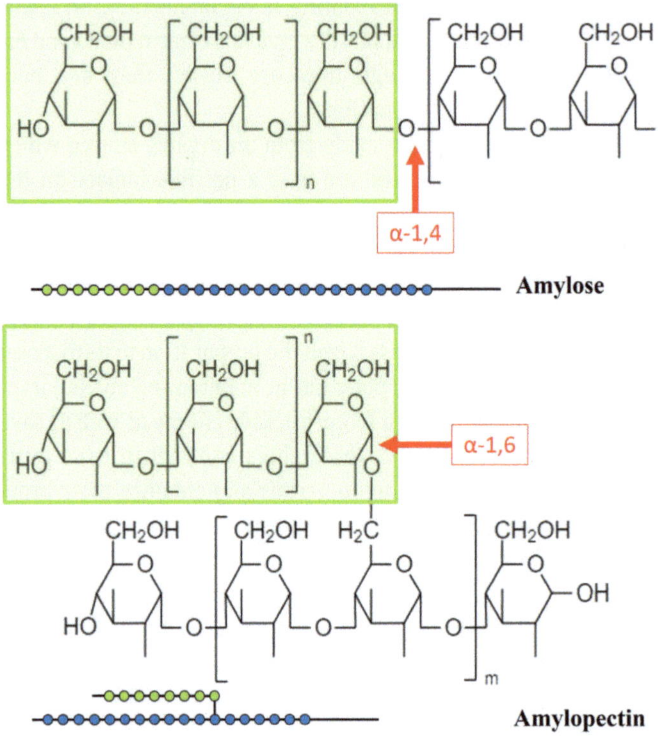

Figure 2.2: Structure amylose and amylopectin [10].

discarded plastics has prompted research towards the production of polymers that disintegrate more quickly in the environment, eventually resulting in total mineralization or bio-assimilation. Biopolymers should be employed in applications where biodegradability and/or the extraction of natural resources provide value, particularly in areas where valuable petroleum-based plastics are used for short-term uses [11].

2.2.2 Advantages and disadvantages of starch as biodegradable packaging materials

In the food packaging industry, biodegradable films and coatings are becoming popular and also triggered the subject of interest for increasing the serviceable life of many food products. It is feasible to maintain the product quality and freshness during the time required for commercialization and consumption conditions by using appropriate materials and packaging technologies [12]. Hence, starch is being suggested in food packaging industry as a biopolymer packaging material due to its biodegradability, low cost and also easy availability/production from common renewable resources [6].

Unfortunately, starch has several disadvantages, such as high hydrophilic tendency (poor moisture barrier) and inferior mechanical qualities than non-biodegradable plastic film that is commonly employed in the food packaging business [11].

One of the biodegradable plastics that can be used to overcome the low mechanical properties of starch is degradable poly-(ε-caprol-actone) (PCL). PCL is a biodegradable synthetic polyester with good mechanical properties that is compatible with a wide range of polymers. It is also one of the most hydrophobic biodegradable polymers that is currently available [13, 14].

2.3 Fabrication of starch-based packaging materials

The pristine starch itself does not have a good processing property, brittle, less compatible for its hydrophilic nature to hydrophobic materials, high viscosity in which reduces the processability and many more. However, starch is a natural and renewable biodegradable material. Thus, there is a need to find suitable processing method for the targeted application. There are various methods that can be used to prepare starch blend or composites into packaging materials. Those methods are synthesis of starch [15], solution casting [16, 17], melt mixing [18], thermoforming [19] and extrusion [20]. Some of the new techniques to prepare starch-based biodegradable material are electrospinning, 3D-printing, reactive extrusion and nanotechnology [21].

2.3.1 Synthesis of starch

Shafqat et al. [22] had used three types of starch; banana peels, rice and corn and potato peel powder, and sawdust as reinforcement filler. The banana peel starch was obtained by washing the feel thoroughly and cut into small pieces and then boiled for 30 min in water. The peels were then collected and dried. And then, it was blended in blender to obtain into paste. The authors had used acetic acid solution (5%), plasticizers (glycerol and sorbitol) and water. In their research, they had varied the compositions of the ingredients and found that bioplastics containing glycerol as a plasticizer exhibited lower values in term of tensile strength and Young's modulus. This could be due to lower mass reported for glycerol, thus enhance the interaction of glycerol and starch.

2.3.2 Film solution casting

Initially, heat and stir a mixture of film-forming ingredients and solvent, which is a film suspension of gelatinized starch, to make a film-forming solution. After that, the solution is poured onto a smooth surface, such as a glass or plastic plate. The solvent is then evaporated by drying the distributed solution in an oven at room temperature. Finally, the prepared films are peeled away from the plate [21]. Furthermore, this method is also quite simple and cheap to implement but it does have some limitations. Biodegradable films that have been successfully prepared from starch by using the film solution casting method are corn starch [23], cassava starch [24], and sweet potato starch [25].

Yurong and Dapeng [16] had used commercially available corn starch, polyvinyl alcohol (PVA), glycerol, water and Ɛ-Polylysine. First, they mixed corn starch, PVA and glycerol in water. Then, they added solution containing Ɛ-polylysine with various concentrations into the starch mixtures. The solution then mechanically stirred continuously at 300 rpm (40 °C) for 30 min, 500 rpm (60 °C) for 30 min, and finally at 500 rpm (95 °C) for an hour. The mixture was then poured into glass plate and then dried in oven for 12 h at 45 °C. The authors reported that both tensile strength and elongation at break increased tremendously with higher content of Ɛ-polylysine. This could be due to bonding formed as of the interaction of amine group of Ɛ-polylysine and hydroxyl group of starch. This shows that addition of Ɛ-polylysine enhanced the flexibility of starch.

2.3.3 Melt mixing

Hu et al. [26] had prepared premixture in which starch mixed with vegetable oil polyol with ratio of 7:3 for 15 min. Then together with the premixture, polylactic acid, polyethylene glycol (PEG), and citric acid (CA) were melt-blended at 170 °C (60 rpm) for 10 min. After that, the blends were pressed using plate vulcanizer. The samples were then prepared into dumbbell-shaped as per according to ISO527-2. In this work, the researchers wanted to study the impact of CA on the properties of different premixture mass ratios in the blends. It was found that the elongation at break increased when the premixture content reached 30 wt%. They mentioned that this could be due to enhanced interaction within the blend's components.

2.3.4 Thermoforming

Mohan and Kanny [27] had prepared thermoformed starch nanocomposites from corn starch and montmorillonite nanoclays. The researchers had prepared various compositions by varying the concentrations of clay in the composites. First, the corn starch was dissolved in water and then stirred at 80 °C (10 min). Then, glycerol was added into the hot solution. Clay was then added into the starch/glycerol solution and then mixed for 45 min. The solution was then poured in a mold and left in oven at 60 °C. The thermoforming was conducted using on a mold having cavity and place over an infrared (IR) chamber. At certain wavelength the IR heating was conducted and as the temperature reaches the targeted temperature, the film was immediately plugged into the chamber to obtain a tray structure and then removed for cooling. In this research, they had studied effect of temperatures on the obtained thermoformed samples. The tensile strength and modulus were reported to be increased but strain of tensile failed for formed sample was reduced. They mentioned this could be absence of moisture in the samples thus made the samples hardened.

2.3.5 Foaming process

Foaming processing is a technique for making foams from starch. Baking, molding, extrusion, and supercritical fluid foaming are the four basic methods for making these foams. Baking foaming entails pouring a mixture of starch, foaming agent, and other ingredients into a mold and heating it in the oven. In the case of molding foaming, the foaming material is placed in a temperature-controlled hydraulic press, where it is heated and pressure changes are used to generate foam. To create gas bubbles in supercritical fluid foaming, supercritical carbon dioxide is commonly used. Under pressure, the foaming ingredient and carbon dioxide are combined. After that, the device is depressurized, allowing the carbon dioxide gas to form bubbles, resulting in a foamed product [21]. Extrusion foaming is a process that includes heating and mixing raw ingredients under pressure before forcing them through a screw extruder to create foam. Dissolved gases generate bubbles that are trapped in the starch matrix when the pressure is released [28].

2.3.6 Extrusion process

Extrusion is an alternative that involved a dry processing method as compared to wet processing method such as solution casting [29]. The initial stage in extrusion processing is to fill the screw extruder with a mixture of film-forming ingredients. Subsequently, under the impacts of heating and shearing, the temperature of the film-forming ingredients in a screw extruder rises to the melting temperature.

This process generally includes three major types: extrusion blowing, extrusion compression molding and extrusion injection molding. To make films, extrusion blowing comprises blowing, pressing, and chilling extruded film-forming material. By hot pressing and chilling the extruded film-forming material, extrusion compression, and injection moldings form the films [21]. Extrusion processing has been used for the preparation of hydroxypropyl starch-based films [30] and cassava starch-based films [31].

McGlashan and Halley [32] had prepared starch nanocomposites using twin-screw extruder. They had inserted starch, polyester and organoclay all together into the extruder. In this research, the starch and polyester ratios were varied and then the properties were studied. They observed that addition of MMT into the starch/polyester exhibited increased tensile strength. In their work, samples with composition of 30 wt% starch showed higher elongation at break and this was due to addition of MMT into the nanocomposites. The improved gelatinization and dispersion of MMT in the starch nanocomposites made it exhibited higher tensile properties especially in elongation at break.

2.3.7 Reactive extrusion

Reactive extrusion is a polymer extrusion processing technology that employs chemical or enzymatic processes to change the polymer, such as crosslinking, grafting, polycondensation and polymerization. In reactive extrusion, the extruder serves as a chemical or bioreactor as well as a processing device [21]. Multiple physical fields, such as considerable shear, temperature, and pressure input, are present in this reactor. To achieve enzymatic reactive extrusion, exogenous enzymes have recently been incorporated into reactive extrusion. The thermo-mechanical action of the extruder is superimposed on the enzyme action in the extrusion phase, increasing production efficiency [33].

To achieve improved efficiency, reactive extrusion can be utilised to change starch in less time and at a cheaper cost. Reactive extrusion has been shown to achieve cassava starch plasticization and succinization [34], grafting of starch and acrylamide monomer [35], cellulose acetate/corn starch as biodegradable food packaging films [36] and starch/polylactic acid as an alternative to petroleum-based materials in studies [31].

2.3.8 Electrospinning

Electrospinning is a method for electrostatically producing small fibers with diameters in the nanometers and microns from polymer solutions or melts. It is widely utilised in the food and biomedical industries [37]. Electrospinning involves charging a polymer solution or melt with a high voltage power supply, which results in droplets of liquid at the needle's tip, which are subsequently dragged out by electrostatic forces to create a Taylor cone. The polymer solution is expelled from the Taylor cone toward the grounded collector when the electrostatic force overcomes the surface tension. Before the jet reaches the collector, the solvent evaporates, causing the polymer chains in the jet to stretch and orient. The jet then freezes into nanofibers, which are deposited on the collector [38].

Because of its biocompatibility, starch offers a lot of potential for electrospinning nanofibers. Starch can be electrospun with various polymers such as polycaprolactone, polyvinyl alcohol, polyepoxyethane, polyglycolide, polylactic acid (PLA), and polyglycolide (ethylene-covinylalcohol) [39]. Electrospinning also gives starch and other biopolymers some unique properties that can be employed in a variety of applications, such as food packaging, drug delivery, and wound dressings [21].

2.3.9 3D printing

Three-dimensional printing (3D printing) is a technology that uses computer software to create 3D models that are then printed layer by layer. This method has been used in a variety of fields, including medical, aerospace, vehicles, cuisine, art, textiles, and architecture [40]. 3D printing has recently been employed in the food sector to manufacture food with complicated geometric shapes, fine textures, and unique nutrient profiles to fit the demands of various people [21].

The study from Kuo et al. [41] on starch/acrylonitrile-butadiene-styrene copolymers have been used to produce specific shapes using 3D-printing. The suitability of starch-based materials for 3D printing is determined by their printability, gelling behavior, mechanical qualities, and heat resistance. Combining starch-based materials with additional ingredients (such as pea protein) [42] or chemical modification can typically increase the printability of starch-based materials (such as ozonation) [43].

2.3.10 Nanotechnology

Nanotechnology is a method for fabrication materials with dimensions of 1–100 nm. Nanomaterials provide us with functional features that are difficult to achieve with traditional materials. As a result, nanotechnology has become widely used in recent years in agriculture, food, medical, automobiles, aerospace, and other industries [44]. The increased emphasis on human health and sustainable development has prompted research on nano-biodegradable biopolymers for use in agriculture, food, and drug delivery [21].

This method can be done by adding nanoscale components to starch matrix and have produced biodegradable nanocomposite films based on starch. Nanomaterials can alter starch's crystallization kinetics, morphology, grain shape, and grain size, increasing the mechanical and barrier properties of starch-based materials [45]. Some studies that shown the enhancement of starch-based materials by addition of variety of nano-materials such cellulose nanofibers [46], montmorillonite nanoclay [45] and bentonite nanoclay [47].

2.4 Additives for starch-based packaging material

Additives play significant roles in determining the qualities of the packaging materials, particularly for starch-based packaging where their brittleness and hydrophilicities are the main concern. The inclusion of additives in the formulation of starch-based packaging is intended to improve the mechanical and the barrier performance's requirements. Nanocellulose, such as cellulose nanocrystal (CNC) and cellulose nanofiber (CNF), has recently been used as an additive to substantially improve the performance of packaging films [48]. The addition of 5–15% CNCs into starch-based film had enhanced the tensile strength and Young's modulus [49]. Furthermore, it was found to be effective in reducing moisture absorption and water vapor permeability [50].

The use of additives can also greatly improve the processing of starch-based packaging materials. By reducing internal friction, processing aid additives increase the productivities of the machineries. Additionally, the use of additives enables manufacturers to tailor the packaging materials for specific applications. For example, the use of essential oils in the production of starch-based films has improved the barrier and optical properties. Furthermore, the active compounds contain in the oils have also imparted the antioxidant and antibacterial properties to the films [51–53].

The following sections discuss several additives used in the production of starch-based packaging materials, such as plasticizers, crosslinkers, antibacterial agents, and antioxidants.

2.4.1 Plasticizer

The main obstacle to the starch-based production process was their brittleness and difficulty to process without the addition of suitable additives, hence plasticizers were added to enhance and improve the processability, workability, and flexibility of the starches [54].

Plasticizers are added during the plasticization process to break the strong inter-and intra-molecular and minimize the interactions of the starch chains to help improve the starch flexibility and chain mobility [55, 56]. Intramolecular and intermolecular hydrogen bonds can be destroyed with the addition of the plasticizers and the double-helical structure of starch molecules are disrupted, reducing their melting points, resulting in a thermoplastic starch [57]. Plasticizer properties, such as their functional groups, molecular weight, thermal stability, and compatibility with polymers, play significant roles in the production of the starch plasticization process [55]. Table 2.1 shows the classification of the common plasticizers used in the plasticization process.

Glycerol is the most common and cheapest plasticizer used in the preparation of starch-based products [57, 59] as it is compatible with amylose. Due to its compatibility with amylose, glycerol is frequently used in starch plasticization as it interferes with amylose packing which decreases the intermolecular forces between the starch [51, 60, 61]. This interference in amylose packing helps to promote better mechanical properties of the starch-based product [62, 63].

Gao et al. [57] incorporated 15 wt% of glycerol with the addition of 15 wt% of water as co plasticizer in the production of arrowroot starch film. The addition of water as a co-plasticizer helped significantly in reducing the cost of production. Meanwhile, Paula

Table 2.1: Classification of the plasticizers [58].

Types	Plasticizers
Polyols	Glycerol (glycerin), propylene glycol, and polyethylene glycols (PEG)
Organic esters	Phthalate esters (diethyl and dibutyl), dibutyl sebacate, citrate esters (triethyl, acetyl triethyl, and acetyl tributyl), and triacetin
Oils/ glycerides	Castor oil, acetylated monoglycerides, and fractionated coconut oil

Guarás et al. [64] used 30 wt% combinations of glycerol and water in the production of corn starch film via an extrusion process. It shows that the addition of water in the plasticization process was able to reduce the temperature profile inside the barrel, thus preventing the thermoplastic starch from thermally degraded during the gelatinization process. Glycerol was also used by Tarique et al. [65] in the production of arrowroot starch biopolymers through the solution casting method. The research revealed that film thickness, moisture content, and solubility in water were proportionately increased with the glycerol content whereby water absorption and density decreased.

Besides glycerol, sorbitol is another type of plasticizer that received much attention due to its effects that exceeded those of glycerol in most categories as complex processing and high content in biocomposites resulting in much more impact [66]. Jha [54] used 30 wt% of sorbitol in the corn starch-chitosan film and found that film had the highest tensile strength but poor in film solubility, water vapor permeability and thermal stability as compared to glycerol. Hazrati et al. [56] applied sorbitol in the production of *Dioscorea hispida* starch film, and found that the moisture content and elongation at break increased while tensile strength decreased with the increasing amount of sorbitol concentrations.

Other types of plasticizers reported in the literatures including isosorbide in the corn starch-filled with poly(3-hydroxybutyrate-co-3-hydroxyvalerate) microparticle [67] and fructose for the corn starch-based bioplastics packaging film [68]. Meanwhile, amide groups-containing plasticizers including urea, formamide, and acetamide are not favored due to their toxicities which are not suitable to be used in food packaging applications [55].

2.4.2 Crosslinker

The use of crosslinker to chemically improve the poor mechanical properties of starch-based packaging is not new. Starch crosslinking involves the formation of intermolecularly or intramolecularly bonding between the polymer chains. During the crosslinking process, the multifunctional reagents that can form ester or ether bonds are added into the formulation. The crosslinking reagents' multifunctional group reacts with the hydroxyl groups of amylose and amylopectin, causing the starch to be less hydrophilic. Epichlorohydrin (ECH), sodium trimetaphosphate (STMP), and sodium tripolyphosphate (STPP) are examples of the crosslinkers used in the preparation of starch-based film [69]. Glutaraldehyde (GA) is also one of the aldehydes that have been widely utilized in the crosslinking of starch [70].

Some crosslinkers such as GA and ECH, are toxic, which limit their use in biomaterials and food-contacted materials. ECH, upon contact with water, could be readily hydrolyzed into 3-chloropropane-1,2-diol, which is a carcinogenic substance that could cause harm to humans' health. Recently, citric acid (CA) has been widely reported as an effective material for improving the properties of native starch in the search for a nontoxic crosslinker. CA contains tri-functional carboxylic acid (COOH), which can interact with the hydroxyl groups (OH groups) of starch to form an ester and establish stronger hydrogen bonds [71]. Studies showed that CA was able to increase the film's tensile strength [72, 73] and reduced water vapor transmittance in starch-based materials [72, 74]. Oxidized sucrose, a novel aldehyde-based green cross-linker, also endowed starch films with substantial improvement in both tensile strength (23 MPa) and elongation (60%) [75].

2.4.3 Antibacterial and antimicrobial agents

Another interesting feature revolving around starch-based packaging is anti-bacterial property. This anti-bacterial characteristic is crucial, especially in starch-based food packaging as good food packaging should prevent unfavorable conditions such as spoilage and microbial contamination. Nanoparticles and nanoclay such as silica dioxide [76] and montmorillonite (MMT) [77] have been added to the starch matrix as an anti-bacterial agent. The starch-based films demonstrated antimicrobial activity against *Staphylococcus aureus, Escherichia coli* and *Candida albicans*. The addition of metal nanoparticles such as silver (Ag) and titanium dioxide (TiO_2) had also improved the antibacterial resistance [77–79]. Interestingly, only a small amount of nanoparticles (<5%) is required to improve the packaging properties.

The utilization of natural biopolymers such as chitosan has imparted the antimicrobial properties onto the starch-based packaging [32]. The presence of protonated amino groups in chitosan inhibited the bacterial growth. Some bacteria including *Bacillus cereus, S. aureus, E. coli, and Salmonella typhimurium*, were found to have an inhibitory zone [80]. The antibacterial properties were evaluated by counting the growth of bacteria in the wrapped cherry tomatoes.

In another study, approximately 98–99% of foodborne bacterial growth and 87–99% of fungal growth have been inhibited when 15 w/v % of citric acid was added into sago starch/poly vinyl acetate (PVA) film [81]. A qualitative assessment of food shelf-life also revealed that the starch-citrate film was more effective than commercial food wrappers at impeding the microbial growth in cake and bread samples (the film extended the shelf-life of the food products; 10 days for cake and 40 days for bread).

Besides that, natural active substances usually extracted from plants like polyphenols, flavonoids, tannins, and alkaloids exhibit antibacterial effects that can effectively inhibit the proliferation of bacteria. These extracts are commonly used in their oil state, known as essential oil (EO). Due to the EO's volatility and lipophilicity, it allows the active compound to penetrate and damage the bacteria's cell wall, which eventually results in cell death [51]. Cinnamon and clove essential oils had proven to be effective against two microorganisms commonly found in bread; *P. commune* and *Escherichia amstelodami* even at lower concentrations [82]. Biodegradable starch film derived from corn/wheat with the addition of lemon EO portrayed antibacterial activity against *E. coli* and *S. aureus* and was enhanced with the presence of surfactants (Tween 80 and Span 80) [83].

2.4.4 Antioxidant or stabilizer

Antioxidants are commonly added into the formulations of many packaging materials, in order to sustain and maintain the shelf life of the packaging for a certain period. The antioxidant activity is manifested by radical scavenging mechanisms including oxygen. As the reaction with radicals stops the chain reactions, it delays the ageing of the polymer [84]. Natural antioxidants such as organic acids, phenolic acids, terpenes, tocopherols, carotenoids, and vitamins have been used in starch films to improve product stability and extend storage time. The most commonly used is polyphenolic compounds as they are effective natural antioxidants due to their reducing properties, which aid in free radical scavenging. Beetroot extract [85], propolis extract [86], and rosemary extract [87] were all reported to be phenolic-rich compounds suitable for starch-based packaging, particularly as bioactive film. A study revealed that starch film incorporating phenolic compounds as a functional food packaging had protected the fatty foods against oxidation, by delaying lipid oxidation approximately 35 days [88].

Water-soluble phenolic compounds (ferulic, *p*-coumaric and protocatechuic acid) extracted from rice straw were also found to be a suitable antioxidant for the starch film as good *in-vitro* antiradical scavenging activity against DPPH* was observed [89]. Meanwhile, potato peel phenolic extract exhibited antioxidant activity (56–85% of $ABTS^{\bullet +}$ inhibition after 7 days) [90]. By reacting directly with the free radical or preventing hydroperoxides from decomposing into free radicals, phenolic compounds can stop propagation reactions induced by free radical intermediates. Not only that, the presence of the aromatic compounds in polyphenol can act as an excellent ultraviolet (UV) barrier as the free radicals are easily formed when exposed to UV light [53]. In conclusion, the addition of various additives in the formulations has driven the development of new starch-based packaging materials.

2.5 Challenge and future prospect on starch based packaging material

Starch-based biodegradable materials play an important role in the food industry, especially in food packaging. Furthermore, the development of more ecologically friendly and sustainable food packaging materials will also be aided by this starch-based biodegradable packaging in the future. However, starch has a few drawbacks such as poor mechanical strength and high water permeability which limits their applications. Hence, some approaches to improve starch-based biodegradable packaging have been investigated, such as processing method and incorporation of additives. In addition, future researchers are recommended to incorporate functional materials into starch-based materials to provide them with new products, such as antibacterial agents, antioxidant agents, sensor technologies, and more to increase their present qualities. As a result, starch-based materials will be able to be used to develop active or smart packaging materials in the future.

2.6 Conclusions

In the future, starch based bioplastic will play a greater role in the packaging industry. The urge to replace synthetic plastic with bioplastic derived from starch is increasing. This is due to the fact that starch possesses desirable characteristics, such as being biodegradable, inexpensive, and easy to obtain from common renewable resources. Nowadays, these characteristics still gaining a lot of attention by researchers to improve starch-based biodegradable materials for future use. This can be seen from the studies on addition of chemical, enzymatic or physical modification of starch. As a result, research into improving starch properties will continue until we achieve the best results from starch-based food packaging.

References

1. The global market for plastic products will be worth $1.2 trillion by 2020 [Online]. Available from: https://resource-recycling.com/resourcerecycling/wp-content/uploads/2017/12/TBRC-Plastic-Products.pdf [Accessed 8 Aug 2022].
2. Gadhave RV, Das A, Mahanwar PA, Gadekar PT. Starch based bio-plastics: the future of sustainable packaging. Wuhan, China: Scientific Research Publishing, Inc.; 2018.
3. Marichelvam MK, Jawaid M, Asim M. Corn and rice starch-based bio-plastics as alternative packaging materials. Fibers 2019;7:32.
4. Appelqvist IAM, Debet MRM. Starch-biopolymer interactions—a review. Food Rev Int 2009;13:163–224.
5. Pokhrel S. A review on introduction and applications of starch and its biodegradable polymers. Int J Environ 2015;4:114–25.
6. Guohua Z, Ya L, Cuilan F, Min Z, Caiqiong Z, Zongdao C. Water resistance, mechanical properties and biodegradability of methylated-cornstarch/poly (vinyl alcohol) blend film. Polym Degrad Stabil 2006;91: 703–11.
7. Tang S, Zou P, Xiong H, Tang H. Effect of nano-SiO2 on the performance of starch/polyvinyl alcohol blend films. Carbohydr Polym 2008;72:521–6.
8. Yu L, Dean K, Li L. Polymer blends and composites from renewable resources. Prog Polym Sci 2006;31: 576–602.
9. Ramesh M, Mitchell JR, Jumel K, Harding SE. Amylose content of rice starch. Starch/Staerke 1999;51:311–3.
10. Malathi AN, Santhosh KS, Nidoni U. Recent trends of Biodegradable polymer: biodegradable films for food packaging and application of nanotechnology in biodegradable food packaging. Bhopal, India: Revista Info Solutions; 2014.
11. Avella M, de Vlieger JJ, Errico ME, Fischer S, Vacca P, Volpe MG. Biodegradable starch/clay nanocomposite films for food packaging applications. Food Chem 2005;93:467–74.
12. Sorrentino A, Gorrasi G, Vittoria V. Potential perspectives of bio-nanocomposites for food packaging applications. Trends Food Sci Technol 2007;18:84–95.

13. Chen DR, Bei JZ, Wang SG. Polycaprolactone microparticles and their biodegradation. Polym Degrad Stabil 2000;67:455–9.
14. Müller RJ, Kleeberg I, Deckwer WD. Biodegradation of polyesters containing aromatic constituents. J Biotechnol 2001;86:87–95.
15. Shafqat A, Al-Zaqri N, Tahir A, Alsalme A. Synthesis and characterization of starch based bioplatics using varying plant-based ingredients, plasticizers and natural fillers. Saudi J Biol Sci 2021;28:1739–49.
16. Yurong G, Dapeng L. Preparation and characterization of corn starch/PVA/glycerol composite films incorporated with ϵ-polylysine as a novel antimicrobial packaging material. E-Polymers 2020;20:154–61.
17. Begum MHA, Hossain MM, Gafur MA, Kabir ANMH, Tanvir NI, Molla MR. Preparation and characterization of polyvinyl alcohol–starch composites reinforced with pulp. SN Appl Sci 2019;1:1091.
18. Hu H, Xu A, Zhang D, Zhou W, Peng S, Zhao X. High-toughness poly(lactic acid)/starch blends prepared through reactive blending plasticization and compatibilization. Molecules 2020;25:5951.
19. Mohan TP, Kanny K. Thermoforming studies of corn starch-derived biopolymer film filled with nanoclays. J Plastic Film Sheeting 2016;32:163–88.
20. McGlashan SA, Halley PJ. Preparation and characterisation of biodegradable starch-based nanocomposite materials. Polym Int 2003;52:1767–73.
21. Cheng H, Chen L, McClements DJ, Yang T, Zhang Z, Ren F, et al. Starch-based biodegradable packaging materials: a review of their preparation, characterization and diverse applications in the food industry. Trends Food Sci Technol 2021;114:70–82.
22. Shafqat A, Al-Zaqri N, Tahir A, Alsalme A. Synthesis and characterization of starch based bioplatics using varying plant-based ingredients, plasticizers and natural fillers. Saudi J Biol Sci 2021;28:1739–49.
23. Gao W, Wu W, Liu P, Hou H, Li X, Cui B. Preparation and evaluation of hydrophobic biodegradable films made from corn/octenyl succinated starch incorporated with different concentrations of soybean oil. Int J Biol Macromol 2020;142:376–83.
24. Junlapong K, Boonsuk P, Chaibundit C, Chantarak S. Highly water resistant cassava starch/poly (vinyl alcohol) films. Int J Biol Macromol 2019;137:521–7.
25. Liu P, Wang R, Kang X, Cui B, Yu B. Effects of ultrasonic treatment on amylose-lipid complex formation and properties of sweet potato starch-based films. Ultrason Sonochem 2018;44:215–22.
26. Hu H, Xu A, Zhang D, Zhou W, Peng S, Zhao X. High-toughness poly (lactic acid)/starch blends prepared through reactive blending plasticization and compatibilization. Molecules 2020;25:5951.
27. Mohan TP, Kanny K. Thermoforming studies of corn starch-derived biopolymer film filled with nanoclays. J Plastic Film Sheeting 2015;32:163–88.
28. Zeng GS, Hu C, Zou S, Zhang L, Sun G. BP neural network model for predicting the mechanical performance of a foamed wood-fiber reinforced thermoplastic starch composite. Polym Compos 2019;40:3923–408.
29. Chen W, Ma S, Wang Q, McClements DJ, Liu X, Ngai T, et al. Fortification of edible films with bioactive agents: a review of their formation, properties, and application in food preservation; 2021. Available from: https://www.tandfonline.com/doi/abs/10.1080/10408398.2021.1881435 [Accessed 10 Aug 2022].
30. Gao W, Liu P, Li X, Qiu L, Hou H, Cui B. The co-plasticization effects of glycerol and small molecular sugars on starch-based nanocomposite films prepared by extrusion blowing. Int J Biol Macromol 2019;133:1175–81.
31. Palai B, Biswal M, Mohanty S, Nayak SK. In situ reactive compatibilization of polylactic acid (PLA) and thermoplastic starch (TPS) blends; synthesis and evaluation of extrusion blown films thereof. Ind Crop Prod 2019;141:111748.
32. McGlashan SA, Halley PJ. Preparation and characterisation of biodegradable starch-based nanocomposite materials. Polym Int 2003;52:1767–73.
33. Xu E, Campanella OH, Ye X, Jin Z, Liu D, BeMiller JN. Advances in conversion of natural biopolymers: a reactive extrusion (REX)–enzyme-combined strategy for starch/protein-based food processing. Trends Food Sci Technol 2020;99:167–80.

34. Fonseca-Florido HA, Soriano-Corral F, Yañez-Macías R, González-Morones P, Hernández-Rodríguez F, Aguirre-Zurita J, et al. Effects of multiphase transitions and reactive extrusion on in situ thermoplasticization/succination of cassava starch. Carbohydr Polym 2019;225:115250.
35. Siyamak S, Laycock B, Luckman P. Synthesis of starch graft-copolymers via reactive extrusion: process development and structural analysis. Carbohydr Polym 2020;227:115066.
36. Herniou-Julien C, Mendieta JR, Gutiérrez TJ. Characterization of biodegradable/non-compostable films made from cellulose acetate/corn starch blends processed under reactive extrusion conditions. Food Hydrocolloids 2019;89:67–79.
37. Kouhi M, Prabhakaran MP, Ramakrishna S. Edible polymers: an insight into its application in food, biomedicine and cosmetics. Trends Food Sci Technol 2020;103:248–63.
38. Liu G, Gu Z, Hong Y, Cheng L, Li C. Electrospun starch nanofibers: recent advances, challenges, and strategies for potential pharmaceutical applications. J Contr Release 2017;252:95–107.
39. Mondragón M, López-Villegas O, Sánchez-Valdés S, Rodríguez-González FJ. Effect of thermoplastic starch and photocrosslinking on the properties and morphology of electrospun poly (ethylene-co-vinyl alcohol) mats. Polym Eng Sci 2020;60:474–80.
40. Liu J, Sun L, Xu W, Wang Q, Yu S, Sun J. Current advances and future perspectives of 3D printing natural-derived biopolymers. Carbohydr Polym 2019;207:297–316.
41. Kuo CC, Liu LC, Teng WF, Chang HY, Chien FM, Liao SJ, et al. Preparation of starch/acrylonitrile-butadiene-styrene copolymers (ABS) biomass alloys and their feasible evaluation for 3D printing applications. Compos B Eng 2016;86:36–9.
42. Chuanxing F, Qi W, Hui L, Quancheng Z, Wang M. Effects of pea protein on the properties of potato starch-based 3D printing materials. Int J Food Eng 2018;14:20170297.
43. Maniglia BC, Lima DC, Matta Junior MD, Le-Bail P, Le-Bail A, Augusto PED. Hydrogels based on ozonated cassava starch: effect of ozone processing and gelatinization conditions on enhancing 3D-printing applications. Int J Biol Macromol 2019;138:1087–97.
44. Sampathkumar K, Tan KX, Loo SCJ. Developing nano-delivery systems for agriculture and food applications with nature-derived polymers. iScience 2020;23:101055.
45. Issa AT, Schimmel KA, Worku M, Shahbazi A, Ibrahim SA, Tahergorabi R. Sweet potato starch-based nanocomposites: development, characterization, and biodegradability. Starch/Staerke 2018;70:1700273.
46. Fourati Y, Magnin A, Putaux JL, Boufi S. One-step processing of plasticized starch/cellulose nanofibrils nanocomposites via twin-screw extrusion of starch and cellulose fibers. Carbohydr Polym 2020;229:115554.
47. Merino D, Mansilla AY, Casalongué CA, Alvarez VA. Effect of nanoclay addition on the biodegradability and performance of starch-based nanocomposites as Mulch films. J Polym Environ 2019;27:1959–70.
48. Zhang W, Zhang Y, Cao J, Jiang W. Improving the performance of edible food packaging films by using nanocellulose as an additive. Int J Biol Macromol 2021;166:288–96.
49. Coelho CCDS, Silva RBS, Carvalho CWP, Rossi AL, Teixeira JA, Freitas-Silva O, et al. Cellulose nanocrystals from grape pomace and their use for the development of starch-based nanocomposite films. Int J Biol Macromol 2020;159:1048–61.
50. Chen QJ, Zhou LL, Zou JQ, Gao X. The preparation and characterization of nanocomposite film reinforced by modified cellulose nanocrystals. Int J Biol Macromol 2019;132:1155–62.
51. Vianna TC, Marinho CO, Marangoni Júnior L, Ibrahim SA, Vieira RP. Essential oils as additives in active starch-based food packaging films: a review. Int J Biol Macromol 2021;182:1803–19.
52. Syafiq R, Sapuan SM, Zuhri MYM, Ilyas RA, Nazrin A, Sherwani SFK, et al. Antimicrobial activities of starch-based biopolymers and biocomposites incorporated with plant essential oils: a review. Polymers 2020;12: 1–26.
53. Cui C, Ji N, Wang Y, Xiong L, Sun Q. Bioactive and intelligent starch-based films: a review. Trends Food Sci Technol 2021;116:854–69.
54. Jha P. Effect of plasticizer and antimicrobial agents on functional properties of bionanocomposite films based on corn starch-chitosan for food packaging applications. Int J Biol Macromol 2020;160:571–82.

55. Montilla-Buitrago CE, Gómez-López RA, Solanilla-Duque JF, Serna-Cock L, Villada-Castillo HS. Effect of plasticizers on properties, retrogradation, and processing of extrusion-obtained thermoplastic starch: a review. Starch/Staerke 2021;73:2100060.
56. Hazrati KZ, Sapuan SM, Zuhri MYM, Jumaidin R. Effect of plasticizers on physical, thermal, and tensile properties of thermoplastic films based on Dioscorea hispida starch. Int J Biol Macromol 2021;185:219–28.
57. Gao W, Zhu J, Kang X, Wang B, Liu P, Cui B, et al. Development and characterization of starch films prepared by extrusion blowing: the synergistic plasticizing effect of water and glycerol. LWT (Lebensm-Wiss & Technol) 2021;148:111820.
58. Foroughi-Dahr M, Mostoufi N, Sotudeh-Gharebagh R, Chaouki J. Particle coating in fluidized beds. In: Reference module in chemistry, molecular sciences and chemical engineering. Amsterdam, Netherlands: Elsevier; 2017:1–17 pp.
59. Paluch M, Ostrowska J, Tynski P, Sadurski W, Konkol M. Structural and thermal properties of starch plasticized with glycerol/urea. J Polym Environ 2022;30:728–40.
60. Farhan A, Hani NM. Characterization of edible packaging films based on semi-refined kappa-carrageenan plasticized with glycerol and sorbitol. Food Hydrocolloids 2017;64:48–58.
61. Nordin N, Othman SH, Rashid SA, Basha RK. Effects of glycerol and thymol on physical, mechanical, and thermal properties of corn starch films. Food Hydrocolloids 2020;106:105884.
62. Abera G, Woldeyes B, Demash HD, Miyake G. The effect of plasticizers on thermoplastic starch films developed from the indigenous Ethiopian tuber crop Anchote (Coccinia abyssinica) starch. Int J Biol Macromol 2020;155:581–7.
63. Ballesteros-Mártinez L, Pérez-Cervera C, Andrade-Pizarro R. Effect of glycerol and sorbitol concentrations on mechanical, optical, and barrier properties of sweet potato starch film. NFS J 2020;20:1–9.
64. Paula Guarás M, Alvarez VA, Ludueña LN. Effect of storage time, plasticizer formulation and extrusion parameters on the performance of thermoplastic starch films. Adv Mater Lett 2019;10:206–14.
65. Tarique J, Sapuan SM, Khalina A. Effect of glycerol plasticizer loading on the physical, mechanical, thermal, and barrier properties of arrowroot (Maranta arundinacea) starch biopolymers. Sci Rep 2021;11:1–17.
66. Rojas-Bringas PM, De-la-Torre GE, Torres FG. Influence of the source of starch and plasticizers on the environmental burden of starch-Brazil nut fiber biocomposite production: a life cycle assessment approach. Sci Total Environ 2021;769:144869.
67. Area MR, Montero B, Rico M, Barral L, Bouza R, López J. Isosorbide plasticized corn starch filled with poly(3-hydroxybutyrate-co-3-hydroxyvalerate) microparticles: properties and behavior under environmental factors. Int J Biol Macromol 2022;202:345–53.
68. Abotbina W, Sapuan SM, Sultan MTH, Alkbir MFM, Ilyas RA. Development and characterization of cornstarch-based bioplastics packaging film using a combination of different plasticizers. Polymers 2021; 13:1–18.
69. Detduangchan N, Sridach W, Wittaya T. Enhancement of the properties of biodegradable rice starch films by using chemical crosslinking agents. Int J Food Res Technol 2014;21:1225–35.
70. Paiva D, Pereira AM, Pires AL, Martins J, Carvalho LH, Magalhães FD. Reinforcement of thermoplastic corn starch with crosslinked starch/chitosan microparticles. Polymers 2018;10:1–14.
71. Reddy N, Yang Y. Citric acid cross-linking of starch films. Food Chem 2010;118:702–11.
72. Ghanbarzadeh B, Almasi H, Entezami AA. Improving the barrier and mechanical properties of corn starch-based edible films: effect of citric acid and carboxymethyl cellulose. Ind Crop Prod 2011;33:229–35.
73. Chen WC, Mohd Judah SNMS, Ghazali SK, Munthoub DI, Alias H, Mohamad Z, et al. The effects of citric acid on thermal and mechanical properties of crosslinked starch film. Chem Eng Trans 2021;83:199–204.
74. Seligra PG, Jaramillo CM, Famá L, Goyanes S. Biodegradable and non-retrogradable eco-films based on starch – glycerol with citric acid as crosslinking agent. Carbohydr Polym 2016;138:66–74.
75. Xu H, Canisag H, Mu B, Yang Y. Robust and flexible films from 100% starch cross-linked by biobased disaccharide derivative. ACS Sustainable Chem Eng 2015;3:2631–9.

76. Zhang R, Wang X, Cheng M. Preparation and characterization of potato starch film with various size of Nano-SiO2. Polymers 2018;10:9–12.
77. Abreu AS, Oliveira M, De Sá A, Rodrigues RM, Cerqueira MA, Vicente AA, et al. Antimicrobial nanostructured starch based films for packaging. Carbohydr Polym 2015;129:127–34.
78. Anaya-Esparza LM, de la Mora ZV, Ruvalcaba-Gómez JM, Romero-Toledo R, Sandoval-Contreras T, Aguilera-Aguirre S, et al. Use of titanium dioxide (TiO2) nanoparticles as reinforcement agent of polysaccharide-based materials. Processes 2020;8:1–26.
79. Liu C, Xiong H, Chen X, Lin S, Tu Y. Effects of nano-TiO2 on the performance of high-amylosestarch based antibacterial films. J Appl Polym Sci 2015;132:42339.
80. Shapi'i RA, Othman SH, Nordin N, Kadir Basha R, Nazli Naim M. Antimicrobial properties of starch films incorporated with chitosan nanoparticles: in vitro and in vivo evaluation. Carbohydr Polym 2020;230:115602.
81. Romainor AN, Chin SF, Lihan S. Antimicrobial starch-based film for food packaging application. Starch/Staerke 2022;74:2100207.
82. Souza AC, Goto GEO, Mainardi JA, Coelho ACV, Tadini CC. Cassava starch composite films incorporated with cinnamon essential oil: antimicrobial activity, microstructure, mechanical and barrier properties. LWT – Food Sci Technol (Lebensmittel-Wissenschaft -Technol) 2013;54:346–52.
83. Song X, Zuo G, Chen F. Effect of essential oil and surfactant on the physical and antimicrobial properties of corn and wheat starch films. Int J Biol Macromol 2018;107:1302–9.
84. Singh P, Saengerlaub S, Wani AA, Langowski HC. Role of plastics additives for food packaging. Pigment Resin Technol 2012;41:368–79.
85. Tran TN, Athanassiou A, Basit A, Bayer IS. Starch-based bio-elastomers functionalized with red beetroot natural antioxidant. Food Chem 2017;216:324–33.
86. De Araújo GKP, De Souza SJ, Da Silva MV, Yamashita F, Gonçalves OH, Leimann FV, et al. Physical, antimicrobial and antioxidant properties of starch-based film containing ethanolic propolis extract. Int J Food Sci Technol 2015;50:2080–7.
87. Piñeros-Hernandez D, Medina-Jaramillo C, López-Córdoba A, Goyanes S. Edible cassava starch films carrying rosemary antioxidant extracts for potential use as active food packaging. Food Hydrocolloids 2017;63:488–95.
88. Rajapaksha SW, Shimizu N. Development and characterization of functional starch-based films incorporating free or microencapsulated spent black tea extract. Molecules 2021;26:3898.
89. Menzel C, González-Martínez C, Vilaplana F, Diretto G, Chiralt A. Incorporation of natural antioxidants from rice straw into renewable starch films. Int J Biol Macromol 2020;146:976–86.
90. Lopes J, Gonçalves I, Nunes C, Teixeira B, Mendes R, Ferreira P, et al. Potato peel phenolics as additives for developing active starch-based films with potential to pack smoked fish fillets. Food Packag Shelf Life 2021;28:100644.

A. S. Norfarhana, R. A. Ilyas*, A. Nazrin, Salit Mohd Sapuan, R. M. O. Syafiq, P. S. Khoo, Abu Hassan Nordin, Abdoulhdi A. Borhana Omran, Dominic C. D. Midhun, H. S. N. Hawanis, Nasmi Herlina Sari, Melbi Mahardika, Mochamad Asrofi and Hairul Abral

3 Nanocellulose: from biosources to nanofiber and their applications

Abstract: Nanocellulose is a product of cellulose, a sustainable and plentiful resource. It's distinctive nanoscale structure makes it a versatile, green and interesting material for a variety of applications. This article describes in detail the biosources of nanocellulose, the types and characteristics of nanocellulose, and the techniques used to produce

***Corresponding author: R. A. Ilyas**, Department of Chemical Engineering, Faculty of Chemical and Energy Engineering, Universiti Teknologi Malaysia, 81310 UTM Johor Bahru, Johor, Malaysia; Centre for Advanced Composite Materials, Universiti Teknologi Malaysia, 81310 UTM Johor Bahru, Johor, Malaysia; Centre of Excellence for Biomass Utilization, Universiti Malaysia Perlis, 02600, Arau, Perlis, Malaysia; and Institute of Tropical Forestry and Forest Products, Universiti Putra Malaysia, Serdang 43400, Selangor, Malaysia, E-mail: ahmadilyas@utm.my. https://orcid.org/0000-0001-6622-2632

A. S. Norfarhana, Abu Hassan Nordin and H. S. N. Hawanis, Department of Chemical Engineering, Faculty of Chemical and Energy Engineering, Universiti Teknologi Malaysia, 81310 UTM Johor Bahru, Johor, Malaysia. https://orcid.org/0000-0001-7546-3833 (A.H. Nordin)

A. Nazrin, Institute of Tropical Forestry and Forest Products, Universiti Putra Malaysia, Serdang 43400, Selangor, Malaysia

Salit Mohd Sapuan, Advanced Engineering Materials and Composites Research Center (AEMC), Faculty of Engineering, Universiti Putra Malaysia, 43400 Seri Kembangan, Selangor, Malaysia

R. M. O. Syafiq, Department of Engineering Design and Manufacture, Faculty of Engineering, University of Malaya, 50603 Kuala Lumpur, Malaysia

P. S. Khoo, Centre for Advanced Composite Materials, Universiti Teknologi Malaysia, 81310 UTM Johor Bahru, Johor, Malaysia

Abdoulhdi A. Borhana Omran, Department of Mechanical and Mechatronic Engineering, Faculty of Engineering, Sohar University, Sohar, P C-311, Oman

Dominic C. D. Midhun, Department of Chemistry, Sacred Heart College, Thevara, Pin 682013, Kochi, Kerala, India

Nasmi Herlina Sari, Mechanical Engineering Department, Faculty of Engineering, University of Mataram, Mataram, West Nusa Tenggara, Indonesia

Melbi Mahardika, Research Center for Biomass and Bioproducts, National Research and Innovation Agency (BRIN), Cibinong 16911, Indonesia

Mochamad Asrofi, Department of Mechanical Engineering, Faculty of Engineering, Universitas Jember, Kalimantan 37, Jember 68121, Indonesia

Hairul Abral, Department of Mechanical Engineering, Universitas Negeri Padang, Padang 25173, Indonesia; and Laboratory of Nanoscience and Technology, Department of Mechanical Engineering, Andalas University, Padang 25163, Indonesia

As per De Gruyter's policy this article has previously been published in the journal Physical Sciences Reviews. Please cite as: A. S. Norfarhana, R. A. Ilyas, A. Nazrin, S. M. Sapuan, R. M. O. Syafiq, P. S. Khoo, A. H. Nordin, A. A. Borhana Omran, D. C. D. Midhun, H. S. N. Hawanis, N. H. Sari, M. Mahardika, M. Asrofi and H. Abral "Nanocellulose: from biosources to nanofiber and their applications" *Physical Sciences Reviews* [Online] 2023. DOI: 10.1515/psr-2022-0008 | https://doi.org/10.1515/9783110773606-003

nanocellulose fibers. The mechanical properties and morphologies of nanocellulose fibers are addressed in depth, along with their prospective applications in sectors, including paper packaging, building materials, composites, biomedicine, energy storage and filtration. In addition, the current state of nanocellulose research, including the opportunities in the field, as well as the future prospects of nanocellulose as a viable and sustainable material for a vast array of applications, are discussed.

Keywords: biosources; composite; nano-fiber; nanocellulose; natural fiber.

3.1 Introduction

Nanocellulose is a natural fiber substance derived primarily either from wood pulp or non-wood pulp [1]. It has received significant attention in recent years as a new class of nanomaterials due to its unique properties and potential uses in a variety of fields. Nanocellulose is produced by mechanically or chemically breaking down cellulose fibers into ultra-small fibers with sizes ranging from a few nanometers to a few hundred nanometers. This produces a material with high mechanical strength, stiffness, and low density, making it ideal for industries such as paper and packaging, building materials, composites, and biomedicine. One of the key benefits of nanocellulose is its biodegradability, which makes it a more environmentally friendly alternative to synthetic materials. Additionally, it can be produced from renewable sources such as wood and agricultural waste, reducing reliance on finite resources. The purpose of this review paper is to provide an overview of the nanocellulose, covering the extraction techniques, its properties, and its potential application in vast field.

3.2 Natural fiber

Natural fibers are favorable because they may be sourced naturally, harvested safely, and are less expensive than glass fibers [2]. Natural fibers can be utilized to make a variety of products, including cushions, mats, rope, brushes, filters, and fish breeding environments [3, 4]. However, both fibers and fruits produce a large amount of waste. As a result, many studies have been conducted with the goal of maximizing the product while reducing waste. Some studies are focusing on bio nanocomposites that use natural fibers as its base due to its abundance and its potential to replace carbon fiber and fiber glass in the future. Natural fiber is a composite of hollow cellulose fibrils bound together by lignin in a hemicellulose matrix.

Natural fibers can be categorized according to their origin either from minerals, animals or plants. Among these classified natural fibers, plant fiber is the most widely used reinforcement biomaterial in biocomposite. According to [5], plant fiber can be classified based on the parts and type of plant that the fibers were extracted. Figure 3.1

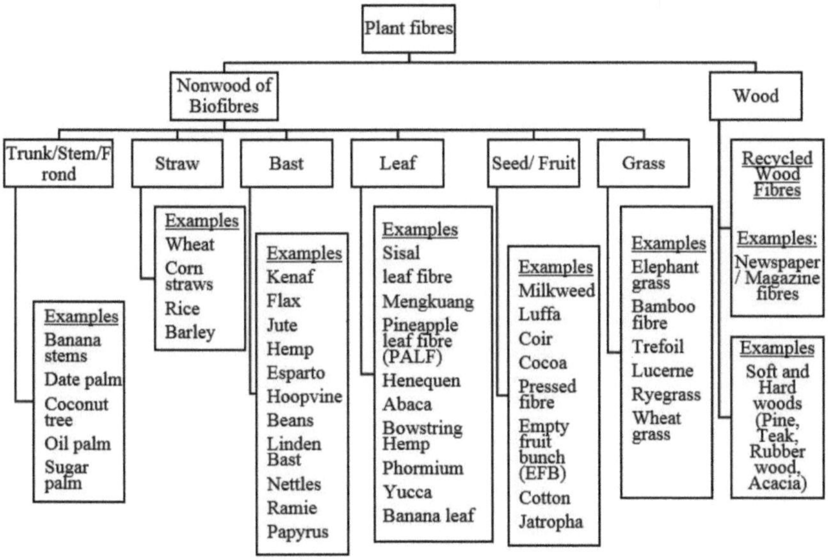

Figure 3.1: Classification of natural (plant) fibers.

depicts the major plant fiber groups, including grass fiber, leaf, stalk, fruit, seed and bast (non-wood fiber), and hardwood and softwood (wood fiber).

3.3 Increased usage of natural fibers as reinforcement for polymer composites

Usage of natural fiber reinforced polymer matrices is important for reducing the destruction of the ecosystem and providing low-cost polymeric matrices incorporated composite. Natural fibers have good mechanical properties, as shown in Table 3.1. Mechanical characteristics of fibrer vary depending on the part of the plant and the height of the plant. This can be seen in Table 3.2, in which sugar palm trunk, sugar palm bunch, sugar palm frond, and ijuk have different mechanical properties. The mechanical characteristics of sugar palm fiber vary with tree height. Higher height sugar palm fiber was shown to have better mechanical characteristics than lower height sugar palm fiber. Table 3.3 displays sugar palm fiber reinforced various polymer composites. The incorporation of sugar palm fiber into polymer matrices may help in addressing the ecosystem problems associated with nonbiodegradable waste that has accumulated in landfill. Heavy continuous dumping of solid waste disposal into the land fill had caused drastically decreasing of landfill area.

Table 3.1: Natural and synthetic fiber mechanical properties [5].

Fiber	Tensile strength (MPa)	Tensile modulus (GPa)	Elongation at break (%)
Abaca	980	–	–
Bagasse	20–290	19.7–27.1	1.1
Banana	355	33.8	5.3
Coie	220	6	15–25
Cotton	400	12	3–10
Flax	800–1500	60–80	1.2–1.6
Hemp	550–900	70	1.6
Henequen	430–580	–	3–4.7
Jute	400–800	10–30	1.8
Kenaf (bast)	295	–	2.7–6.9
Oil palm (empty fruit bunch)	248	3.2	2.5
Pineapple	170–1627	82	1–3
Ramie	500	44	2
Sisal	600–700	38	2–3
Sugar palm (frond)	421.4	10.4	9.8
E-glass	2000–3500	70	2.5
S-glass	4570	86	2.8

Table 3.2: Mechanical properties of fiber derived from different regions of the sugar palm tree [6].

Fibers	Sugar palm frond	Sugar palm bunch	Sugar palm trunk	Ijuk
Tensile strength (MPa)	421.4	365.1	198.3	276.6
Tensile modulus (GPa)	10.4	8.6	3.1	5.9
Elongation at break (%)	9.8	12.5	29.7	22.3

Table 3.3: Mechanical properties of sugar palm fiber at various tree heights [4].

Height (m)	1	3	5	7	9	11	13	15
Tensile strength (MPa)	15.5	81	149	201	266	292	279	270
Tensile modulus (GPa)	0.49	1.15	1.97	2.76	3.22	3.34	3.37	2.68
Elongation at break (%)	5.75	12.54	27.75	28.32	24.68	23.08	21	18.80
Toughness (Mj/m^3)	0.58	7.36	33.58	46.09	50.64	52.46	45.21	35.71

3.4 Downsides of natural fibers as reinforcement for polymer composites

3.4.1 Inconsistent fiber properties

Sugar palm offers numerous advantages, as described earlier in the preceding part; however, they also have some disadvantages that limit their utilisation in the composite industry [7]. The reasons for the inconsistency in sugar palm fiber properties were (1) the growing environment of the plant (i.e. soil properties, light intensity, and average annual rainfall in a region falls) (2) plant maturity (3) the part of the plant from which the fibers are extracted, and (4) the fiber harvesting and processing method. The inconsistency of sugar palm fiber properties would represent the non-uniformity of the mechanical properties of the fibers. To address this issue, fibers from diverse areas of a single tree must be blended [8].

3.4.2 Hydrophilicity of natural fibers

High moisture content of natural fibers had restricted their applications in industry that cover outdoor applications, (i.e. construction building material and automotive parts). Hydrophilic properties of SPF lead to ease rotting and microbial attack. These disadvantages would mostly be affected during period of delivery, long-term storage, and composite processing. The hygroscopic properties of cellulose in SPF allows it to absorb and swell when exposed to water. Swelling affects not just mechanical and physical properties, but also the stability of composite dimensions [9, 10].

3.4.3 Poor fiber – matrix adhesion

One of the most difficult difficulties confronting the usage of natural fiber is its low fiber-matrix adhesion properties. Natural fibers, in general, possess hydrophilic properties, which contrast with the thermoplastic/thermoset polymer properties, which are hydrophobic in nature. This behavior would have a negative impact on the compatibility of these two materials. Strong bonding is required at the interface between natural fibers and polymer matrices in order for the matrix to transfer load to the fiber via shear stresses. Thus, inadequate fiber-matrix adhesion would result in low mechanical characteristics of natural fiber reinforced polymer composites [5].

3.4.4 Low thermal stability

Another limitation for the use of natural fiber reinforced composites is their low thermal stability. Most natural fibers can only endure processing temperatures below 200 °C; over that temperature, the fibers shrink and degrade. When natural fiber is exposed to high temperatures, it undergoes recrystallization, drying, depolymerization, oxidation, decarboxylation, and hydrolysis, which alters its chemical and physical structure. Preventing such processing flaws demands restricting the processing temperature, pressure, and time ranges [8].

3.4.5 Mitigating the drawbacks of sugar palm fibers for improved reinforcing

It is important to overcome the shortcomings of sugar palm fibers in order to maximize their maximal potential as an alternative for man-made synthetic fiber. This may improve the characteristics of SPF reinforced polymer composites and indirectly broaden their application scope. Poor adhesion between the fiber and matrix, resulting in low mechanical characteristics, is a major concern for the majority of the parts linked to SPF reinforced polymer composites. Therefore, it is critical to execute surface modification treatment to improve the incompatibility of SPF and polymer matrix. Many investigations have been carried out in order to increase the mechanical characteristics of biocomposite by improving the SPF surface through surface modification treatment.

3.5 Nanocellulose

3.5.1 Types of nanocellulose

There are three forms of nanocelluloses: bacterial nanocellulose (BNCs), nanocrystalline cellulose (NCCs), and nano-fibrillated cellulose (NFCs) [11]. Table 3.4 shows a comparison of the three major nanocelluloses. Bottom-up or top-down approaches can be used to create these nanocelluloses [12]. NCCs and NFCs are typically isolated and extracted from crops, higher plants, and by-products of forestry and agriculture using a top–down approach such as enzymatic, physical, or chemical techniques. Figure 3.2 depicts cellulosic fiber structures, emphasizing the origin of NCCs from cellulosic plant sources, cellulose fibers, and microfibril bundles with crystalline and amorphous regions. In contrast, bacterial nanocelluloses (BNCs) are produced by a particular family of bacteria and are cultivated in a culture medium as microfibrils through a bottom–up approach. Additionally, there are methods that can be used to produce NCCs by hydrolyzing BNCs into bacteria nanocrystals using acid hydrolysis which is proven to have similar characteristics as NCCs. According to [13], one of the advantages of extraction of plant-based cellulose nanofibers compared to bacterial cellulose is that the nanocellulose can be extracted into thinner fiber. Typically, NCCs are separated from cellulose-source material in two stages: the pre-treatment process (which includes partial and complete elimination of matrix materials such as lignin, hemicellulose, and cellulosic fiber separation) and the chemical treatment process, or hydrolysis, which is critical for removing the amorphous regions of the cellulose polymer.

3.5.2 Preparation of nanocrystalline cellulose

There are numerous methods for obtaining NCCs from plants. Hydrolysis treatment is one of the most used methods of isolating NCCs (Table 3.5). Plant raw materials must be

Table 3.4: A comparison of the various types of nanocellulose [14].

	Nanocrystalline cellulose (NCCs)	**Nanofibrillated cellulose (NFCs)**	**Bacterial nanocellulose (BNCs)**
Images	 Transmission electron microscope (100 nm)	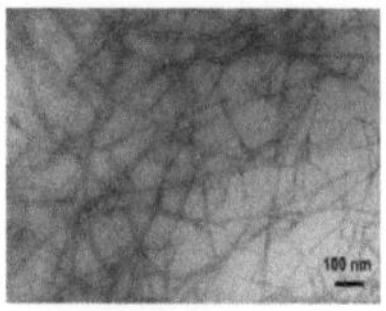 Transmission electron microscope (100 nm)	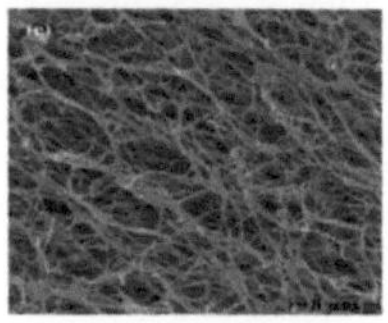 Scanning microscope (8 µm)
Synonyms	Crystallites, cellulose nanocrystals, nanowhiskers, whiskers, rod-like cellulose microcrytals	Microfibrillated cellulose, microfibrils, nanofibrils, nanofiber cellulose	Bacterial cellulose, microbial cellulose, biocellulose
Common source	Cotton, wood fibers (e.g. Pine, spruce and eucalyptus), softwood wood flour, branch-bark of mulberry, flax, hemp, wheat straw, rice straw, ramie, tunicin, avicel, mcc, algae, bacterial cellulose, pea hull fiber, tunicate, valonia, kraft wood, palm oil, pineapple leaf fibers, grass, swede root, chardonnay grape skins, coconut fiber, kenaf fiber, rice husk, sabdariffa fibers, and corncob.	Cotton, cotton linter, hardwood and softwood pulp (e.g. Acacia pulp, eucalyptus pulp), sugar beet pulp, recycled pulp, date palm tree (rachis/leaflets), mengkuang leaves, pea hull, bagasse, flax, hemp, curaua, ramie, sisal, bamboo, sugarcane bagasse, kenaf, wheat straw, rice straw, cassava, coconut husk fiber, rice husk, algae (valonia), tunicate cellulose, bacterial cellulose, tunicate cellulose, banana, capimdourado, mulberry, *luffacylindrica*, and tunicate.	Alcohols, low molecular weight sugar, and bacteria species such as agrobacterium, gluconacetobacter, pseudomonas, rhizobium, and sarcin.
Formation process	Chemical (acid hydrolysis) or enzymatic treatment	Chemical, TEMPO-mediated oxidation, mechanical shearing process integration via high pressure homogenization and enzymatic hydrolysis	Bacterial synthesis and cultivation in glucose, oxygen and phosphate-rich aqueous culture mediums
Average size	Diameter: 5–80 nm Length: 100–250 nm	Diameter: 5–70 nm Length: 100–250 nm (from plant); 100 nm to several micrometer (from cellulose of tunicates, algae, bacteria)	Diameter: 20–100 nm Length: 1.0–5.0 µm Nanofiber networks of various sorts
Tensile strength	2–6 GPa	2–4 GPa	200–300 MPa
Young's modulus	50–143 GPa	15–150 GPa	15–35 GPa
Films visual properties	Iridescent and transparent	Transparent	Transparent

Table 3.4: (continued)

	Nanocrystalline cellulose (NCCs)	Nanofibrillated cellulose (NFCs)	Bacterial nanocellulose (BNCs)
Eco- toxicity	Low toxicity Some proinflammatory cytokines	Low toxicity Some pulmonary inflammation	Low toxicity
Cost	\$1/g (dry) \$5/g (slurry)	\$2/g (dry) \$3/g (slurry)	\$1.6/g

exposed to a highly acidic environment while being sonicated during hydrolysis treatment. This phase begins with the hydrolysis of cellulose, which results in nanocrystalline cellulose (NCCs) with rod-like shapes. Concentrated sulfuric acid is a mineral acid that is often employed in the separation of NCCs from raw sources. Surface OH groups esterification in cellulose occurs during the sulfuric acid hydrolysis process, resulting in a large number of negatively charged sulfate groups on the surface of NCCs. This technique also improves NCC characteristics by minimizing NCC flocculation and agglomeration in an aqueous media [16]. However, there is one disadvantage: NCCs have only modest thermostability. Fortunately, this can be avoided by adding sodium hydroxide (NaOH), which acts as a nanoparticle neutralizer. Most experiments employed H_2SO_4 and HCl for the preparation of NCCs, although there are some studies that used HBr and H_3PO_4 for this process [17]. A number of studies have postulated a convergence between the impact of experimental circumstances on the physical properties of the resulting nanocrystals. The research revealed that the acid concentration, acid-to-fiber ratio, temperature, and time of the hydrolysis process are all important in defining the size and shape of the produced NCCs. The length and size of the nanocrystals were reduced as the hydrolysis period and acid to fiber ratio were increased. Another way for extracting NCCs from natural fiber is to combine acid hydrolysis with mechanical approaches such as homogenization.

Figure 3.3 depicts three phases in the acid hydrolysis mechanism:

I. Formation of conjugated acid via processes involving oxygen protons and glycoside acid.
II. C–O bond breakdown and conjugated acid segregation into cyclic carbonium ions.
III. After the addition of water, the proton and free sugar are released.

Recently, researchers have begun to investigate the potential use of agricultural or forest wastes as NCCs sources. As a result, numerous local sources are being employed to investigate the potential of NCCs in certain technologies.

3.5.3 Distinct characteristics of NCCs for reinforcement

NCCs are extremely useful for reinforcement and have the potential to be employed as a filler in composite materials. Therefore, understanding the extraction process, and the

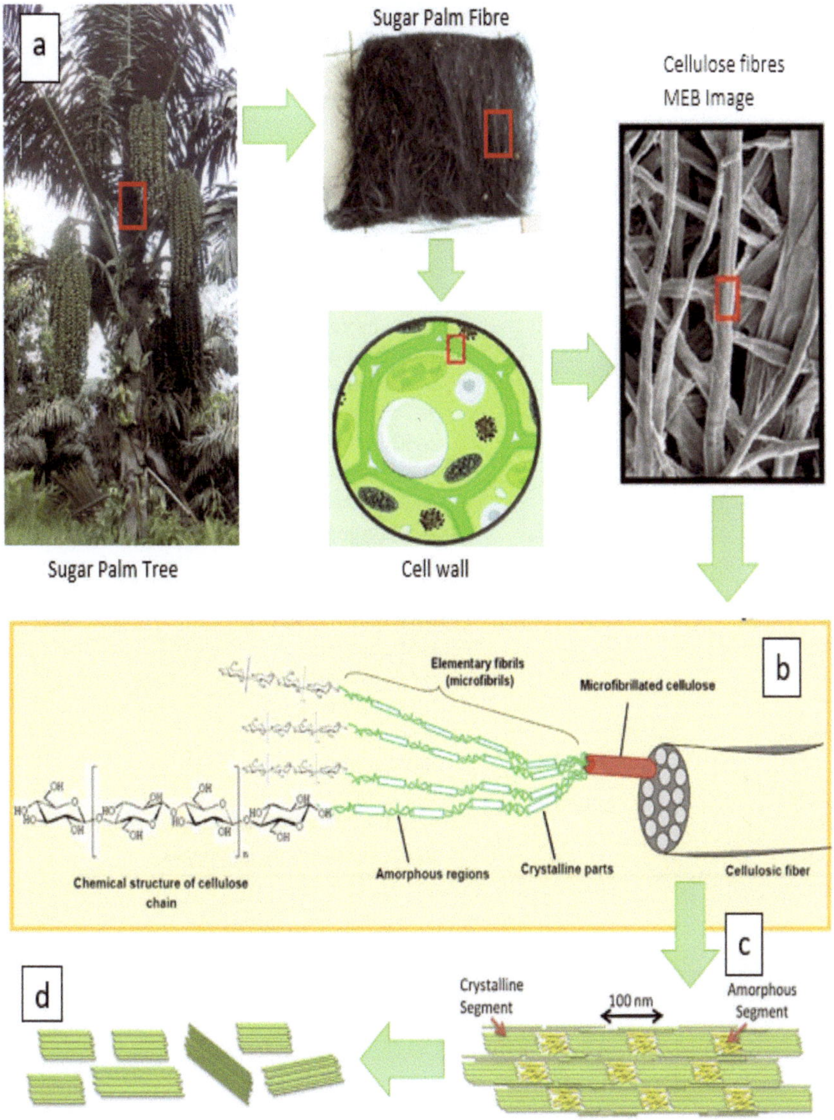

Figure 3.2: Fiber structures illustrating the origin of NCCs from (a) cellulosic plant sources, (b) cellulose fibers, (c) microfibrils bundles with amorphous and crystalline areas, and (d) nanocrystalline cellulose (NCCs) [15].

structural, mechanical, and chemical characteristics of NCCs, is critical. The structure and characteristics of NCCs are generally influenced by a number of controlled conditions under which they are extracted utilising various methods. NCCs are also recognized to be good reinforcing agents because of their high surface area, abundant hydroxyl

Table 3.5: Various acid hydrolysis techniques for NCC isolation from various sources [17].

Sources	Process
Ramie	H_2SO_4 hydrolysis
MCC	H_2SO_4 hydrolysis
Industrial bioresidue	H_2SO_4 hydrolysis
Grass fibers	H_2SO_4 hydrolysis
Rice husk	H_2SO_4 hydrolysis
Cotton Whatman filter paper	H_2SO_4 hydrolysis
Bacterial cellulose	H_2SO_4 hydrolysis
Cotton (cotton wool)	H_2SO_4 hydrolysis
Industrial bioresidue (Sludge)	H_2SO_4 hydrolysis
Eucalyptus kraft pulp	H_2SO_4 hydrolysis
Cotton linters	HCl hydrolysis
Rice straw	H_2SO_4 hydrolysis
Soy hulls	H_2SO_4 hydrolysis
Mengkuang leaves	H_2SO_4 hydrolysis
Sugarcane bagasse	H_2SO_4 hydrolysis
Algae	H_2SO_4 hydrolysis
Mulberry	H_2SO_4 hydrolysis
Banana fiber	$H_2C_2O_4$ hydrolysis
Sesame husk	H_2SO_4 hydrolysis
Sisal fiber	H_2SO_4 hydrolysis
Bamboo	H_2SO_4 hydrolysis
Colored cotton	H_2SO_4 hydrolysis
Tunicate	H_2SO_4 hydrolysis
Coconut husk	H_2SO_4 hydrolysis
Wood pulp	TEMPO-oxidation followed by HCl hydrolysis
Curaua fiber	H_2SO_4, H_2SO_4/HCl, HCl hydrolysis

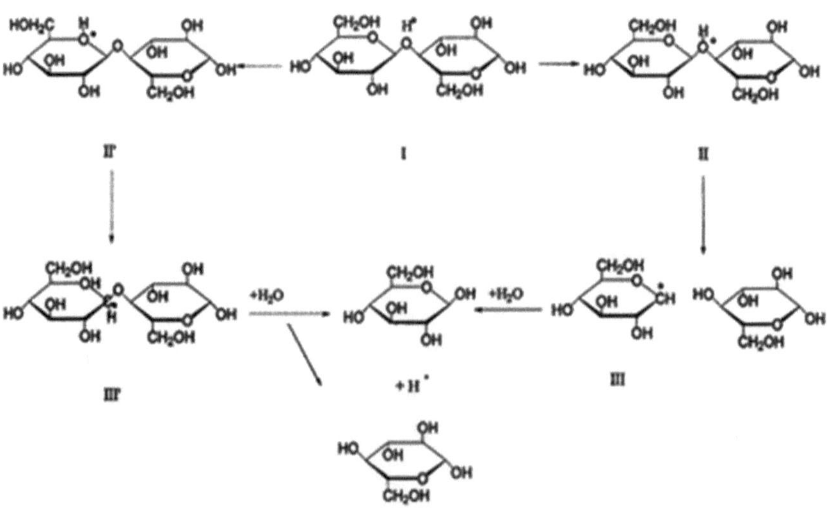

Figure 3.3: Mechanism of hydrolysis of acid [18].

groups, high crystallinity, large specific surface area, high strength and stiffness, low weight, biodegradability, high aspect ratio, and excellent mechanical and thermal properties [19].

3.5.3.1 Abundant surface hydroxyl groups

In nature, NCCs are noted for their renewability and availability. Despite their abundance, NCCs have multiple surface OH groups all around the nanocrystals cellulose. According to [20]; the strong hydrogen bonds that hold the OH groups of NCCs together are the main features of the cellulose network that can be exploited for polymer reinforcement. [21], discovered many hydroxyl functional groups or hydrogen bonded networks on the surface of NCCs using the Fourier transform infrared spectroscopy (FTIR) technique. These hydroxyl functional groups or hydrogen bonded networks correspond to the hydroxyl stretching vibration at the 3700–3000 cm^{-1} region. The findings of this study confirm the notion that the more active sites for hydrogen bonding generated through fiber dimension reduction, the greater the reinforcing effect of cellulose fibers with polymer. The comparison of nano-fibrillated cellulose (NFCs) and nanocrystalline cellulose (NCCs) has received a lot of attention. The results reveal that NCCs have greater reinforcing ability than NFCs. This is due to the lack of a hydroxyl group on the amorphous regions of NFCs, as illustrated in Figure 3.4 [13].

3.5.3.2 Large specific surface area

Previous research has shown that NCCs have a significant surface area due to the nanodimension of structural elements ranging from more than 100 $m^2\,g^{-1}$ to many hundreds of $m^2\,g^{-1}$ [19]. Thus, the increased surface area of NCCs would affect the interaction of the NCCs with the polymer matrix via the hydrogen bond. Increasing the aspect ratio diameter of NCCs increases surface contact with the polymer matrix, which improves chain mobility and surface reactivity of the polymer, as well as improved molecular level dispersion, and hence the characteristics of the composite [17].

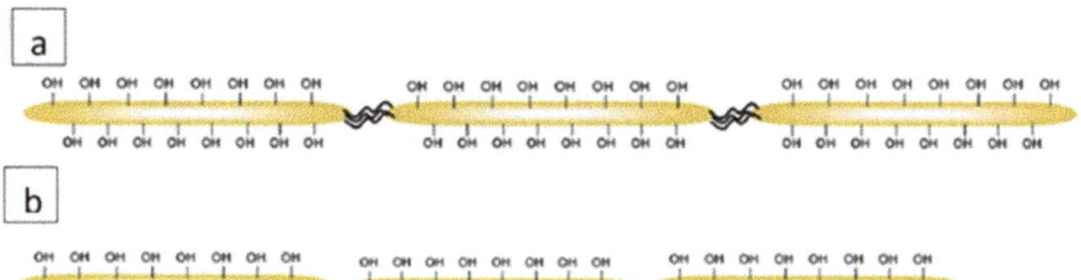

Figure 3.4: Surface-abundant hydroxyl groups of (a) Nano-fibrillated cellulose and (b) Nanocrystalline cellulose [22].

3.5.3.3 High aspect ratio

One important requirement of the nanofiller composite reinforcement is the filler must have a high aspect ratio and length to width ratio. When the aspect ratio of the nano-element is 30 or higher, the reinforcement effect is stronger. NCCs have been reported to have an aspect ratio of up to 100 depending on the technique of isolation and the source of NCCs used [23]. According to Reddy et al. [15], a high aspect ratio of NCCs could improve the characteritics of nano-composite materials due to nano-enabled functional properties including percolation and tangling. Dalmas et al. [20], mentioned that fillers with a high aspect ratio have good malleability for producing a multifarious intertwined micro-structure with more reinforcement with polymer chains, resulting in greater restriction of polymer chain motions.

3.5.3.4 High crystallinity

Figure 3.5 displays the fundamental chemical structure of a cellulose chain, demonstrating the remarkable crystallinity of NCCs with three hydroxyl groups on each monomer. As a result of intra and inter-molecular hydrogen bonding among the cellulose chains, a very compact system of cellulose crystalline packing is formed. According to Jonoobi et al. [17], as the closely packed crystalline structural region of NCCs bears inter-chain hydrogen bonding, it becomes inaccessible to most water uptake at composite and organic solvent equilibrium. However, due to the fewer organized regions of nano-fibrils where the chains are wider apart and loosely packed compared to NCCs, NFCs are more accessible for interaction with other molecules such as water.

Moreover, the high packing crystallinity matrix of NCCs may improve polymer structure, stiffness and rigidity. Follain et al. [25], reported that the high crystallinity of NCCs makes it difficult for penetrant molecules such as gas or water in polymer to penetrate, hence indirectly strengthening the barrier characteristics of hydrophilic polymer.

3.5.3.5 High thermal resistance

Cellulosic materials experience rapid thermal deterioration from low to moderate temperatures, which are at 400 °C and below. The highly crystalline structure of NCCs is the

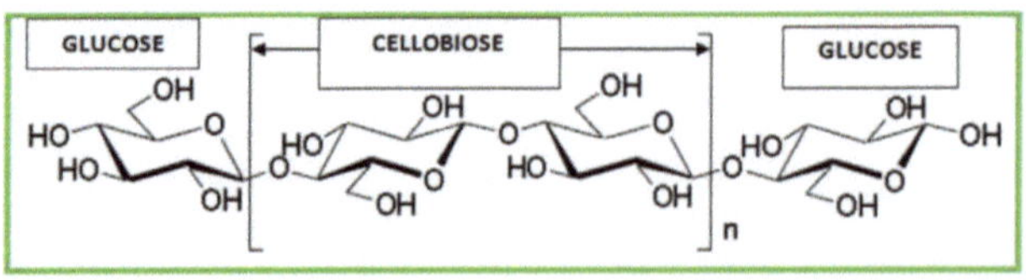

Figure 3.5: The basic chemical structure of cellulose made up of cellobiose repeat units [24].

primary element in their exceptional thermal resistance. Cellulose chains are packed in a highly ordered manner with compact crystals with the strong stabilizing hydrogen bonding [20]. According to Dalmas et al. [20], the formation of highly crystalline structures is thought to generate high thermal resistance due to inter-chain hydrogen bonding in the tightly packed crystalline structure region that cannot be burned or destroyed by high temperature, preventing cellulose from melting. As a result of their great thermal stability, NCCs have been shown to improve the thermal characteristics of polymers. The emergence of NCCs is imperative for reducing heat loss and thermal expansion of polymers [20].

3.5.3.6 Good mechanical properties

The mechanical characteristics of fillers are critical in the selection of materials for reinforcement. The crystalline regions of NCCs, which include an abundance of interchain hydrogen bonds, provide the reinforced polymer material with excellent stiffness and strength. According to Savadekar and Mhaske [23], NCCs have a high Young's modulus and a high elastic modulus of up to 150 GPa and 145 GPa, respectively. This is due to the highly ordered crystal generated by strong hydrogen bonding stabilisation. NCCs are anticipated to have five times the axial Young's modulus and mechanical strength of magnesium alloy or mild steel. The mechanical characteristics of NCCs are mostly determined by the cellulose used. Each form of cellulose has a unique crystalline structure [24]. NCCs can achieve higher properties for sustainable green nanocomposite due to their fibrous structure, which can vary and adjust along a matrix axis.

3.6 Nanofabrillated cellulose (NFCs)

Cellulose (also known as sugar palm cellulose) is composed of a linear chain of ringed glucose molecules in a flat ribbon-like conformation. Two anhydroglucose rings $(CH_6H_{10}O_5)_n$ were consolidated to generate a replication unit (Figure 3.6a); with n ranging from 10,000 to 15,000 depending on the source of material. These two anhydroglucose rings were linked by the β1–4 glucisiduc bond, in which oxygen was covalently bonded to C1 of one glucose ring and C4 of the adjacent ring (1 → 4 linkage) [26].

Furthermore, the interchain hydrogen bonding between the oxygen of the adjacent ring molecules (O → C4) and the hydroxyl groups results in the linkage being stabilised and the cellulose chain being linear. Microfibrils with diameters of 5–50 nm and lengths of several microns were generated throughout the biosynthesis process from the aggregation of elementary fibrils of cellulose chains. The intra- and inter-chain hydrogen bonding network contributes to the cellulose fibril's strong axial stiffness and stability. As demonstrated in Figure 3.6b, cellulose fibrils are made up of both highly organised (crystalline) and disordered (armorphous) regions [26].

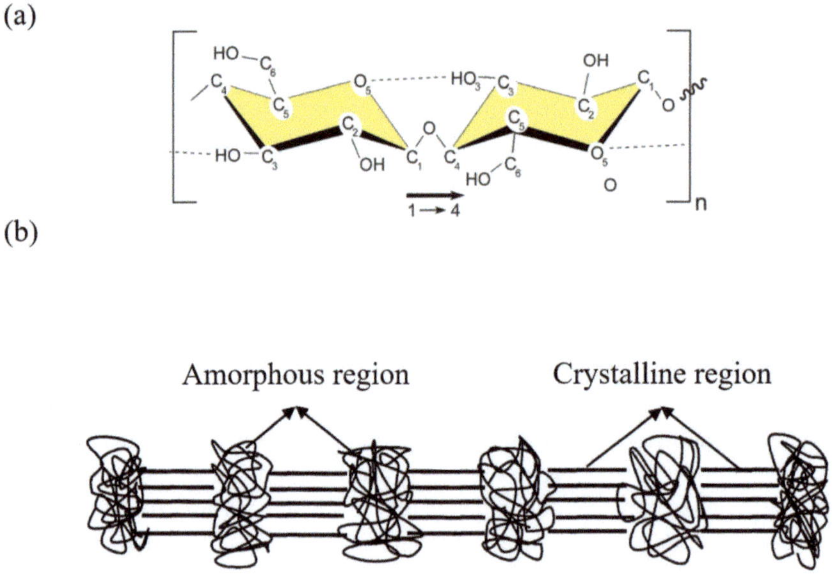

Figure 3.6: Diagram of (a) a single cellulose chain replication unit [26] and (b) crystalline and amorphous region structure6.1 structure of nanofabrillated cellulose (NFCs).

Nanofibrillated cellulose can be isolated from cellulose using one of three methods: (1) mechanical processes (such as grinding, high-pressure homogenization, and milling), (2) chemical processes (such as TEMPO oxidation), or (3) a combination of mechanical and chemical processes. Nanofibrillated cellulose has a nanometer diameter (2–10 nm) and several micrometer-long entangled fibrils ribbon-like made up of both crystalline and amorphous domains (Figure 3.7). However, their measurements vary depending on the plant origin. Synonyms for "microfibril" include "nano-fiber" and "nano-fibrillated cellulose" [14].

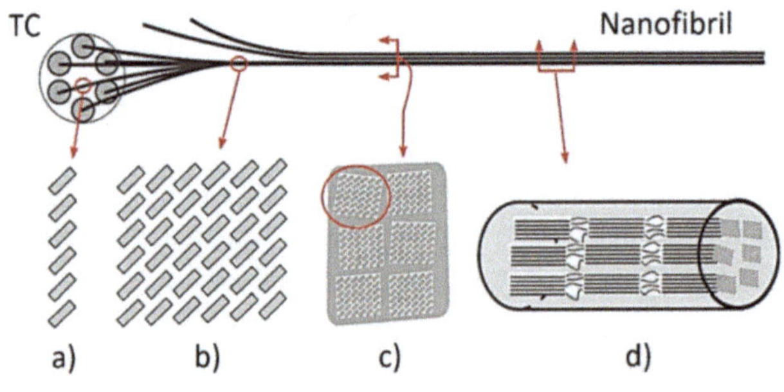

Figure 3.7: Wood nanofibril synthesis (a) a single cellulose subunit, (b) an elementary fibril cross-section, (c) a nanofibril cross-section, and (d) a nanofibril lateral section showing the layout of crystalline and amorphous regions [26].

3.6.1 Extraction of nanofabrillated cellulose (NFCs)

In the process of extracting cellulose fibers from cellulosic raw materials, two steps occurred. The first stage is the purification and homogenization of the source material. These treatments, which vary based on the type of cellulosic raw material, are used to achieve a more consistent reaction in following treatments. Pretreatments for plants and woods, for example, require the isolation of individual cellulose fibers through the partial or complete removal of matrix components (lignin, hemicellulose, waxes, etc.). Whereas tunicate, algal, and bacterial cellulose require the isolation of mantel and protein matrix removal, alga wall matrix removal requires culturing and purifying processes, and bacteria and other media removal requires culturing and washing processes, respectively [27].

In the second stage, the purified cellulose materials are next disintegrated into micro or nanofibrillar components. Mechanical techniques such as grinders/refiners, high intensity ultrasonic treatment, high pressure homogenizer, cryocrushing, and microfluidazation can be used to extract NFCs from cellulose fibrils. These techniques created a strong shear stress, causing the cellulose macro/microfibrillar structure to transverse cleave along the longitudinal axis, resulting in the disintegration of long cellulose nanofibrils, known as nanofabrillated cellulose [26]. Different mechanical methods are used to extract the micro/nanographs of nanofibrillated cellulose, as shown in Figure 3.8 and Table 3.6 [27]. Typically, cellulose fibers are mechanically treated in multiple passes during the high pressurise homogenization (HPH) process. This is done to generate cellulose nanofibers that are smaller and more homogeneous in diameter. This treatment, however, produced mechanical damage to the nanocrystalline cellulose region, which resulted in a reduced proportion of fiber crystallinity [27]. After a period of time, a filtration step was introduced to separate the partially fabrillated fractions and larger unfibrillated cellulose. Moreover, the NFCs produced by these techniques are then chemically treated in order to either enhance the NFCs or chemically functionalize the surface of the nanoparticle. To simplify the process of disintegrating cellulose microfibrils into smaller and thinner NFCs (Figure 3.8), three processing processes can be utilized partially removing all matrix material (lignin, hemicellulose, etc.) that is reinforced with cellulose, chemical treatment, and never using dried materials. These processes were taken in order to decrease the interfibrillar hydrogen bonds within the cellulose microfiber.

When the source material dries, the gaps between the microfibril bundles compact. Therefore, when the material source is dried, the material agglomerates and is difficult to separate because hydrogen bonds form between the microfibrils. Rehydration or reswelling of dry materials does not displace all of the hydrogen bonds between the fibrils, resulting in the same "weakening" effect [28]. Likewise, if the removal of matrix material was inadequate (e.g., just a portion of the hemicelluloses were removed), these matrix elements can hinder the coalescence of the microfibril bundles during the drying

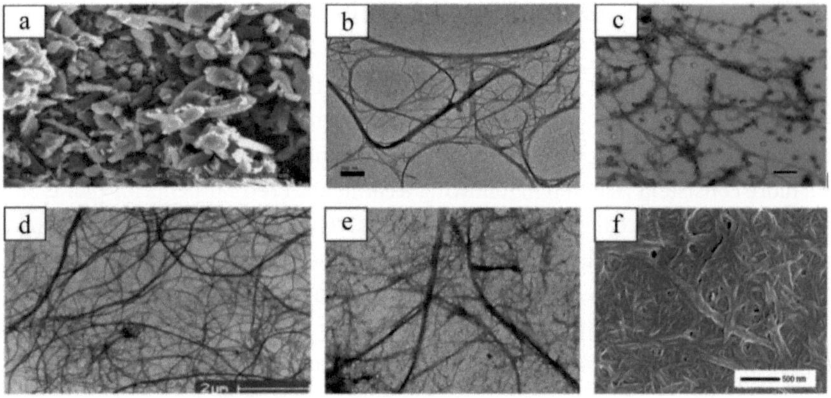

Figure 3.8: SEM micrographs of (a) microcrystalline cellulose (MCC), (b) TEM micrographs of SPNFCs produced by high pressurized homogenization through 5 passes at 500 MPa, (c) NFCs produced by microfluidization through 15 passes at 150 MPa, (d) NFCs produced by microgrinding for 6 h at 1500 rpm, (e) bleached frozen pulp after cryocrushing and treatment in a homogenizer, and (f) ultrasonication derived nanocellulose [27].

Table 3.6: Different mechanical methods to extract NFCs.

Methods	Descriptions
High pressurized homogniza-tion (HPH)	The HPH process involves putting the cellulose slurry into the vessel at high pressure using a very small nozzle. Shear rates in the stream are caused by high pressure, velocity, shear, and impact force on fluid, which reduces the size of the fiber from micro to nanoscale.
Microfluidization	This process employs an intensifier pump to raise pressure and an interation chamber to defibrillate the fibers via applying impact and shear force to the channel wall and clashing streams. The fibrillation mechanism in micro-fluidization is identical to HPH.
Grinding	The pulp slurry is passed between the rotating and static grind stones throughout the grinding operation. The mechanism of disintegration in the grinder is to break down the cell wall and hydrogen bond of the fiber by shear forces, which indirectly reduces the size of the fiber from individualization of pulp to nanoscale fibers.
Cryocrushing	The cryocrushing process involves immersing cellulosic fibers in water in liquid nitrogen and then crushing them with a mortar and pestle. The mechanism of fibrillation in this technique is to separate nanofibers by breaking down the cell wall of the frozen fiber by high impact force on the ice crystal.
High intensity ultrasonication (HIUS)	The HIUS process involves the isolation of cellulose pulp by hydrodynamic forces of ultrasound utilizing powerful oscillating power and high intensive waves is part of the HIUS process. This mechanism causes small gas bubbles to develop, expand, and implode as molecules absorb ultrasonic energy.

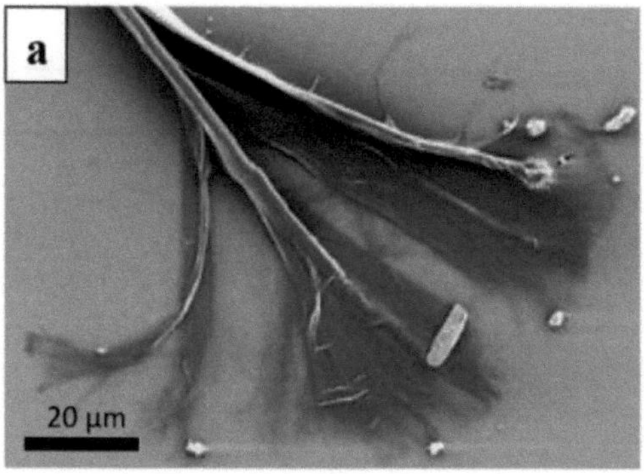

Figure 3.9: TEMPO-mediated oxidation of cellulose fibers to form microfibrils and nanofibrils [28].

phase and facilitate the material's subsequent fibrillation. Besides, conveying a charge on the surface of a microfibril by oxidation (2,2,6,6-tetramethyl-piperidinyl-1-oxyl radical (TEMPO) area selective oxidation) or adsorption of charged polyelectrolytes (e.g. carboxymethyl cellulose treatment) would increase the interfabrillar repulsive forces [28]. Figure 3.9 depicts the TEMPO-mediated oxidation of cellulose fibers into microfibrils.

3.7 Nanocellulose reinforced starch-based composites

The physical and chemical properties of the polymer matrix, as well as the interaction between the nanoparticle and the polymer matrix, influence the properties of nanocomposite [29, 30]. Due to its wide surface area with abundant surface hydroxyl group, crystallinity, high aspect ratio, thermal resistance, and mechanical characteristics, nanosize reinforced composite offers several benefits over microsize reinforced composite [21]. Bionanocomposite (TPS/NCCs or NFCs) are completely biodegradable and fully dissolve in optimal conditions when compared to conventional polymers. The biodegradation of nanocomposite is influenced by a variety of parameters, including the type of organism, polymer characteristics, and pretreatment type. Bacteria and fungi are among the microorganisms responsible for the breakdown of both synthetic and natural plastic. Table 3.7 covers some of the most recent, scientific, and industrial NFC reports. Moreover, NFCs are widely employed as filler in thermoplastic polymer matrices to make highly durable nanocomposite for load-bearing and stress-transfer in thermoplastic at a low cost. Furthermore, these nanocomposites reinforcement were applied in food packaging for reduced water sensitivity, thermal stable starch, and continuous papermaking [11, 31, 32].

Owing nanocellulose's high aspect ratio and mechanical strength, there is a growing interest in utilizing it as a loading-bearing material in the development of innovative and inexpensive biodegradable materials. Previous researchers [23, 33, 34] have extensively reported on the effect of nanocellulose in starch-based matrix. They discovered that including nanocellulose into the starch matrix increased the mechanical characteristics. This is attributable to a combination of variables, including strong adhesive bonding

Table 3.7: Recent advances in NFCs-thermoplastic composites [11].

Nanofiller	Polymer component	Manufacturing technique	Applications
NFCs	Starch	Solution casting	Thermally stable, reduces water sensitivity, increases mechanical strength
	Maize amylopectin	Solution casting	Continuous papermaking
	Polyacrylamide	Solution casting	Good mechanical, optical, thermal, and oxygen barrier properties in films
	PVA	Solution casting	Flexible displays, optical devices, food packaging, and automobile windows
	Carboxymethyl cellulose	Solution casting	Edible coatings and packaging materials
	Polyethylene	Extrusion	High performance cellulosics, environmentally friendly HDPE, evaluation of cotton filler in LDPE
	Polyethylene glycol	PEG-g-CNF ribbons via stretching hydrogel	Ultra-high tensile strength and modulus for optoelectronic and medical devices
	Amorphous dialcohol cellulose	Oxidation + reduction of CNF surface	Barrier film
	Polyvinyl amine	Layer by layer	Self-healing polymer film
	Poly(butylene adipate-co- terephthalate)	Injection molding	Low weight and excellent performance for defense, infrastructure and energy

between nanocellulose and matrix, good fiber dispersion, low porosity, and the absence of agglomeration [35].

In this section, starch is used as a polymer, and the solution casting process is generally used in the examined data [36]. Tables 3.8 and 3.9 summarized the mechanical characteristics, source and preparation of NFCs and NCCs, as well as the manufacturing procedure of various nano-fibrillated cellulose reinforced starch-based nanocomposite. The sources of fibers utilised in the production of nanocrystalline cellulose (NCCs) and nano-fibrillated cellulose (NFCs) differ between the current study and the previous research indicated in Tables 3.8 and 3.9. Table 3.8 shows that super masscolloider, ultrasonification, and high pressurised homogenization techniques are used to produce nano-fibrillated cellulose. Potato starch, corn starch, and mango puree are among the starches utilized for NFC reinforcement. Meanwhile, NFCs used for starch reinforcement are derived from a variety of plants, including softwood wood flour, kenaf, wheat, rice straw, wheat straw, and bamboo fiber. Based on Table 3.8, it is possible to conclude that the solution casting method is the most often utilized technique for reinforcing NFCs and starch biopolymers. According to [37], the solution casting method was used to maintain the uniform dispersion of the nano-fiber in the nanocomposites film. Based on the mechanical data collected in Table 3.8, the highest tensile strength of 38 MPa was achieved by [38] through the reinforcement of kenaf NFCs and corn starch. In the meantime, Hietala et al. [39], achieved the highest Young's modulus of 1317 MPa by reinforcing

Table 3.8: Examples of polymers based on starch, NFC nanocomposites, and their mechanical characteristics.

Starch-based polymers	NFCs Sources	NFCs Preparation	Manufacturing technique	Tensile strength (MPa)	Young's modulus (MPa)	References
Maize starch	Kenaf	Super masscolloider	Solution-casting	2.35	53.6	[40]
Corn starch	Kenaf	Super masscolloider	Solution-casting	38.0	141.0	[38]
Mango puree	Wheat	–	Solution-casting	8.76	322.05	[41]
Potato starch	Softwood wood flour	Super masscolloider	Solution-casting	17.5	1317.0	[39]
Maize starch	Wheat straw	High pressurize homogenizer/ 15 min	Solution-casting	6.75	220	[42]
Potato starch	Rice straw	Ultrasonication	Solution-casting	5.01	160	[43]
Corn starch	Bamboo fiber	–	Solution-casting	11.2	12.4	[44]

Table 3.9: Examples of polymers based on starch, NCC nanocomposites, and their mechanical characteristics.

Starch-based polymers	NCCs Sources	Isolation chemical/ time of NCCs	Manufacturing technique	Tensile strength (MPa)	Young's modulus (MPa)	References
Pea starch	Hemp	64 wt% H_2SO_4/4 h	Solution-casting	3.9–11.5	31.9–823.9	[45]
Pea starch	Flax	64 wt% H_2SO_4/4 h	Solution-casting	3.9–11.9	31.9–498.2	[46]
Pea starch	Bamboo	50 wt% H_2SO_4/48 h	Solution-casting	2.5–12	20.4–210.3	[47]
Maize starch	Tunicin	55 wt% H_2SO_4/20 min	Solution-casting	0.24–20	51–315	[48]
Maize starch	Waxy maize starch	H_2SO_4/5 days	Solution-casting	1–15	11–320	[49]
Maize starch	Tunicate	–	Solution-casting	42	208–838	[50]
Wheat starch	Cottonseed linter	64 wt% H_2SO_4/4 h	Solution-casting	2.5–7.8	36–301	[51]
Plasticized starch	Cotton cellu-lose powders	H_2SO_4	Solution-casting	–	–	[52]
Wheat starch	Ramie	64 wt% H_2SO_4/4 h	Solution-casting	2.8–6.9	56–480	[53]
Potato starch	MCC	64 wt% H_2SO_4/2 h	Solution-casting	13.7	460	[54]
Wheat starch	MCC	36.5 wt% HCl	Solution-casting	3.15–10.98	–	[55]

Table 3.9: (continued)

Starch-based polymers	NCCs Sources	Isolation chemical/ time of NCCs	Manufacturing technique	Tensile strength (MPa)	Young's modulus (MPa)	References
Potato starch	Cotton linter	64 wt% H_2SO_4/1 h	Solution-casting	4.93	–	[56]
Potato starch	Potato peel waste	64 wt% H_2SO_4/90 min	Solution-casting	–	460	[57]
Maize starch	Sugarcane bagasse	64 wt% H_2SO_4/3 h	Solution-casting	17.4	520	[58]

softwood wood flour NFCs with potato starch. The lowest tensile strength measured was 2.35 MPa using kenaf NFCs and maize starch biopolymer reinforcement.

Natural fibers from different plants are utilised to extract NCCs via delignification, mercerization, and hydrolysis. Treatments such as delignification (using sodium chlorite, $NaClO_2$) and mercerization (using sodium hydroxide, NaOH) are used to eliminate the amorphous structure in fibers such as lignin and hemicellulose, respectively, which also splits the fibers into smaller fibrils known as cellulose microfibrils. Following that, the cellulose microfibrils were hydrolyzed (HCl and H_2SO_4 at varied concentrations, reaction times, and rotating speeds) to eliminate the amorphous region within the cellulose [59].

According to Table 3.9, the majority of the researchers used sulfuric acid (H_2SO_4) to produce nanocrystalline cellulose. Each author employed a different concentration of H_2SO_4, however the majority of them used 64 wt.% H_2SO_4. This could be related to the esterification of surface OH groups in cellulose during the sulfuric acid hydrolysis process, which results in a significant number of negatively charged sulfate groups on the surface of NCCs [60]. Besides, this process improves NCC characteristics by minimising flocculation and agglomeration of NCCs in an aqueous solution, making it an effective option for reinforcement with starch bio-based polymer [16]. Wheat, pea, maize, and potato starch are among the starches utilised to enhance NCCs. Meanwhile, NCCs used for starch reinforcement are derived from various types of plants, including flax, hemp, tunicin, tunicate, bamboo, ramie, waxy maize starch, cottonseed linter, cotton cellulose powders, microcrystalline cellulose (MCC), cotton linter, sugarcane bagasse, and potato peel waste. Solution casting is the most common approach used to reinforce NCCs within starch biopolymers. This method was used to ensure that the nanofiber dispersion in the nanocomposites film was uniform [37]. Based on Table 3.9, the highest tensile strength and Young's modulus achieved by [50] from the reinforcement of tunicate NCCs and maize starch are 17.4 MPa and 838 MPa, respectively. The lowest tensile strength measured was 4.93 MPa using cotton linter NFCs and potato starch biopolymer reinforcement. As a result, diverse types of starches and NFCs with reinforcing properties led to distinct mechanical properties.

3.8 Potential application of nanocellulose

Nanocellulose fibers have numerous potential applications, ranging from flexible optoelectronics to tissue regeneration scaffolds [61]. Cellulose is a biosynthetic material derived by animals, microorganisms, or plants. The word 'nanocellulose' refers to isolated cellulose with a nano-scale dimension [11]. Table 3.10 depicts various polymer components, manufacturing methods, and uses. Nanocellulose has demonstrated potential in a variety of applications, including:

1. Paper and packaging: Nanocellulose has been utilized to increase the strength, stiffness, and optical characteristics of paper products [62–67].
2. Building materials: Nanocellulose has been employed as a reinforcement material in the production of lightweight and strong building materials such as cement composites [68].
3. Composites: Nanocellulose has been utilized in composites as a reinforcement material to improve mechanical properties such as strength, stiffness, and toughness [1, 69].
4. Biomedicine: Nanocellulose has been used in various biomedical applications, including wound dressing, drug delivery systems, and tissue engineering [70–74].
5. Energy storage: Nanocellulose has been used as a scaffold material in energy storage devices like supercapacitors [75–80].

Table 3.10: Polymer-reinforced nanocelluloses, manufacturing techniques and applications [11].

Polymer component	Manufacturing technique	Applications
Methylcellulose	Hydrogel by aqueous dispersion	Thermoreversible and tunable nanocellulose-based hydrogels
Plasticized starch	Solution casting	Transparent materials
Starch	Blending Solution casting	Surface-sized paper, air permeable, resistant Food packaging
PVA	Solution casting	Stretchable photonic devices Wound diagnosis/biosensor scaffolds Conductive materials
Plasticized PLA	Twin-screw extruder	Packaging, film blowing
Maleic-anhydride grafted PLA	Electrospinning	Bone tissue engineering
Cellulose esterified with lauryl chloride	Solution casting and thermopressing	Interface melting
PC	Master batch melt extrusion process	Optical devices
PC based polyurethane blend	Solution casting	Smart actuators and sensors
Ethylene-co-vinyl acetate rubber	Solution mixing and vulcanization	Transparent, rubbery materials
PU	Solution casting	High temperature biomedical devices

6. Filtration: Nanocellulose can be used to create ultra-fine filters with a high surface area and pore size, making it suitable for a range of applications, including water filtration and air purification [81–84].

3.9 Conclusions

The hierarchical nature of cellulose enables for the extraction of nano-fibers from this naturally occurring polymer. Mechanical shearing can liberate more or fewer individual microfibrils, which can subsequently be longitudinally cut using a strong acid hydrolysis procedure to dissolve the amorphous domains. These nanoparticles' remarkable mechanical characteristics, reinforcing ability, abundance, low density, and biodegradability make them ideal for the production of polymer nanocomposites. With a Young's modulus ranging from 100 to 130 GPa and a surface area of several hundred square meters per grams, cellulose nanoparticles offer fascinating new possibilities. Despite the abundance of nanocellulose in nature and its numerous outstanding properties, extracting it from lignocellulosic biomass and cellulosic materials remains challenging. The high aggregation of lignin, hemicelluloses, and other substances in plant cell walls necessitates significant pretreatment to remove non-cellulosic materials, which is time-consuming, entails the use of hazardous chemicals, and generates wastewater. Traditional extraction procedures, such as acid hydrolysis, mechanical processing, and enzymatic hydrolysis, all have limitations, such as significant energy consumption, extended reaction times, and the production of vast amounts of acid waste water. The goal of nanocellulose research is to develop novel extraction processes that are more efficient and produce less or no waste. The yield and characteristics of the nanocellulose produced should also outperform those achieved using traditional methods.

Acknowledgement: The research has been carried out under the programme, Research Excellence Consortium (JPT (BPKI) 1000/016/018/25 (57)), provided by the Ministry of Higher Education Malaysia (MOHE).

References

1. Sabaruddin FA, Paridah MT, Sapuan SM, Ilyas RA, Lee SH, Khalina A, et al. The effects of unbleached and bleached nanocellulose on the thermal and flammability of polypropylene-reinforced kenaf core hybrid polymer bionanocomposites. Polymers 2020;13:116.
2. Suriani MJ, Ilyas RA, Zuhri MYM, Khalina A, Sultan MTH, Sapuan SM, et al. Critical review of natural fiber reinforced hybrid composites: processing, properties, applications and cost. Polymers 2021;13:3514.
3. Azman MA, Asyraf MRM, Khalina A, Petru M, Ruzaidi CM, Sapuan SM, et al. Natural fiber reinforced composite material for product design: a short review. Polymers 2021;13:1917.
4. Ishak MR, Sapuan SM, Leman Z, Rahman MZA, Anwar UMK, Siregar JP. Sugar palm (arenga pinnata): its fibres, polymers and composites. Carbohydr Polym 2013;91:699–710.
5. Sanyang ML, Sapuan SM, Jawaid M, Ishak MR, Sahari J. Recent developments in sugar palm (arenga pinnata) based biocomposites and their potential industrial applications: a review. Renew Sustain Energy Rev 2016;54:533–49.

6. Sahari J, Sapuan SM, Ismarrubie ZN, Rahman MZ. Physical and chemical properties of different morphological parts of sugar palm fibres. Fibres Text East Eur 2012;91:21–4.
7. Norizan MN, Abdan K, Ilyas RA, Biofibers SP. Effect of fiber orientation and fiber loading on the mechanical and thermal properties of sugar palm yarn fiber reinforced unsaturated polyester resin composites. Polimery 2020;65:34–43.
8. Koronis G, Silva A, Fontul M. Green composites: a review of adequate materials for automotive applications. Compos B Eng 2013;44:120–7.
9. Burgueño R, Quagliata MJ, Mohanty AK, Metha G, Drzal LT, Misra M. Hybrid biofiber-based composites for structural cellular plates. Composites, Part A 2005;36:581–93.
10. Nurazzi NM, Sabaruddin FA, Harussani MM, Kamarudin SH, Rayung M, Asyraf MRM, et al. Mechanical performance and applications of CNTs reinforced polymer composites—a review. Nanomaterials 2021;11: 2186.
11. Abitbol T, Rivkin A, Cao Y, Nevo Y, Abraham E, Ben-Shalom T, et al. Nanocellulose, a tiny fiber with huge applications. Curr Opin Biotechnol 2016;39:76–88.
12. Norfarhana AS, Ilyas RA, Ngadi N. A review of nanocellulose adsorptive membrane as multifunctional wastewater treatment. Carbohydr Polym 2022;291:119563.
13. Brinchi L, Cotana F, Fortunati E, Kenny JM. Production of nanocrystalline cellulose from lignocellulosic biomass: technology and applications. Carbohydr Polym 2013;94:154–69.
14. Klemm D, Kramer F, Moritz S, Lindström T, Ankerfors M, Gray D, et al. Nanocelluloses: a new family of nature-based materials. Angew Chem, Int Ed 2011;50:5438–66.
15. Reddy MM, Vivekanandhan S, Misra M, Bhatia SK, Mohanty AK. Biobased plastics and bionanocomposites: current status and future opportunities. Prog Polym Sci 2013;38:1653–89.
16. Beck-Candanedo S, Roman M, Gray DG. Effect of reaction conditions on the properties and behavior of wood cellulose nanocrystal suspensions. Biomacromolecules 2005;6:1048–54.
17. Jonoobi M, Oladi R, Davoudpour Y, Oksman K, Dufresne A, Hamzeh Y, et al. Different preparation methods and properties of nanostructured cellulose from various natural resources and residues: a review. Cellulose 2015;22:935–69.
18. Xiang Q, Lee YY, Pettersson PO, Torget RW. Heterogeneous aspects of acid hydrolysis of α-cellulose. In: Biotechnology for fuels and chemicals. Totowa, NJ: Humana Press; 2003:505–14 pp.
19. Silvério HA, Neto WPF, Dantas NO, Pasquini D. Extraction and characterization of cellulose nanocrystals from corncob for application as reinforcing agent in nanocomposites. Ind Crop Prod 2013;44:427–36.
20. Dalmas F, Cavaille JY, Gauthier C, Chazeau L, Dendievel R. Viscoelastic behavior and electrical properties of flexible nanofiber filled polymer nanocomposites. Influence of processing conditions. Compos Sci Technol 2007;67:829–39.
21. Majeed K, Jawaid M, Hassan A, Abu Bakar A, Abdul Khalil HPS, Salema AA, et al. Potential materials for food packaging from nanoclay/natural fibres filled hybrid composites. Mater Des 2013;46:391–410.
22. Islam MT, Alam MM, Zoccola M. Review on modification of nanocellulose for application in composites. Int J Innov Res Technol Sci Eng 2013;2:5444–51.
23. Savadekar NR, Mhaske ST. Synthesis of nano cellulose fibers and effect on thermoplastics starch based films. Carbohydr Polym 2012;89:146–51.
24. Siqueira G, Bras J, Dufresne A. Cellulosic bionanocomposites: a review of preparation, properties and applications. Polymers 2010;2:728–65.
25. Follain N, Belbekhouche S, Bras J, Siqueira G, Marais S, Dufresne A. Water transport properties of bio-nanocomposites reinforced by Luffa cylindrica cellulose nanocrystals. J Membr Sci 2013;427:218–29.
26. Moon RJ, Martini A, Nairn J, Simonsen J, Youngblood J. Cellulose nanomaterials review: structure, properties and nanocomposites. Chem Soc Rev 2011;40:3941.
27. Khalil HPSA, Bhat AH, Yusra AFI. Green composites from sustainable cellulose nanofibrils: a review. Carbohydr Polym 2012;87:963–79.

28. Johnson RK, Zink-Sharp A, Renneckar SH, Glasser WG. A new bio-based nanocomposite: fibrillated TEMPO-oxidized celluloses in hydroxypropylcellulose matrix. Cellulose 2009;16:227–38.
29. Asyraf MRM, Ishak MR, Syamsir A, Nurazzi NM, Sabaruddin FA, Shazleen SS, et al. Mechanical properties of oil palm fibre-reinforced polymer composites: a review. J Mater Res Technol 2022;17:33–65.
30. Sharma S, Sudhakara P, Omran AAB, Singh J, Ilyas RA. Recent trends and developments in conducting polymer nanocomposites for multifunctional applications. Polymers 2021;13:2898.
31. Chen Y, Liu C, Chang PR, Cao X, Anderson DP. Bionanocomposites based on pea starch and cellulose nanowhiskers hydrolyzed from pea hull fibre: effect of hydrolysis time. Carbohydr Polym 2009;76:607–15.
32. Diyana ZN, Jumaidin R, Selamat MZ, Ghazali I, Julmohammad N, Huda N, et al. Physical properties of thermoplastic starch derived from natural resources and its blends: a review. Polymers 2021;13:1–20.
33. Ilyas RA, Sapuan SM, Ishak MR, Zainudin ES, Atikah MSN. Nanocellulose reinforced starch polymer composites: a review of preparation, properties and application. In: Proceeding: 5th international conference on applied sciences and engineering (ICASEA, 2018), Capthorne Hotel. Cameron Highlands, Malaysia: Global Academic Excellence (M) SDN BHD; 2018:325–41 pp.
34. Ilyas RA, Sapuan SM, Ishak MR, Zainudin ES. Development and characterization of sugar palm nanocrystalline cellulose reinforced sugar palm starch bionanocomposites. Carbohydr Polym 2018;202: 186–202.
35. Abral H, Dalimunthe MH, Hartono J, Efendi RP, Asrofi M, Sugiarti E, et al. Characterization of tapioca starch biopolymer composites reinforced with micro scale water hyacinth fibers. Starch/Staerke 2018;70:1–8.
36. Tarique J, Sapuan SM, Khalina A, Sherwani SFK, Yusuf J, Ilyas RA. Recent developments in sustainable arrowroot (maranta arundinacea Linn) starch biopolymers, fibres, biopolymer composites and their potential industrial applications: a review. J Mater Res Technol 2021;13:1191–219.
37. Khan A, Khan RA, Salmieri S, Tien CL, Riedl B, Bouchard J, et al. Mechanical and barrier properties of nanocrystalline cellulose reinforced chitosan based nanocomposite films. Carbohydr Polym 2012;90: 1601–8.
38. Babaee M, Jonoobi M, Hamzeh Y, Ashori A. Biodegradability and mechanical properties of reinforced starch nanocomposites using cellulose nanofibers. Carbohydr Polym 2015;132:1–8.
39. Hietala M, Mathew AP, Oksman K. Bionanocomposites of thermoplastic starch and cellulose nanofibers manufactured using twin-screw extrusion. Eur Polym J 2013;49:950–6.
40. Karimi S, Paridah MT, Dufresne A, Karimi A, Abdulkhani A. A comparative study on characteristics of nanocellulose reinforced thermoplastic starch biofilms prepared with different techniques. Nord Pulp Pap Res J 2014;29:41–5.
41. Azeredo HMC, Mattoso LHC, Wood D, Williams TG, Avena-Bustillos RJ, McHugh TH. Nanocomposite edible films from mango puree reinforced with cellulose nanofibers. J Food Sci 2009;74:31–5.
42. Kaushik A, Singh M, Verma G. Green nanocomposites based on thermoplastic starch and steam exploded cellulose nanofibrils from wheat straw. Carbohydr Polym 2010;82:337–45.
43. Nasri-Nasrabadi B, Behzad T, Bagheri R. Preparation and characterization of cellulose nanofiber reinforced thermoplastic starch composites. Fibers Polym 2014;15:347–54.
44. Llanos JHR, Tadini CC. Preparation and characterization of bio-nanocomposite films based on cassava starch or chitosan, reinforced with montmorillonite or bamboo nanofibers. Int J Biol Macromol 2017;107: 371–82.
45. Cao X, Chen Y, Chang PR, Stumborg M, Huneault MA. Green composites reinforced with hemp nanocrystals in plasticized starch. J Appl Polym Sci 2008;109:3804–10.
46. Cao X, Chen Y, Chang PR, Muir AD, Falk G. Starch-based nanocomposites reinforced with flax cellulose nanocrystals. Express Polym Lett 2008;2:502–10.
47. Liu D, Zhong T, Chang PR, Li K, Wu Q. Starch composites reinforced by bamboo cellulosic crystals. Bioresour Technol 2010;101:2529–36.
48. Anglès MN, Dufresne A. Plasticized starch/tunicin whiskers nanocomposite materials. 2. Mechanical behavior. Macromolecules 2001;34:2921–31.

49. Angellier H, Molina-Boisseau S, Dole P, Dufresne A. Thermoplastic Starch–Waxy maize starch nanocrystals nanocomposites. Biomacromolecules 2006;7:531–9.
50. Mathew AP, Thielemans W, Dufresne A. Mechanical properties of nanocomposites from sorbitol plasticized starch and tunicin whiskers. J Appl Polym Sci 2008;109:4065–74.
51. Lu Y, Weng L, Cao X. Biocomposites of plasticized starch reinforced with cellulose crystallites from cottonseed linter. Macromol Biosci 2005;5:1101–7.
52. Yang S, Tang Y, Wang J, Kong F, Zhang J. Surface treatment of cellulosic paper with starch-based composites reinforced with nanocrystalline cellulose. Ind Eng Chem Res 2014;53:13980–8.
53. Lu Y, Weng L, Cao X. Morphological, thermal and mechanical properties of ramie crystallites—reinforced plasticized starch biocomposites. Carbohydr Polym 2006;63:198–204.
54. Kvien I, Sugiyama J, Votrubec M, Oksman K. Characterization of starch based nanocomposites. J Mater Sci 2007;42:8163–71.
55. Chang PR, Jian R, Zheng P, Yu J, Ma X. Preparation and properties of glycerol plasticized-starch (GPS)/ cellulose nanoparticle (CN) composites. Carbohydr Polym 2010;79:301–5.
56. Noshirvani N, Ghanbarzadeh B, Fasihi H, Almasi H. Starch-PVA nanocomposite film incorporated with cellulose nanocrystals and MMT: a comparative study. Int J Food Eng 2016;12:37–48.
57. Chen D, Lawton D, Thompson MR, Liu Q. Biocomposites reinforced with cellulose nanocrystals derived from potato peel waste. Carbohydr Polym 2012;90:709–16.
58. Slavutsky AM, Bertuzzi MA. Water barrier properties of starch films reinforced with cellulose nanocrystals obtained from sugarcane bagasse. Carbohydr Polym 2014;110:53–61.
59. Dufresne A. Nanocellulose: a new ageless bionanomaterial. Mater Today 2013;16:220–7.
60. Rashid ESA, Gul A, Yehya WAH, Julkapli NM. Physico-chemical characteristics of nanocellulose at the variation of catalytic hydrolysis process. Heliyon 2021;7:e07267.
61. Kadier A, Ilyas RA, Huzaifah MRM, Harihastuti N, Sapuan SM, Harussani MM, et al. Use of industrial wastes as sustainable nutrient sources for bacterial cellulose (BC) production: mechanism, advances, and future perspectives. Polymers 2021;13:1–51.
62. Abral H, Ariksa J, Mahardika M, Handayani D, Aminah I, Sandrawati N, et al. Transparent and antimicrobial cellulose film from ginger nanofiber. Food Hydrocolloids 2020;98:105266.
63. Ilyas RA, Sapuan SM, Ibrahim R, Abral H, Ishak MR, Zainudin ES, et al. Sugar palm (arenga pinnata (wurmb.) merr) cellulosic fibre hierarchy: a comprehensive approach from macro to nano scale. J Mater Res Technol 2019;8:2753–66.
64. Ilyas RA, Sapuan SM, Atiqah A, Ibrahim R, Abral H, Ishak MR, et al. Sugar palm (arenga pinnata [wurmb.] merr) starch films containing sugar palm nanofibrillated cellulose as reinforcement: water barrier properties. Polym Compos 2020;41:459–67.
65. Ilyas RA, Sapuan SM, Ibrahim R, Abral H, Ishak MR, Zainudin ES, et al. Thermal, biodegradability and water barrier properties of bio-nanocomposites based on plasticised sugar palm starch and nanofibrillated celluloses from sugar palm fibres. J Biobased Mater Bioenergy 2020;14:234–48.
66. Nazrin A, Sapuan SM, Zuhri MYM, Ilyas RA, Syafiq R, Sherwani SFK. Nanocellulose reinforced thermoplastic starch (TPS), polylactic acid (PLA), and polybutylene succinate (PBS) for food packaging applications. Front Chem 2020;8:1–12.
67. Nazrin A, Sapuan SM, Zuhri MYM, Tawakkal ISMA, Ilyas RA. Water barrier and mechanical properties of sugar palm crystalline nanocellulose reinforced thermoplastic sugar palm starch (TPS)/Poly(Lactic acid) (PLA) blend bionanocomposites. Nanotechnol Rev 2021;10:431–42.
68. Hisseine OA, Wilson W, Sorelli L, Tolnai B, Tagnit-Hamou A. Nanocellulose for improved concrete performance: a macro-to-micro investigation for disclosing the effects of cellulose filaments on strength of cement systems. Constr Build Mater 2019;206:84–96.
69. Omran AAB, Mohammed AABA, Sapuan SM, Ilyas RA, Asyraf MRM, Koloor SSR, et al. Micro- and nanocellulose in polymer composite materials: a review. Polymers 2021;13:231.

70. Khalil HPSA, Adnan AS, Yahya EB, Olaiya NG, Safrida S, Hossain MS, et al. A review on plant cellulose nanofibre-based aerogels for biomedical applications. Polymers 2020;12:1–26.
71. Hossain AS, Uddin MM, Veettil VN, Fawzo M. Nano-cellulose based nano-coating biomaterial dataset using corn leaf biomass an innovative biodegradable plant biomaterial. Data Brief 2018;17:162–8.
72. Jorfi M, Foster EJ. Recent advances in nanocellulose for biomedical applications. J Appl Polym Sci 2015;132: 1–19.
73. Lin N, Dufresne A. Nanocellulose in biomedicine: current status and future prospect. Eur Polym J 2014;59: 302–25.
74. Tayeb AH, Amini E, Ghasemi S, Tajvidi M. Cellulose nanomaterials — binding properties and applications: a review. Molecules 2018;23:1–24.
75. Chen W, Yu H, Lee SY, Wei T, Li J, Fan Z. Nanocellulose: a promising nanomaterial for advanced electrochemical energy storage. Chem Soc Rev 2018;47:2837–72.
76. Erlandsson J, Durán VL, Granberg H, Sandberg M, Larsson PA, Wågberg L. Macro- and mesoporous nanocellulose beads for use in energy storage devices. Appl Mater Today 2016;5:246–54.
77. Guo M, Xu P, Lv J, Gong C, Zhang Z, Li C. Engineering nanocellulose/graphene hybrid aerogel for form-stable composite phase change materials with high phase change enthalpy for energy storage. Diamond Relat Mater 2022;127:109131.
78. Li Z, Liu J, Jiang K, Thundat T. Carbonized nanocellulose sustainably boosts the performance of activated carbon in ionic liquid supercapacitors. Nano Energy 2016;25:161–9.
79. Lv P, Lu X, Wang L, Feng W. Nanocellulose-based functional materials: from chiral photonics to soft actuator and energy storage. Adv Funct Mater 2021;31:2104991.
80. Shaghaleh H, Xu X, Wang S. Current progress in production of biopolymeric materials based on cellulose, cellulose nanofibers, and cellulose derivatives. RSC Adv 2018;8:825–42.
81. Bai L, Wu H, Ding J, Ding A, Zhang X, Ren N, et al. Cellulose nanocrystal-blended polyethersulfone membranes for enhanced removal of natural organic matter and alleviation of membrane fouling. Chem Eng J 2020;382:122919.
82. Cheng R, Kang M, Zhuang S, Shi L, Zheng X, Wang J. Adsorption of Sr(II) from water by mercerized bacterial cellulose membrane modified with EDTA. J Hazard Mater 2019;364:645–53.
83. Huang Y, Yang P, Yang F, Chang C. Self-supported nanoporous lysozyme/nanocellulose membranes for multifunctional wastewater purification. J Membr Sci 2021;635:119537.
84. Karim Z, Mathew AP, Grahn M, Mouzon J, Oksman K. Nanoporous membranes with cellulose nanocrystals as functional entity in chitosan: removal of dyes from water. Carbohydr Polym 2014;112:668–76.

Salit Mohd Sapuan*, Moklis Muhammad Harussani, Aleif Hakimi Ismail, Noorashikin Soh Zularifin Soh, Mohamad Irsyad Mohamad Azwardi and Vasi Uddin Siddiqui

4 Development of nanocellulose fiber reinforced starch biopolymer composites: a review

Abstract: In the last few years, there are rising numbers for environmental waste due to factors such as plastic based food packaging that really need to get enough attention in order to prevent the issue from becoming worse and bringing disaster to society. Thus, the uses of plastic composite materials need to be reduced and need to be replaced with materials that are natural and have low degradation to preserve nature. Based on the statistics for the global, the production of plastic has been roughly calculated for passing 400 million metric tons every year and has a high probability of approaching the value of 500 million metric tons at the year of 2025 and this issue needs to be counteracted as soon as possible. Due to that, the increasing number for recent development of natural biopolymer, as an example starch, has been investigated as the substitution for the non-biodegradable biopolymer. Besides, among all biodegradable polymers, starch has been considered as promising substitution polymer due to its renewability, easy availability, and biodegradability. Apart from that, by the reinforcement from the nanocellulose, starch fiber has an increasing in terms of mechanical, barrier and thermal properties. In this review paper, we will be discussing the up-to-date development of nanocellulose fiber reinforced starch biopolymer composites throughout this century.

Keywords: biopolymer; food packaging; nanocellulose; reinforcement; starch.

4.1 Introduction

According to the [1]; as shown in Figure 4.1 and Table 4.1, the total forest area in the year 2008 was 17, 680, 508 ha covering 53.6% of the total area of Malaysia. Meanwhile,

***Corresponding author: Salit Mohd Sapuan**, Department of Mechanical and Manufacturing Engineering, Advanced Engineering Materials and Composites (AEMC) Research Centre, Universiti Putra Malaysia (UPM), Serdang 43400, Selangor, Malaysia, E-mail: sapuan@upm.edu.my
Moklis Muhammad Harussani, Energy Science and Engineering, Department of Transdisciplinary Science and Engineering, School of Environment and Society, Tokyo Institute of Technology, Meguro 152-8552, Tokyo, Japan
Aleif Hakimi Ismail, Noorashikin Soh Zularifin Soh, Mohamad Irsyad Mohamad Azwardi and Vasi Uddin Siddiqui, Department of Mechanical and Manufacturing Engineering, Advanced Engineering Materials and Composites (AEMC) Research Centre, Universiti Putra Malaysia (UPM), Serdang, Selangor 43400, Malaysia

As per De Gruyter's policy this article has previously been published in the journal Physical Sciences Reviews. Please cite as: S. M. Sapuan, M. M. Harussani, A. H. Ismail, N. S. Zularifin Soh, M. I. Mohamad Azwardi and V. U. Siddiqui "Development of nanocellulose fiber reinforced starch biopolymer composites: a review" *Physical Sciences Reviews* [Online] 2023.
DOI: 10.1515/psr-2022-0007 | https://doi.org/10.1515/9783110773606-004

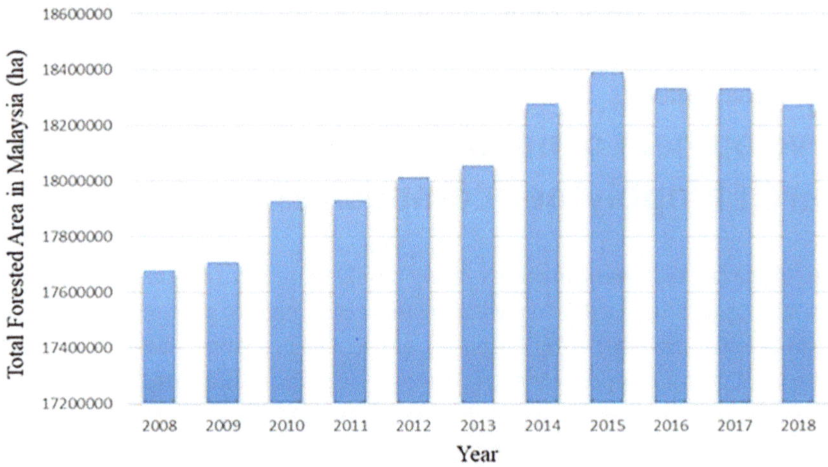

Figure 4.1: Total forest area in Malaysia (ha) from 2008 to 2010 [1].

a decade later, which is in 2018, the total forest area shows about 3.35%, increasing equivalent to 18, 273, 487 ha or 55.4% of the total land area of Malaysia. With this abundant number of forests in Malaysia, an attractive alternative to developing natural fiber and biopolymers, which is produced by living organisms, primarily plants, can be adapted for replacing non-biodegradable materials [2, 3]. The utilization of natural fiber and biopolymer, for example, cellulose and starch, offer a myriad of possibilities such as low-cost, sustainable, and ecological materials for producing environmentally friendly materials that could be used on a wide scale in industrial applications [2, 4]. Cellulose is a polymer that serves as the primary structural support for trees and other plants. It is developed from animals, algae, and a few bacteria but is most abundant in various plants such as cotton, flax, jute, sugarcane, and cereals [5, 6]. Meanwhile, nanocellulose extracted from cellulose possesses unique properties such as broad availability, biodegradability, economical, nano-scale effects, and excellent specific strength [7]. The application of nanocellulose includes food packaging, filtration, wound dressing and numerous other applications [8–11].

Nanocellulose can be divided into three: Nanocrystalline cellulose (NCC), nanofibrillated cellulose (NFC) and bacterial nanocellulose (BNC) [12, 13]. Both NCC and NFC are generated from plants, while BNC is mostly generated from bacteria. Meantime, starch is a carbohydrate commercially generated from a range of natural sources, including barley, wheat, maize, and pea [14, 15]. It also possesses unique properties such as nanocellulose [5]. Due to its tremendous potential, starch was not only utilized as food but is now being employed in many applications, including paper, medicines, textiles, and more [15]. Due to the remarkable capabilities of nanocellulose when in contact with polymer matrices, it is often used as a reinforcing phase in polymer matrices [16]. Table 4.2 summarizes the recent research on nanocellulose-starch polymer composite,

Table 4.1: Total forest area in Malaysia (ha) from 2008 to 2010 [1].

Year	Total forested area in Malaysia (ha)
2008	17,680,508
2009	17,706,821
2010	17,926,964
2011	17,931,424
2012	18,013,251
2013	18,056,155
2014	18,277,602
2015	18,389,686
2016	18,332,583
2017	18,332,583
2018	18,273,487

Table 4.2: Recent research, nanocellulose fiber and starch biopolymer sources, and research focus on nanocellulose-starch composites.

Recent Research	Nanocellulose fiber source	Starch biopolymer source	Research focus	Reference
Plasticized starch/ramie nanocrystallites (PS/RN) composite	Ramie	Wheat	Morphological, thermal, and mechanical properties	[17]
Plasticized starch/tunicin whiskers nanocomposite	Tunicin	Maize	Structural analysis	[18]
Plasticized starch/flax cellulose nanocrystals (PS/FCNs) composite	Flax	Pea	Morphology, thermal behavior, mechanical properties, and water sensitivity	[8]
Thermoplastic starch/ rice straw cellulose nanofibers composite	Rice straw	Potato	Mechanical and dynamic mechanical properties, morphology, humidity absorption, and transparency	[19]
Plasticized starch/kenaf cellulose nanofiber (TPS/ CNFs)	Kenaf	Corn	Morphology, water absorption, water vapor permeability rate, tensile, dynamic mechanical analysis, and fungal degradation properties	[20]
Plasticized starch/ bamboo nanofiber	Bamboo	Corn	Mechanical properties, crystallinity, molecular changes, and morphology	[21]

the source of nanocellulose fiber and starch biopolymer, and research focuses on studies of reinforcement nanocellulose-starch composites. The advantages gained from reinforcing nanocellulose fiber in starch biopolymer composite will surely give numerous promising applications. Thus, this article will discuss further recent developments of nanocellulose fiber reinforced in starch biopolymer composites.

4.2 Nanocellulose

4.2.1 Overview

Nanocellulose is a cellulose-derived nanofiber with a diameter of less than 100 nm. Plants, bacteria, and animals are the primary sources of cellulose; however, most research focused on extracting nanocellulose from plants [22]. This is due to the fact that nanocellulose extracted from plants may be reduced to a finer fiber, and roughly 35–50% of cellulose is generated by plants [23–25]. A variety of cellulosic sources have been used to extract nanocellulose, including sugarcane, areca nuts, bamboo, Ramie, and raw cotton [26–30]. An enzymatic hydrolysis, mechanical extraction, and chemical acid hydrolysis are three methods for extracting the nanocellulose [22, 31, 32]. The acid hydrolysis method is the most popular way for obtaining nanocellulose compared to other extraction methods, with sulfuric acid being the most frequently applied acid in the process [33–35]. Crystalline and amorphous regions distinguish cellulose as ordered and disordered. Around 40–70% of the cellulose is covered by crystalline regions, depending on the cellulose origin and the isolation method in which the chain molecules are organized to make cellulose stiff, elastic, and strong [6, 12]. As well as being resistant to chemical, mechanical, and enzymatic treatments, it is more stable compared to amorphous materials [6, 36]. Meanwhile, the amorphous regions contribute to the bulk material's elasticity and plasticity [12]. Due to its ability to hydrolyze into the amorphous regions of cellulose, sulfuric acid is the most common acid to be used in acid hydrolysis. This results in the nanocellulose having excellent distributed ability in an aqueous medium as well as a high stiffness [22, 6, 25, 35–37]. Ranby and Ribi first reported this method in 1950 [37].

4.2.2 Nanocellulose properties and performance

Nanocellulose offers unique properties such as low density (1.6 g/cm^3), lightweight, and low toxicity. Its biocompatibility and biodegradability provide exceptional crystallinity, stiffness (220 GPa elastic modulus), and tensile strength (10 GPa). Furthermore, it possesses exceptional specific strength, great physical, mechanical, thermal, and barrier properties, a high surface area with nano-scale effects, and relatively reactive surfaces with typical shape, foldability, and enhanced optical transparency [7, 12, 13, 38–44]. Thus, nanocellulose has been utilized to enhance the performance of products and develop materials with its novel features [37]. Nanocellulose is widely used in reinforcement applications as polymer nanocomposites [32]. Polymer nanocomposites consist of polymers as the matrix and nanocellulose as the filler or reinforcing materials [45]. Numerous studies observed that the mechanical, thermal, and barrier characteristics of the final nanocomposites can be enhanced by incorporating nanocellulose into the polymeric matrix even at extremely low concentrations [41]. The presence of strong intermolecular and intermolecular hydrogen bonds with the polymer matrix, in addition to the high aspect ratio of nanocellulose, which contributes to entanglements within the nanocomposite, are responsible for the increased mechanical capabilities of nanocomposites [46]. Strong hydrogen bonds have been formed between the nanocellulose and the polymer matrix of the nanocomposite, which has led to an improvement in the material's thermal characteristics. In addition to this, the presence of nanocellulose may enhance the barrier properties of the nanocomposite. This is because, when nanocellulose is distributed uniformly throughout the polymer matrix, a significant interfacial connection between the nanocellulose and the polymer matrix forms. This results in a reduction in chain segmental mobility and penetrant diffusivity [47]. These nanocellulose-based materials may also be used in various other industries, including packaging, building, furniture, biomedical applications, nanocomposite materials, and textiles [25, 45].

Nanocellulose also has been used in surface modification. There are three methods in performing surface modification; 1) Nanocrystalline cellulose synthesis, 2) physical adsorption, or 3) chemical modification [48]. Due to the enormous surface area and abundance of hydroxyl groups of nanocellulose, the most frequently used surface modification approach is a direct chemical modification [40, 43]. Fabrication of amphiphilic surfaces, which protect both polar and nonpolar fluids, is one of the primary applications of nanocellulose surface modification [49]. Using glass microfiber (GM) and hydrophilic polytetrafluoroethylene (PTFE) filter paper [50] developed amphiphobic nanocellulose-modified papers that could repel polar and nonpolar liquids in a variety of environments. This type of innovative anti-wetting nanocellulose-modified paper may be used in regions that need the antimicrobial, anti-corrosive, and self-cleaning surfaces such as in protective coatings, drug delivery, tissue engineering, and 3D bioprinting [25, 48, 49, 51, 52]. There are three varieties of nanocellulose, as shown in Figure 4.2, such as cellulose nanocrystals (CNCs) or nanocrystalline cellulose (NCC), nanofibrillated cellulose, and bacterial nanocellulose (BC) [12, 13].

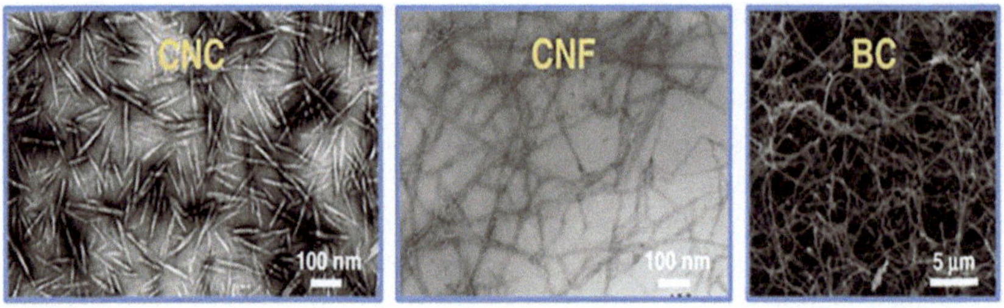

Figure 4.2: Three varieties of nanocellulose (CNC, CNF, and BC) under electron microscope [39].

4.2.3 Nanocrystalline cellulose (NCC)

The other names for nanocrystalline cellulose are cellulose nanocrystals (CNC) and cellulose nanowhiskers (CNW). The source of the NCC can be obtained from by-products of forestry and agriculture, higher plants and crops such as ramie, flax, bamboo, sugarcane bagasse, and many more [8, 17, 53–55]. In plant-based sources such as cellulose fibrils, NCC can be extracted through chemical processes such as acid hydrolysis [12, 40, 41]. Figure 4.3 illustrates the acid hydrolysis procedure used to obtain NCC from cellulose fibrils. The amorphous cellulose elements are hydrolyzed and removed during acid hydrolysis, whereas the crystalline elements are retained [12, 43]. As a result of this process, NCC is in high stiffness as it contains high crystallinity, around 54–88%, with short-rod-like fibrils or whisker form as shown in Figure 4.4 [12, 25, 43, 56]. Table 4.3 shows the NCC's diameter, length, tensile strength, and Young's Modulus (YM).

According to research on NCC conducted by Kovacs et al. [59], NCC is safe for the environment and does not pose any threat to health. Aside from that, an article said that NCC is not cytotoxic and does not trigger an immune response [16]. In addition, NCC is rich in hydroxyl groups, has a large specific surface area, possesses strength and stiffness, is biodegradable, and possesses exceptional mechanical and thermal properties. These characteristics combine to make NCC an excellent material for use as either a reinforcing agent or filler in composite materials [54, 60–62]. The following should be emphasized to maximize the performance of NCCs as filler in polymer composites: NCC must be properly extracted, the plant source must have high cellulose

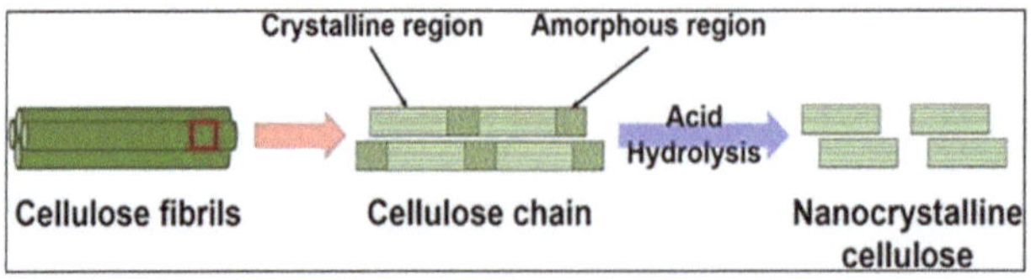

Figure 4.3: NCC extraction using acid analysis [25].

Figure 4.4: NCCs under transmission electron microscope of 100 nm [57].

Table 4.3: The properties of NCCs [12, 25, 43, 56, 58].

Diameter (nm)	Length (nm)	Tensile strength (GPa)	Young's modulus (GPa)
2–20	100–500	2–6	50–143

content, and the plant source's growing conditions must be optimal [58]. There are several potential applications of NCC in various industries.

One of the applications of NCC is food packaging in the packaging industry [63]. This has been studied by Cao et al. [8] in plasticised starch/flax cellulose nanocrystals (PS/FCNs) composite films. They discovered that the nanocomposite films exhibited a high degree of interfacial bonding between PS and FCNs, which increases glass transition temperature and improved vibration and thermal shock resistance. Apart from that, the PS/FCNs nanocomposite film showed a remarkable improvement in tensile strength and YM as the FCNs volume increased from 0-30% in water uptake (wt). As demonstrated in Figure 4.5, the study also demonstrated that water absorption reduced as the FCNs concentration increased during conditioning at 98% relative humidity (RH), demonstrating their greater water resistance. These outstanding characteristics make them ideal for food packaging applications.

Other applications of NCC are antimicrobial wound dressing in the drug delivery industry. Antibacterial wound dressings have emerged as effective solutions for reducing wound bacterial transmission of infection, hence improving the healing process [64]. This application has been reported by [65] in a developing antibacterial wound dressing film using NCC as a drug delivery mechanism for curcumin. Polyvinyl

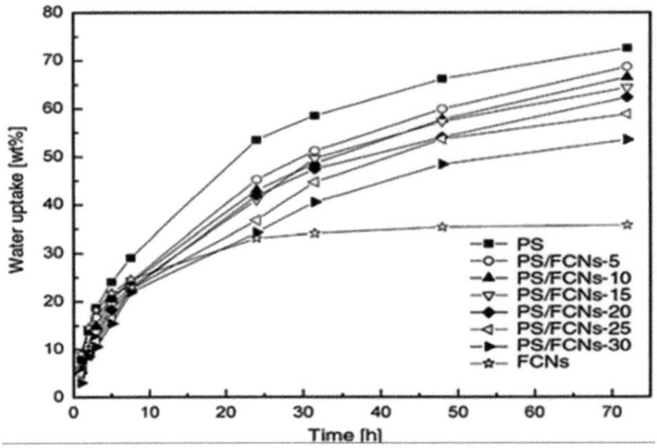

Figure 4.5: Water absorption of PS and composite films during conditioning in 98% RH [8].

alcohol (PVA) was used as a polymer matrix to enhance the films' mechanical characteristics even at modest NCC loadings. They discovered that curcumin loaded NCC films are soft and flexible, demonstrate considerable inhibitory effect against both positive and negative bacteria, and have a high launderability due to the curcumin's strong binding to the cellulose nanocrystals. The controlled release is critical to the film's long-term antibacterial activity, and the created film shows it maintains a steady drug concentration in a wound without rapidly increasing the concentration of the drug. It delivers a therapeutically effective dose by continuously delivering the drug into the wound to stabilize the drug level. This condition favors the application as a wound dressing in which the NCC in the film serves as a diffusion barrier [66].

4.2.4 Nanofibrillated cellulose (NFC)

The other name of nanofibrillated cellulose is microfibrillated cellulose (MFC), cellulose nanofiber (CNF) or cellulose nanofibrils (CNF). The process of extracting NFC from its source is analogous to that of NCC [41, 67]. The extraction of NFC can be extracted via the mechanical extraction process [39, 68]. The mechanical process isolates the cellulose fibrils by cleaving the cellulose fibers longitudinally with a high shear force [12, 69]. The most common mechanical extraction methods include high-pressure homogenization, ultrasonication, and grinders and refiners [25, 43]. Figure 4.6 illustrates the mechanical process procedure used to obtain NFC from cellulose fibrils. The force is applied towards the cellulose fiber during the mechanical process, resulting in the cleavage longitudinally [43]. NFC is made up of both aggregated and individual nanofibrils made up of alternating crystalline and amorphous cellulose domains as a result of this processes and 100% of cellulose chemical composition, with a soft and elongated chains morphology caused by the entanglement of the cellulose chains as shown in

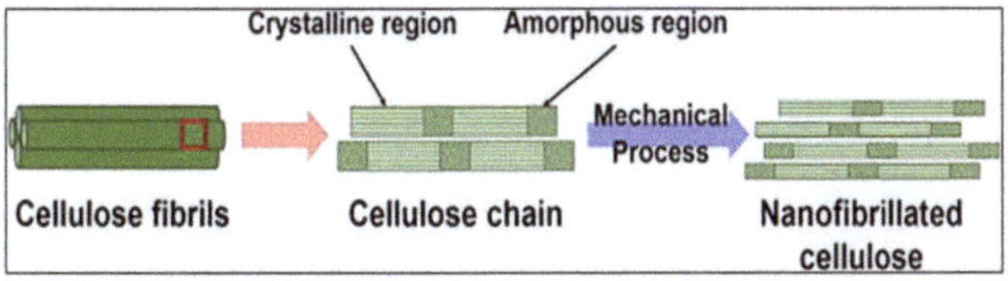

Figure 4.6: NFC extraction using mechanical processes [25].

Figure 4.7 [12, 40, 70]. Due to their amorphous nature, NFC is more elastic than NCC [67]. Table 4.4 shows the diameter, length, tensile strength, and YM of NFC.

NFC contains a number of promising properties, including excellent hydrophilicity, biocompatibility, high concentration of hydroxyl groups, and extensive specific surface area [40, 56, 67, 71]. There are several potential applications of NFC in various industries. One of the applications is cement mortar in the construction industry. Ardanuy et al. [72] has developed a cement mortar made of NFC extracted from sisal pulp in his study. According to his research, sisal pulp was the most suited reinforcement agent compared to cotton fiber for the brittle cementitious matrix due to sisal fiber being longer and narrower, resulting in a higher aspect ratio. For the composite preparation, the ratio of cement, sand and water were set to 1:1:0.67, and

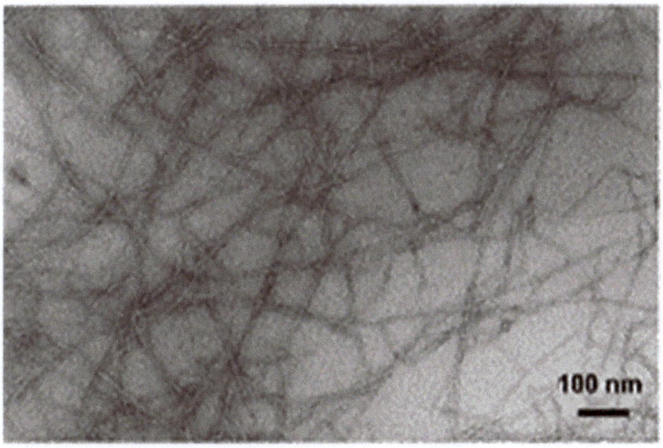

Figure 4.7: NFC under transmission electron microscope of 100 nm [57].

Table 4.4: The properties of NFC [16, 39, 56, 67, 68].

Diameter (nm)	Length (nm)	Tensile strength (GPa)	Young's modulus (GPa)
1–100	500–2000	2–4	15–150

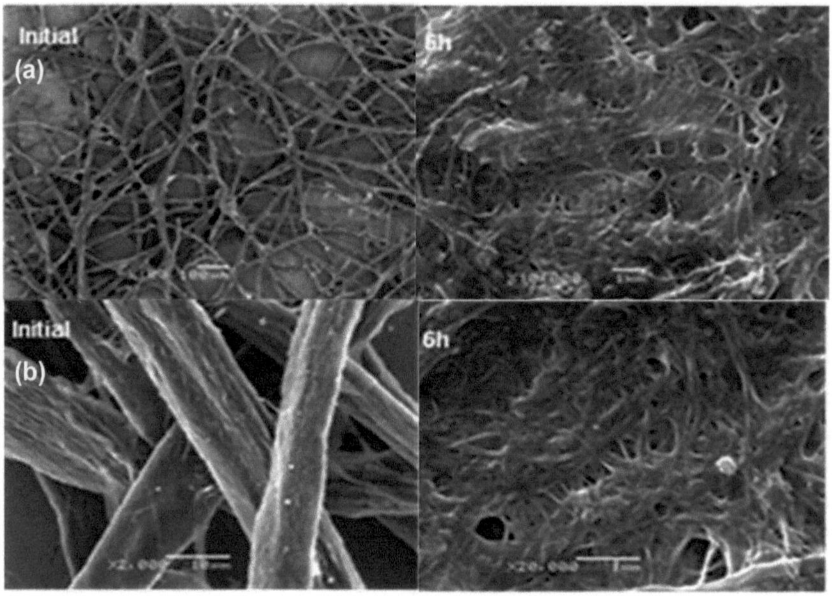

Figure 4.8: SEM micrographs of the initial fiber and 6 h refinement at low (a) and high (b) magnifications [72].

the reinforcement amount of NFC was set at 3.3 wt%, which is the maximal amount permitted in the preparation of the uniform mixtures. NFC was prepared by subjecting sisal pulp to a high-intensity refining process. The influence of mechanical processes on the microstructure of the fiber was investigated using a scanning electron microscope (SEM), as shown in Figure 4.8. It was revealed that initial sisal fiber has a diameter of 10–20 m, and further refining up to 6 h results in highly branched fibers with a diameter of 25–250 nm. These NFC may have a higher strength and aspect ratio than the original fibers, enhancing the reinforcing capabilities and making these pulps suitable for composites manufacturing. Additionally, their large specific surface area may enhance matrix connectivity, resulting in increased fiber-matrix bonding effectiveness.

As indicated in Table 4.5, the mechanical characteristics of cement mortar composites reinforced with cellulose sisal fiber and NFC were investigated using flexural testing. Comparing the cement mortar composite made of cellulose fiber and the one reinforced with NFC, the composite treated with NFC demonstrated a considerable improvement in both bending modulus and its capacity to withstand force or pressure. This is due to NFC having a high specific surface area, which enhances fiber-matrix connection. However, it leads to an embrittlement of the composite and is confirmed by having a lower value of fracture energy than cellulose sisal fiber. This brittle failure is due to the NFC's poor crack bridging ability, as its nanoscale length is insufficient to inhibit matrix crack formation. To solve the following problem, the refining time must

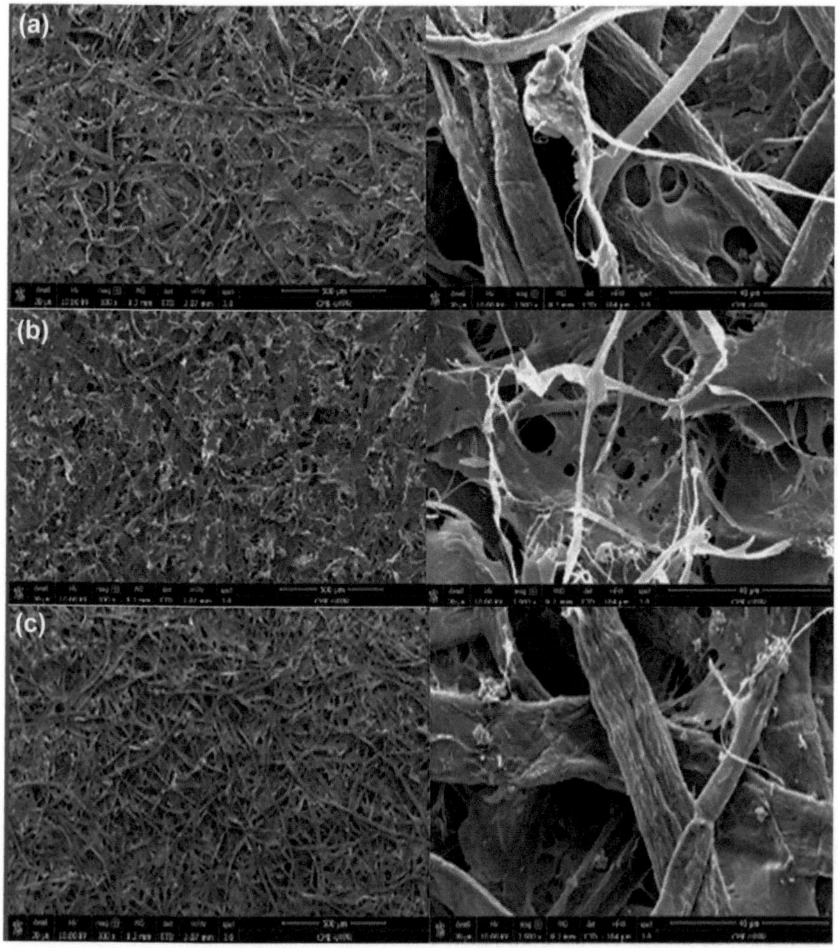

Figure 4.9: SEM micrograph of recycled cardboard (a), newspaper (b), and writing paper (c) fibers at low (left) and high (right) magnification [73].

Table 4.5: Mechanical properties of cement mortar composites [38].

Composites	Flexural Modulus (GPa)	Flexural Strength (MPa)	Fracture Energy (J)
Cellulose fibers reinforced cement mortar composites	2.4 ± 0.1	10.3 ± 0.06	759 ± 78
NFC reinforced cement mortar composites	4.1 ± 0.1	14.0 ± 1.5	357 ± 48

be optimized to obtain longer NFC fibers that retain their flexural qualities while increasing ductility.

Another application of NFC is as a paper additive. According to Viana et al. [73], the physical and mechanical properties of recycled papers can be improved by adding NFC. This has been proved in their study in which three types of pulp paper produced from recycled cardboard, newspaper and writing paper were investigated with different reinforcement percentages of NFC (0, 5 and 10 wt%). The findings showed that the thickness was reduced while the density was enhanced with the addition of NFC into the recycled cardboard and papers. The enhancement in density is related to the formation of tiny fibers. The smaller the fibers are, the greater the surface area of interaction between the adjacent cellulose fibers. This results in a greater number of hydrogen bridge links, thus generating a denser network and increasing the paper's strength and stiffness. By comparing the three types of pulp paper, recycled writing paper shows the best physical and mechanical properties. As shown in Figure 4.9, the SEM micrograph of all types of recycled cardboard and papers were observed. It was revealed that the recycled cardboard's fibers formed mostly fine and damaged with varying lengths, while the newspaper fibers have high quantities of fine, incomplete and severely damaged fibers. This indicates that the material has low physical and mechanical resistance. Meanwhile, the recycled printing/writing paper fibers have low fine quantities, uniform lengths and less structural damage compared to other fibers. This generates a denser network and increases the paper's mechanical properties. Other than that, it also shows the improvement in the tensile (96%), burst (101%), and tear (104%) resistances with the addition of 10 wt% of NFC, compared to normal papers. Nevertheless, the addition of NFC improves fiber morphology, resulting in a more compact, uniform, and durable structure for all recycled cardboard and papers.

4.2.5 Bacterial nanocellulose (BNC)

The other names for bacterial nanocellulose are biocellulose (BC) and bacterial cellulose (BC) [41, 54]. BNC can be isolated from a variety of bacterial species, including gluconacetobacter, agrobacterium, pseudomonas, rhizobium, and sarcin; however, the gluconacetobacter species is the one that has received the most attention from the researchers [41, 43, 58, 74]. BNC is produced and cultivated in an aqueous medium containing glucose, oxygen, and phosphate [16]. BNC has the shape of a twisting ribbon with a high surface area per unit, as shown in Figure 4.10 [75]. BNC showed a high degree of purity, and it is the most refined nanofibers since it is produced in a pure form that does not require any laborious or time-consuming treatment to eliminate impurities or pollutants such as pectin, hemicellulose, or lignin [70, 71, 76, 77].

BNC is chemically similar to nanocellulose that plants produce. However, due to its 3D nanofibrillar network, it displays unique physical characteristics such as high crystallinity, water absorption capacity, tensile strength, porosity, degree of polymerization,

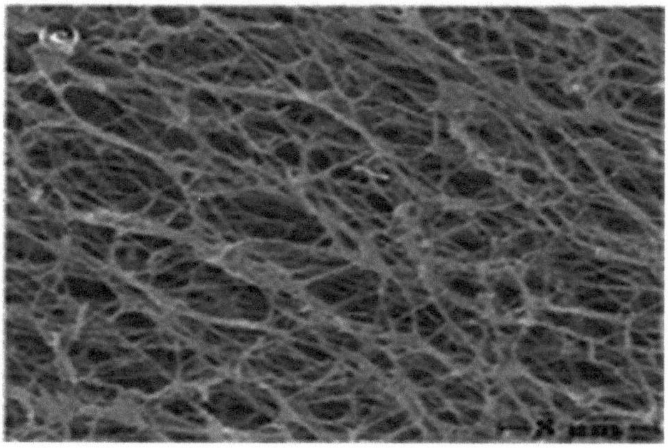

Figure 4.10: BNC under scanning microscope of 8 μm [78].

mechanical strength, and biodegradable [11, 79–81] (Table 4.6). There are several potential applications of BNC in various industries. BNC has a number of applications in the biomedical industry, including wound dressing for the recovery of damaged skin [10, 11]. Wound dressings regenerate skin tissue by absorbing exudate throughout the healing process and removing it from the wound area after recovery [70, 82, 83]. Multilayer fermentation may be used to create BNC for skin tissue repair, demonstrating minimal cytotoxicity and high multiplication of human adipose-tissue-derived stem cells [70]. Experiments on animals have proven that tissue regenerates and capillary development in the wound region more rapidly on a BNC biomaterial dressing than on a commercial dressing [83, 84]. This is shown in research conducted by [85]; in which they generate BNC/collagen composites by immersing *Acetobacter xylinum* in a collagen solution and then freeze-drying them. Figure 4.11 (b, d) shows the SEM micrograph of collagen molecules coated the BNC surface and penetrating inside BNC, forming a smooth layer of BNC/collagen composite. Additionally, it has a porous network structure with a large surface area required for cellular adhesion and vascularization and stimulating cellular ingrowth in biomedical applications.

Other than that, BNC is also used in sheet masks in the cosmeceutical industry. Amorim et al. [86] developed a moisturizing and anti-inflammatory sheet mask using a propolis extract integrated into a BNC film formed in a modified medium. Propolis extract is a natural waxy material generated by bees where it contains polyphenols several flavonoids and has an infinite number of biological properties. Thus, propolis is used in medical treatment to help in the reduction of inflammation. The high-water content of BNC, on the other hand, assists in skin hydration and improves the texture of affected skin. The result shows that the composite film helps to moisten and treat acne-related inflammations.

4.3 Starch biopolymers

4.3.1 Polymers

Biopolymers are polymers extracted from living creatures, and microbes. They are plentiful renewable resources that are often utilized to make environmentally friendly bioplastics [87]. A biopolymer is a polymer that can be degraded by microbes, moisture, and heat. There is no specific standard for biodegradation, according to the literature. Agricultural waste starch may be utilized to make biopolymers [88]. Biopolymers are mass-produced commercially for a wide range of applications. A multitude of factors influence biopolymer breakdown capabilities, including polymer type, chemical makeup, and environmental conditions. Protein, starch, cellulose, deoxyribonucleic acid (DNA), ribonucleic acid (RNA), lipids, collagen, carbohydrates, and other biopolymers are examples.

4.3.1.1 Biopolymers

Based on their origin, biodegradable polymers are classified into two types: bio generated or natural polymers and synthetic polymers. Since natural biopolymers are formed from biological systems. In contrast, synthetic polymers are created from natural resources such as petroleum, charcoal, and fossil fuels. There are two categories of biopolymers based on degradation: non-biodegradable and biodegradable. Polymers derived from plants and biopolymers derived from fossil fuels are two more forms of biopolymers [89]. Biopolymers are categorized into three types depend on the type of their functional group: Polysaccharides, such as glucose (for example, cellulose found in plants), proteins, such as amino acids (such as methemoglobin found in muscle cells), and nucleic acids, such as nucleotide (DNA, genetic material of a given organism), are the three types of carbohydrates. According to their intended use, biopolymers can be classified as bioplastics, biosurfactants, bio detergents, bioadhesives, bioflocculants, and so on. Because of their low cost, extensive availability, high moisture absorption, high thermal stability, favorable mechanical properties, resistance to degradation, and biocompatibility, biopolymers have been put to use in a diverse array of applications [90]. The hydrophilic nature of biopolymers, poor mechanical qualities, and a low lasting degradation ratio in moist settings are some of the biopolymers' shortcomings, despite the fact that these materials are widely used and have many benefits [91].

Ordinary packaging materials have transformed daily life in a wide range of applications. Over 50 percent of all plastics are reused for single-use purposes (they are only used for a few moments) based on the claim. However, these polymers persist on the planet for at least several centuries before entirely disintegrating; moreover, some of them do not degrade at all [92]. The use of petroleum resources in the manufacture of these plastic polymers raises knowledge about potential resilience, people and animal welfare, and health of the economy (landfill sites). This plastic generated major environmental challenges such as the buildup and agglomeration of plastic rubbish in landfills, as well as physical problems for flora and fauna. This is due to plastic entrapment or consumption, as well as the chemical leaching process of plastic into rivers, seas, and water storage basins. It may later result in the spread of chemical compounds to live animals, which can lead to infection.

Natural biopolymers such as polyols, polyhydroxyalkanoates, enzymes, and polysaccharide-based plastics have been investigated as viable replacements for regular plastics in order to address the increasing ecological problem caused by nonbiodegradable plastics [58]. As illustrated in Figure 4.12, a variety of biopolymers derived from various sources, including agriculture and agri, microbial, and synthesized biopolymers, are used to improve biodegradability. Starch is the only biodegradable polymer that is renewable, cost-effective, physically pleasing, and completely biodegradable with no adverse effects.

4.3.2 Starch

Semi-permeable to water vapor, oxygen, carbon dioxide, and the components of food additives, starch-based polymers are nontoxic and odorless, transparent, and biodegradable [93]. Starch is one of the viable options for replacing non-biodegradable plastic due to its accessibility, environmental friendliness, cost, and biodegradability.

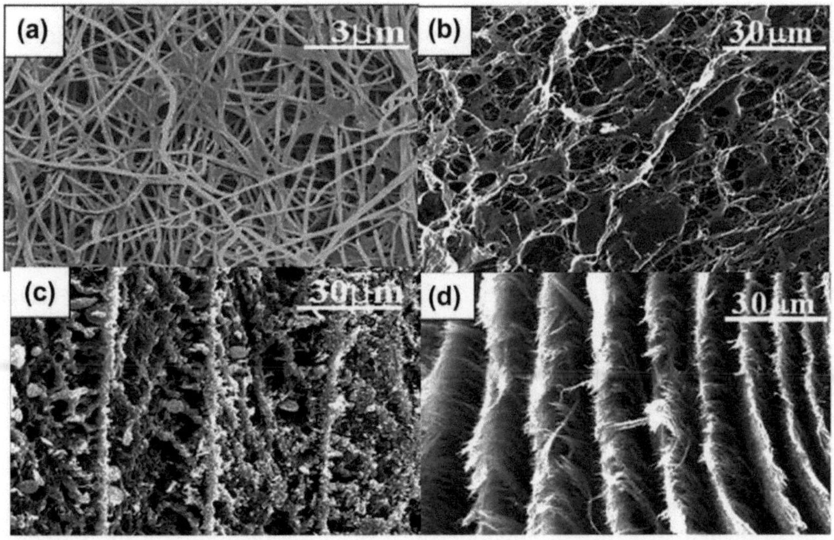

Figure 4.11: SEM micrograph of the initial BNC (a: surface; c: cross section) and BNC/collagen composite (b: surface; d: Cross section) [85].

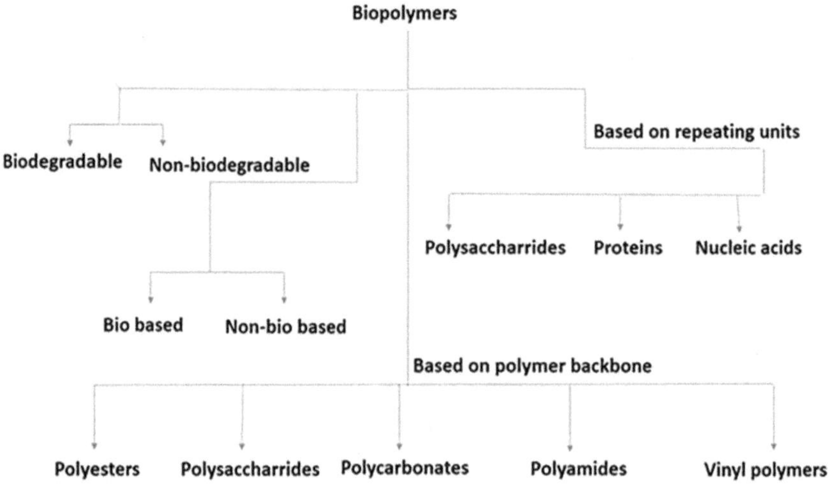

Figure 4.12: Main categories of biopolymers [89].

Furthermore, these biopolymers have been thoroughly studied for their prospective application in the manufacture of products such as food and fruit packaging films, plastic shopping bags, flushable linter, pharmaceutical delivery system, and technology [94]. Starches are also hydrophilic and can be found in nature as partially and separate crystalline tiny grains connected by an enormous microsphere structure of mixed substances. Starch is composed of both morphology and longitudinal carbohydrates known as amylopectin and amylose, which are derived from different agricultural sources. Figure 4.13 shows schematic representation of biopolymers from different sources [16]. The molecular and biochemical of amylopectin and amylose are depicted in Figure 4.14 (a) and (b), respectively.

4.3.2.1 Amylose and amylopectin

Amylose is made up of α-1, 4 connected glucose units in a linear configuration. Amylopectin, on the other hand, has a highly branching structure formed of short α-1,4 chains joined by α-1,6 links. Amylose molecules are typically made up from 200–20,000 glucose monomers arranged in a helical. This occurred as a result of the distinct attachment orientations in between glycosidic bonds. Amylopectin is composed of 30 short glucose unit side chains connected per 20–30 monosaccharides along the strand. Amylopectin molecules can include approximately 2 million molecules of glucose [94]. With its horizontal nature, amylose behaves similarly to synthetic polymers. Meanwhile, the branched structure of amylopectin lessens the flexibility of the polymer chains. It also prevents them from becoming orientated tightly enough to demonstrate large quantities of hydrogen bonding. In most situations, flour includes between 70 and

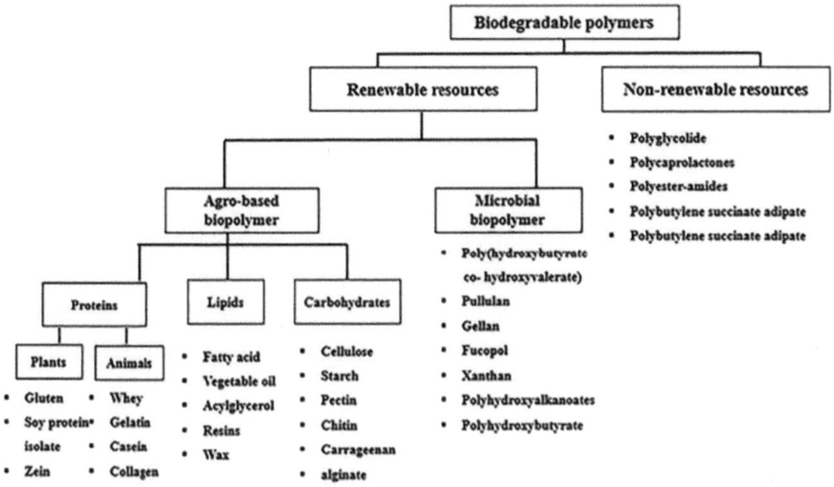

Figure 4.13: Schematic representation of biopolymers from different sources [54].

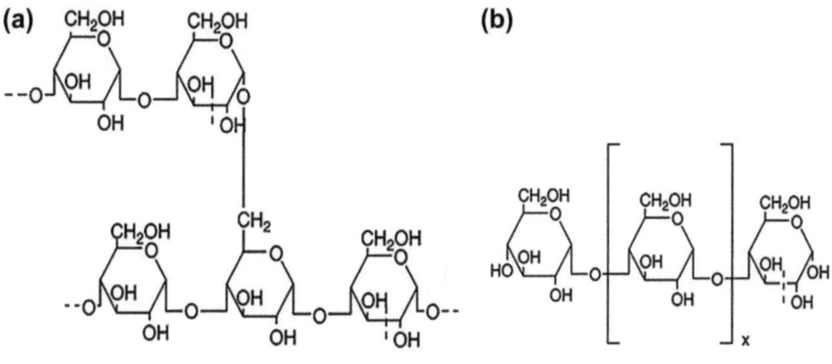

Figure 4.14: Structure in chemical of (a) amylopectin and (b) amylose [95].

85 percent amylopectin and 15 to 30 percent amylose [96]. Due to the obvious large distribution of numerous hydrogen bonds starch processing is slowed, resulting in a melting heat that is higher compared to the decomposition temperature. As a result, modifiers (water, glycerin, and methanol) are being used to enhance free volume and, as a result, lower softening and melting temperatures [94]. Thermoplastic starch (TPS) is a kind of starch that has been thermally and/or chemically plasticized to replace any need for synthetic polymers. It is manufactured when the molecular structure of starch is disrupted, as when heating starch granules induces swelling and under some conditions, amorphous parts undergo a non-reversible transition in the addition of a softener [94].

4.4 Nanocellulose reinforced starch biopolymer composites

4.4.1 Starch biopolymer composites

In the last 10 years, much emphasis has been placed on the use of organic rigid additives as reinforcements in starch-based thermoplastics in order to increase water sensitivity and mechanical characteristics of starch biopolymer at the same time lowering the utilization of non-petroleum-based polymer and synthetic resin. This is related to the rigidity, low moisture reactivity, and low tensile strength of starch-based thermoplastics. As a result, plant-based fillers have been bonded to starch-based thermoplastics to create composite materials or bioplastics that improve water resistance as well as mechanical strength. A composite material is one that is composed of more than one constituent material that has been blended to produce significant chemical and physical behavior, resulting in a material having distinct characteristics than the separate capacity [97–100]. Wooden (trees) and vein arteries are examples of natural composites. Wooden is made up of cellulose-reinforced fibrils that are held on by a lignocellulosic materials structure [101, 102]. In the early 1990s, car industry researchers produced composites that included nanosized clay support within polymer (i.e., terra cotta polymer nanocomposites), resulting in considerable increase in rigidity, high stiffness, plus thermal expansion when compared to pure polymer [103].

4.4.2 Nanocellulose

Nanocellulose is a remarkable substance that is made up of nano-scaled cellulose microfibrils and nanocrystals. It can be produced from cotton linters, plant cell walls, or microorganisms by the use of mechanical, enzymatic, or chemical processes. It is possible to produce nanocellulose by synthesizing any form of cellulose, including that found in viruses, scum, and fungi, as well as in flora and fauna. Agricultural waste and wood, on the other hand, have been examined as crucial materials for nano-cellulose extraction. Nanocellulose is a cellulose derivative with a width of 100 nm and a magnitude of a few micrometers. The three primary types of nanomaterials are cellulose nanofibers, cellulose nanocrystals, and bacterial cellulose, which are titled after the process and synthesis conditions that determine the quality, content, and characteristics of nanocellulose [104]. The synthesis of CNCs and CNFs involves the dissolution of cellulose fibers into nanoparticles, whereas the formation of BC involves the inoculation of bacteria. The lengths of NCC and NCF are 50 nm–0.5 m and 50 nm–3 m, respectively, with diameters of 3–15 nm and 5–50 nm, whereas the length of bacterial

Table 4.6: The properties of BNC [54, 109].

Diameter (nm)	Length (nm)	Tensile strength (GPa)	Young's modulus (GPa)
20–100	20–100	200–300	15–35

cellulose is 200 nm–3 m [105]. Nanocellulose is characterized by a wide range of desirable properties, including a high modulus of elasticity (110–220 GPa), the elastic ability (7.5–7.7 GPa), a tunable aspect ratio, a sizable specific surface area, easy surface functionalization, tunable crystallinity, a high degree of polymerization, and a high resistance to chemicals [106]. Nanocellulosic sustainable materials are gaining traction in a variety of industries, including those that deal with transportation, packaging, energy devices, and aerospace, due to their exceptional chemical and physical qualities. Additionally, the use of fossil fuels and the impact of synthetic materials on the climate are both important factors to take into account. Therefore, making the switch to these eco-friendly nanomaterials will be crucial over the next years. Recent advances in engineering, biomaterials, and other cutting-edge applications have brought to light the importance and adaptability of various celluloses, especially nanocellulose. These chemicals have several practical applications, including water treatment, human and animal health care. The effectiveness of nanocellulose-based products is dependent on the precision with which nanocellulose dimensions are made, making it crucial to understand the dimensions of nanocellulose created from various plants and NFs [107, 108]. When compared to microcomposite, nanocomposites have improved characteristics (reduced resistance, greater impact resistance, etc.). Table 4.7 shows the comparison of different characteristics against each variety of nanocellulose.

Table 4.7: Comparison of different characteristics against each variety of nanocellulose [105].

Properties	NFC	CNC	BNC
Dimensions and aspect ratio	Dia - (1–10 μm); very high A.R.	Dia - (5–30 μm); low A.R.	Dia - (20–100 μm); low A.R.
Chemical requirements	Corrosive reagents	Corrosive reagents	Noncorrosive chemicals
Sustainability	Not very sustainable	Not very sustainable	Green approach
Cost	Low cost	Low cost	High cost
Energy	High energy process	High energy process	Green process

4.4.3 Mechanical and physical properties of nanocellulose reinforced starch biopolymer composites

Starch is a ubiquitous, low-cost, extremely simple biopolymer. Starch films are a potential option for the packaging of food due to their improved physical properties, as well as their biodegradability and long-term sustainability. However, the brittle nature of starch films, as well as their low water resistance, limits their applicability. Polymers and plasticizers can be used to alleviate these drawbacks [110, 111]. Because of its high crystallinity, huge surface area, improved mechanical strength, and electrically active sites, nanocellulose may have use in the packaging of food, the manufacture of pharmaceuticals, and the treatment of medical conditions. It is critical to fortify nanocellulose derived from several starch film sources for packing [110]. The well-akin diffusion and spread of nanocellulose in the starch biopolymeric matrix, according to Siqueira et al. [112], is crucial because it impacts physical performance. Table 4.8 reveals

Table 4.8: Starch-based polymer, NFC nanocomposites, and their mechanical performance [113].

Starch based Polymers	NFC Source	NFC Preparation	Manufacturing Technique	Tensile strength (MPa)	Young's Modulus (MPa)	References
Cassava starch	Cassava bagasse	Hydrolyzed in 6.5 M H_2SO_4/40 min	Solution casting	4.8	84.3	[114]
Mango puree	Wheat	-	Solution casting	8.76	322.05	[115]
Maize starch	Wheat straw	High-pressurize homogenizer/15 min	Solution casting	6.75	220	[105]
Maize starch	Cotton cellulose	Hydrolyzed in 6.5 M H_2SO_4/75 min	Twin screw extruder and hot press	0.35	3.12	[116]
Potato starch	Scotswood wood flour	Supermasscolloider	Solution casting	17.5	1317.0	[117]
Potato starch	Rice straw	Ultrasonication	Solution casting	5.01	160	[19]
Maize starch	Kenaf	Supermasscolloider	Solution casting	2.35	53.6	[118]
Corn starch	Kenaf	Supermasscolloider	Solution casting	38.0	141.0	[20]
Corn starch	Bamboo fiber	-	Solution casting	11.2	12.4	[21]
Sugar palm starch	Sugar palm fiber	High-pressure homogenizer, 500 bar	Solution casting	10.68	121.26	[58]
Yam bean starch	Water hyacinth	Ultrasonic crusher treatment	Solution casting	11.47	443	[119]

that the bionanocomposites made from the cellulose nanofiber reinforced maize starch had the highest tensile strength (38.0 MPa).

The nanocomposites constructed from grain cellulose nanofiber reinforced mango pureed had the highest YM (322.05 MPa). These findings were carried out by Azeredo et al. [115]. Bionanocomposite was developed by putting some wheat CNF from FMC Biopolymer as a hardening filler to mangoes mush-based edible films at a particular loading of 36 g/100 g. Before being casted and allowed to cure for 16 h at 42% relative humidity and 22 °C, the film forming dispersions were vacuumed to reduce air pockets. The study discovered that adding CNF to mango puree biopolymer films considerably enhanced their mechanical, vapor permeability, and thermal characteristics. Corn starch had the highest tensile strength (38.0 MPa). Interestingly, the addition of CNF successfully increased the tensile properties and melting point of films. This occurred as a result of the development of a nanofibrillar network inside the polymer matrices [115]. Despite some accumulating agglomeration, the results show that the incorporation of CNF enhanced the mechanical properties of plain maize starch biopolymer at loadings more than 2.5 wt percent. This is due to the authors' manufacturing process, which is a twin screw extruder. Table 4.8 shows the mechanical performance of starch-based polymers and NFCs nanocomposites.

Numerous CNC enhanced starch-based nanocomposites' mechanical performance is shown in Table 4.9. The majority of researchers and scientists, according to the data, in order to remove nanofiber from naturally available fiber, they used a sulfuric acid (H_2SO_4) hydrolysis method. During the hydrolysis process, natural fiber will be subjected to a very extreme ph. For 5 min, the mixture was stirred in a preheated autoclave reactor with scalded water at 160 °C. The liquid was then sucked to eliminate air bubbles before being cast in a Teflon mold. Because of the deep relationship between the tunicin CNC, sticky wheat corn starch biopolymer matrix, moisture, and softener, as well as the CNC's ability to create a three-dimensional stiff network, the nanocomposites have high mechanical strength scalded water. On the other hand, the mechanical properties of the nanocomposite films increased proportionally with the nanofiller concentration, which demonstrates an effective stress transmission between the biopolymer matrix and CNC nanofiller [120, 121].

For 5 min, the mixture was warmed in a boiling water evaporator reactor at 160 °C. Following that, the solution was sucked to eliminate any air bubbles before being summoned in a Teflon mold. When compared to a clean starch biopolymer, the inclusion of CNC at a 25% concentration increased the mechanical performance of the composite from 0.24 to 20 MPa. Compared to the spotless starch biopolymer film, the nanocellulose strengthening reduced the dispersibility of starches biopolymer, resulting in a stiffer, fewer stretchy, and more fracture conferring resistance nanocomposite. Furthermore, Ilyas et al. [16] observed that the enforcement of 0.5 wt% sugar palm nanocellulose significantly enhances the strength of nanocomposites film (tensile strength and YM increased by 11.466 and 178.828 MPa, respectively). Nanocellulose has a low population density, an availability of hydroxyl groups, a large density, a high

Table 4.9: Starch-based polymer, CNCs nanocomposites, and their mechanical performance [113].

Starch based Polymers	NCCs Sources	Isolation Chemical/Time of NCCs	Manufacturing Technique	Tensile strength (MPa)	Young's Modulus (MPa)	References
Pea starch	Hemp	64 wt% H_2SO_4/4 h	Solution casting	3.9–11.5	31.9–823.9	[8]
Pea starch	Flax	64 wt% H_2SO_4/4 h	Solution casting	3.9–11.9	31.9–498.2	[8]
Pea starch	Bamboo	50 wt% H_2SO_4/48 h	Solution casting	2.5–12	20.4–210.3	[122]
Maize starch	Waxy maize starch	H_2SO_4/5 days	Solution casting	1–15	11–320	[123]
Maize starch	Tunicin	55 wt% H_2SO_4/20 min	Solution casting	0.24–20	51–315	[124]
Wheat starch	Cotton seed linter	64 wt% H_2SO_4/4 h	Solution casting	2.5–7.8	36–301	[17]
Potato starch	Potato peel waste	64 wt% H_2SO_4/90 min	Solution casting	-	460	[125]
Sugar palm starch	Sugar palm fiber	64 wt% H_2SO_4/45 min	Solution casting	11.5	178	[55]
Tapioca starch	Water hyacinth	HCl	Solution casting	5.8	403	[119]

hardness, a high surface to volume ratio, and a large surface area, making it excellent as an element for strengthening in starch biopolymers, according to previous published works [54, 58]. Moreover, the starch biopolymer's ability to physically interface with the nanofibers led to the improved mechanical characteristics of nanocomposites. When nanocellulose is coupled with a starch biopolymer, the nanocellulose and the starch biopolymer generate a new robust interfacial bonding. Table 4.9 demonstrates the mechanical characteristics of starch-based polymers, CNCs nanocomposites, and their mechanical performance.

4.5 Recent development of nanocellulose fiber reinforced starch biopolymer composites

4.5.1 Food packaging

In the last several decades, we are seeing a spike in popularity in food packaging innovation in response to the growing environmental effect of waste, customer demand for conducive foods that have the characteristics as an example prolonged shelf life, and eco-friendly concerns about restricted natural resources. Numerous materials, including plastic, paper, paperboard, steel, and aluminum, that have been utilized to develop packaging materials globally, with more than two-thirds would be used mainly in the food industry [113]. According to present study, the manufacturing and utilization of petroleum-based food packaging materials have developed in popularity because of their abundance and outstanding mechanical properties [126]. Apart from that, research by Fonseca-García et al. [127] stated that global plastic use presently reaches 700 million tons annually and is anticipated to achieve one billion in 2021. As a basis, we would see that the packaging industry is the main aspect responsible for the growth of plastic waste, and the concern must be tackled seriously in order to protect the environment. Furthermore, films exhibiting moderate water vapor permeability (WVP) are appropriate for food packaging utilization, together with minimization and inhibition of humidity movement across the environment and the food. Hence, decreasing the WVP of sugar palm starch (SPS) films is crucial for their remarkable prospective utilizations [54].

Looking at the latest development of nanocellulose fiber reinforced starch biopolymer composites, it was initially invented from starch biopolymer, which is nontoxic, transparent, flavorless, and semi-permeable to humidity, oxygen, carbon dioxide, and flavor enhancers components, but also has some drawbacks, which include poor degradability, constrained power, and a top level for water absorption capacity of starch films, which restrict its functionality to be applicable to most industry [113]. Various approaches, such as starch manipulation and the insertion of strengthening compounds into the structure of the films, can drastically raise their total attributes for food applications and it can be changed using a combination of physical, chemical, enzymatic, and/or mixed ways [128]. As a consequence of this innovation, it has a big potential to be implemented in a variety of applications, as example food packaging and additives. According to Bangar et al. [113] analysis, starch bio-composites enhanced with nanocellulose may be utilized further to prolong the longevity and sustain the conditions of food. Similarly, nanocomposites have been used in food packaging to expand the shelf life of food during processing and handling [129]. A fresh study by Montero et al. [130] discovered that the use of potato starch film enhanced with 5% cellulose nanoparticles enhanced the solidity, thermal stability, and resistance to water of the film, revealing that this could be a more reasonable replacement for short-term applications in the packaging sector. This conclusion was confirmed by the findings of experiments performed on thermoplastic starch (TPS) and TPS/CNC films, which were then analyzed. Also Syafri et al. [131] had synthesized bengkuang starch and hyacinth nanocellulose bio composites proceeded with 15–30 min vibration time for ultrasonic incubation. Both thermal stability and moisture resistance of biocomposites were boosted by ultrasonication vibration, highlighting its promising application as green technology packing bags. The ultrasonic bath's kinetic energy lowered the idle OH bonding to the matrix. The optimum findings were achieved using biocomposite samples that were pulsates for 60 min. Furthermore, extending the vibration duration lowered the biodegradability properties of biocomposites in the soil. The soil burial test identified that biocomposites which have been through a vibration period tend to be degraded at a moderate speed than sampled biocomposites. Yet, it was revealed that the biodegradation rate fell as the vibration period was boosted.

According to the most recent investigations executed by Souza et al. [132]; the melting temperatures of the dispersion emulsion films were relatively low. This was because the polyphenols found in essential oils reacted with the amylopectin chains, thereby shifting the coupling forces between the amorphous matrix and the crystallites. Based on the result, its water vapor permeability ratings decreased significantly. This discovery, nevertheless, may be intriguing since water can operate as a medium for essential oils, delivering the essential oils active components and hence boosting the film's biological movements. In summary, starch-based films that also include a nanocellulose-stabilized dispersion emulsion and include wood essential oil might be used as materials for packaging if the mechanical, thermal, and permeability factors were taken into consideration. According to the research by Chen et al. [133]; they utilize TEMPO-mediated oxidation and highly pressurized homogenization, nanocellulose was obtained using three types of non-wood fibers (bamboo, cotton linter, and sisal). The solution casting process was used to generate starch-based composite films incorporating multiple types of nanocellulose at varying concentrations starting from 0 wt% and ending at 10 wt%. Although bamboo nanocellulose had the lowest crystallinity degree and has outstanding potential for the food packaging industry, it is notable that the bamboo nanofibers with the highest aspect ratio reinforced the starch films the most. By using a field emission scanning electron microscope (FESEM), researchers were able to confirm that the nanofibers of sugar palm nanocrystalline cellulose (SPNCCs) were uniformly dispersed throughout the SPS matrix and improved the films' mechanical, thermal, and moisture barrier properties. It was astonishing to discover that elevating the nanofillers concentration from 0 to 1.0 wt percent led to an enhancement in YM and tensile strength from 54 to 178.83 MPa and 4.80–11.47 MPa, correspondingly, in the SPNCCs filler SPS nanocomposite films [58]. As an outcome, it has a large potential as a replacement material for food packaging in place for synthetic materials, which are damaging to the environment if used in their pure form.

Bagde and Nadanathangam [134] had studied the corn starch films mixed with bacteriocins immobilised on the surface of NCC. Both bacteriocins have been obtained from *P. acidiclactic* and *E. faecium*, and NCC was synthesized with cotton linters in a biomedical technique. At 27 °C, bacteriocin immobilized NCC (BIN) films remained unchanged for four weeks, however bacteriocin integrated films disintegrated after two weeks. The deterioration rate for NCC films was the greatest, preceded by BIN and bacteriocin films. It was revealed that the combination of NCC and BIN films boosted the solubility of water by 36% and 41%–46%, respectively. A massive increase of about 6% and 39%–69% in the tensile strength of NCC-integrated films and BIN was observed. Food packaging would certainly benefit from the use of BIN integrated starch films in the future prospects. Montoya et al. [135] discovered that enclosing mango slices in thermoplastic starch (TPS) or TPS/BC (BC-bacterial cellulose) films conserved their physical characteristics in the storage, showing that these films have a wide potential for being applied as food packaging materials to sustain minimally processed mangoes. The TPS/BC films have the most successful in sustaining the fruit's physical and chemical qualities for 5 days at a relative humidity of 75% and a temperature of 5 °C. The findings of this study indicate that biodegradable TPS/BC films provide an innovative alternative for maintaining minimally processed fruit within limited storage time.

In the investigation by Noorbakhsh-Soltani et al. [136], they explored the combination of nanocellulose with gelatin and starch as it was the main objective for this research. The results demonstrate that boosting the nanocellulose content until 10% enhances the tensile strength at breakage to 8121 MN/m^2 and also diminish the elongation at break (EAB) significantly. Furthermore, enhancing the chitosan level from 5% to 30% can extend the lifespan of food until the range of 15 days. Apart from this research, there was also research conducted by Slavutsky and Bertuzzi [55] which involved water barrier qualities of starch and starch/CNC films that had been produced from sugarcane bagasse. It has been shown by the result that reinforced starch/CNC films had a smaller attraction for molecules of water compared to the starch films. The incorporation of CNC lowered the surface hydrophilicity and better water vapor barrier qualities, highlighting that the combination for the research has incredible potential power as a revolutionary biomaterial for food packaging and preservation Cao et al. [8] examined research involving a suspension of cellulose nanocrystals which was created using flax fibers by acid hydrolysis and employed to strengthen the plastic starch containing within 5% and 30% by composition for the goal of casting nanocomposite composites. It's indeed important to mention that when the flax cellulose nanocrystals proportion was elevated from 0% until 30%, the tensile strength rose beginning at 3.9 MPa up to 11.9 MPa while YM of the nanocomposites rose from 31.9 MPa till 498.2 MPa, significantly. Moreover, the starch-based nanocomposites exhibit additional water resistance, making them ideal for utilization in food packaging.

Based on new findings by Nazrin et al. [137], a variation of PLA/TPS composition blend nanocomposites were formed by melt blending and compression molding with various amounts of TPS which are in range of 20%–80% and the same amount of sugar palm nanocellulose fibers which is 0.5%. Approximately 60% of the TPS amount, YM went up beginning at 1.19 GPa and ended at 1.56 GPa. There's also a rise in density, water absorption, as well as thickness swelling when further TPS content was included in there. It can be linked to the dominant functional group of hydroxyls. PLA60TPS40 can be characterized as having good operational attributes in food packaging. Furthermore, according to the recent research by Yuan and Chen [138], the composite antibacterial films starch – based were apparently made utilizing the casting technique containing nanocellulose for the reinforcement agent while polyhexamethylene biguanide (PHMB) as the antibacterial force. When nanocellulose was supplied for 5%

while PHMB was 6% for their composition, the starch nanocellulose polyhexamethylene biguanide (SNP) composite film displayed superb qualities. TGA findings for the SNP composite film revealed high levels of thermal stability in the film with PHMB added, revealing that the SNP composite material with PHMB presented high resistance to heat. The incorporation of PHMB and nanocellulose improves the mechanical performance, hinting that composite films should be exploited as an impressive food packaging material.

According to Jayaraj et al. [139], digestible films generated from natural potato starch via solution casting process with nanocellulose made of rice husk concentration ranging from 0% to 15%. By comparison to the authentic films, the nanocellulose supplied had minimizing water vapor transmission rate, enhanced mechanical strength, and greater transparency. Hence, nanocellulose found to be a good ingredient for creating nanocomposite edible films with superior film characteristics than native control films and could be applied effectively for edible packaging. After that, research conducted by Owi et al. [140], bionanocomposites for food packaging were created utilizing low-cost tapioca starch-based films. To conduct the assessment, two crosslinkers, *Citrus aurantifolia* (LJ) and industrialized citric acid (CA), were mixed with glycerol using the solvent casting process in one timeframe. Acid hydrolysis was employed to extract various amounts of nanocellulose (NC) from empty oil palm fruit bunches (OPEFB). The extra amount of NC boosted the tensile strength of the TS bionanocomposite films; meanwhile the addition of LJ boosted their flexibility and break elongation. By incorporating LJ and NC into TS films, the percentage of water solubility was dropped, showing that food packaging films that have been generated contain a high level of water resistance. Apart from that, formulation F11 which included in the research made by Bruna et al. [141], reached the upper threshold with a nearly 47% rise in elongation and a 94% reduction in elongation as a consequence, compared to the original starch. Moreover, the presence of nanocellulose in the starch polymeric matrix boosted the boundary qualities for the films, as example water vapor permeability and water activity. So, the introduction of coconut cellulose inside films composed of cassava starch and plasticized with glycerol might drastically affect their mechanical and boundary characteristics.

The findings indicate that introducing sisal cellulose nanofibers in glycerol-plasticized cassava starch-based films can optimize mechanical, barrier, and thermal properties. Applying cellulose nanofibers and raising nanofiber percentages boosted young modulus and ultimate tensile force. Meanwhile, their EAB, water swelling degree, and water vapor permeability were minimized. These findings are in agreement with the general concepts of food preservation packaging. Bio-based nanocomposite films featuring outstanding barrier and tensile characteristics have been established using cassava starch as the polymer matrix, glycerol as the plasticizer, and cellulose nanofibers as reinforcing agents. By doing so, this research will cooperate to massively generate new ecologically friendly packaging for food [142]. According to research by Salehudin et al. [143], they were combining nanocellulose from oil palm empty fruit bunches with starch polymer. The input of nanocellulose in reinforced starch-based packaging ranged from 2%–10% every mass of starch. Applying 2% cellulose nanofiber elevated the film's tensile strength to 468 MPa, opposed to 366 MPa for the original starch film. The tensile strength decreased with cellulose nanofiber incorporation exceeding 2%. YM follows the identical progress as tensile strength, with 2% of cellulose nanofiber content providing the best results. It also dropped the elongation percentage of nanobiocomposite films by approximately 2%. Considering the low elongation percentage, a 2% content of cellulose nanofiber addition increases the mechanical characteristics of starch-based biopolymers making them suitable candidates for food packaging application.

Depending on the outcomes of the research conducted by Costa et al. [144], biodegradable films composed of starch and glycerol and incorporating CNC derived from licuri and alcoholic extracts of red propolis (ERP) may be a viable option to synthetic packaging films. Carbon nanocrystals have an extraordinary structure that maximizes mechanical strength while minimizing moisture activity and water-vapor permeability, two key properties for food packaging. The native film that combined with the content of 0.70% ERP and 0.50% concentration of CNC prevented the development of coagulase-positive staphylococci in cheese and delayed the oxidation of butter. Further exploration is required to determine the efficiency of these films on a variety of food matrices, taking into account storage circumstances, product characteristics, and possible interactions with the films.

In the up-to-date research by Othman et al. [145], they used the solvent casting process to generate corn starch (CS) films reinforced by nanocellulose fiber and thymol contents starting from 0.1% to 0.5% weight of thymol/volume of solution. The crystallinity of the film reduces while the thymol concentration rises. The tensile strength and YM of the CS films were improved, while the EAB, oxygen permeability, and water vapor permeability were lowered. The incorporation of NCF to CS films boosted their thermal stability. Thymol's plasticizing effect reduced the thermal strength and YM of the CS/NCF bionanocomposite films while boosting the EAB. Thymol raised thermal stability but lowered the barrier properties of the film. Greater thymol concentrations demonstrated a more significant influence on mechanical, thermal, and barrier properties. Finally, the combination of NCF and thymol increased the flexibility and thermal stability of the CS films which makes it the best candidate for replacing synthetic material for food packaging [146]. In the previous research, the target of this research was to develop sustainable biodegradable films using biopolymers, primarily tapioca starch (TS). Microcrystalline cellulose (MCC) was applied as a filler in TS/MCC composite films prepared by solvent casting at content varying from 0 wt% to 10 wt% to strengthen the films' constrained properties. 3 wt% MCC content was discovered to be the optimal filler concentration for the finest mechanical and barrier properties. MCC improved the films' thermal properties significantly. The TS/MCC films proposed in this research have improved properties and have a high potential to be utilized as food packaging materials. Based on the research by Zhou et al. [147], they conducted research for biodegradable poly (vinyl alcohol) (PVA), corn starch, and CNF, were combined to produce mechanically durable and long-lasting composites via a melt processing process. Thermal analysis proves that the PVA composite's processing window has been greatly widened, according to the findings of the study. The addition of 10% CS and 10% CNF significantly improved the composites' tensile strength, modulus, and PVA, as well as other biodegradable plastics, can benefit from our study as we target an economical cost, straightforward production method, and equivalent performance with universal polymers, which is of major importance for the vast uses of biodegradable plastics in many fields. According to the findings of the study that was conducted by Asrofi et al. [119], the bionanocomposites were derived from a yam bean (Pachyrhizus spp.) starch matrix, and the nanocellulose was derived from a water hyacinth fiber (WHF). Both of these materials were formed through the use of the solution casting method. Modifications were made so that the amount of nanocellulose suspension in the matrix ranged from 0 wt percent up to 1 wt percent. When nanocellulose was included into the material, both the tensile strength and the modulus of elasticity saw significant increases. This study found that a concentration of nanocellulose equal to one weight percent produced the best results in terms of tensile strength as well as the modulus of elasticity. The crystallinity index of the bionanocomposite was improved to its maximum value with the introduction of nanocellulose at a quantity of less than 1%. Thermal stability and moisture resistance were also improved as nanocellulose concentration was enhanced. As a result, yam bean starch reinforced with water hyacinth fiber nanocellulose bionanocomposites may be excellent for food packaging applications.

Nordin et al. [148] published on the effective production of microcellulose fibers using ultrasonication and NaOH/urea treatment (MCF). This straightforward, safe, and economical method of preparation warrants additional investigation. Thermal strength and EAB parameters were enhanced by 1 wt percent MCF at their optimal values. Due to MCF agglomeration inside the film matrix, the TS and EAB of the composite films decreased at 5% and 10% MCF concentrations, respectively. However, because MCF lacked thermal resistance, it degraded at a lower temperature than native TPS films. This finding may aid future research, particularly in the production of microcellulose fibers and their use as reinforcement in bio-based polymers for food packaging. Based on the findings by Yang et al. [149], they utilize starch-based composite films with the content of NCC starting with 0.1 wt% and ending with 0.5 wt% The outcomes showed that NCC reinforced composite films operated similarly to non-Newtonian fluids. These combination rheological properties are based on the NCC concentration content. It enhanced the starch-based composite films significantly related to thermal stability. Surface-sized paper's mechanical and air permeable qualities proved to be best at 0.3 wt % NCC concentration. In the research of Xu et al. [150], CNC helps to make the films extra durable and water permeable. Grape pomace extract (GPE) extracted from Viognier, a white grape variety, has massive effects on the cumulative phenolic content and antibacterial activity. *L. monocytogenes* and *S. aureus* can grow on the surface of ready-to-eat deli meat products after they've been injected intravenously with GPE. The nanocomposite films with GPE reinforced with tapioca starch biocomposite films showed potential in blocking this growth during refrigeration storage for *S. aureus* (97.5% and 97.6%) and *L. monocytogenes* (54.7% and 62.7%) for both growth inhibition. It can be concluded that this combination was found to have a superior antibacterial response and should be used in antimicrobial food packaging films.

4.5.2 Food additives

According to the most recent findings by Yuan and Chen [138], the SNP composite film exhibits exceptional antibacterial effect when the PHMB dosage is more than 6 wt%. By implementing a biodegradable antibacterial coating, this innovation widens the prospects for food preservation. Furthermore, *in vitro* antibacterial experiments revealed that combination of SNP and PHMB dosage of 6% composite films inhibited *E. coli* 100%, with a bacteriostatic loop radius exceeding 10 mm. The best formula of the SNP composite film is a combination of 5 wt% nanocellulose and 6 wt% PHMB after the result demonstrates that it achieves the optimum value for physical and antibacterial properties. Beside this [53], investigates the influence of CNC on the gelatinization and pasting capabilities of normal maize starch (NMS), waxy maize starch (WMS), and sweet potato cellulose starch (SPS). It has been shown that the incorporation of CNC to native maize, waxy maize, and sweet potato starches inhibits both short-term and long-term retrogradation, as evidenced by the finding for this study. In order to generate starch-based goods with extended shelf-lives, CNC can be added as an excellent ageing inhibitor. According to Ji et al. [151], higher concentration of CNC reduced the activities of α-amylase and glucoamylase. CNC affected the secondary structure of α-amylase and glucoamylase, as determined by Fourier transform infrared (FTIR) spectroscopy, and circular dichroism (CD). After incorporating CNC into maize, potato, and pea starches, the proportions of quickly and moderately digestible starch in the cooked starches reduced, whereas the resistant starch content was higher. The findings of this research may provide understanding on the interactions between amylases that are economically relevant and nanoparticles that have potential use in the food industry as preservatives agents.

Then, according to Liu and Kong [152], high concentrations of TEMPO-CNF and CNC dramatically inhibited glucose transmission, especially at 0.36% TEMPO-CNF and 2%–4% CNC content. CNF and TEMPO-CNF significantly reduced the content of free fatty acid produced during intestinal digestion of milk fat by 30%. CNC at a concentration of 0.3% impeded the transport of free amino nitrogen throughout the milk protein intestinal's digestion. Even though the nanocellulose attracted significant amounts of Fe and Zn, it did not collect any other minerals such as Ca, Mg, Zn, Cu, or Ag. The outcome of this research indicates that all three types of nanocellulose may impact digestion and nutrient absorption when consumed as zero-calorie fiber. Additionally, according to Corral et al. [153], the presence of BNC elevated the baking performance of wheat flours by modifying the mass's viscoelastic properties. It can be contended that the incorporation of BNC improved nutrient content and BNC could be used as a performance stimulant in bread-making. Based on the newest research by Bagde and Nadanathangam [134], they employed bacteriocins from broth cultures of *P. acidilactici* and *E. faecium* to dissolve on the surface of CNC and fortify the starch film. Antibacterial films were made from corn starch and fortified with bacteriocin immobilised crystalline nanocellulose (BIN). The presence of bacteriocins reduced the water solubility of starch films in range of 19%–23%. Furthermore, bacteriocin incorporated films decayed in 2 weeks, but BIN-incorporated films maintained in a period of 4 weeks. The BIN-coated corn starch film seems to be an antibacterial packaging film with enhanced mechanical performance. The tensile strength of corn starch film was raised 69% while its antibacterial activity was boosted 57%. Bacteriocin decreased significantly corn starch film biodegradability rate. Khairuddin et al. [154] revealed the antibacterial activities of hydroxyethyl cellulose (HEC)/wheatstarch films containing thymol which is essential oil at dosages ranging from 0.5% to 2.5% (w/v). Starting with concentration of 1.6%–2.5% thymol, it had a small impact on physical, chemical, and thermal qualities. Microbial inhibition analysis revealed that a film with 1.5% (w/v) thymol had the highest inhibitory effects against all chosen bacteria, including *E. coli*, *B. subtilis*, and *A. niger*. It can be concluded from this research that nanocellulose reinforced with starch biopolymer can be used as food additives agent to preserve the quality of food.

4.5.3 Paper industry

Research published by Sharma et al. [155] suggests that highly substituted cationic starch can be used to compact paper, defend it from bacteria, and make it look denser. In the long term, CNF could make paper even more useful. It could make it thin, strong, flexible, air-resistant, and easier to print. In fact, when cationic starches are incorporated, they do not enhance as much as when they are manufactured from petroleum-based polymers like PVA due to the fact that they permit a much greater concentration of polysaccharide. Using cationic starches with higher levels of quality should be attempted to see if they work more effectively in future research.

4.5.4 Biomedical industry

As for biomedical industry [156], evolved BNC-pectin bionanocomposite to BNC-pectin-CMC and BNC-pectin-starch appropriate as probiotic entrapment matrix for preserving *B. coagulans* vitality upon drying and gastrointestinal digestion. After evaluating various matrix formulations, it was determined that the optimal matrix composition comprised of 15% concentration of pectin, 60% and 25% for the content of BNC, and CMC. A scientific explanation was that starch coating increased heat protection when compared to CMC coating, which enhanced the limitation qualities to drying and digesting. As a result, it is hypothesized that entrapment matrices offer probiotics benefits over pure bacteria and maintain probiotic viability for the period of preparation and digestion. Zhai et al. [157] altered nanocellulose by cross-linking it with borax and incorporating it into cationized high amylose maize starch (CS) films. The relative crystallinity of borax modified nanocellulose (BMNC) was slightly better than that of nanocellulose, and heat stability was increased. To advance the tensile strength and resistance for water of CS films, borax modified nanocellulose was adopted. The cross-linking of borax improved the dispersion of nanocellulose in CS films. Mechanical and water resistance of CS/BMNC films were stronger than CS/nanocellulose films. Its tensile strength was raised greatly when compared to the standard CS film, while its water-vapor permeability was reduced. The novel boron ester bonding process used to prepare BNC may result in the production of aesthetically starch films. Satyamurthy et al. [158] reported that nanocellulose can be synthesized from non-spinnable short staple cotton fibers by method of microbial hydrolysis. This substrate has been used to enhance the starch/polyvinyl alcohol biocomposite film being used to immobilize the urease enzyme. Nanocellulose was used as a carrier for enzymes and as a film reinforcer. The incorporation of 1.5% nanocellulose enhanced the tensile strength of the film by a factor of two. The disclosed method successfully conjugated urease on nanocellulose at a rate of 68%. The research reveals that the nanocomposite film may boost the effectiveness and storage of urea biosensors.

4.5.5 Pharmaceutical industry

According to Meneguin et al. [159], these research results, blended with the advancement of Methotrexate (MTX) release rates in colonic pH, prove the potential of nanocomposites films as an advancement of technological channel for specialized pharmaceutical demands, in example increasing the oral bioavailability of badly water-soluble drugs and evolving colon-specific drug delivery systems. Nanocellulose incorporated with big concentration amylose starch films were dropped by 4 ml of MTX. The recorded observation demonstrated that nanocomposite films had the capability to enhance MTX emission by 40.84%–64.13% during the initial 15 min of evaluation, highlighting a significant method for treating the oral bioavailability challenges related to poorly soluble medications. Other aspects as an example pH, mucus film thickness, and colon transportation period are major contributors to the mucoadhesion of generated nanocomposite films, thereby enhancing colon specific MTX release.

According to the research conducted by Patil et al. [160], gelatinized maize starch and cellulose nanofibers are used to create nanocomposites. Dimethyl phthalate (DMP) introductory distribution is strongly limited by CNF; however, this total distribution is massively improved. This offers for the controlled release of contained substances and guaranteeing that they may be used in applications effectively. They identified that the introductory distribution of dimethyl phthalate was slowed significantly by the addition of CNFs to starch. Even so, the whole distribution was significantly greater, and 85%–90% of the drug was discharged in one week.

4.5.6 Electronic appliances

According to the research by Hazrol et al. [161], the influence of sugar palm nanocrystalline cellulose (SPNCC) tested with (0.00–0.10 wt%) on the electrical resistance, resistivity, and conductivity of (SPS)/SPNCC nanocomposite films was investigated in this research. The experiments were performed effectively using the four-probe approach, and the electrical parameters were measured using Ohm's law, resistivity, and conductivity equations. The resistivity values of SPS/SPNCC films were determined in between 3.1×10^2 to 1.5×10^4 (Ω cm). Conductivity values of polymer composites dropped as current supply grew. The resistivity of the polymer composites rose when the SPNCC content was escalated. It is reasonable to conclude that sugar palm nanocomposite has the capability to be an excellent insulator for electrical equipment [161].

4.5.7 Sport industry

Based on the research by Zhen [162], nanocellulose has been used as a reinforcement agent as it was introduced to a TPS polymer matrix. Composite materials were formed with the relevant solvent by solution casting technology. The inclusion of nanocellulose invented the composite films to be extra durable, water-absorbing, and powerful. Once the mass fraction of nanocellulose in composite films reaches 8%, both tensile strength and elongation to break the composite films reach their highest standards. Afterward, they initiate to drop off. There is a lot of ability for nanocellulose to boost the mechanical properties of biopolymers.

4.6 Conclusions

In conclusion, from the present-day research that utilized the development of nanocellulose reinforced with starch biopolymer composites, it can be concluded that the combination of both sides really improved the properties of the films. Apart from that, the reinforcement of starch-based film with nanocellulose has been explored and implemented into various types of applications such as food packaging, food additives, pharmaceutical industry, paper industry, biomedical industry, electronic appliances, and sport industry. This review has the main objectives to spread knowledge and collect the data related to these biocomposites which can benefit further study in the future. Besides that, there are also some improvements for the related properties of these biocomposites in order to replace the newest material that have been used such as petro-based packaging that bring harmful effects to the environment. It needs to be finished in the future by the researchers in order to save the environment due to plastic packaging waste.

4.7 Challenges and future recommendation

From naturally occurring cellulose, nanocrystalline cellulose is a newly developed renewable nanomaterial that shows significant promise in textiles, enhanced cellulose papers, improved plastics, and biomedical and biological applications. It is also quite interesting from a practical perspective to use NCC as reinforcement in composite structures. The use of nanocellulose in starch biocomposite sheets opens up new possibilities for creating innovative biodegradable packaging with enhanced barrier, mechanical, and tensile qualities. Nanocellulose has been widely investigated over the last 20 years, but significant obstacles remain, including surface and end-reduction changes, environmentally friendly extraction at lower cost with reduced energy consumption, and upscaling manufacturing. However, eliminating various components to increase performance and recycling challenges remain key concerns with nanocellulosic materials. In order to meet the present challenges, upcoming research on nanocellulose-based materials will need to combine a wide range of methods and approaches, including biomimetics, numerical models, AI, and machine learning-based algorithms. For homogeneous and on-demand nanocellulose use, it is important to research novel production approaches like advanced composite material manufacturing or 3D printing. Manufacturing nanocellulose that is technologically viable for extensive production requires the exploration of novel technologies based on new techniques. As a result of these characteristics, scientists will be able to create improved functionalized nanocellulose-starch materials and composites in the future. Furthermore, these materials should have a very small effect on the environment. With the development of low-cost commercial nanocellulose sources, it is conceivable to develop new applications and improve old ones. This is an especially important topic for the future because it can be employed in a diverse range of industries that demand advanced materials. As a result, further research is required to cover present gaps by transferring laboratory experiments to industrial or commercial manufacturing, establishing the feasibility of the finished materials, and introducing them to the market. These research gaps will allow for the study of sustainable nanocellulose for high-performance and advanced applications. Despite the problems stated above, the next generation of nanocellulose-based products will surely improve the life quality of individuals. As a result, the reinforcement of nanocellulose and starch biopolymers has enormous potential to partially or completely replace synthetic polymers and can be used for a wide range of low-and high-cost packaging, industrial, agricultural mulch, and medical applications.

References

1. Ministry of Energy and Natural Resources; 2020. Total Forested Areas in Malaysia (1990 – 2018): Ketsa. Available from: https://www.ketsa.gov.my/en-my/KetsaCore/Forestry/Pages/Total-Forested-Areas-in-Malaysia.aspx.
2. Gironès J, López J, Mutjé P, Carvalho A, Curvelo A, Vilaseca F. Natural fiber-reinforced thermoplastic starch composites obtained by melt processing. Compos Sci Technol 2012;72:858–63.
3. Gowthaman N, Lim H, Sreeraj T, Amalraj A, Gopi S. Advantages of biopolymers over synthetic polymers. In: Biopolymers and Their Industrial Applications. Amsterdam, Netherlands: Elsevier; 2021:351–72 pp.
4. Kargarzadeh H, Mariano M, Huang J, Lin N, Ahmad I, Dufresne A, et al. Recent developments on nanocellulose reinforced polymer nanocomposites: a review. Polymer 2017;132:368–93.
5. Havstad MR. Biodegradable plastics. In: Plastic waste and recycling. Cambridge, Massachusetts: Academic Press; 2020:97–129 pp.
6. Trache D, Tarchoun AF, Derradji M, Hamidon TS, Masruchin N, Brosse N, et al. Nanocellulose: from fundamentals to advanced applications. Front Chem 2020;8:392.
7. Salas C, Nypelö T, Rodriguez-Abreu C, Carrillo C, Rojas OJ. Nanocellulose properties and applications in colloids and interfaces. Curr Opin Colloid Interface Sci 2014;19:383–96.
8. Cao X, Chen Y, Chang PR, Muir AD, Falk G. Starch-based nanocomposites reinforced with flax cellulose nanocrystals. Express Polym Lett 2008;2:502–10.
9. Cruz-Tato P, Ortiz-Quiles EO, Vega-Figueroa K, Santiago-Martoral L, Flynn M, Díaz-Vázquez LM, et al. Metalized nanocellulose composites as a feasible material for membrane supports: design and applications for water treatment. Environ Sci Technol 2017;51:4585–95.
10. Czaja WK, Young DJ, Kawecki M, Brown RM. The future prospects of microbial cellulose in biomedical applications. Biomacromolecules 2006;8:1–12.
11. Iguchi M, Yamanaka S, Budhiono A. Bacterial cellulose – a masterpiece of nature's arts. J Mater Sci 2000; 35:261–70.
12. Dufresne A. Chapter 1. Nanocellulose: potential reinforcement in composites. In: Green chemistry series. London: The Royal Society of Chemistry; 2012:1–32 pp.
13. Lee KY, Aitomäki Y, Berglund LA, Oksman K, Bismarck A. On the use of nanocellulose as reinforcement in polymer matrix composites. Compos Sci Technol 2014;105:15–27.
14. Gozzo A, Glittenberg D. Chapter 9.2. Starch: a versatile product from renewable resources for industrial applications. In: Green chemistry series. London: The Royal Society of Chemistry; 2009:238–63 pp.
15. Laycock BG, Halley PJ. Starch applications: state of market and new trends. In: Starch polymers. Amsterdam, Netherlands: Elsevier; 2014:381–419 pp.
16. Ilyas RA, Sapuan SM, Sanyang ML, Ishak MR, Zainudin ES. Nanocrystalline cellulose as reinforcement for polymeric matrix nanocomposites and its potential applications: a review. Curr Anal Chem 2018;14: 203–25.
17. Lu Y, Weng L, Cao X. Morphological, thermal and mechanical properties of ramie crystallites—reinforced plasticized starch biocomposites. Carbohydr Polym 2006;63:198–204.
18. Anglès MN, Dufresne A. Plasticized starch/tunicin whiskers nanocomposites. 1. Structural analysis. Macromolecules 2000;33:8344–53.
19. Nasri-Nasrabadi B, Behzad T, Bagheri R. Preparation and characterization of cellulose nanofiber reinforced thermoplastic starch composites. Fibers Polym 2014;15:347–54.

20. Babaee M, Jonoobi M, Hamzeh Y, Ashori A. Biodegradability and mechanical properties of reinforced starch nanocomposites using cellulose nanofibers. Carbohydr Polym 2015;132:1–8.
21. Llanos JH, Tadini CC. Preparation and characterization of bio-nanocomposite films based on cassava starch or chitosan, reinforced with montmorillonite or bamboo nanofibers. Int J Biol Macromol 2018;107: 371–82.
22. Das S, Ghosh B, Sarkar K. Nanocellulose as sustainable biomaterials for drug delivery. Sensor Int 2022;3: 100135.
23. Brinchi L, Cotana F, Fortunati E, Kenny J. Production of nanocrystalline cellulose from lignocellulosic biomass: technology and applications. Carbohydr Polym 2013;94:154–69.
24. Costa AFS, Almeida FCG, Vinhas GM, Sarubbo LA. Production of bacterial cellulose by gluconacetobacter hansenii using corn steep liquor as nutrient sources. Front Microbiol 2017;8:2027.
25. Phanthong P, Reubroycharoen P, Hao X, Xu G, Abudula A, Guan G. Nanocellulose: extraction and application. Carbon Resour Convers 2018;1:32–43.
26. Chandra J, George N, Narayanankutty SK. Isolation and characterization of cellulose nanofibrils from arecanut husk fibre. Carbohydr Polym 2016;142:158–66.
27. Kusmono, Affan MN. Isolation and characterization of nanocrystalline cellulose from ramie fibers via phosphoric acid hydrolysis. J Nat Fibers 2020;19:1–12.
28. Morais JPS, Rosa MDF, de Souza Filho MDSM, Nascimento LD, do Nascimento DM, Cassales AR. Extraction and characterization of nanocellulose structures from raw cotton linter. Carbohydr Polym 2013;91:229–35.
29. Wulandari WT, Rochliadi A, Arcana IM. Nanocellulose prepared by acid hydrolysis of isolated cellulose from sugarcane bagasse. IOP Conf Ser Mater Sci Eng 2016;107:012045.
30. Yu M, Yang R, Huang L, Cao X, Yang F, Liu D. Preparation and characterization of bamboo nanocrystalline cellulose. Bioresources 2012;7:1802–12.
31. Nazrin A, Sapuan SM, Zuhri MYM. Mechanical, physical and thermal properties of sugar palm nanocellulose reinforced thermoplastic starch (TPS)/Poly (lactic acid) (PLA) blend bionanocomposites. Polymers 2020;12:2216.
32. Ribeiro RSA, Pohlmann BC, Calado V, Bojorge N, Pereira N. Production of nanocellulose by enzymatic hydrolysis: trends and challenges. Eng Life Sci 2019;19:279–91.
33. Bondeson D, Mathew A, Oksman K. Optimization of the isolation of nanocrystals from microcrystalline cellulose by acid hydrolysis. Cellulose 2006;13:171–80.
34. Dong XM, Revol JF, Gray DG. Effect of microcrystallite preparation conditions on the formation of colloid crystals of cellulose. Cellulose 1998;5:19–32.
35. Gupta GK, Shukla P. Lignocellulosic biomass for the synthesis of nanocellulose and its eco-friendly advanced applications. Front Chem 2020;8:601256.
36. Nasir M, Hashim R, Sulaiman O, Asim M. Nanocellulose. In: Cellulose-Reinforced Nanofibre Composites. Sawston, Cambridge: Woodhead Publishing; 2017:261–76 pp.
37. Nascimento DM, Nunes YL, Figueirêdo MCB, de Azeredo HMC, Aouada FA, Feitosa JPA, et al. Nanocellulose nanocomposite hydrogels: technological and environmental issues. Green Chem 2018;20:2428–48.
38. Abdul Khalil H, Bhat A, Ireana Yusra A. Green composites from sustainable cellulose nanofibrils: a review. Carbohydr Polym 2012;87:963–79.
39. Abitbol T, Rivkin A, Cao Y, Nevo Y, Abraham E, Ben-Shalom T, et al. Nanocellulose, a tiny fiber with huge applications. Curr Opin Biotechnol 2016;39:76–88.
40. Dufresne A. Nanocellulose: a new ageless bionanomaterial. Mater Today 2013;16:220–7.
41. Fahma F, Febiyanti I, Lisdayana N, Arnata I, Sartika D. Nanocellulose as a new sustainable material for various applications: a review. Arch Mater Sci Eng 2021;2:49–64.
42. Lin N, Huang J, Dufresne A. Preparation, properties and applications of polysaccharide nanocrystals in advanced functional nanomaterials: a review. Nanoscale 2012;4:3274.
43. Moon RJ, Martini A, Nairn J, Simonsen J, Youngblood J. Cellulose nanomaterials review: structure, properties and nanocomposites. Chem Soc Rev 2011;40:3941.

44. Wang B, Sain M. Isolation of nanofibers from soybean source and their reinforcing capability on synthetic polymers. Compos Sci Technol 2007;67:2521–7.
45. Mondal S. Review on nanocellulose polymer nanocomposites. Polym-Plast Technol Eng 2017;57:1377–91.
46. Luong ND, Korhonen JT, Soininen AJ, Ruokolainen J, Johansson LS, Seppälä J. Processable polyaniline suspensions through in situ polymerization onto nanocellulose. Eur Polym J 2013;49:335–44.
47. Bhanvase B, Sonawane S. Ultrasound assisted in situ emulsion polymerization for polymer nanocomposite: a review. Chemical Engineering and Processing. Process Intensif 2014;85:86–107.
48. Tortorella S, Vetri Buratti V, Maturi M, Sambri L, Comes Franchini M, Locatelli E. Surface-modified nanocellulose for application in biomedical engineering and nanomedicine: a review. Int J Nanomed 2020; 15:9909–37.
49. Hayase G, Kanamori K, Hasegawa G, Maeno A, Kaji H, Nakanishi K. A superamphiphobic macroporous silicone monolith with marshmallow-like flexibility. Angew Chem Int Ed 2013;52:10788–91.
50. Phanthong P, Guan G, Karnjanakom S, Hao X, Wang Z, Kusakabe K, et al. Amphiphobic nanocellulose-modified paper: fabrication and evaluation. RSC Adv 2016;6:13328–34.
51. Liu K, Tian Y, Jiang L. Bio-inspired superoleophobic and smart materials: design, fabrication, and application. Prog Mater Sci 2013;58:503–64.
52. Si Y, Guo Z. Superhydrophobic nanocoatings: from materials to fabrications and to applications. Nanoscale 2015;7:5922–46.
53. Cui S, Li M, Zhang S, Liu J, Sun Q, Xiong L. Physicochemical properties of maize and sweet potato starches in the presence of cellulose nanocrystals. Food Hydrocolloids 2018;77:220–7.
54. Ilyas R, Sapuan S, Ishak M, Zainudin E. Development and characterization of sugar palm nanocrystalline cellulose reinforced sugar palm starch bionanocomposites. Carbohydr Polym 2018;202:186–202.
55. Slavutsky AM, Bertuzzi MA. Water barrier properties of starch films reinforced with cellulose nanocrystals obtained from sugarcane bagasse. Carbohydr Polym 2014;110:53–61.
56. Lavoine N, Desloges I, Dufresne A, Bras J. Microfibrillated cellulose – its barrier properties and applications in cellulosic materials: a review. Carbohydr Polym 2012;90:735–64.
57. Saito T, Kuramae R, Wohlert J, Berglund LA, Isogai A. An ultrastrong nanofibrillar biomaterial: the strength of single cellulose nanofibrils revealed via sonication-induced fragmentation. Biomacromolecules 2012; 14:248–53.
58. Ilyas RA, Sapuan SM, Ridzwan MI, Zainudin ES, Atikah M. Nanocellulose reinforced starch polymer composites: a review of preparation, properties and application. In: 5th International Conference on Applied Sciences and Engineering Application (ICASEA 2018). Cameron Highlands, Malaysia; 2018.
59. Kovacs T, Naish V, O'Connor B, Blaise C, Gagné F, Hall L, et al. An ecotoxicological characterization of nanocrystalline cellulose (NCC). Nanotoxicology 2010;4:255–70.
60. Islam MT, Alam MM, Zoccola M. Review on modification of nanocellulose for application in composites. Int J innov Res Sci Eng Technol 2013;2:5444–51.
61. Silvério HA, Flauzino Neto WP, Dantas NO, Pasquini D. Extraction and characterization of cellulose nanocrystals from corncob for application as reinforcing agent in nanocomposites. Ind Crop Prod 2013;44: 427–36.
62. Zhang W, Yang X, Li C, Liang M, Lu C, Deng Y. Mechanochemical activation of cellulose and its thermoplastic polyvinyl alcohol ecocomposites with enhanced physicochemical properties. Carbohydr Polym 2011;83:257–63.
63. Criado P, Hossain FMJ, Salmieri S, Lacroix M. Nanocellulose in food packaging. In: Composites materials for food packaging. Beverly, MA: Scrivener Publishing; 2018:297–329 pp.
64. Simões D, Miguel SP, Ribeiro MP, Coutinho P, Mendonça AG, Correia IJ. Recent advances on antimicrobial wound dressing: a review. Eur J Pharm Biopharm 2018;127:130–41.
65. Tong WY, bin Abdullah AYK, Binti Rozman NAS, bin Wahid MIA, Hossain MS, Ring LC, et al. Antimicrobial wound dressing film utilizing cellulose nanocrystal as drug delivery system for curcumin. Cellulose 2017; 25:631–8.

66. Lu P, Hsieh YL. Preparation and properties of cellulose nanocrystals: rods, spheres, and network. Carbohydr Polym 2010;82:329–36.
67. Hasan N, Rahman L, Kim SH, Cao J, Arjuna A, Lallo S, et al. Recent advances of nanocellulose in drug delivery systems. J Pharm Investig 2020;50:553–72.
68. Nechyporchuk O, Belgacem MN, Bras J. Production of cellulose nanofibrils: a review of recent advances. Ind Crop Prod 2016;93:2–25.
69. Abdul Khalil H, Davoudpour Y, Islam MN, Mustapha A, Sudesh K, Dungani R, et al. Production and modification of nanofibrillated cellulose using various mechanical processes: a review. Carbohydr Polym 2014;99:649–65.
70. Lin N, Dufresne A. Nanocellulose in biomedicine: current status and future prospect. Eur Polym J 2014;59: 302–25.
71. Klemm D, Schumann D, Udhardt U, Marsch S. Bacterial synthesized cellulose — artificial blood vessels for microsurgery. Prog Polym Sci 2001;26:1561–603.
72. Ardanuy RM, Claramunt JB, Arévalo PR, Parés SF, Aracri E, Vidal LT. Nanofibrillated cellulose (NFC) as a potential reinforcement for high performance cement mortar composites. Bioresources 2012;7:3883–94.
73. Viana LC, Potulski DC, Muniz GIBD, Andrade ASD, Silva ELD. Nanofibrillated cellulose as an additive for recycled paper. Cerne 2018;24:140–8.
74. Gomes FP, Silva NH, Trovatti E, Serafim LS, Duarte MF, Silvestre AJ, et al. Production of bacterial cellulose by Gluconacetobacter sacchari using dry olive mill residue. Biomass Bioenergy 2013;55:205–11.
75. Jozala AF, de Lencastre-Novaes LC, Lopes AM, de Carvalho Santos-Ebinuma V, Mazzola PG, Pessoa A Jr., et al. Bacterial nanocellulose production and application: a 10-year overview. Appl Microbiol Biotechnol 2016;100:2063–72.
76. Chen G, Wu G, Chen L, Wang W, Hong FF, Jönsson LJ. Comparison of productivity and quality of bacterial nanocellulose synthesized using culture media based on seven sugars from biomass. Microb Biotechnol 2019;12:677–87.
77. Huang Y, Huang X, Ma M, Hu C, Seidi F, Yin S, et al. Recent advances on the bacterial cellulose-derived carbon aerogels. J Mater Chem C 2021;9:818–28.
78. Heßler N, Klemm D. Alteration of bacterial nanocellulose structure by in situ modification using polyethylene glycol and carbohydrate additives. Cellulose 2009;16:899–910.
79. Czaja W, Krystynowicz A, Bielecki S, Brownjr R. Microbial cellulose—the natural power to heal wounds. Biomaterials 2006;27:145–51.
80. Darder M, Aranda P, Ruiz-Hitzky E. Bionanocomposites: a new concept of ecological, bioinspired, and functional hybrid materials. Adv Mater 2007;19:1309–19.
81. Mahsuli T, Suryanto H, Larasati A, Muhajir M. Mechanical properties of bacterial nanocellulose membrane from pineapple peel waste after homogenization process. In: International Conference On Biology And Applied Science (ICOBAS); 2019.
82. Fu L, Zhang J, Yang G. Present status and applications of bacterial cellulose-based materials for skin tissue repair. Carbohydr Polym 2013;92:1432–42.
83. Sionkowska A, Mężykowska O, Piątek J. Bacterial nanocelullose in biomedical applications: a review. Polym Int 2019;68:1841–7.
84. Fu L, Zhou P, Zhang S, Yang G. Evaluation of bacterial nanocellulose-based uniform wound dressing for large area skin transplantation. Mater Sci Eng C 2013;33:2995–3000.
85. Zhijiang C, Guang Y. Bacterial cellulose/collagen composite: characterization and first evaluation of cytocompatibility. J Appl Polym Sci 2011;120:2938–44.
86. Amorim JDP, Junior CJGS, Costa AFS, Nascimento HA, Vinhas GM, Sarrubo LA. Biomask, a polymer blend for treatment and healing of skin prone to acne. Chem Eng Trans 2020;79:205–10.
87. Alam M. A., Sapuan S. M., Ya H. H., Hussain P. B., Azeem M., Ilyas R. A. Application of biocomposites in automotive components: a review. Sawston, Cambridge: Woodhead Publishing; 2021:1–17.

88. Hussein FH, Al-Khafaji YF. Biopolymers, biocomposites, and their types. In: Advanced green materials. Sawston, Cambridge: Woodhead Publishing; 2021:43–60 pp.
89. Ashraf MA, Batool S, Ahmad M, Sarfraz M, Noor WSAWM. Biopolymers as biofilters and biobarriers. In: Biopolymers and biotech admixtures for eco-efficient construction materials. Sawston, Cambridge: Woodhead Publishing; 2016:387–420 pp.
90. Harries KA, Sharma B, editors. Nonconventional and vernacular construction materials: characterisation, properties and applications. Sawston, Cambridge: Woodhead Publishing; 2019.
91. Wirawan R., Sapuan S. M. Sugarcane bagasse-filled poly (Vinyl Chloride) composites: a review. In: Natural Fibre Reinforced Vinyl Ester and Vinyl Polymer Composites. Sawston, Cambridge: Woodhead Publishing; 2018:157–68 pp.
92. Mishra RK, Ha SK, Verma K, Tiwari SK. Recent progress in selected bio-nanomaterials and their engineering applications: an overview. J Sci Adv Mater Devices 2018;3:263–88.
93. Shah U, Naqash F, Gani A, Masoodi FA. Art and science behind modified starch edible films and coatings: a review. Compr Rev Food Sci Food Saf 2016;15:568–80.
94. Sanyang ML, Sapuan SM, Jawaid M, Ishak MR, Sahari J. Recent developments in sugar palm (Arenga pinnata) based biocomposites and their potential industrial applications: a review. Renew Sustain Energy Rev 2016;54:533e549.
95. Pérez S, Baldwin PM, Gallant DJ. Structural features of starch granules I. In: Starch. Cambridge, Massachusetts: Academic Press; 2009:149–92 pp.
96. Reddy MM, Misra M, Mohanty AK. Bio-based materials in the new bio-economy. Chem Eng Prog 2012;108: 37–42.
97. Atiqah A, Jawaid M, Sapuan SM, Ishak MR, Ansari MNM, Ilyas RA. Physical and thermal properties of treated sugar palm/glass fibre reinforced thermoplastic polyurethane hybrid composites. J Mater Res Technol 2019;8:3726e3732.
98. Ilyas R, Sapuan S, Norrrahim MNF, Yasim-Anuar TAT, Kadier A, Kalil MS, et al. Nanocellulose/starch biopolymer nanocomposites: processing, manufacturing, and applications. In: Advanced processing, properties, and applications of starch and other bio-based polymers. Amsterdam, Netherlands: Elsevier; 2020:65–88 pp.
99. Mazani N, Sapuan SM, Sanyang ML, Atiqah A, Ilyas RA. Design and fabrication of a shoe shelf from kenaf fiber reinforced unsaturated polyester composites. In: Lignocellulose for future bioeconomy. Amsterdam, Netherlands: Elsevier; 2019:315–32 pp.
100. Nurazzi NM, Khalina A, Sapuan SM, Ilyas RA, Rafiqah SA, Hanafee ZM. Thermal properties of treated sugar palm yarn/glass fiber reinforced unsaturated polyester hybrid composites. J Mater Res Technol 2020;9: 1606–18.
101. Asyraf MRM, Ishak MR, Sapuan SM, Yidris N, Ilyas RA. Woods and composites cantilever beam: a comprehensive review of experimental and numerical creep methodologies. J Mater Res Technol 2020;9: 6759–76.
102. Jaafar J, Siregar JP, Oumer AN, Hamdan MHM, Tezara C, Salit MS. Experimental investigation on performance of short pineapple leaf fiber reinforced tapioca biopolymer composites. BioResources 2018; 13:6341–55.
103. Takagi H, Asano A. Effects of processing conditions on flexural properties of cellulose nanofiber reinforced "green" composites. Compos Appl Sci Manuf 2008;39:685e689.
104. Mehanny S, Abu-El Magd EE, Ibrahim M, Farag M, Gil-San-Millan R, Navarro J, et al. Extraction and characterization of nanocellulose from three types of palm residues. J Mater Res Technol 2021;10:526e37.
105. Kaushik M, Moores A. Review: nanocelluloses as versatile supports for metal nanoparticles and their applications in catalysis. Green Chem 2016;18:622–37.
106. Onuaguluchi O, Banthia N. Plant-based natural fibre reinforced cement composites: a review. Cement Concr Compos 2016;68:96e108.

107. Shah N, Fehrenbach J, Ulven CA. Hybridization of hemp fiber and recycled-carbon fiber in polypropylene composites. Sustain Times 2019;11:3163.
108. Khalid MY, Arif ZU. Novel biopolymer-based sustainable composites for food packaging applications: a narrative review. Food Packag Shelf Life 2022;33:100892.
109. Norrrahim MNF, Nurazzi NM, Jenol MA, Farid MAA, Janudin N, Ujang FA, et al. Emerging development of nanocellulose as an antimicrobial material: an overview. Mater Adv 2021;2:3538–51.
110. Faradilla RHF, Lee G, Rawal A, Hutomo T, Stenzel MH, Arcot J. Nanocellulose characteristics from the inner and outer layer of banana pseudo-stem prepared by TEMPO-mediated oxidation. Cellulose 2016;23: 3023–37.
111. Punia S. Barley starch: structure, properties and in vitro digestibility – a review. Int J Biol Macromol 2020; 155:868–75.
112. Siqueira G, Bras J, Dufresne A. Cellulose whiskers versus microfibrils: influence of the nature of the nanoparticle and its surface functionalization on the thermal and mechanical properties of nanocomposites. Biomacromolecules 2009;10:425e432.
113. Bangar SP, Whiteside WS. Nano-cellulose reinforced starch bio composite films-A review on green composites. Int J Biol Macromol 2021;185:849–60.
114. Teixeira EDM, Pasquini D, Curvelo AA, Corradini E, Belgacem MN, Dufresne A. Cassava bagasse cellulose nanofibrils reinforced thermoplastic cassava starch. Carbohydr Polym 2009;78:422–31.
115. Azeredo HMC, Mattoso LHC, Wood D, Williams TG, Avena-Bustillos RJ, McHugh TH. Nanocomposite edible films from mango puree reinforced with cellulose nanofibers. J Food Sci 2009;74:N31–5.
116. Teixeira EDM, Lotti C, Corrêa AC, Teodoro KB, Marconcini JM, Mattoso LH. Thermoplastic corn starch reinforced with cotton cellulose nanofibers. J Appl Polym Sci 2011;120:2428–33.
117. Hietala M, Mathew AP, Oksman K. Bionanocomposites of thermoplastic starch and cellulose nanofibers manufactured using twin-screw extrusion. Eur Polym J 2013;49:950–6.
118. Karimi S, Tahir PM, Dufresne A, Karimi A, Abdulkhani A. A comparative study on characteristics of nanocellulose reinforced thermoplastic starch biofilms prepared with different techniques. Nord Pulp Pap Res J 2014;29:41–5.
119. Asrofi M, Abral H, Kasim A, Pratoto A, Mahardika M, Hafizulhaq F. Characterization of the sonicated yam bean starch bionanocomposites reinforced by nanocellulose water hyacinth fiber (WHF): the effect of various fiber loading. J Eng Sci Technol 2018;13:2700–15.
120. Herrera N, Salaberria AM, Mathew AP, Oksman K. Plasticized polylactic acid nanocomposite films with cellulose and chitin nanocrystals prepared using extrusion and compression molding with two cooling rates: effects on mechanical, thermal and optical properties. Compos Appl Sci Manuf 2016;83:89e97.
121. Mathew AP, Thielemans W, Dufresne A. Mechanical properties of nanocomposites from sorbitol plasticized starch and tunicin whiskers. J Appl Polym Sci 2008;109:4065e4074.
122. Liu D, Zhong T, Chang PR, Li K, Wu Q. Starch composites reinforced by bamboo cellulosic crystals. Bioresource technology. Bioresour Technol 2010;101:2529–36.
123. Angellier H, Choisnard L, Molina-Boisseau S, Ozil P, Dufresne A. Optimization of the preparation of aqueous suspensions of waxy maize starch nanocrystals using a response surface methodology. Biomacromolecules 2004;5:1545–1551.
124. Anglès MN, Dufresne A. Plasticized starch/tunicin whiskers nanocomposite materials. Mechanical behavior. Macromolecules 2001;34:2921e2931.
125. Chen D, Lawton D, Thompson MR, Liu Q. Biocomposites reinforced with cellulose nanocrystals derived from potato peel waste. Carbohydr Polym 2012;90:709–16.
126. Park JH, Koo MS, Cho SH, Lyu MY. Comparison of thermal and optical properties and flowability of fossil-based and bio-based polycarbonate. Macromol Res 2017;25:1135–44.
127. Fonseca-García A, Jiménez-Regalado EJ, Aguirre-Loredo RY. Preparation of a novel biodegradable packaging film based on corn starch-chitosan and poloxamers. Carbohydr Polym 2021;251:117009.

128. Bruni GP, Oliveira JP, Fonseca LM, Silva FT, Dias ARG, da Rosa Zavareze E. Biocomposite films based on phosphorylated wheat starch and cellulose nanocrystals from rice, oat, and Eucalyptus husks. Starch Staerke 2020;72:1900051.
129. Youssef AM, El-Sayed SM. Bionanocomposites materials for food packaging applications: concepts and future outlook. Carbohydr Polym 2018;193:19–27.
130. Montero B, Rico M, Rodríguez-Llamazares S, Barral L, Bouza R. Effect of nanocellulose as a filler on biodegradable thermoplastic starch films from tuber, cereal and legume. Carbohydr Polym 2017;157: 1094–104.
131. Syafri E, Sudirman M, Yulianti E, DeswitaAsrofi M, Abral H, Sapuan S, et al. Effect of sonication time on the thermal stability, moisture absorption, and biodegradation of water hyacinth (Eichhornia crassipes) nanocellulose-filled bengkuang (Pachyrhizus erosus) starch biocomposites. J Mater Res Technol 2019;8: 6223–31.
132. Souza AG, Ferreira RR, Paula LC, Mitra SK, Rosa DS. Starch-based films enriched with nanocellulose-stabilized Pickering emulsions containing different essential oils for possible applications in food packaging. Food Packag Shelf Life 2021;27:100615.
133. Chen Q, Liu Y, Chen G. A comparative study on the starch-based biocomposite films reinforced by nanocellulose prepared from different non-wood fibers. Cellulose 2019;26:2425–35.
134. Bagde P, Nadanathangam V. Mechanical, antibacterial and biodegradable properties of starch film containing bacteriocin immobilized crystalline nanocellulose. Carbohydr Polym 2019;222:115021.
135. Montoya R, Zuluaga R, Castro C, Vélez L, Gañán P. Starch and starch/bacterial nanocellulose films as alternatives for the management of minimally processed mangoes. Starch Staerke 2018;71:1800120.
136. Noorbakhsh-Soltani S, Zerafat M, Sabbaghi S. A comparative study of gelatin and starch-based nano-composite films modified by nano-cellulose and chitosan for food packaging applications. Carbohydr Polym 2018;189:48–55.
137. Nazrin A, Sapuan SM, Zuhri MYM, Ilyas RA, Syafiq R, Sherwani SFK. Nanocellulose reinforced thermoplastic starch (TPS), polylactic acid (PLA), and polybutylene succinate (PBS) for food packaging applications. Front Chem 2020;8:213.
138. Yuan Y, Chen H. Preparation and characterization of a biodegradable starch-based antibacterial film containing nanocellulose and polyhexamethylene biguanide. Food Packag Shelf Life 2021;30:100718.
139. Jayaraj JJ, Renugadevi K, Prakash P, Harish M, Kumar RD. Effect of nanocellulose extracted from rice husk on the film properties of native starch based edible films. In: 3rd international conference on frontiers in automobile and mechanical engineering (FAME 2020); 2020.
140. Owi WT, Ong HL, Sam ST, Villagracia AR, Tsai CK, Akil HM. Unveiling the physicochemical properties of natural Citrus aurantifolia crosslinked tapioca starch/nanocellulose bionanocomposites. Ind Crop Prod 2019;139:111548.
141. Bruna ASM, João HDOR, Lindaiá SC, Ingrid LL, Josiane DVB, Joyce BA, et al. Characterization of cassava starch films plasticized with glycerol and strengthened with nanocellulose from green coconut fibers. Afr J Biotechnol 2017;16:1567–78.
142. Santana JS, do Rosário JM, Pola CC, Otoni CG, de Fátima FerreiraSoares N, Camilloto GP, et al. Cassava starch-based nanocomposites reinforced with cellulose nanofibers extracted from sisal. J Appl Polym Sci 2016;134:44637.
143. Salehudin MH, Salleh E, Muhamad II, Mamat SNH. Starch-based biofilm reinforced with empty fruit bunch cellulose nanofibre. Mater Res Innovat 2014;18:S6–322.
144. Costa SS, Druzian JI, Machado BAS, de Souza CO, Guimarães AG. Bi-functional biobased packing of the cassava starch, glycerol, licuri nanocellulose and red propolis. PLoS One 2014;9:e112554.
145. Othman SH, Nordin N, Azman NAA, Tawakkal ISMA, Basha RK. Effects of nanocellulose fiber and thymol on mechanical, thermal, and barrier properties of corn starch films. Int J Biol Macromol 2021;183:1352–61.
146. Othman SH, Majid NA, Tawakkal ISMA, Basha RK, Nordin N, Shapi'i RA. Tapioca starch films reinforced with microcrystalline cellulose for potential food packaging application. Food Sci Technol 2019;39:605–12.

147. Zhou P, Luo Y, Lv Z, Sun X, Tian Y, Zhang X. Melt-processed poly (vinyl alcohol)/corn starch/nanocellulose composites with improved mechanical properties. Int J Biol Macromol 2021;183:1903–10.
148. Nordin N, Othman S, Kadir Basha R, Abdul Rashid S. Mechanical and thermal properties of starch films reinforced with microcellulose fibres. Food Res 2018;2:555–63.
149. Yang S, Tang Y, Wang J, Kong F, Zhang J. Surface treatment of cellulosic paper with starch-based composites reinforced with nanocrystalline cellulose. Ind Eng Chem Res 2014;53:13980–8.
150. Xu Y, Rehmani N, Alsubaie L, Kim C, Sismour E, Scales A. Tapioca starch active nanocomposite films and their antimicrobial effectiveness on ready-to-eat chicken meat. Food Packag Shelf Life 2018;16:86–91.
151. Ji N, Liu C, Li M, Sun Q, Xiong L. Interaction of cellulose nanocrystals and amylase: its influence on enzyme activity and resistant starch content. Food Chem 2018;245:481–7.
152. Liu L, Kong F. In vitro investigation of the influence of nano-cellulose on starch and milk digestion and mineral adsorption. Int J Biol Macromol 2019;137:1278–85.
153. Corral ML, Cerrutti P, Vázquez A, Califano A. Bacterial nanocellulose as a potential additive for wheat bread. Food Hydrocolloids 2017;67:189–96.
154. Khairuddin N, Muhamad II, Wan Abd Rahman WA, Siddique BM. Physicochemical and thermal characterization of hydroxyethyl cellulose – wheat starch based films incorporated thymol intended for active packaging. Sains Malays 2020;49:323–33.
155. Sharma M, Aguado R, Murtinho D, Valente AJ, Mendes De Sousa AP, Ferreira PJ. A review on cationic starch and nanocellulose as paper coating components. Int J Biol Macromol 2020;162:578–98.
156. Khorasani AC, Shojaosadati SA. Starch-and carboxymethylcellulose-coated bacterial nanocellulose-pectin bionanocomposite as novel protective prebiotic matrices. Food Hydrocolloids 2017;63:273–85.
157. Zhai X, Gao S, Xiang Y, Wang A, Li Z, Cui B, et al. Cationized high amylose maize starch films reinforced with borax cross-linked nanocellulose. Int J Biol Macromol 2021;193:1421–9.
158. Satyamurthy P, Nadanathangam V. Nanocellulose as functional filler in starch/polyvinyl alcohol film for preparation of urea biosensor. Curr Sci 2018;114:897.
159. Meneguin AB, Ferreira Cury BS, dos Santos AM, Franco DF, Barud HS, da Silva Filho EC. Resistant starch/pectin free-standing films reinforced with nanocellulose intended for colonic methotrexate release. Carbohydr Polym 2017;157:1013–23.
160. Patil MD, Patil VD, Sapre AA, Ambone TS, Torris AT, A, Shukla PG, et al. Tuning controlled release behavior of starch granules using nanofibrillated cellulose derived from waste sugarcane bagasse. ACS Sustainable Chem Eng 2018;6:9208–17.
161. Hazrol M, Sapuan S, Ilyas R, Othman M, Sherwani S. Electrical properties of sugar palm nanocrystalline cellulose, reinforced sugar palm starch nanocomposites. Polimery 2020;65:363–70.
162. Zhen WJ. Study on nanocellulose/starch composites. Adv Mater Res 2011;187:544–7.

Dang Mao Nguyen*, Julia Buchner, Thien Huu Tran, DongQuy Hoang, Thi My Hanh Diep and Quoc-Bao Bui

5 Highly functional nanocellulose-reinforced thermoplastic starch-based nanocomposites

Abstract: Starch/nanocellulose nanocomposite is of interest because of its potential applications in the field of biodegradable food packaging and biomedical applications thanks to its safe, biodegradable, fabricated by simple traditional methods, and cheap. The starch reinforced by nanocellulose significantly improved the physicochemical properties, especially the mechanical properties, thermal stability as well as barrier properties, compared to the starch matrix. With outstanding advantages compared to polymer nanocomposites derived from petroleum, the starch/nanocellulose composite is considered a potential agent for biodegradable food packaging and biomedical technology.

Keywords: biodegradable films, nanocellulose, starch/nanocellulose film, thermoplastic starch

5.1 Introduction

The current fossil fuel crisis leads to a shortage of petroleum-based plastic materials. Moreover, current climate change is causing serious problems in energy security and affecting the quality of life due to carbon dioxide (CO_2) emissions and the greenhouse effect. New regulations are enacted to improve the quality of life by reducing the use of petroleum-based materials by developing biodegradable materials for various applications such as bio-insulating materials, increasing the use of bio-based materials in

***Corresponding author: Dang Mao Nguyen**, Laboratoire Innovation Matériau Bois Habitat (LIMBHA), Ecole Supérieure du Bois, 7 Rue Christian Pauc, 44306 Nantes, France, E-mail: mao.nguyen@esb-campus.fr. https://orcid.org/0000-0002-0304-9227
Julia Buchner, Laboratoire Innovation Matériau Bois Habitat (LIMBHA), Ecole Supérieure du Bois, 7 Rue Christian Pauc, 44306 Nantes, France, E-mail: julia.buchner@esb-campus.fr
Thien Huu Tran, VNUK – Institute for Research and Executive Education, The University of Da Nang, Da Nang, Vietnam, E-mail: thien.tran180316@vnuk.edu.vn
DongQuy Hoang, Faculty of Materials Science and Technology, University of Science, Ho Chi Minh City, 700000, Vietnam; and Vietnam National University, Ho Chi Minh City, 700000, Vietnam, E-mail: htdquy@hcmus.edu.vn
Thi My Hanh Diep, Vietnam National University, Ho Chi Minh City, 700000, Vietnam; and Center Research for Natural Resources Conservation, University of Science, Ho Chi Minh City, Vietnam, E-mail: mhg286@gmail.com
Quoc-Bao Bui, Sustainable Developments in Civil Engineering Research Group, Faculty of Civil Engineering, Ton Duc Thang University, Ho Chi Minh City, Vietnam, E-mail: buiquocbao@tdtu.edu.vn

As per De Gruyter's policy this article has previously been published in the journal Physical Sciences Reviews. Please cite as: D. M. Nguyen, J. Buchner, T. Huu Tran, D. Hoang, T. M. Hanh Diep and Q.-B. Bui "Highly functional nanocellulose-reinforced thermoplastic starch-based nanocomposites" *Physical Sciences Reviews* [Online] 2023. DOI: 10.1515/psr-2022-0038 | https://doi.org/10.1515/9783110773606-005

construction, especially developing high-performance biodegradable food packaging and biomedical technology. One of the recent approaches that has gained wide interest is to fabricate biodegradable films derived from starch for food packaging applications. Thermoplastic starch is of great interest due to its safe for health, biodegradable, cheap, and renewable [1–4].

However, there are still some disadvantages that need to be improved such as starch has a high crystallinity content, so it must be plasticized during processing, and high hydrophilicity, leading to easily absorbing vapour moisture to swelling, dissolution, and spoilage during use at high humidity. The water and gas molecules are easily dispersed through the starch film. In addition, the mechanical properties of starch-based films are very low compared to those of petroleum-based films. To improve the above disadvantages, thermoplastic starch is combined with highly compatible biological agents to complement and optimize the properties of each ingredient. On the other way, the starch film was reinforced by inorganic components at nanometer size, these reinforcements have significantly improved the disadvantages of starch film such as mechanical and barrier properties (air or water vapour). In previous reports, the use of reinforcement phases at nanometer such as montmorillonite (MMT), silicon dioxide (SiO_2), and graphene oxide (GO) has significantly improved the physicochemical properties of polymer nanocomposites thanks to their superior properties compared to the polymer matrix [5, 6, 7]. It is also known to be the reinforcer at nanometers and extracted from agricultural by-products. Nanocellulose is considered a potential reinforcer for starch films and is also of strong interest due to its superior properties compared to the starch matrix. In addition, it has a high surface area, good compatibility with the starch matrix, and high crystallinity. The combination of starch and nanocellulose has also been studied before, although the properties of starch/nanocellulose materials have been improved compared to the starch matrix, they still have certain limitations such as the antibacterial ability (poor or no antibacterial) [8]. Therefore, the modification of starch/nanocellulose films by antibacterial agents has been studied and they showed significant improvement in the bacteria treatment.

This chapter aims to synthesize the reinforcing effect of nanocellulose on starch matrix materials. Some outstanding properties of starch/nanocellulose materials are evaluated and a better understanding of the behaviour and mechanism of improving the properties of the starch matrix. Some properties of starch/nanocellulose materials were synthesized and analysed such as mechanical properties, barrier properties, and medical application. In addition, processing methods to create starch/nanocellulose materials are also listed to better understand the diversity in processing and manufacturing this material.

5.2 Starch/nanocellulose preparation methods

Depending on various applications, cellulose/starch materials are prepared by different methods. The commonly used methods are summarized in Table 5.1. Among them, the

Table 5.1: The processing methods and mechanical properties of starch/nanocellulose materials.

Materials	Nanocellulose sources	Processing method	Tensile strength [MPa]	Young's modulus [MPa]	Elongation at break [%]	Elastic modulus [MPa]	Reference
Nanocomposite films from pumpkin (*Cucurbita maxima*) starch (PSF), Cellulose nanocrystal (CNC), or cellulose nanofiber (CNF)	Commercial cellulose nanocrystal (CNC) and cellulose nanofiber (CNF)	Solvent casting	3.95 (24)	–	58.83 (6.33)	–	[9]
PSF/CNC 1 wt%			6.23 (30.32)	–	61.75 (6.67)	–	
PSF/CNC 2 wt%			5.51 (28.76)	–	55.63 (5.89)	–	
PSF/CNF 10 wt%			4.56 (25.96)	–	59.63 (6.88)	–	
PSF/CNF 15 wt%			5.08 (27.15)	–	57.10 (6.13)		
Nanocomposite films from starch/pectin (RS/P) reinforced with nanocellulose	Cellulose nanofibers (BCN) or cellulose nanofibrils (CNF) are obtained by grinding plant cellulose	Solvent casting	–	–	5.68	–	[10]
RS/P-0.5 BCN							
RS/P-1.0 BCN			–	–	2.43		
RS/P-2.0 BCN			–	–	4.75		
RS/P-3.0 BCN			–	–	5.29		
RS/P-0.5 CNF			–	–	3.32		
RS/P-1.0 CNF			–	–	5.23		
RS/P-2.0 CNF			–	–	8.84		
RS/P-3.0 CNF			–	–	5.61		
RS/P				–	0.96		
Sugar palm based nanocellulose/starch/cinnamon essential oil (EO) composite films	Sugar palm based nanocellulose (CNF)	Solvent casting method	4.53	–	18.1	112.49	[13]
TPS/CNF							
TPS/CNF + 0.8% EO			4.81		17.3	122.75	
TPS/CNF + 1.2% EO			4.94		16.4	122.44	

Table 5.1: (continued)

Materials	Nanocellulose sources	Processing method	Tensile strength [MPa]	Young's modulus [MPa]	Elongation at break [%]	Elastic modulus [MPa]	Reference
TPS/CNF + 1.6% EO			5.08		15.6	128.84	
TPS/CNF + 2% EO			5.27		13.9	130.52	
Thermoplastic starch (TPS) nanocomposite films from potato starch and cellulose nanofibers (CNFs)	Cellulose nanofibers (CNFs) from *Chrysopogon zizanioides* roots.	Solution casting	1.77	15.82	31.36		[16]
TPS							
TPS/CNF 1 wt%			3.14	24.62	38.82		
TPS/CNF 2 wt%			3.63	24.70	53.01		
TPS/CNF 3 wt%			4.62	42.35	29.82		
TPS/CNF 4 wt%			3.98	50.45	29.86		
Thermoplastic corn starch (TPS)/nanocellulose (CNC)/essential oils (ho wood, cardamom or cinnamom) films.	Cellulose nanocrystals (CNC) from eucalyptus biomass are prepared by the ball mill method	Casting	6.4	–	39.8	223.7	[12]
TPS							
TPS/2 w% Ho wood			12.9	–	26.9	479.5	
TPS/5 w% Ho wood			10.8		29.5	491.7	
TPS/2 w% cardamom			2.4		9.1	130.2	
TPS/5 w% cardamom			1.7		44.7	71.4	
TPS/2 w% cinnamom			5.7		12.8	361.2	
TPS/5 w% cinnamom			5.1		32.4	208.6	
Poly (vinyl alcohol) (PVA)/corn starch/nanocellulose (CNF) composites	Cellulose nanofibrils (CNFs) commercial product	Melt processing	~22	~950	~5.5	–	[18]
PVA/CNF/Starch [100/0/10]							
PVA/CNF/starch [100/0/20]			~20	~900	~2.5		

Table 5.1: (continued)

Materials	Nanocellulose sources	Processing method	Tensile strength [MPa]	Young's modulus [MPa]	Elongation at break [%]	Elastic modulus [MPa]	Reference
PVA/CNF/starch [100/10/10]			28.19	1572.54	10.72		
PVA/CNF/starch [100/10/20]			~24	~1000	~7.5		
PVA/CNF/starch [100/20/10]			~27	~1000	~10		
PVA/CNF/starch [100/20/20]			~24	~750	~5		
Nanocomposite films from high amylose maize starch (CS)/nanocellulose (NC).	Nanocellulose (NC) was extracted by enzymatic hydrolysis-assisted high-pressure homogenization from cotton fiber	Solution casting method	Increasing until 6 wt% NC	Increasing until 6 wt% NC	Decreasing until 6 wt% NC		[14]
CS							
CS/NC 2 wt%							
CS/NC 4 wt%							
CS/NC 6 wt%							
CS/NC 8 wt%							
CS/NC 2 wt% + borax			Higher than without borax	Higher than without borax	Higher with borax		
CS/NC 4 wt% + borax							
CS/NC 6 wt% + borax			7.15				
CS/NC 8 wt% + borax							
Cassava starch (CS)/lignocellulose nanofibers (LCNF) films	Lignocellulose nanofibers (LCNF) were produced from cassava bagasse	Casting	5.3		48.7		[19]
CS/LCNF 0.65 wt%							
CS/LCNF 1.3 wt%			6.6		44.4		
Thermoplastic starch (TPS)/cellulose nanofibers nanocompos tes	Cellulose nanofiber (CNF) was prepared from softwood wood flour	Twin-screw extrusion	8.8 ± 0.7	455.3 ± 39.2	23.0 ± 5.5		[20]
TPS 100 wt%							

Table 5.1: (continued)

Materials	Nanocellulose sources	Processing method	Tensile strength [MPa]	Young's modulus [MPa]	Elongation at break [%]	Elastic modulus [MPa]	Reference
TPS/CNF 95/5 wt%			11.8 ± 0.6	622.9 ± 34.8	8.9 ± 2.9		
TPS/CNF 90/10 wt%			16.4 ± 1.1	914.9 ± 125.0	8.8 ± 2.0		
TPS/CNF 85/15 wt%			16.8 ± 02.1	1056.0 ± 99.9	3.0 ± 0.9		
TPS/CNF 80/20 wt%	Softwood wood flour		17.5 ± 2.1	1317.0 ± 134.7	1.9 ± 0.4		
Thermoplastic cassava starch (TPS)/cellulose cassava bagasse nanofibrils (CBN)	Cellulose nanofibrils (CBN) was produced from cassava bagasse	Melting mixer processing	1.8		29.8	16.8	[21]
TPS 100 wt%							
TPS/CBN 95/5 wt%			1.5		76.7	23.7	
TPS/CBN 90/10 wt%			2.8		76.5	25.8	
TPS/CBN 80/20 wt%			2.6		55.0	27.3	

solution casting method is one of the most commonly used approaches at the laboratory scale to form films with homogeneous structures. The starch-based films reinforced by nanocellulose may be obtained by this technique based on starch gelatinization (wet). This technique is closely related to coating application on various surfaces such as coating for food. However, it is considered a highly energy-consuming method because it involves the gelatinization of an aqueous starch suspension and it is not a continuous process. Furthermore, it has many limitations such as the size and number of films produced, the long preparation time and a large amount of solvent used. In the literature, this technique was an effective method for preparing food-preserving films such as pumpkin starch/nanocellulose composite films [9]. It was also used to prepare the starch/pectin free-standing films reinforced nanocellulose for colonic methotrexate release that enhanced their barrier properties and low oral bioavailability of methotrexate [10], the PVA/(ligno)nanocellulose films to improve the mechanical, barrier and antioxidant properties [11] as well as food packaging from starch/nanocellulose-stabilized Pickering emulsions [12]. The nanocellulose/starch films combined with plasticizers and additives were fabricated by solution casting method [13]. The raw component in different ratios were mixed together by simple stirring and sonication to produce a homogeneous film. The solution suspension was cooled and placed under a vacuum to remove air bubbles. This solution was then poured into a Petri dish to undergo the solution casting process. The film was peeled off the Petri dish after being placed in an overnight oven at 40 °C. The film was stored at 23 ± 2 °C under a relative humidity (RH) of 53 ± 1% in a controlled room for 7 days. The nanocellulose modified by borax cross-linking was developed and incorporated into a cationic high amylose corn starch films by solution casting method to increase their applicability [14]. The new films have significantly improved their tensile strength and water resistance due to the good dispersion of nanocellulose in the starch films. In addition, this method was used to prepare functional films such as eco-friendly and food-preservative packaging with improved barrier and controlled antimicrobial properties. These nanocomposites were fabricated from cellulose nanofibrils and dextran-coated silver nanoparticles [15], from cellulose-rich *Chrysopogon zizanioides* nanofibers and potato starch with exceptional tensile strength, tensile modulus, thermal stability, and crystallinity [16], and from PVA/banana *pseudostem* fibers reinforced with nanocellulose filler [17].

In contrast to the solution method, the melting process method is an alternative to handling insoluble polymers. This is a very useful method in making polymer blends or composites containing thermoplastic polymers by melting the polymers with heat while preserving the original properties of the polymers when cooled. Moreover, this method is simple and quick, so it has been widely applied to large-scale industrial production. In the melting process, the polymers are melted to form a viscous liquid, then the reinforcing fillers are mixed into the matrix polymers through a high shear mixer or in an extruder. The final products are obtained through injection molding, compression molding, or extrusion. The disadvantage of this method is the undesired thermal degradation of products under high shear conditions. Therefore, the processing conditions must be optimized to eliminate undesirable defects. In the literature, the melt-processing method was used to fabricate sustainable materials from corn starch and cellulose nanofibrils incorporated with biodegradable poly (vinyl alcohol) [18]. The different materials with various component contents were internally mixed using a thermal mixer at 150 °C for 10 min with a rotating speed of 30 rpm. The mixture was then shredded into small pieces at room temperature before being shaped into three-dimensional samples by compression molding at 160 °C for 10 min and 10 MPa. On the other hand, the nanocomposites-based thermoplastic starch and cellulose nanofibers were prepared by melt processing using twin-screw extrusion to improve the mechanical properties and moisture sensitivity of thermoplastic starch [20]. The water vapour was removed from materials during melt-processing through two atmospheric vents and vacuum ventilation, which were made on the extruder body. Furthermore, the parameters processing including temperature, the retention time of the sample during extrusion, and screw speed were optimized before processing. The melt processing was also a wide method used in the preparation of starch/nanocellulose nanocomposites from laboratory scale to industrial scale [21, 22].

5.3 Mechanical properties

One of the most prominent properties of inorganic nanofillers is the significant enhancement of the mechanical properties of polymer nanocomposites at the appropriate filler content. Nanocellulose was discovered as a renewable and sustainable nanofiller, synthesized from agricultural by-products. It is used to reinforce matrix polymers to improve the physical and mechanical properties through the linkages between the functional groups between fillers and polymer matrix to create a cross-linking network inside polymer nanocomposites, leading to significantly improving some properties of polymer nanocomposites, especially the mechanical properties. In addition, the morphology and size of the nanocellulose dispersed into the matrix also greatly affect the mechanical properties of the material, which is related to the number of functional groups present on its surface as well as the change in crystal structure and percentage of crystallinity of the matrix polymer through crystal nucleation during the melt-processing of polymer nanocomposite. The nanocomposites with strongly polar matrix polymers such as starch, the nanocellulose exhibit a significant influence on the mechanical properties of their polymer nanocomposites. The mechanical properties of starch/nanocellulose (nano) composites are listed in Table 5.1.

The film nanocomposites were prepared from pumpkin starch and cellulose nanocrystal and cellulose nanofiber. The film showed an enhancement of the highest tensile strength of 30.32 MPa with 1 wt% cellulose nanocrystal. Moreover, the pumpkin starch/cellulose nanocrystal/cellulose nanofiber composite films were expected to be especially used in food packaging [9]. For the starch matrix, nanocellulose has shown a significant role in enhancing the mechanical properties of this composite. The enhancement of the mechanical properties was explained in relation to the high mechanical properties of nanocellulose compared with starch [10]. Furthermore, the active functional groups (hydroxyl) on the surface of nanocellulose interacted with starch functional groups to form the cross-linking networks in the composites. The large contact surface area of nanocellulose also facilitates the favourable interactions between starch and nanocellulose. The mechanical properties of thermoplastic starch films reinforced by cellulose nanofibers were improved with increasing their content) [16, 23]. The nanocomposite film with 3 wt% nanofiller showed an exceptional tensile strength at 16% and tensile modulus (~167%) compared to the neat starch film. It was confirmed that the reinforcing role of cellulose nanofibers with the starch matrix is due to the intermolecular hydrogen bonding interaction and good interfacial adhesion between the starch matrix and reinforcers [16]. Moreover, with its small size, nanocellulose can fill in the holes and defects within the starch matrix, leading to a more compact structure for the films. Moreover, a high surface area of nanocellulose and hydrogen-bonded network between nanocellulose and starch could improve the rigidity part, which helps to transfer the stress efficiently due to the higher mechanical properties of the nanocellulose [16]. Moreover, the influence of nanocellulose on the enhancement of mechanical properties of starch-based nanocomposites was due to the formation of a rigid network of nanofibers, leading to reducing the mobility of macromolecular chains and making them difficult to slide [24, 25]. The external impact force was transferred from the starch matrix to the rigid network nanocellulose/starch, which absorbed deformation and dissipated the stress [26, 27].

5.4 Barrier properties

In the literature, it is reported that inorganic fillers not only enhance the physicomechanical properties of the polymer matrix but can also be used to significantly improve the barrier performance for polymer nanocomposites. Nanocellulose is one of the most inorganic fillers of interest today, it is increasingly preferred for inclusion in matrix polymers to improve barrier properties. This is due to the fact that it possesses many outstanding features compared to other fillers such as sustainable materials deriving from agricultural by-products, which are extracted through simple methods. Especially, when nanocellulose was dispersed into the polymer matrix at manometer size, it exhibits an excellent barrier role even for both, hydrophobic or hydrophilic matrix polymers such as starch. When nanocellulose was incorporated into hydrophilic matrix resins such as starch, nanocellulose also exhibits effective barrier properties for the matrix through strong hydrogen bonding interactions between the functional groups on the surface of nanocellulose and the functional groups of the polymer matrix. This impedes the diffusion of water molecules within the material and thereby reducing the absorption and transmission rate of water. The role of nanocellulose in increasing barrier performance of the matrix polymer is to create more tortuous pathways for permeation of water and gas molecules instead of these molecules passing through directly within the materials (F.V [28]. Moreover, with the surface containing many active functional groups, the nanocellulose is compatible with the polar polymer matrix to create a change in the crystallinity of the polymer. The crystallinity of the polymer matrix also contributes to the improvement of barrier properties.

The starch film was usually reported as unsatisfactory with the water vapour barrier properties, compared to conventional polymers films [14]. In order to improve the water vapour barrier property and broaden its applications, nanocellulose was used as a promising potential nanofiller [20]. The barrier performance of starch/nanocellulose materials are synthesed in Table 5.2.

So can be for example be seen, that the addition of nanofiller may enhance the thermal stability of the material however reduces the water vapour permeability [10, 14].

Water vapour permeability (WVP) is an important value used to evaluate the transmission of water vapour through materials, to check the moisture penetration of materials, and to understand the macromolecular interaction of materials and water. The WVP value depends on the solubility, permeability, and diffusivity of water molecules in the membrane. Thermoplastic starches are not completely amorphous, they have a crystalline part called amylose and an amorphous part called amylopectin. Therefore, different types of starches have different WVP values because they differ in structure [30].

The addition of nanocellulose to the starch film significantly improved the water vapour resistance of the film or significantly reduced the WVP value because of the low water absorption capacity of nanocellulose and high structural arrangement compared to starch, so, they exhibit a good water vapour barrier for starch/nanocellulose nanocomposites. As reported in the literature, the highest water vapour resistance was obtained at 6% nanocellulose reinforced to the starch film [14]. At this appropriate concentration, nanocellulose was able to evenly disperse into the starch matrix, thereby filling the voids inside the film, increasing the tortuous pathways of water molecules, this means increasing the average free paths the water molecule has to take to get to the other end of the materials, leading to limiting permeability penetration, dissolution, and diffusion of water across the film [31–33]. Furthermore, the reinforcement of the nanocellulose into the thermoplastic starch matrix causes the material to become dense and less porous, and the materials become stiffer, leading to the migration of water molecules through the materials becoming more difficult. The good interaction and bonding ability between the starch matrix and nanocellulose also contribute to limiting the water diffusivity of the nanocomposite film compared to the neat starch film.

However, as the nanocellulose content continued to increase, the water vapour resistance of the film decreased or the WVP value of the film increased slightly [34, 35].

The WVP value also contributes to the assessment of the hydrophilic properties of the starch/nanocellulose films, the results obtained help predict the solubility and swelling in the water of the films. Furthermore, the WVP value of starch/nanocellulose films is not much affected by their thickness and cellulose concentration [10]. The water vapour

Table 5.2: The barrier properties of starch/nanocellulose composites.

Materials	Nanocellulose sources	Moisture diffusion coefficient (D) [cm^2/s]	Water vapour permeability [g mm m^{-2} day^{-1} kPa^{-1}/*h^{-1}*instead day/ **different unit*]	Reference
Nanocomposite films from resistant starch/pectin (RS/P) reinforced with nanocellulose	Cellulose nanofibers (BCN) or cellulose nanofibrils (CNF) are obtained by grinding plant cellulose		3.95×10^{-5}	[10]
RS/P-0.5 BCN				
RS/P-1.0 BCN			3.27×10^{-5}	
RS/P-2.0 BCN			4.01×10^{-5}	
RS/P-3.0 BCN			4.65×10^{-5}	
RS/P-0.5 CNF			4.57×10^{-5}	
RS/P-1.0 CNF			4.61×10^{-5}	
RS/P-2.0 CNF			4.85×10^{-5}	
RS/P-3.0 CNF			3.5×10^{-5}	
RS/P			19.1×10^{-5}	
Biodegradable films from hemicellulose-based bulgur bran/nanocellulose	Commercial cellulose nanocrystal (CNC) and cellulose nanofiber (CNF)		2.12*	[29]
Film + no nanocellulose				
Film + 10 wt%			2.02*	
CNC + 10 wt% CNF				
Thermoplastic corn starch (TPS)/nanocellulose (CNC)/essential oils (ho wood, cardamom, or cinnamom) films.	Cellulose nanocrystals (CNC) from eucalyptus biomass are prepared by the ball mill method		2.6**g mm m^{-2} s^{-1} Pa^{-1}	[12]
TPS				
TPS/2 wt% Ho wood			2.8**	
TPS/5 w% Ho wood			2.9**	
TPS/2 w% cardamom			2.9**	
TPS 5 w% cardamom			2.6**	
TPS/2 w% cinnamom			2.6**	
TPS/5 w% cinnamom			2.7**	
Corn starch/bacteriocin immobilized crystalline nanocellulose films	Whisker-shaped crystalline nanocellulose (CNC) was prepared from cotton linters by bio-mechanical process		1.9*	[8]
Corn starch film				
Starch/CNC film			1.78	

Table 5.2: (continued)

Materials	Nanocellulose sources	Moisture diffusion coefficient (D) [cm^2/s]	Water vapour permeability [g mm m^{-2} day^{-1} kPa^{-1}/*h^{-1}*instead day/ **different unit]*	Reference
Starch/bacteriocin from *P.acidilactici*)			1.72	
Starch/bacteriocin from *P.faecium*			1.72	
Thermoplastic starch (TPS) films from tuber, cereal, and legume reinforced by nanocellulose	Cellulose nanocrystals (CNC)	2.01×10^{-10}		[30]
TPS film				
Potato TPS/CNF 2 wt%		0.44×10^{-10}		
Potato TPS/CNF 5 wt%		0.42×10^{-10}		
Corn TPS film		5.25×10^{-10}		
Corn TPS/CNF 2 wt%		1.91×10^{-10}		
Corn TPS/CNF 5 wt%		1.43×10^{-10}		
TPS		9.19×10^{-10}		
Pea TPS/CNF 2 wt%		1.18×10^{-10}		
Pea TPS/CNF 5 wt%		1.15×10^{-10}		
Nanocomposite films from high amylose maize starch (CS)/nano-cellulose (NC).	Nanocellulose (NC) was extracted by enzymatic hydrolysis-assisted high-pressure homogenization from cotton fibre		2.8**g mm m^{-2} s^{-1} Pa^{-1}	[14]
CS				
CS/NC 2 wt%			2.53**	
CS/NC 4 wt%			2.5**	
CS/NC 6 wt%			2.16**	
CS/NC 8 wt%			2.21**	
CS/CN 2 wt% + borax			2.16**	
CS/CN 4 wt%+ borax			1.94**	
CS/CN 6 wt% + borax			1.77**	
CS/CN 8 wt% + borax			1.80**	
Thermoplastic starch (TPS)/cellulose nano-fibers nanocomposites	Cellulose nanofiber (CNF) was prepared from softwood wood flour	1.66×10^{-9}		[20]
TPS 100 wt%				
TPS/CNF 95/5 wt%		1.50×10^{-9}		
TPS/CNF 90/10 wt%		1.43×10^{-9}		
TPS/CNF 85/15 wt%		1.31×10^{-9}		
TPS/CNF 80/20 wt%		1.27×10^{-9}		

Table 5.2: (continued)

Materials	Nanocellulose sources	Moisture diffusion coefficient (D) [cm^2/s]	Water vapour permeability [g mm m^{-2} day^{-1} kPa^{-1}/*h^{-1}*instead day/* ***different unit]*	Reference
Cassava starch (CS)/ lignocellulose nanofibers (LCNF) films	Lignocellulose nanofibers (LCNF) were produced from cassava bagasse		0.041	[19]
CS 100 wt%				
CS/LCNF 0.65 wt%			0.032	
CS/LCNF 1.3 wt%			0.047	

diffusivity coefficient of starch films decreased bitterly in the presence of nanocellulose, but this effect was also highly dependent on the nature of the starch species [30].

5.5 Medical applications

In previous publications, in addition to significantly improving the physical and mechanical properties of polymer nanocomposites, nanocellulose is also known as a potential candidate for biomedical applications. In particular, nanocellulose modified by biological agents are used as encapsulation and drug delivery agents in the pharmaceutical industry and improving drug release at appropriate pH value [36, 37]. One of the applied studies of nanocellulose was in combination with other biological agents to increase its ability against colorectal and prostate cancer cells [38]. In addition, nanocellulose is coupled with other biological agents such as poly (ethyl ethylene phosphate) to deliver drugs to cancer cells [39]. Nanocellulose was used as a carrier for immobilization of silver nanoparticles (AgNPs) which showed increased inhibitory effect on the growth of *Escherichia coli* (Gram negative) and *Bacillus subtilis* (Gram positive) compared with AgNPs only. Furthermore, nanocellulose were also shown to be effective in inhibiting the infectivity of alphaviruses in Vero cells B [40]. In addition, nanocellulose has also been studied and applied in biomedical engineering through bone tissue regeneration and improved adhesion of bone implants and injectable tissue supports, reconstruction of calvarial bone defects in mice, and marginal nanocellulose has gradually become a good candidate for bone grafting [41]. Nanocellulose is also used to make biosensors for medical applications thanks to its surface-active functional groups. These biological sensors are used to detect organic compounds such as amino acids and type 2 diabetes serum. It is combined with biologically active substances to detect immunity in mice through the detection of related diseases [42].

In the literature, the starch/nanocellulose composites were modified by polyhexamethylene biguanide (PHMB) as antimicrobial films for *E. coli* by the agar well method [42]. PHMB showed remarkable antibacterial activity for starch/nanocellulose films at PHMB contents from 6% to 12%. This confirmed that the film was not able to inhibit bacteria at low PHMB concentrations (<6%) and was inhibitory stable at high concentrations (>12%). Furthermore, the *E. coli* antibacterial ability of the films were also tested by the liquid immersion film method. The results also confirmed the effect of inhibition of *E. coli* at PHMB content of at least 6%. The antimicrobial ability of sugar palm starch/sugar palm nanocrystalline cellulose/cinnamon essential oil (EO) films was tested using the agar disc method [13]. It was reported that at least 2.0 wt% of EO showed the most effective inhibition against bacteria such *B. subtilis, S. aereus,* and *E. coli.* In addition, the starch/bacteriocin immobilized crystalline nanocellulose film also exhibited antibacterial activity against *E. coli* and *S. aureus*) [8].

On the other hand, nanocellulose and bio-thermoplastic are two biopolymers that have extraordinary qualities that are useful in the medical field. Different sizes of cellulose, such as micro and nanocellulose, can be generated depending on the intended function [43]. NCC and BNC demonstrate the useful qualities in medicine [43]. NCC is frequently used to create nanometer-thick biofilms. According to [44]; NCC can be used in vascular grafts. NCC-nanocomposite fibrin's structure makes it appropriate for variations in mechanical strength and elasticity that mimic blood vessels. NCC is additionally utilised in tissue engineering and antimicrobial [8, 45]. In addition, bacterial nanocellulose (BNC), which is directly generated by bacteria, nanocellulose of plant origin also exhibits extraordinary features. Bacteria nanocellulose is produced in essentially pure form, up to 90% free of lignin and hemicellulose, by the terminal complex of bacteria. BNC is regarded as a non-cytotoxic, non-genotoxic, and highly biocompatible substance due to its exceptional purity, attracting interest in a variety of fields, particularly medicine [46]. developed novel magnetic materials containing base bacterial cellulose ligands and magnetite nanoparticles (Fe_3O_4NPs) for antimicrobial and cytotoxic activities, as well as chemotherapy in cancer treatment, using a multi-step procedure. Biodegradable thermoplastics, including polylactic acid and polycaprolactone, are found to have many important roles in the medical field. PLA is commonly used in drug delivery, tissue engineering, and regenerative medicine [47]. Also, PCL shows excellent biocompatibility for tissue engineering and medical implants [48]. Otherwise, thermoplastic starch and nanocellulose can be reinforced polylactic acid (PLA) and polycaprolactone (PCL) to enhance its chemical and physical properties that directly improve their applications in medication [47–49]. Consequently, several advantages were discovered in applying biopolymers in medical aspect, due to its biocompatibility. The development of biocomposite materials containing cellulose has improved in terms of mechanical, thermal, and aging resistance. However, the properties of biopolymers have some challenges to widen their application, included poor toughness, low melting point.

Besides the medical sector, nanocellulose based materials fit to a broader range of engineering applications such as automotive industry, optically transparent materials, coating films, tissue technology, biomimetic materials, aerogels, sensors, three-dimensional printing, filtration, scaffolding, etc. [50, 51].

5.6 Challenges and future recommendations

Nanocellulose has shown a number of specific applications as mentioned in this paper and in numerous other references. However, it will make sense as more and more practical applications appear on an industrial scale. In fact, the application of nanocellulose for industrial products is limited, possibly because of the limited quantity and stability of products on an industrial scale. Therefore, the big challenge is to bring this product to practical applications and expand it to more applications at the industrial scale.

5.7 Conclusions

Nanocellulose could be prepared from different biomass by-products with different methods to obtain desired morphological, size and properties. Cellulose has shown a similar tendency improvement in the characteristic properties of almost various starches. In particular, the barrier properties of the starch films were significantly improved, especially the water vapour transmission when the nanocellulose content was increased into the starch films. Moreover, the high crystalline properties of nanocellulose also contribute to the barrier efficiency of starch films. In addition, nanocellulose also significantly improves the mechanical and physical properties of starch films because of its superior rigidity compared to neat starch. The starch/nanocellulose film combines with other antibacterial agents to create the food packaging against bacterial activity. Finally, starch/cellulose mixtures are also investigated for biomedical applications.

References

1. Builders PF, Arhewoh MI. Pharmaceutical applications of native starch in conventional drug delivery: pharmaceutical applications of native starch .... Starch Staerke 2016;68:864–73.
2. Caetano Kdos S, Lopes NA, Costa TMH, Brandelli A, Rodrigues E, Flôres SH, et al. Characterization of active biodegradable films based on cassava starch and natural compounds. Food Packag Shelf Life 2018;16: 138–47.
3. Vaezi K, Asadpour G, Sharifi SH. Effect of coating with novel bio nanocomposites of cationic starch/cellulose nanocrystals on the fundamental properties of the packaging paper. Polym Test 2019;80:106080.
4. Yang LC, Kim H, Beak S-E, Song KB. Utilization of foxtail millet starch film incorporated with clove leaf oil for the packaging of *queso blanco* cheese as a model food. Starch Staerke 2018;70:1700171.

5. Mallakpour S, Zahra K. Surface modified SiO2 nanoparticles by thiamine and ultrasonication synthesis of PCL/SiO2-VB1 NCs: morphology, thermal, mechanical and bioactivity investigations. Ultrason Sonochem 2018;41:527–37.
6. Mao ND, Jeong H, Nguyen TKN, Nguyen TML, Do TVV, Thuc CNH, et al. Polyethylene glycol functionalized graphene oxide and its influences on properties of poly(lactic acid) biohybrid materials. Compos B Eng 2019;161:651–8.
7. Nguyen DM, Vu TT, Grillet A-C, Ha Thuc H, Ha Thuc CN. Effect of Organoclay on morphology and properties of linear low density polyethylene and Vietnamese cassava starch biobased blend. Carbohydr Polym 2016; 136:163–70.
8. Bagde P, Nadanathangam V. Mechanical, antibacterial and biodegradable properties of starch film containing bacteriocin immobilized crystalline nanocellulose. Carbohydr Polym 2019;222:115021.
9. Zhang L, Zhao J, Zhang Y, Li F, Jiao X, Li Q. The effects of cellulose nanocrystal and cellulose nanofiber on the properties of pumpkin starch-based composite films. Int J Biol Macromol 2021;192:444–51.
10. Meneguin AB, Cury BSF, dos Santos AM, Franco DF, Barud HS, da Silva Filho EC. Resistant starch/pectin free-standing films reinforced with nanocellulose intended for colonic methotrexate release. Carbohydr Polym 2017;157:1013–23.
11. Espinosa E, Bascón-Villegas I, Rosal A, Pérez-Rodríguez F, Chinga-Carrasco G, Rodríguez A. PVA/(Ligno) Nanocellulose biocomposite films. Effect of residual lignin content on structural, mechanical, barrier and antioxidant properties. Int J Biol Macromol 2019;141:197–206.
12. Souza AG, Ferreira RR, Paula LC, Mitra SK, Rosa DS. Starch-based films enriched with nanocellulose-stabilized pickering emulsions containing different essential oils for possible applications in food packaging. Food Packag Shelf Life 2021;27:100615.
13. Syafiq R, Sapuan SM, Zuhri MRM. Antimicrobial activity, physical, mechanical and barrier properties of sugar palm based nanocellulose/starch biocomposite films incorporated with cinnamon essential oil. J Mater Res Technol 2021;11:144–57.
14. Zhai X, Gao S, Xiang Y, Wang A, Li Z, Cui B, et al. Cationized high amylose maize starch films reinforced with borax cross-linked nanocellulose. Int J Biol Macromol 2021;193:1421–9.
15. Lazić V, Vera V, Peršin Z, Stoiljković M, Ratnayake IS, Ahrenkiel PS, et al. Dextran-coated silver nanoparticles for improved barrier and controlled antimicrobial properties of nanocellulose films used in food packaging. Food Packag Shelf Life 2020;26:100575.
16. Dominicdos Santos Rosa CDMD, Camani PH, Kumar AS, Neenu KV, Sabura Begum PM, Dinakaran D, et al. Thermoplastic starch nanocomposites using cellulose-rich chrysopogon zizanioides nanofibers. Int J Biol Macromol 2021;191:572–83.
17. Srivastava KR, Dixit S, Pal DB, Mishra PK, Srivastava P, Srivastava N, et al. Effect of nanocellulose on mechanical and barrier properties of PVA–banana pseudostem fiber composite films. Environ Technol Innovat 2021;21:101312.
18. Zhou P, Luo Y, Lv Z, Sun X, Tian Y, Zhang X. Melt-processed poly (vinyl alcohol)/corn starch/nanocellulose composites with improved mechanical properties. Int J Biol Macromol 2021;183:1903–10.
19. Travalini AP, Lamsal B, Magalhães WLE, Demiate IM. Cassava starch films reinforced with lignocellulose nanofibers from cassava bagasse. Int J Biol Macromol 2019;139:1151–61.
20. Hietala M, Mathew AP, Oksman K. Bionanocomposites of thermoplastic starch and cellulose nanofibers manufactured using twin-screw extrusion. Eur Polym J 2013;49:950–6.
21. Teixeira Ede M, Pasquini D, Curvelo AAS, Corradini E, Belgacem MN, Dufresne A. Cassava bagasse cellulose nanofibrils reinforced thermoplastic cassava starch. Carbohydr Polym 2009;78:422–31.
22. Fourati Y, Albert M, Putaux J-L, Boufi S. One-step processing of plasticized starch/cellulose nanofibrils nanocomposites via twin-screw extrusion of starch and cellulose fibers. Carbohydr Polym 2020;229:115554.
23. Almasi H, Azizi S, Amjadi S. Development and characterization of pectin films activated by nanoemulsion and pickering emulsion stabilized marjoram (Origanum majorana L.) essential oil. Food Hydrocolloids 2020;99:105338.

24. Muscat D, Adhikari R, McKnight S, Guo Q, Adhikari B. The physicochemical characteristics and hydrophobicity of high amylose starch–glycerol films in the presence of three natural waxes. J Food Eng 2013;119:205–19.
25. Siqueira G, Bras J, Dufresne A. Cellulose whiskers versus microfibrils: influence of the nature of the nanoparticle and its surface functionalization on the thermal and mechanical properties of nanocomposites. Biomacromolecules 2009;10:425–32.
26. Chi K, Catchmark JM. Improved eco-friendly barrier materials based on crystalline nanocellulose/chitosan/carboxymethyl cellulose polyelectrolyte complexes. Food Hydrocolloids 2018;80:195–205.
27. Kaushik A, Singh M, Verma G. Green nanocomposites based on thermoplastic starch and steam exploded cellulose nanofibrils from wheat straw. Carbohydr Polym 2010;82:337–45.
28. Ferreira FV, Dufresne A, Pinheiro IF, Souza DHS, Gouveia RF, Mei LHI, et al. How do cellulose nanocrystals affect the Overall properties of biodegradable polymer nanocomposites: a comprehensive review. Eur Polym J 2018;108:274–85.
29. Kocabaş DS, Akçelik ME, Bahçegül E, Özbek HN. Bulgur bran as a biopolymer source: production and characterization of nanocellulose-reinforced hemicellulose-based biodegradable films with decreased water solubility. Ind Crop Prod 2021;171:113847.
30. Montero B, Rico M, Rodríguez-Llamazares S, Barral L, Bouza R. Effect of nanocellulose as a filler on biodegradable thermoplastic starch films from tuber, cereal and legume. Carbohydr Polym 2017;157:1094–104.
31. Estevez-Areco S, Guz L, Famá L, Candal R, Goyanes S. Bioactive starch nanocomposite films with antioxidant activity and enhanced mechanical properties obtained by extrusion followed by thermo-compression. Food Hydrocolloids 2019;96:518–28.
32. Liu W, Wang Z, Liu J, Dai B, Hu S, Hong R, et al. Preparation, reinforcement and properties of thermoplastic starch film by film blowing. Food Hydrocolloids 2020;108:106006.
33. Maliha M, Herdman M, Brammananth R, McDonald M, Coppel R, Werrett M, et al. Bismuth phosphinate incorporated nanocellulose sheets with antimicrobial and barrier properties for packaging applications. J Clean Prod 2020;246:119016.
34. Savadekar NR, Mhaske ST. Synthesis of nano cellulose fibers and effect on thermoplastics starch based films. Carbohydr Polym 2012;89:146–51.
35. Shankar S, Rhim J-W. Preparation of nanocellulose from micro-crystalline cellulose: the effect on the performance and properties of agar-based composite films. Carbohydr Polym 2016;135:18–26.
36. Akhlaghi SP, Berry RC, Tam KC. Surface modification of cellulose nanocrystal with chitosan Oligosaccharide for drug delivery applications. Cellulose 2013;20:1747–64.
37. Letchford J, Wasserman B, Wadood Hamad Y, Burt H. The use of nanocrystalline cellulose for the binding and controlled release of drugs. Int J Nanomed 2011;6:321–30.
38. García-García D, Balart R, Lopez-Martinez J, Ek M, Moriana R. Optimizing the yield and physico-chemical properties of pine cone cellulose nanocrystals by different hydrolysis time. Cellulose 2018;25:2925–38.
39. Wang Y, Zhang H, Wu Y, Yang J, Zhang L. Preparation and properties of natural rubber/rectorite nanocomposites. Eur Polym J 2005;41:2776–83.
40. Zoppe JO, Ruottinen V, Ruotsalainen J, Rönkkö S, Johansson L-S, Hinkkanen A, et al. Synthesis of cellulose nanocrystals carrying tyrosine sulfate mimetic ligands and inhibition of alphavirus infection. Biomacromolecules 2014;15:1534–42.
41. Yang X, Bakaic E, Todd H, Cranston ED. Injectable polysaccharide hydrogels reinforced with cellulose nanocrystals: morphology, rheology, degradation, and cytotoxicity. Biomacromolecules 2013;14:4447–55.
42. Yuan Y, Chen H. Preparation and characterization of a biodegradable starch-based antibacterial film containing nanocellulose and polyhexamethylene biguanide. Food Packag Shelf Life 2021;30:100718.
43. Omran AAB, Mohammed AABA, Sapuan SM, Ilyas RA, Asyraf MRM, Koloor SSR, et al. Micro-and nanocellulose in polymer composite materials: a review. Polymers 2021;13:231.

44. Brown EE, Hu D, Abu Lail N, Zhang X. Potential of nanocrystalline cellulose–fibrin nanocomposites for artificial vascular graft applications. Biomacromolecules 2013;14:1063–71.
45. Lalia BS, Samad YA, Hashaikeh R. Nanocrystalline cellulose-reinforced composite mats for lithium-ion batteries: electrochemical and thermomechanical performance. J Solid State Electrochem 2013;17:575–81.
46. Chaabane L, Chahdoura H, Mehdaoui R, Snoussi M, Beyou E, Lahcini M, et al. Functionalization of developed bacterial cellulose with magnetite nanoparticles for nanobiotechnology and nanomedicine applications. Carbohydr Polym 2020;247:116707.
47. Ilyas RA, Zuhri MYM, Aisyah HA, Asyraf MRM, Hassan SA, Zainudin ES, et al. Natural fiber-reinforced polylactic acid, polylactic acid blends and their composites for advanced applications. Polymers 2022;14: 202.
48. Ilyas RA, Zuhri MYM, Faiz Norrrahim MN, Mohamad Misenan MS, Jenol MA, Samsudin SA, et al. Natural fiber-reinforced polycaprolactone green and hybrid biocomposites for various advanced applications. Polymers 2022;14:182.
49. Nazrin A, Sapuan SM, Zuhri MYM, Ilyas RA, Syafiq R, Sherwani SFK. Nanocellulose reinforced thermoplastic starch (TPS), polylactic acid (PLA), and polybutylene succinate (PBS) for food packaging applications. Front Chem 2020;8:213.
50. Ansari F, Ding Y, Berglund LA, Dauskardt RH. Toward sustainable multifunctional coatings containing nanocellulose in a hybrid glass matrix. ACS Nano 2018;12:5495–503.
51. Ferreira FV, Caio Otoni G, De France KJ, Barud HS, Lona LMF, Cranston ED, et al. Porous nanocellulose gels and foams: breakthrough status in the development of scaffolds for tissue engineering. Mater Today 2020; 37:126–41.

A. Nazrin, A. S. Norfarhana, R. A. Ilyas*, S. M. Sapuan, A. Khalina, R. M. O. Syafiq, M. Y. S. Hamid, C. S. Hassan, I. Idris, P. S. Khoo, A. H. Nordin, H. S. N. Hawanis and M. L. Sanyang

6 Sugar palm (*Arenga pinnata*) thermoplastic starch nanocomposite films reinforced with nanocellulose

Abstract: The growing consciousness about global environmental concerns, particularly landfills, in conjunction with the rapid use of petroleum-based plastics, is a key factor behind the use of natural and biodegradable polymers in short-life applications like food packaging, container, and tray. Sugar palm stem is a biomass that has proven the potential to produce biodegradable polymers such as sugar palm starch. Nevertheless, their applications were limited due to their low tensile strength and excessive hydrophilicity. Plasticization using polyols, reinforcement with sugar palm fiber, cellulose, microcrystalline cellulose, or nanocellulose, blending with thermoplastic polymer, and addition of essential oils has been used to maximize the functional qualities of the starch biopolymer. As the content of plasticizers grew, the glass transition temperature and water absorption

***Corresponding author: R. A. Ilyas**, Institute of Tropical Forestry and Forest Products, Universiti Putra Malaysia, Serdang 43400, Selangor, Malaysia; Department of Chemical Engineering, Faculty of Chemical and Energy Engineering, Universiti Teknologi Malaysia, 81310 UTM Johor Bahru, Johor, Malaysia; Centre for Advanced Composite Materials, Universiti Teknologi Malaysia, 81310 UTM Johor Bahru, Johor, Malaysia; and Centre of Excellence for Biomass Utilization, Universiti Malaysia Perlis, 02600, Arau, Perlis, Malaysia, E-mail: ahmadilyas@utm.my. https://orcid.org/0000-0001-6622-2632

A. Nazrin and A. Khalina, Institute of Tropical Forestry and Forest Products, Universiti Putra Malaysia, Serdang 43400, Selangor, Malaysia

A. S. Norfarhana, M. Y. S. Hamid, A. H. Nordin and H. S. N. Hawanis, Department of Chemical Engineering, Faculty of Chemical and Energy Engineering, Universiti Teknologi Malaysia, 81310 UTM Johor Bahru, Johor, Malaysia. https://orcid.org/0000-0001-7546-3833 (A.H. Nordin)

S. M. Sapuan, Advanced Engineering Materials and Composites Research Center (AEMC), Universiti Putra Malaysia, Faculty of Engineering, 43400 Seri Kembangan, Selangor, Malaysia

R. M. O. Syafiq, Department of Engineering Design and Manufacture, Faculty of Engineering, University of Malaya, 50603 Kuala Lumpur, Malaysia

C. S. Hassan, Faculty of Engineering, Technology and Built Environment, UCSI University, Kuala Lumpur, Malaysia

I. Idris, School of Chemical Engineering, College of Engineering, Universiti Teknologi MARA (UiTM), 40450 Shah Alam, Selangor, Malaysia

P. S. Khoo, Centre for Advanced Composite Materials, Universiti Teknologi Malaysia, 81310 UTM Johor Bahru, Johor, Malaysia

M. L. Sanyang, School of Agriculture and Environmental Science; University of The Gambia, P.O. Box 3530, MDI Road, Kanifing, The Gambia; and Directorate of Research and Consultancy, University of The Gambia, P.O. Box 3530, MDI Road, Kanifing, The Gambia

As per De Gruyter's policy this article has previously been published in the journal Physical Sciences Reviews. Please cite as: A. Nazrin, A. S. Norfarhana, R. A. Ilyas, S. M. Sapuan, A. Khalina, R. M. O. Syafiq, M. Y. S. Hamid, C. S. Hassan, I. Idris, P. S. Khoo, A. H. Nordin, H. S. N. Hawanis and M. L. Sanyang "Sugar palm (*Arenga pinnata*) thermoplastic starch nanocomposite films reinforced with nanocellulose" *Physical Sciences Reviews* [Online] 2023. DOI: 10.1515/psr-2022-0031 | https://doi.org/10.1515/9783110773606-006

ability decreased. Furthermore, the addition of sugar palm nanocellulose to sugar palm starch improves the performances of sugar palm starch-based films as a packaging material. Addition of essential oil contributes to antibacterial properties and slightly improved tensile strength of the film. A comprehensive understanding on the interaction of starch-based biodegradable polymer and nanocellulose constituents for enhancing the physico-chemical properties of starch-based films is prerequisite for researchers in the design of industrial products with enhanced functional attributes. To address the knowledge gap, more studies including the reinforcement of new types of biodegradable polymer and nanocellulose derived from natural sources should be conducted in order to continually populate the database for research purposes.

Keywords: *Arenga pinnata*, cellulose, lignocellulose, nanocellulose, nanocomposite, sugar palm, thermoplastic starch films

6.1 Introduction

The disposal of plastics packaging had been a challenging issue in the past decades due to its versatile characteristic. The versatile criteria of packaging plastic are essential to preserve food quality during transportation, storage, and avoiding external factors such as moisture, microorganism, odor, and gases that might spoil the food quality [1, 2]. Nowadays, due to the active lifestyle of modern society, food producers are facing difficulties in developing food packaging system to adapt with it. Ready-made foods are much more convenience and preferable, which required protective function to maintain the consumable freshness for customer satisfaction. Plastic packaging offers excellent protection against the food spoilage, but it poses threat to the environment as it is not biodegradable and difficult to be reuse or recycle [3]. Thus, this paper highlights the importance of utilizing sugar palm starch bio-based plastics instead of fossil-based plastics in various applications. Sugar palm plant is yet another one of the potential plants for the development of bio-based packaging materials [4].

Starch-based bioplastics exhibit several shortcomings such as high-water sensitivity, poor mechanical properties, low water vapor barrier property, and low freeze stability during fabrication process [5]. Thus, the addition of nanocellulose is an alternative to improve these characteristics [6–9]. Although the incorporation of nanocellulose and plasticizing agents is known to improve the mechanical and barrier properties of starch-based polymers, the study of the interaction of raw materials from different sources with different concentrations is important to try to achieve a biodegradable material with physical properties on par with traditional polymers. The lack of comprehensive literatures on the interaction between sugar palm starch and nanocellulose constituents for enhancing the physico-chemical properties of biopolymers is a major challenge for researchers and its holistic understanding should be considered a prerequisite in the design of industrial products with enhanced functional attributes. Thus, in this paper, the effects of nanocellulose incorporation on the properties of sugar palm thermoplastic starch-based polymers are discussed, based on studies previously conducted by several authors.

6.2 Type of packaging materials

The concept of food storage had brought forth various type of packaging materials such as glass, metal, paper and plastic. Each type of material offers its own unique characteristic that is compatible with the type of food that will be stored. Nevertheless, plastic-based packaging is the best adapted in food industry due to its good processability, good aesthetic, and decent physicochemical properties [10]. In comparison with glass and metal, plastic is affordable, flexible, lightweight, and portable. Meanwhile, paper has limited application in food packaging because of its low durability and high-water sensitivity. Paper is dominantly derived from wood of multispecies trees, thus, the exploitation of wood for packaging will lead to major deforestation. Therefore, plastic strives over other materials in the food packaging industry.

6.3 Sugar palm starch

Currently, the raw material resources for the starch-based polymer are mainly extracted from cereals (rice, wheat, etc.), legumes (green pea, bean, etc.), roots (cassava, yam, etc.), and tubers (potatoes, sweet potatoes, etc.). Condemnation and criticism were imposed toward the utilization of such materials as a matrix in composite development since underdeveloped nation of low livelihood regard these resources as a main staple [11]. Thus, researchers and scientists are prioritizing the usage of non-food resources as polymeric matrix to develop biopolymer. One of them is sugar palm starch (SPS), which can be obtained from a multipurpose sugar palm (*Arenga pinnata*) tree. Sugar palms are famous for its brown sugar block and black fiber. Since sugar palms are mainly cultivated for these purposes, the unproductive trees are cut and the trunks are processed to acquire starch similar to the production of sago starch. A single tree can yield an average of 50–100 kg of starch. Nevertheless, there are certain limitations to plastic starch biopolymer, such as its high hydrophilic nature and weak mechanical characteristics [12]. As a result, as compared to synthetic thermoplastic polymers, plastic starch biopolymer has limited application as starch-based films, particularly for packaging [12–15].

In fact, the degree of polymerization efficiency of starch biopolymer was rely on the starch's amylose content, whereby the amylose's linear structure closely resembling the behavior of conventional synthetic polymers [15, 16]. SPS, fortunately, has higher amylose content (37 %) compared to cassava, potato, corn, sago, wheat, and maize, based on Table 6.1 [17]. It was reported by Sahari et al. [18] that higher amylose content (37.60 %) in SPS holds the potential for the production of superior biopolymer compared to other starches. Its counterpart, amylopectin, is defined as branched polysaccharide component composed of hundreds of small chains built of α-D-glucopyranosyl residue with $(1 \rightarrow 4)$

Table 6.1: The chemical composition of various starches.

Starch	Density	Ash (%)	Amylose (%)	Water content (%)
Wheat	1.44	0.2	26–27	13
Tapioca	1.45–1.46	0.2	17	13
Maize	1.5	0.1	26–28	12–13
Potato	1.54–1.55	0.4	20–25	18–19
Sago	–	0.2	24–27	10–20
SPS	1.54	0.2	37.60	15

links. These are connected by (1 → 6)-α-linkages, of which 5–6 % are found at the branch sites. The molecular weight measures around 10^7–10^9 with a low intrinsic viscosity of 120–190 mL/g owing to its extensively branched molecular structure. Table 6.1 indicates the chemical composition of commercial starches in comparison with SPS [19].

6.4 Sugar palm fiber

Sugar palm fiber (SPF) can be extracted from the sugar palm stem (mostly made of starch) after the sugar palm fruits and sap have been processed for culinary goods. Sugar palm stem is a by-products remaining in the field after these principal products have been harvested [14]. SPF is a woven fiber that naturally entwined around the sugar palm trunk from the bottom to the top part of the tree. It is known for its durability and resilience to salinity. Besides, SPF does not require any auxiliary process and is ready upon harvest. At different height of sugar palm tree, the fibers are categorised into black sugar palm fibers (Ijuk), sugar palm bunch (SPB), sugar palm frond (SPF), and sugar palm trunk (SPT), as shown in Figure 6.1. On this account, numerous products had been made out of SPF such as brooms, brushers, cushions, filters, mats, roofs, and ropes [20]. The fiber is black and measures up to 1.19 m in length, around 94–370 µm in diameter and a density of 1.26 kg/m^3. It was stated that these fibers have good tensile properties making it a decent reinforcing agent in enhancing composite material [21]. Bachtiar et al. [21] reported that SPF has the tensile strength of 190.29 MPa, tensile modulus of 3.69 GPa and elongation at break of 19.6 %. Table 6.2 displays the chemical compositions of sugar palm fiber harvested from various heights [20]. It can be observed that as the tree ages, the chemical composition of the fiber fluctuates, which can be observed from the bottom to the top of the tree. At 1 m height, the fiber was characterized to have lower content of cellulose, hemicellulose and lignin (37.3 %, 4.71 %, and 17.93 %, respectively) in comparison with the fiber obtained from the area of live palm frond. An increment in cellulose, hemicelluloses, and lignin contents (53.41–56.8 %, 7.45–7.93 %, and 20.45–24.92 %, respectively) was reported as the fiber get higher but only up to 5 m where beyond this height (5 m), it remained unchanged.

Figure 6.1: Sugar palm tree.

Table 6.2: The chemical compositions of sugar palm fiber harvested from various heights.

Height (m)	1	3	5	7	9	11	13	15
Holocellulose (%)	43.4	56.36	64.28	65.55	64.8	63.75	62.42	57.41
Cellulose (%)	37.3	49.36	55.28	56.55	56.8	55.75	54.42	53.41
Hemicellulose (%)	4.71	6.11	7.36	7.68	7.93	7.92	7.89	7.45
Lignin (%)	17.93	18.941	20.89	20.45	23.6	22.96	24.27	24.92
Ash (%)	30.92	14.04	5.8	4.23	2.06	4.09	3.98	4.27
Extractive (%)	2.49	2.019	1.71	1.41	1.35	1.48	1.21	0.85
Moisture content (%)	5.36	8.64	7.92	8.37	8.19	7.72	8.12	8.7

6.4.1 Cellulose fiber

There are two processes involved in extracting cellulose fiber from sugar palm fiber (SPF) namely delignification and mercerization. The first process involves chlorination or bleaching, performed to eliminate the lignin from SPF, producing holocellulose. A clear physical appearance of SPF was acquired with the elimination of waxy layer on the outer surface. After chemical treatment, the size of the bleached fibers was reduced from 212.01 ± 2.17 µm to 121.80 ± 10.57 µm. Later, holocellulose is treated using alkaline (NaOH) to obtain α-cellulose. This process separates the fiber bundles into individual fiber measuring around 11.87 ± 1.04 μm with smoother and groovier surface.

6.4.2 Microcrystalline cellulose

Microcrystalline cellulose (MCC) is isolated from α-cellulose in the form of purified partially depolymerized cellulose. MCC powder can be attributed as a white, odorless, tasteless material made up of crystalline and amorphous regions [22]. MCC are commonly utilized in pharmaceutical industry to control viscosity in producing powders and tablets as compaction and disintegration agents [23].

6.4.3 Nanocellulose fiber

The downsizing of fiber into nanoscale is term as nanocellulose consisting of a collection of cellulose particles having at least one dimension (diameter, length, or thickness) ranging from 1 to 100 nm. Nanocellulose possesses unique appealing characteristics such as high surface area, high aspect ratio of 100, high crystallinity, high mechanical strength, and low density, which can advance the material science industry to a higher level. Cao et al. [24] stated that the pure nanocrystalline cellulose domain has a theoretical Young modulus of 150 GPa, which came closer to the value of steel, 200 GPa. Moreover, nanocellulose can be derived from various plant resources (e.g., sugar palm, oil palm, etc.) making it an easily available, biodegradable and renewable material. Nanocellulose is packed with abundant of hydroxyl groups on its surface. These hydroxyl groups offer active sites for the establishment of hydrogen bonding through the physical interlocking between the nonpolar matrices with the –OH groups of hydrophilic polymer matrices.

There has been an immense attraction towards plant-based nanocrystalline celluloses (NCCs) in material science in response to its outstanding intrinsic features including nano-dimension and high surface area (100 $m^2 g^{-1}$). Ilyas et al. [25] conducted study on the isolation and characterization of nanocrystalline cellulose from sugar palm fibers (*Arenga pinnata*). Primarily, sugar palm fibers underwent delignification and mercerization to obtain cellulose. Subsequently, 60 wt% of concentrated sulfuric acid was utilized to isolate sugar palm nanocrystalline celluloses (SPNCCs) from the cellulose. At each stage, the chemical composition of sugar palm fibers was analysed. Structural analysis was carried out by Brunauer–Emmett–Teller (BET), X-ray diffraction (XRD) and Fourier transform infrared spectroscopy (FT-IR). Morphological analysis of extracted cellulose and isolated nanocrystalline cellulose (NCCs) was investigated by using field emission scanning electron microscopy (FESEM), atomic force microscopy (AFM), and transmission electron microscopy (TEM). The thermal stability of sugar palm fibers at different stages of treatment was investigated by thermogravimetric analysis (TGA). Delignification and mercerization processes were responsible in removing the lignin and hemicellulose, respectively, from the extracted cellulose. The isolated SPNCCs were found to have lengths and diameters of 130 ± 30 nm and 9 ± 1.96 nm, respectively, as shown in Figure 6.2.

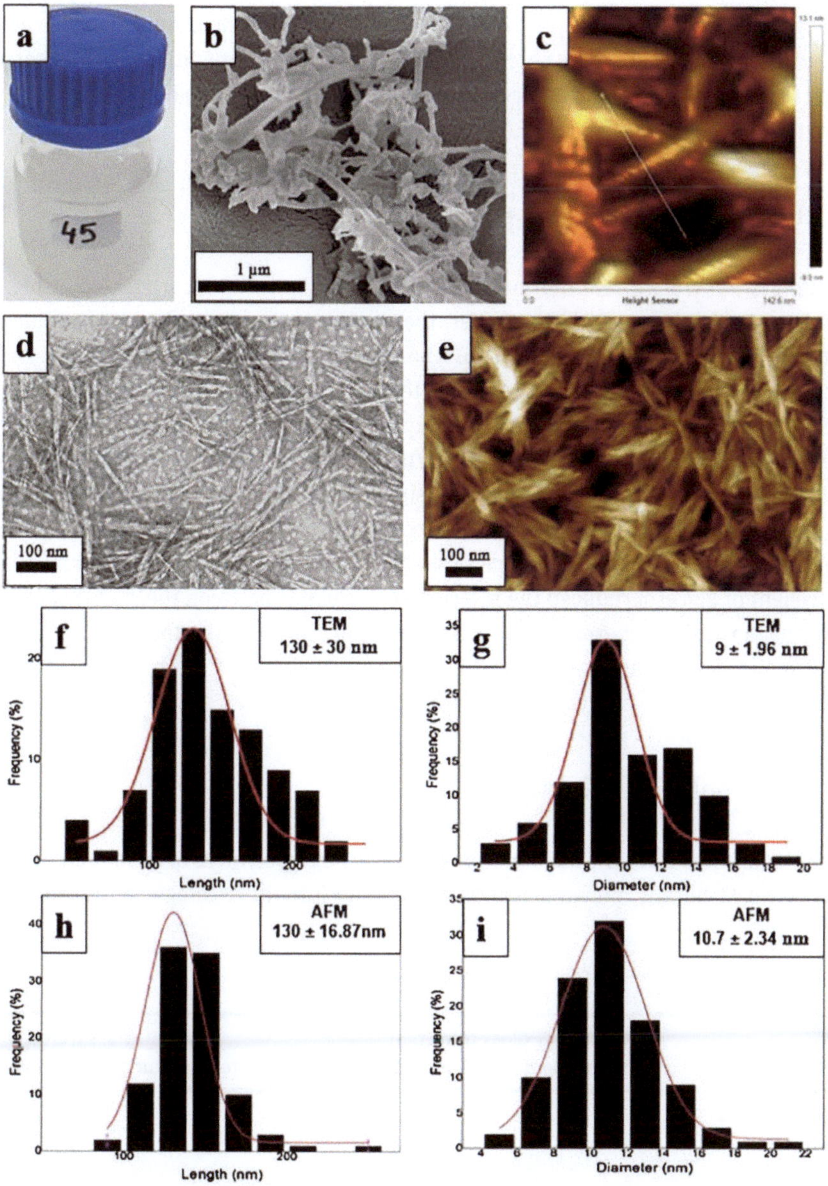

Figure 6.2: Morphology and size of nanocrystalline cellulose. (a) Aqueous suspension (2 wt%), (b) FESEM micrograph, (c) height measurement of NCCs using atomic force microscopy (AFM), (d) transmission electron (TEM) micrograph, (e) atomic force microscopy (AFM) and their length (f and h) and diameter, and (g and i) histograms of nanocrystalline cellulose extracted from sugar palm fibres [25].

6.5 Modification and reinforcement of sugar palm starch films

Starch-based polymers have limited application due to their strong hydrophilic properties. Long-term exposure to high humidity deteriorates the mechanical properties rendering it useless for packaging application. Therefore, various modifications had been executed such as plasticization, fiber reinforcement, and polymer blending in an effort to improve the functional properties of the biopolymer.

6.5.1 Plasticization of sugar palm starch films

Starch itself is not truly a real thermoplastic unless plasticizer is incorporated at high temperature. In the presence of plasticizer, thermoplastic starch has higher flexibility and better processability. Depending on the type and concentration of plasticizer, the end product might yield lower glass transition temperature (T_g) and increased tensile strength. The frequently used plasticizer is polyols such as glycerol, sorbitol, and many more. As glycerol tends to migrate while sorbitol easily crystallizes from the starch matrix, combining both plasticizers helps to stabilize their intermolecular interaction (plasticizer to plasticizer) preventing the occurrence [26]. Sahari et al. reported that the increment in glycerol loading introduced into SPS resulted in lower T_g. The small molecular weight of glycerol reduced the intermolecular bonds between the starch chains, thus altering the overall cohesion force. This interaction also facilitates the chain mobility of starch molecules promoting the flexibility and extensibility of the biopolymer [18]. Table 6.3 indicates the T_g of glycerol plasticized SPS [18].

6.5.2 Sugar palm starch blend

Polymer blending involves the combination of two or more polymers to offset individual polymer flaws. Generally, polymer blends will end up into three categories, which are

Table 6.3: The glass transition temperature of glycerol plasticized SPS.

Sample	Glass transition temperature, T_g (onset)	Glass transition temperature, T_g (midpoint)
Native SPS	237.91	242.14
SPS/15% Glycerol	225.68	229.26
SPS/20% Glycerol	206.44	217.90
SPS/30% Glycerol	189.57	187.65
SPS/40% Glycerol	176.71	177.03

immiscible or heterogeneous, miscible or homogenous and compatible blends. Immiscible polymer blends have separate phases and respective T_g. Polymers of similar chemical structures are miscible with each other thus merged into a single-phase structure with one T_g. Meanwhile, compatible blends are immiscible on a molecular level, but continuous and uniform dispersion of each polymer established sufficiently strong interactions between the component polymers.

Studies conducted by Nazrin et al. [27, 28], thermoplastic sugar palm starch (TPS) was blended with poly lactic acid (PLA) in effort to reduce the water sensitivity. Various compositions were experimented, PLA60TPS40 (PLA:TPS, 60:40) recorded lower water vapor permeability (WVP) compared to PLA70TPS30 (PLA:TPS, 70:30). The improvement in WVP rate was not entirely dependent on the PLA content but the compatibility of the two materials in establishing homogenous and uniform distribution of microstructure hindering the gases from passing through. The harmonized combination of PLA60TPS40 correlated with its higher thermal degradation temperature at 50 % weight loss (324 °C) compared to PLA70TPS30 (320 °C).

6.5.3 Sugar palm starch bilayer films

Another approach adapted to widen the application of starch-based polymer is multilayering polymer films. It is difficult for an individual polymer to have all the required properties for the intended application. In order to compensate for the drawbacks, polymer films are stacked up to create a multilayer film composed of polymers of different characteristics. Sanyang et al. reported that SPS and polylactic acid (PLA) bilayer film showed better tensile strength and water permeability compared to SPS films. FTIR results showed no additional characteristic peaks apart from the peaks revealed in pristine SPS and PLA. The findings concluded that the SPS-PLA bilayer films were compounded through physical interaction rather than chemical interaction indicating their immiscibility [29].

6.5.4 Fiber reinforced sugar palm starch biocomposites

It is a common practice to incorporate fibers into matrices to offset their drawbacks. Sugar palm fiber (SPF) is well known for its versatility, utilized in various products. Ibrahim et al. [30] fabricated hybrid composite films from corn starch (CS) reinforced with cornhusk fiber (CHF) and SPF. The scanning electron microscopy (SEM) indicated excellent interfacial adhesion of fibers and matrix as reflected by improved tensile strength and Young modulus. A noteworthy enhancement was observed after the introduction of SPF into the hybrid composites. At 6 % of SPF loading, the tensile strength was 19.09 MPa, higher than neat corn starch films (6.8 MPa) and CS/CHF composite films (12.84 MPa). The combination of cellulosic fibers established a complex network

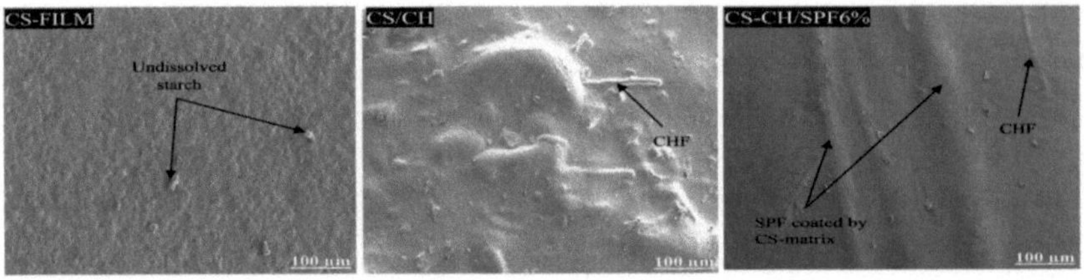

Figure 6.3: The SEM images of surface of neat CS, CS/CH composite, and CS/CH-SPF hybrid composite films [30].

dominating the mobility of TPS molecules, which holds the structural system of the hybrid composites thus facilitates the stress transfer. Figure 6.3 illustrates the SEM of neat CS, CS/CHF composite and CS/CHF-SPF hybrid composite films.

Jumaidin et al. [31] prepared hybrid composites made of TPS/agar reinforced by seaweed (Sw)/SPF. The hybrid composites with a 50:50 ratio of Sw/SPF yield the highest tensile strength (17.74 MPa) and flexural strength (31.24 MPa). A higher onset degradation temperature (259 °C) was recorded for 25:75 Sw/SPF hybrid composites compared to the singular seaweed composites (253 °C).

6.5.5 Cellulose reinforced sugar palm starch biocomposites

The extraction of sugar palm cellulose from SPF reduces the size by four times, providing more effective molecular bonding between starch matrix and fiber due to higher surface area. According to Sanyang et al. [19], the incorporation of 1 wt% of sugar palm cellulose (SPC) into the neat SPS films enhanced the tensile strength from 7.79 MPa to 10.5 MPa and tensile modulus from 20.11 MPa to 31.38 MPa. Good compatibility between SPC and SPS matrices formed adequate interfacial adhesion enhancing the tensile properties. However, the composite films became stiffer and less elastic due to the reduction in molecular mobility with the introduction of SPC. Remarkably, WVP was significantly improved by 63.53 % compared to the neat SPS films. The dispersed SPC established a tortuous path in the composite films limiting the diffusion of water molecules to pass through it. In food packaging application, it is crucial to minimize the WVP to avoid moisture transfer that might quicken the food spoilage.

6.5.6 Microcrystalline cellulose reinforced sugar palm starch

There have been number of documentations on the reinforcement of MCC in various polymer matrices. Sartika et al. [22] used Schweizer reagent to dissolve the MCC extracted from SPF to enhance avocado seed starch-based films. The ratio of starch and MCC was 7:3, 8:2, and 9:1 whereas glycerol was incorporated at the variation of 0.1 %–0.4 %. The optimum tensile strength of 20.874 MPa was recorded at the starch MCC ratio of 7:3 with only 0.1 % of glycerol. Though, higher content of MCC induced stiffer composites with elongation at break of 6.22 %. Through these results, a good compatibility between SPS and MCC can be achieved due to their polar group (–OH) around the plasticizer chain. Plus, SPS and MCC from the same botanical origin have the same chemical composition might further enhance the affinity.

6.5.7 Sugar palm nanocellulose reinforced sugar palm starch

Nanocellulose can be found abundantly in natural plants [32]. These bio-based materials have been adapted in many applications, from medical treatment of scaffold tissue regeneration to optoelectronic devices of blue laser, LED traffic lights, photo diodes, optical fiber, solar cells, and telecommunication laser. The physical and chemical properties of the polymer matrix and the interaction formed between the nanoparticles and polymer matrix determine the properties of the nanocomposites. The use of nano-sized materials instead of micro-sized particles presents huge advantages owing to their large surface area, abundant surface hydroxyl groups, high aspect ratio, high mechanical strength, high thermal resistance, high crystallinity, and low defect rate of reinforcing part.

Ilyas et al. [33] conducted a study on the development and characterization of sugar palm nanocrystalline cellulose reinforced sugar palm starch bionanocomposites (Figure 6.4). The isolation of sugar palm nanocrystalline cellulose (SPNCCs) from sugar palm fiber (SPF) underwent sodium chlorite ($NaClO_2$) and sodium hydroxide (NaOH) treatments and subsequently acid hydrolysis. Using solution casting method, sugar palm starch (SPS) was incorporated with sorbitol/glycerol and reinforced with various loading of nanofiller SPNCCs (0–1.0 wt%) to develop bionanocomposites films. Characterization

Figure 6.4: Sugar palm nanocrystalline cellulose reinforced sugar palm starch bionanocomposites.

of the resulting fibers and nanocomposites were performed, which includes morphology (FESEM and TEM), footprint, crystallinity (XRD), light transmittance, biodegradability, mechanical, physical, thermal (TGA, DSC, and DMA), and water barrier properties. The length (*L*), diameter (*D*), and *L*/*D* values of the SPNCCs were 130 ± 30.23, 8.5 ± 1.82 nm, and 15.3, respectively. In comparison to neat SPS films, SPS/SPNCCs nanocomposite films demonstrated higher tensile strength, Young's modulus, thermal, water-resistance, and crystallinity. The improvement in tensile strength and moduli verified the impact of SPNCCs reinforcement into the bionanocomposites with the optimum loading of 0.5 %.

Ilyas et al. [33] reported an increment of WVP after SPS film was reinforced with sugar palm crystalline nanocellulose (SPNCC). By altering the SPNCC into smaller size, a more complex structure of nanocomposite was formed further obstructing the diffusion of water molecules as recorded by SPS/SPNCC-30, SPS/SPNCC-45, and SPS/SPNCC-60 with the improvement of 11.17 %, 14.72 % and 19.94 %, respectively. In term of biodegradation rate, neat SPS films experience 61.94 % weight loss followed by SPS/SPNCC-30 (56.88 %), SPS/SPNCC-45 (55.74 %), and SPS/SPNCC-60 (52.61 %) after 7 days being buried in compost soil. As water is the main medium for the microbial degradation activity, higher moisture is more prone to the microbial invasion thus leading towards more rapid decomposition of the biomaterial. Furthermore, the presence of crystalline region in the nanocomposites was reported to resist degradation compared to starch [34]. Figure 6.5 elucidates the image of neat SPS and SPNCC reinforced SPS films before and after being buried for 7 days.

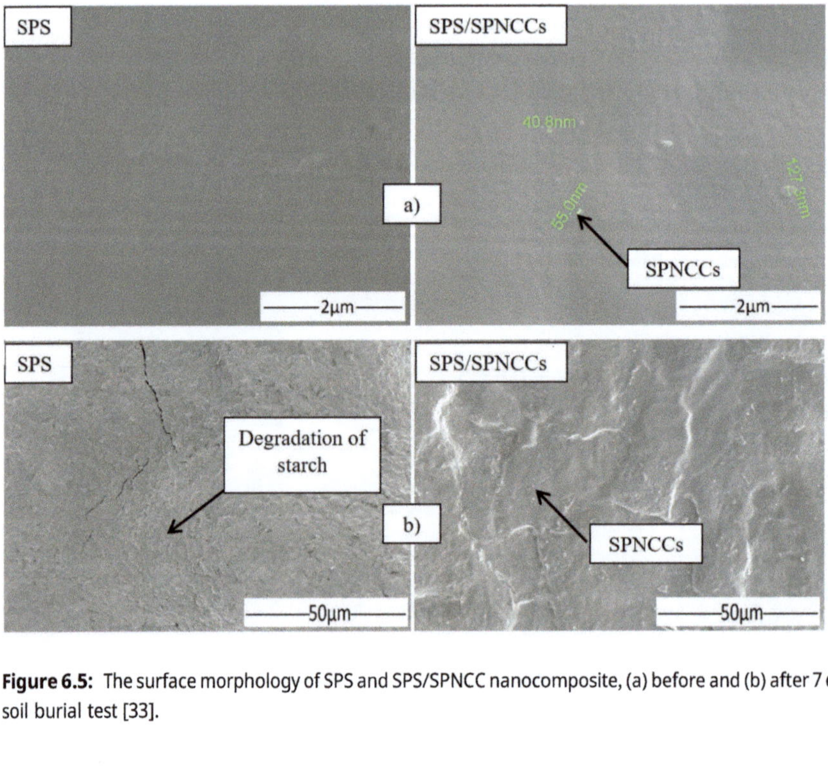

Figure 6.5: The surface morphology of SPS and SPS/SPNCC nanocomposite, (a) before and (b) after 7 days of soil burial test [33].

The demand for antibacterial active packaging is escalating due to the needs to improve food quality and shelf life. Furthermore, plastic waste is one of the world's concerns for human health. Biocomposites have attracted attention among the researchers, due to their environmental friendliness and sustainable nature [35]. Starch is an alternative to substitute petroleum based plastic film [36]. There are several limiting factors in using starch, which are fragile and brittle nature of the starch. Plasticizers were added to the starch film-forming solutions to help overcome the unplasticized starch films' fragile and brittle nature. Besides, cinnamon essential oil was incorporated as an antibacterial agent in the biocomposite films to form antibacterial active packaging. In order to transform into the high performance of antibacterial active packaging, cinnamon essential oil incorporated sugar palm nanocellulose reinforced sugar palm starch biopolymer composites were casted by using solution-casting method to investigate the properties of biocomposite films. The characterizations of biocomposite films was performed using field emission scanning electron microscopic (FESEM), thermal gravimetric analysis (TGA), Fourier transform infrared spectroscopic (FTIR), density, antibacterial activity, and seal and tear strength. The effect of various types of plasticizer with different concentrations (glycerol, sorbitol, and (glycerol + sorbitol) at 0–4.5 %wt) on the mechanical, physical, water barrier, and biodegradable properties were evaluated. It shows water absorption was increased respective to the plasticizer concentration. Besides, water vapor permeability (WVP) and solubility of the different concentration plasticizer used in the biopolymer shows an increasing trend due to high water content. Water content for sample glycerol (G-plasticized), sorbitol (S-plasticized), and mixture of glycerol and sorbitol (GS-plasticized) were shown increasing trend, which were from 14.86-30.30 %, 12.36–15.00 %, and 13.58–35.04 % respective to their concentration. Film of higher plasticizer content showed greater thickness and moisture content values. In contrast, density of plasticized films was decrease as plasticizer content was increased. Three types of essential oil, i.e. cinnamon, eucalyptus, and rosemary oils were tested for antibacterial activity, where cinnamon essential oil (CEO) showed the highest inhibition zone. CEO was chosen as antibacterial agent in the biocomposite films. The effect of various concentration of CEO (0–2.0 wt%) on the mechanical, morphological, flammability, thermal, and antibacterial activity were evaluated. The effectiveness of antibacterial packing is determined by the diameter of the zone of inhibition; the wider the diameter, the more sensitive the bacterium to the packaging. In attempt to evaluate the antibacterial property, agar disc method (ADM) was deployed to observe the inhibition zone of the films on the Gram-positive bacteria, Gram-negative bacteria, and yeast. The result suggested that the optimum amount of CEO used (2.0 wt%) displayed positive result of inhibition response. An observation towards the antimicrobial activity of the films incorporated with 2.0 wt% showed inhibition zones of 7.85 mm, 6.63 mm, and 7.43 mm for *Bacillus subtilis*, *Staphylococcus aureus*, and *Escherichia coli*, respectively. The cross-sections of the CEO-containing films images through scanning electron microscopy elucidate the presence of micro-porous spots associated with the partial evaporation occurred on the cryo-fractured surface due to the vacuum environment. As higher loading of CEO

Figure 6.6: Antimicrobial activity of sugar palm based nanocellulose/starch biocomposite films incorporated with cinnamon essential oil Syafiq et al. [37].

incorporated, greater number and size of the micro-porous holes were observed. The thermal stability was significantly improved by the fusion of CEO. SPS/SPNCC/CEO biopolymer composite films indicated a remarkable enhancement in thermal stability (sample 5, Tmax = 296.25 °C) compared to the unfilled compound (sample 1, Tmax = 289.18 °C). FESEM micrograph shows the films were slightly yellowish and transparent films. Overall, cinnamon essential oil/sugar palm nanocellulose/starch biocomposite films have good potential for active food packaging applications (Figure 6.6).

Meanwhile, Syafiq et al. [37, 38] found out that the addition of cinnamon essential oil (CEO) into SPS/SPNCC not only showed antimicrobial effect but slightly improved tensile strength. At 0.8 wt% of CEO concentration, the bioactive film evaluation towards Gram-positive bacteria namely *B. subtilis*, *S. aureus* and *E. coli* demonstrated an inhibition zone of 1.63 mm, 0.84 mm, and 1.32 mm, respectively. The increment in CEO concentration significantly enhanced the thermal stability of the film. The occurrence is associated with the rearrangement of polymer molecules due to the insertion of CEO and with higher concentration, the crystallinity index also increased parallel with thermal stability.

6.5.8 Sugar palm nanocellulose reinforced sugar palm starch/ polylactic acid blend biocomposites

In recent years, the escalating of petroleum-based plastic wastes had urged researchers to develop alternative biodegradable packaging plastics derived from bioresources. Traditional plastics are non-biodegradable associated with environment pollutions endangering human and animal life. In study conducted by Nazrin et al. [27], sugar palm crystalline nanocellulose (SPCNC) reinforced thermoplastic sugar palm starch (TPS)/poly (lactic acid) (PLA) blend biocomposites were prepared using melt blending and compression molding (Figure 6.7). Prior to melt blending process, the dispersion of SPCNC at a fixed loading (0.5 %) within TPS was achieved via sonication which later proceed with solution casting method to produce TPS biocomposite films. The PLA and TPS were blended into 5 different ratios (PLA/TPS; 20:80, 40:60, 60:40, 70:30, and 80:20) using Brabender mixer. The effectiveness of SPCNC reinforcement based on the compatibility of each composition was evaluated using scanning electron microscopic (SEM) analysis, Fourier transform infrared spectroscopy (FTIR), thermal gravimetric analysis (TGA), X-ray diffraction (XRD) analysis, dynamic mechanical analysis (DMA), limiting oxygen index (LOI), tensile, flexural, impact strength, thickness swelling, water solubility, water absorption, water vapor permeability, flammability, and soil and aqueous environment biodegradation test. XRD patterns of all blends biocomposite samples elucidate amorphous scattering background due to the obstruction of PLA side groups to form hydrogen bond between starch molecule chains. Among all samples, PLA40TPS60 and PLA60TPS40 had the good compatibility as indicated by the minor agglomeration and crack lines through SEM images. FTIR spectra exhibited lower wavenumber shift in the O–H band compared to other samples. Thickness swelling recorded minor reduction from PLA40TPS60 (11.41 %) to PLA60TPS40 (10.72 %). Both samples were classified as HB in UL94 rating flammability test with PLA40TPS60 (19.2 %) had slightly higher LOI value compared to PLA60TPS40 (18.8 %). The PLA40TPS60 had the highest improvement in tensile, flexural, and impact strength, which are 75.47 %, 72.63 %, and 33.13 % respectively. At PLA60TPS40, the tensile, flexural and impact strength were improved by 20.89 %, 25.20 %, and 12.09 %, respectively. Storage modulus (E′) exhibited notable increment at PLA60TPS40 (53.2 %) while minor increment at PLA70TPS30 (10 %) and PLA80TPS20 (0.6 %). At 25 % weight loss, PLA60TPS40 recorded a degradation temperature of 307 °C while PLA70TPS30 was 298 °C. The water vapor permeability was reduced up to 1.02×10^{-11} g/m s Pa for PLA60TPS40 but increased at PLA70TPS30 (1.1×10^{-11} g/m s Pa). For the PLA60TPS40, significant reduction (46 %–69 %) was recorded in maximum water uptake in all mediums while soil degradation rate experienced insignificant increment (7.92 %). The PLA60TPS40 indicated water solubility value of 39.43 % corresponded to the proportion of TPS content. PLA60TPS40 displayed well dispersed TPS within PLA phase correspondingly to SPCNC distribution resulting insignificant reduction in mechanical, water barrier and thermal properties. In effort to overcome plastics

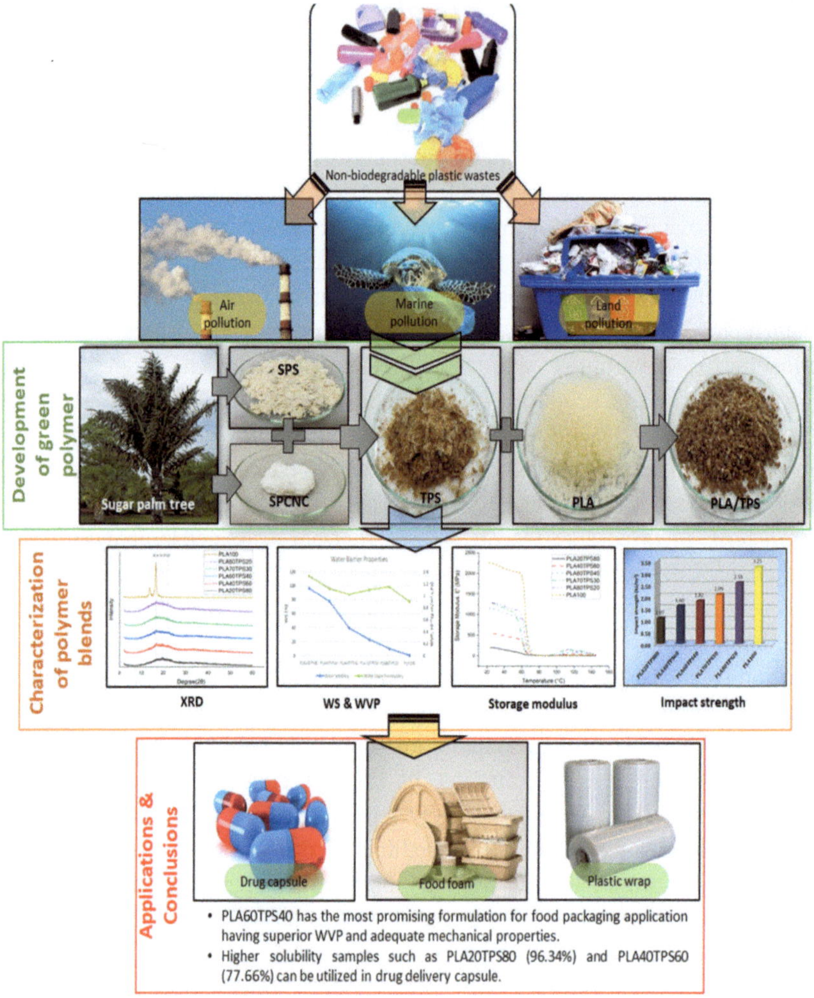

Figure 6.7: Sugar palm crystalline nanocellulose reinforced thermoplastic sugar palm starch (TPS)/ poly(lactic acid) (PLA) blend bionanocomposites [27].

pollution and reduce the over-reliance on depleting fossil resources, biodegradability, and low-cost material are prioritized through maximizing their functional properties.

6.6 Conclusions

For years, cellulose fiber was dominantly extracted from wood and cotton as raw material resource in textile, paper and pharmaceutical industries. The diversified application of cellulose into food, energy, biomedical, and even water treatment industries intensified the production rate making it necessary to promote an alternative resource rather than solely depending on forest woods. Hence, the utilization of cellulose from sugar palm fiber could promote the usage of lignocellulosic waste into renewable materials to satisfy the demand of raw materials and preserving those forest woods. The versatility of SPF showed great potential as matrix reinforcement for starch-based packaging products. Plasticization using polyols, SPS reinforcing with sugar palm fiber, cellulose, microcrystalline cellulose, or nanocellulose, blending with thermoplastic polymer such as polylactic acid, and addition of essential oil were all used to enhance the functional qualities of the starch biopolymer. Plasticizers of varying types and concentrations affected the physico-chemical properties of starch-based films in various ways. The final product may have a lower glass transition temperature (T_g), lower moisture content, and higher tensile strength. As the content of plasticizers grew, the glass transition temperature and water absorption ability decreased. Furthermore, the addition of sugar palm nanocellulose to SPS improves the performances of sugar palm starch-based films as a packaging material. Nanocellulose reinforcement will not resist natural degradation when exposed in decomposition environment, but it will maintain its essential functional properties for packaging application.

6.7 Challenges & future perspective

Sugar palm starch-films research is currently focused on tray, container, and food packaging applications. Sugar palm starch with a high amylose concentration has the potential to be a biopolymer that can be utilised in place of non-biodegradable synthetic thermoplastic polymers. Sugar palm starch is naturally weak and brittle; however, plasticizers such as glycerol and sorbitol substantially improved their characteristics. Aside from glycerol and sorbitol, several polyols have yet to be discovered. Furthermore, the use of essential oils such as cinnamon oil has been shown to provide antibacterial characteristics, somewhat better tensile strength, and greatly improved thermal stability to SPS composites. Researchers could continue to modify sugar palm starch with other types and concentrations of plasticizers, as well as essential oils, to broaden the functional potential of sugar palm starch.

Food packaging is a key aspect in preserving and maintaining the food quality. Recent advancements in smart food packaging and system are paving the way for more precise information to be sent not only to suppliers but, more significantly, to consumers in order for them to be alert with expiry date of food and minimise food waste. In fact, novel packaging methods can be developed to remind the quality of the food being packaged and allow for the detection of the degree of food deterioration, resulting in advantages in terms of food quality and safety. The development of novel and bio-based smart food packaging shall meet the demands of consumers while raising consciousness about the environmental problems due to excessive use of petroleum-based polymers. However, the major constraint of this product is the raising in food packaging costs, as well as the fact that current food packaging are incompatible with digital platforms.

References

1. Sanyang ML, Ilyas RA, Sapuan SM, Jumaidin R. Sugar palm starch-based composites for packaging applications. In: Jawaid M, Swain S, editors. Bionanocomposites for Packaging Applications, 1st ed. Cham, Switzerland: Springer International Publishing; 2018:125–47 pp.
2. Sharma S, Sudhakara P, Omran AAB, Singh J, Ilyas RA. Recent trends and developments in conducting polymer nanocomposites for multifunctional applications. Polymers 2021;13:2898.
3. Azman MA, Asyraf MRM, Khalina A, Petrů M, Ruzaidi CM, Sapuan SM, et al. Natural fiber reinforced composite material for product design: a short review. Polymers 2021;13:1917.
4. Nurazzi NM, Khalina A, Chandrasekar M, Aisyah HA, Rafiqah SA, Ilyas RA, et al. Effect of fiber orientation and fiber loading on the mechanical and thermal properties of sugar palm yarn fiber reinforced unsaturated polyester resin composites. Polimery 2020;65:115–24.
5. Asrofi M, Sapuan SM, Ilyas RA, Ramesh M. Characteristic of composite bioplastics from tapioca starch and sugarcane bagasse fiber: effect of time duration of ultrasonication (Bath-Type). Mater Today Proc 2021;46: 1626–30.
6. Bangar SP, Harussani MM, Ilyas RA, Ashogbon AO, Singh A, Trif M, et al. Surface modifications of cellulose nanocrystals: processes, properties, and applications. Food Hydrocolloids 2022;130:107689.
7. Nazrin A, Sapuan SM, Zuhri MYM, Tawakkal ISMA, Ilyas RA. Flammability and physical stability of sugar palm crystalline nanocellulose reinforced thermoplastic sugar palm starch/poly (lactic acid) blend bionanocomposites. Nanotechnol Rev 2022;11:86–95.
8. Norfarhana AS, Ilyas RA, Ngadi N. A review of nanocellulose adsorptive membrane as multifunctional wastewater treatment. Carbohydrate Polym 2022;291:119563.
9. Syafri E, JamaluddinSari NH, Mahardika M, Amanda P, Ilyas RA. Isolation and characterization of cellulose nanofibers from Agave gigantea by chemical-mechanical treatment. Int J Biol Macromol 2022; 200:25–33.
10. Rhim JW, Park HM, Ha CS. Bio-nanocomposites for food packaging applications. Prog Polym Sci 2013;38: 1629–52.
11. Asyraf MRM, Ishak MR, Syamsir A, Nurazzi NM, Sabaruddin FA, Shazleen SS, et al. Mechanical properties of oil palm fibre-reinforced polymer composites: a review. J Mater Res Technol 2022;17:33–65.
12. Atikah MSN, Ilyas RA, Sapuan SM, Ishak MR, Zainudin ES, Ibrahim R, et al. Degradation and physical properties of sugar palm starch/sugar palm nanofibrillated cellulose bionanocomposite. Polimery/ Polymers 2019;64:27–36.
13. Abral H, Basri A, Muhammad F, Fernando Y, Hafizulhaq F, Mahardika M, et al. A simple method for improving the properties of the sago starch films prepared by using ultrasonication treatment. Food Hydrocolloids 2019;93:276–83.
14. Ilyas RA, Sapuan SM, Ibrahim R, Abral H, Ishak MR, Zainudin ES, et al. Effect of sugar palm nanofibrillated cellulose concentrations on morphological, mechanical and physical properties of biodegradable films based on agro-waste sugar palm (Arenga pinnata (Wurmb.) Merr) starch. J Mater Res Technol 2019;8: 4819–30.
15. Jumaidin R, Sapuan SM, Jawaid M, Ishak MR, Sahari J. Characteristics of thermoplastic sugar palm starch/ agar blend: thermal, tensile, and physical properties. Int J Biol Macromol 2016;89:575–81.
16. Zou W, Yu L, Liu X, Chen L, Zhang X, Qiao D, et al. Effects of amylose/amylopectin ratio on starch-based superabsorbent polymers. Carbohydrate Polym 2012;87:1583–8.
17. Sahari J, Sapuan SM, Zainudin ES, Maleque MA. Physico-chemical and thermal properties of starch derived from sugar palm tree (Arenga pinnata). Asian J Chem 2014;26:955–9.
18. Sahari J, Sapuan SM, Zainudin ES, Maleque MA. Thermo-mechanical behaviors of thermoplastic starch derived from sugar palm tree (Arenga pinnata). Carbohydrate Polym 2013;92:1711–6.

19. Sanyang ML, Sapuan SM, Jawaid M, Ishak MR, Sahari J. Effect of sugar palm-derived cellulose reinforcement on the mechanical and water barrier properties of sugar palm starch biocomposite films. Bioresources 2016;11:4134–45.
20. Ishak MR, Sapuan SM, Leman Z, Rahman MZAA, Anwar UMKK, Siregar JP. Sugar palm (Arenga pinnata): its fibres, polymers and composites. Carbohydrate Polym 2013;91:699–710.
21. Bachtiar D, Sapuan SM, Zainudin ES, Khalina A, Dahlan KZM. The tensile properties of single sugar palm (Arenga pinnata) fibre. IOP Conf Ser Mater Sci Eng 2010;11:012012.
22. Sartika M, Lubis M, Harahap MB, Afrida E, Ginting MHS. Production of bioplastic from avocado seed starch as matrix and microcrystalline cellulose from sugar palm fibers with schweizer's reagent as solvent. Asian J Chem 2018;30:1051–6.
23. Ventura-Cruz S, Tecante A. Nanocellulose and microcrystalline cellulose from agricultural waste: review on isolation and application as reinforcement in polymeric matrices. Food Hydrocolloids 2021;118: 106771.
24. Cao X, Chen Y, Chang PR, Muir AD, Falk G. Starch-based nanocomposites reinforced with flax cellulose nanocrystals. Express Polym Lett 2008;2:502–10.
25. Ilyas RA, Sapuan SM, Ishak MR. Isolation and characterization of nanocrystalline cellulose from sugar palm fibres (Arenga Pinnata). Carbohydrate Polym 2018;181:1038–51.
26. Sanyang M, Sapuan S, Jawaid M, Ishak M, Sahari J. Effect of plasticizer type and concentration on tensile, thermal and barrier properties of biodegradable films based on sugar palm (Arenga pinnata) starch. Polymers 2015;7:1106–24.
27. Nazrin A, Sapuan SM, Zuhri MYM, Tawakkal ISMA, Ilyas RA. Water barrier and mechanical properties of sugar palm crystalline nanocellulose reinforced thermoplastic sugar palm starch (TPS)/poly(lactic acid) (PLA) blend bionanocomposites. Nanotechnol Rev 2021;10:431–42.
28. Nazrin A, Sapuan SM, Zuhri MYM, Tawakkal ISMA, Ilyas RA. Flammability and physical stability of sugar palm crystalline nanocellulose reinforced thermoplastic sugar palm starch/poly(lactic acid) blend bionanocomposites. Nanotechnol Rev 2021;11:86–95.
29. Sanyang ML, Sapuan SM, Jawaid M, Ishak MR, Sahari J. Development and characterization of sugar palm starch and poly(lactic acid) bilayer films. Carbohydrate Polym 2016;146:36–45.
30. Ibrahim MIJ, Sapuan SM, Zainudin ES, Zuhri MYM. Preparation and characterization of cornhusk/sugar palm fiber reinforced Cornstarch-based hybrid composites. J Mater Res Technol 2020;9:200–11.
31. Jumaidin R, Sapuan SM, Jawaid M, Ishak MR, Sahari J. Thermal, mechanical, and physical properties of seaweed/sugar palm fibre reinforced thermoplastic sugar palm Starch/Agar hybrid composites. Int J Biol Macromol 2017;97:606–15.
32. Sabaruddin FA, Paridah MT, Sapuan SM, Ilyas RA, Lee SH, Abdan K, et al. The effects of unbleached and bleached nanocellulose on the thermal and flammability of polypropylene-reinforced kenaf core hybrid polymer bionanocomposites. Polymers 2020;13:116.
33. Ilyas RA, Sapuan SM, Ishak MR, Zainudin ES. Development and characterization of sugar palm nanocrystalline cellulose reinforced sugar palm starch bionanocomposites. Carbohydrate Polym 2018;202: 186–202.
34. Alvarez VA, Ruseckaite RA, Vázquez A. Degradation of sisal fibre/Mater Bi-Y biocomposites buried in soil. Polym Degrad Stabil 2006;91:3156–62.
35. Nurazzi NM, Sabaruddin FA, Harussani MM, Kamarudin SH, Rayung M, Asyraf MRM, et al. Mechanical performance and applications of CNTs reinforced polymer composites—a review. Nanomaterials 2021;11: 2186.
36. Diyana ZN, Jumaidin R, Selamat MZ, Ghazali I, Julmohammad N, Huda N, et al. Physical properties of thermoplastic starch derived from natural resources and its blends: a review. Polymers 2021; 13:1–20.

37. Syafiq R, Sapuan SM, Zuhri MRM. Antimicrobial activity, physical, mechanical and barrier properties of sugar palm based nanocellulose/starch biocomposite films incorporated with cinnamon essential oil. J Mater Res Technol 2021;11:144–57.
38. Syafiq RMO, Sapuan SM, Zuhri MRM. Effect of cinnamon essential oil on morphological, flammability and thermal properties of nanocellulose fibre–reinforced starch biopolymer composites. Nanotechnol Rev 2020;9:1147–59.

R. M. O. Syafiq, S. M. Sapuan*, M. Y. M. Mohd Zuhri, S. H. Othman and R. A. Ilyas

7 Morphological, water barrier and biodegradable properties of sugar palm nanocellulose/starch biopolymer composites incorporated with cinnamon essential oils

Abstract: In the past few decades, there has been increased interest in the use of natural fibers as reinforcement in bioplastic polymer composites because it is biodegradable. This is a result of the drawbacks of biodegradable polymer-based materials, which are brittle, intractable, and poorly water-sensitive. Natural fibers are chosen because they may be obtained organically, can be collected safely, and can be less expensive. In this work, cinnamon essential oil (CEO)/sugar palm nanocellulose/starch biopolymer composites were casted to investigate the morphological, water barrier and biodegradable properties. From the research, it shows water absorption increased regarding the plasticizer concentration. Besides, water vapor permeability (WVP) and solubility of the different concentration plasticizer used in the biopolymer shows an increasing trend due to high water content. All films degrade completely after the 12th day indicating the biodegradability of the film. Furthermore, seal strength for the

***Corresponding author: S. M. Sapuan**, Laboratory of Biocomposite Technology, Institute of Tropical Forestry and Forest Products (INTROP), Universiti Putra Malaysia, UPM, 43400 Serdang, Selangor, Malaysia; and Department of Mechanical and Manufacturing Engineering, Advanced Engineering Materials and Composites Research Centre (AEMC), Universiti Putra Malaysia, UPM, 43400 Serdang, Selangor, Malaysia, E-mail: sapuan@upm.edu.my

R. M. O. Syafiq, Laboratory of Biocomposite Technology, Institute of Tropical Forestry and Forest Products (INTROP), Universiti Putra Malaysia, UPM, 43400 Serdang, Selangor, Malaysia

M. Y. M. Mohd Zuhri, Laboratory of Biocomposite Technology, Institute of Tropical Forestry and Forest Products (INTROP), Universiti Putra Malaysia, UPM, 43400 Serdang, Selangor, Malaysia; and Department of Mechanical and Manufacturing Engineering, Advanced Engineering Materials and Composites Research Centre (AEMC), Universiti Putra Malaysia, UPM, 43400 Serdang, Selangor, Malaysia. https://orcid.org/0000-0002-1069-7345

S. H. Othman, Department of Process and Food Engineering, Universiti Putra Malaysia, UPM, 43400 Serdang, Selangor, Malaysia

R. A. Ilyas, Laboratory of Biocomposite Technology, Institute of Tropical Forestry and Forest Products (INTROP), Universiti Putra Malaysia, UPM, 43400 Serdang, Selangor, Malaysia; Department of Chemical Engineering, Faculty of Chemical and Energy Engineering, Universiti Teknologi Malaysia, 81310 Johor Bahru, Johor, Malaysia; Center of Advanced Centre for Advanced Composite Materials (CACM), Universiti Teknologi Malaysia, 81310 Johor Bahru, Malaysia; and Centre of Excellence for Biomass Utilization, Universiti Malaysia Perlis, 02600, Arau, Perlis, Malaysia. https://orcid.org/0000-0001-6622-2632

As per De Gruyter's policy this article has previously been published in the journal Physical Sciences Reviews. Please cite as: R. M. O. Syafiq, S. M. Sapuan, M. Y. M. Mohd Zuhri, S. H. Othman and R. A. Ilyas "Morphological, water barrier and biodegradable properties of sugar palm nanocellulose/starch biopolymer composites incorporated with cinnamon essential oils" *Physical Sciences Reviews* [Online] 2023. DOI: 10.1515/psr-2022-0029 | https://doi.org/10.1515/9783110773606-007

lower concentrations of plasticizer shows the higher strength, while GS4.5 cannot be sealed due to high water contents. The images show the compatible films with slightly yellowish and transparent films. The variation of plasticizers did not affect the antibacterial activity of CEO inside the film forming solution. Overall, cinnamon essential oil (CEO)/sugar palm nanocellulose/starch biopolymer composites are good potential to enhance their suitability for food packaging applications.

Keywords: biodegradable films; cinnamon essential oil (CEO); soil burial; sugar palm nanocellulose; water barrier properties.

7.1 Introduction

Edible films made from natural materials have gotten a lot of attention recently, rising consumer awareness of environmental conservation and food safety forced this research area to be grown [1–5]. Due to their advantages, it can produce suitable source, environmentally friendly, more economical, lightweight, and simplified degradable. Besides, by minimizing negative reaction such as enzymatic, physical, and chemical reactions can cause the films improve the food shelf life [6–9]. As a natural high molecular weight polysaccharide, starch has the advantages of low cost, biocompatibility, and ease of conversion into thermoplastic polymers [10–12]. It can also be employed as a carrier for bioactive chemicals to improve packaging effectiveness, and it has a lot of potential for food packaging material development [13–15]. At room temperature, natural starch is insoluble in water and has a propensity to retrograde. It has been reported that starch edible films still minor utilization as food packaging [16–20]. Nonetheless, there are limiting factor associated with natural polymer matrices that are associated with water absorption due to bundle of hydroxyl group in their structure [8, 21]. Moisture build-up in biocomposites can result in physical degradation, such as dimensional changes and thickness swelling, as well as a reduction in the film's mechanical properties. The fibers' adherence to the matrix may be reduced as a result of the thickness swelling. Water absorption alters the characteristics of biocomposites in an unfavorable effect, so that so, water absorption were carried out to ensure that the quality of bionanocomposite materials.

Sugar palm fiber gained attention from researchers due to its easy availability, which can readily isolated the lignin and cellulose from its fiber components [22–25]. An environmental-friendly biocomposites can be obtained from this single tree. Sugar palm starch (SPS) and sugar palm nanocellulose (SPNCC) were obtained from the trunk of sugar palm tree [8, 9]. Plasticizers are low-volatility chemicals that are added to biopolymer materials to change their functional properties by increasing their mechanical properties, barrier properties, flexibility, elasticity, stiffness, and brittleness [12, 26–28]. Plasticizing hydrophilic polymers with polyols such as glycerol and sorbitol has been reported to be particularly effective [12, 18, 19, 29, 30]. Plant essential oils (EO) contain functional and active substances so which encourages the researcher to incorporate EO into the polymer matrix [8–10, 31]. Essential oils and their active components offer a wide range of biological effects, including antioxidant, antibacterial, and antifungal properties, and they were considered "generally recognized as safe" by the U.S. Food and Drug Administration, FDA (GRAS). Although most EO is regarded safe, their usage as food packaging polymer matrix is limited to their strong aroma. To produce the bioactive benefits to the nanocomposite films, however, a sufficient amount of EO was required. Rather than simply applying the EO to the food product, incorporation EO to the biopolymer films allows the active components to diffuse slowly, causing the most bioactive effect at a low concentration of EO. As a result, unfavorable odors emitted to the food would be reduced [32]. The volatile oil is derived from the bark, leaves, and fruit of cinnamon also known as cinnamon essential oil (CEO), and its main constituents are cinnamaldehyde and eugenol [8, 9, 32, 33]. Due to its hydrophobicity, to achieve a homogenous distribution of oil particles and preserve the homogenous of the film matrix it is necessary to dissolve the CEO in surfactants such as Twee 80 that contained high hydrophilic-lipophilic balance (HLB) [8, 34].

In this study, SPS/SPNCC reinforced with different plasticizers and concentrations also incorporated with (2 wt%) of CEO were produced by solution casting method. The goal of this work is to study the properties of cinnamon essential oil (CEO)/sugar palm nanocellulose/starch biopolymer composites as a potential to enhance their suitability for food packaging applications.

7.2 Materials and methods

7.2.1 Materials

Sugar palm fiber and starch were extracted from sugar palm trees in Kuala Jempol, Negeri Sembilan, Malaysia. Plasticizers: (Glycerol [99% purity, food-grade], D-sorbitol [with 99% purity]), CEO, and emulsifier Tween 80 were supplied by Evergreen Sdn. Bhd., Semenyih, Malaysia.

7.2.2 Methods

7.2.2.1 SPS extraction and preparation methods: A chainsaw was used to remove SPS from within the stem of a mature sugar palm tree. The starch powder was then washed by adding water to the mixture, and the starch was extracted from the mixture using a specially developed extractor machine for sugar palm. The mixture was then filtered through a (<300 μm) sieve, with the fiber remaining at the top of the sieve and the starch granules flowing into the container with the water. Because starch is denser than water, it was separated from the water by gently pouring the water until it reached the level of the starch. Wet starch was used to extract fiber residues that are by-products. The wet starch was then sun-dried for 30 min before being oven-dried for 24 h at 120 °C [35].

Table 7.1: Denotations of the film's sample.

Denotation of the film	Formulation				
	Sorbitol (%)	Glycerol (%)	CEO (%)	Tween 80 (%)	Starch/SPNCC (%)
G1.5	0	1.5	2	0.05	10/0.05
G3.0	0	3.0	2	0.05	10/0.05
G4.5	0	4.5	2	0.05	10/0.05
S1.5	0	1.5	2	0.05	10/0.05
S3.0	0	3.0	2	0.05	10/0.05
S4.5	0	4.5	2	0.05	10/0.05
GS1.5	0.75	0.75	2	0.05	10/0.05
GS3.0	1.5	1.5	2	0.05	10/0.05
GS4.5	2.25	2.25	2	0.05	10/0.05

7.2.2.2 Fabrication of SPS/SPNCC incorporated CEO nanocomposite films: The denotations of sugar palm-based films made using the conventional solution-casting technique is shown in Table 7.1. To evaluate the effect of each plasticizer on SPS films, glycerol (G), sorbitol (S), and their 1:1 combination (GS) were utilized as plasticizers. The following are the steps in the film preparation process: Initially, a 10% (w/w) aqueous dispersion of gelatinized sugar palm starch and 0.05% (w/w) sugar palm nanocellulose (SPNCC) were prepared by sonicating them in a sonicator machine for dispersed the SPNCC. The film-forming solution was then heated in a hot water bath at 95 ± 2 °C for 15 min with steady stirring. By dissolving the starch granules, this step helps to achieve homogenous dispersion. Following that, the various plasticizers were added to the dispersion at 0, 1.5, 3.0, or 4.5% (w/w). The heating process was maintained at 95 ± 2 °C for another 15 min. Next, CEO and emulsifier (Tween 80) were added into the dispersions at 2% and 0.05% (w/w, starch basis), respectively. Before being cast in glass petri-dishes, the film forming-solutions were let to cool. The smooth and flat surface of the film is due to the glass petri-dishes used as a casting surface. To allow evaporation, the freshly cast films were heated in a 40 °C oven. All of the films were made in triplicate. Films were peeled off the casting surfaces after 24 h drying and stored in desiccators with relative humidity (53 ± 1%).

7.2.2.3 Film thickness and density of SPS/SPNCC incorporated CEO nanocomposite films: A digital micrometer with 0.001 mm sensitivity (Mitutoyo Co., Japan) was used to measure the thickness of the film. For each sample, thickness measurements were taken from five separate film sections. The average of the measurements for each sample was calculated.

A densimeter was used to determine the density of the films (Mettler-Toledo (M) Sdn. Bhd). To prevent water uptake by the hydrophilic film samples, the immersing liquid utilized in this research was Xylene rather than distilled water. The density of the liquid must be lower than that of the film in order for the film to not float. As a result, due to its lower density, Xylene is far more suited than water. Before immersing the film sample in the liquid, it was weighed (m). (V) represents the amount of liquid displaced after immersing the film in the liquid. The density (ρ) was calculated using Eq. (7.1). The test was carried out four times and got the average.

$$\rho = m/V \tag{7.1}$$

7.2.2.4 Swelling index: Film samples (20 × 20 mm) were immersed in distilled water at room temperature (23 ± 2 °C) to determine the swelling index of each film. The film will be taken from the distilled water and weighed on a regular basis until it reaches a constant mass. The test performed in triplicate. The swelling index was determined using Eq. (7.2):

$$\text{Swelling index } (\%) = (M_{\text{Swollen}} - M_{\text{Initial}}/M_{\text{Initial}}) \times 100 \tag{7.2}$$

where M_{Initial} represents the initial while M_{Swollen} represents the swollen mass of the film sample. The method used for this testing was adopted from [36].

7.2.2.5 Water vapor permeability (WVP): The film samples were dried at room temperature in a desiccator with a relative humidity of 50%. The WVP test was investigated in accordance with standards ASTM E96 [37], with minor alterations in accordance with Sanyang et al. [38]. The experiment was carried out three times. The cup's 300 mm aperture was filled with 20 g of silica gel at first. Circular film samples were cut and placed over the cup's mouth, with about 3 mm of vacuum remaining at the top. Then it was carefully sealed. The test cup was then weighed before being placed in a constant relative humidity chamber (25 °C, relative humidity 75%). The mass gain values of the cups were measured on a regular basis until they reached equilibrium. The mass increments in the test cups were recorded, and WVP was calculated using Eq. (7.3).

$$\text{WVP} = (m \times d)/(A \times t \times P) \tag{7.3}$$

where m (g) is the test cup's weight increment, d (mm) is the film thickness, A (m^2) is the exposed film area, t (s) is the permeation time, and P (Pa) is the water vapor partial pressure across the films. The result expressed in g mm s^{-1} m^{-2} Pa^{-1}.

7.2.2.6 Tear and thermo-seal resistance strength: The ASTM D-1938 [39] standard method was used to assess the tear propagation resistance of the films, with some modifications [40]. Ten film samples were used to be examined in this experiment. Samples (25 × 75 mm) with a 12.5 mm slit were prepared according to the standard test method. Tensile tests were conducted using an Instron 3365 universal machine (Instron, USA) with a constant crosshead speed of 250 mm/min and a 50 mm gap between grips. The tests were performed at room temperature.

The thermo-sealer (PFS-200, 110 V–220 V, 300 W) was used to seal the packaging. Thermo-seals were produced using an impulse-wire thermo-sealer. This thermo-sealer was chosen because it allows cooling the seal before the jaws are opened, which enhanced the strength seal [41]. Furthermore, it is the most widely use method for flexible synthetic materials, and it involves heating a film sample to its fusion temperature with a short, intense electric pulse. Then, thermos-seal strength was evaluated in an INSTRON 3365 machine (Instron, USA) by tensile assay at a constant crosshead speed of 200 mm/min and the gap between grips 25 mm. according to ASTM F88-00 [42] standard method, two sample films were been sealed by a thermo-sealer at a wide range of 25.4 mm. For each type of film sample, 10 samples were been used to get an average value.

7.2.2.7 Water solubility: Each film yielded samples (30 mm × 10 mm) that were dried in a 105 °C oven for 24 h. The starting dry matter of each film was determined by weighing the samples (W_i). At temperature 23 ± 2 °C, each sample was submerged in 50 mL of distilled water for 6 h. The beaker was agitated using a magnetic stirrer periodically. The insoluble portion of the film sample was separated from the distilled water and dried for 24 h at 105 °C. The weight of the solubilized dry matter was determined by reweighing the oven-dried samples (W_o). Eq. (7.4) was used to determine the water solubility of each film.

$$\text{Solubility} (\%) = (W_i - W_o/W_i) \times 100 \tag{7.4}$$

7.2.2.8 Soil burial: Biodegradability was measured by the weight loss of the bionanocomposite buries in compost soil under moisture-controlled circumstances. A total of 30 × 10 mm samples were buried 100 mm into the prepared soil, which was moistened with distilled water on a regular basis. Each sample was retrieved from the compost soil, rinsed with water, and dried in a 60 °C oven to a uniform weight after being buried for 24, 48, 72, and 168 h. The following equation, Eq. (7.5) was used to calculate the weight loss:

$$\text{Weight loss}\,(\%) = (W_o - W_t/W_o) \times 100 \tag{7.5}$$

where, W_0 = weight before being buried, W_t = weight after being buried

7.2.2.9 Morphological properties: Using a field emission electron microscope (FEI NOVA NanoSEM 230, Czech Republic), the morphology of the films was examined. To avoid the unwanted charge, the entire sample was coated with gold using an argon plasma metallizer (sputter coater K575X) (Edwards Limited, Crawley, UK). A 1 kV acceleration voltage was used to conduct the FESEM experiment.

7.3 Results and discussion

7.3.1 Film thickness, density, and water content of SPS/SPNCC incorporated CEO nanocomposite films

Table 7.2 shows film thickness, density and water content of different plasticizer and concentrations. Depending on the plasticizer type, increasing the plasticizer concentration resulted in a significant rise in film thickness. Plasticizers may play a role in dissolving and organising intermolecular polymer chain networks, resulting in larger free volumes and hence thicker films. A similar finding of the plasticizer concentration effect on the film thickness was described by Sanyang et al. [43], Syafiq et al. [8], Razavi et al. [44], Hazrol et al. [30], and Ilyas et al. [36]. Moreover, using different plasticizer types revealed a significant effect on films' thickness, as presented in Table 7.2.

Table 7.2: Table of film thickness, density and water content of different plasticizer and concentrations.

Samples	Thickness (mm)	Density (g/cm^3)	Water content (%)
G1.5	0.104	2.04	14.86
G3.0	0.16	1.93	15.48
G4.5	0.204	1.65	30.30
S1.5	0.118	2.08	13.24
S3.0	0.16	1.98	15.00
S4.5	0.206	1.86	12.36
GS1.5	0.12	2.23	13.58
GS3.0	0.178	1.89	29.27
GS4.5	0.202	1.82	35.04

S-plasticized films were thicker than G-plasticized films, with GS-plasticized films being the thickest at 1.5 wt%. Since the formulation of the film-forming solution was constant, the differences in film thickness may be attributed to their molar masses. At constant concentration, G-plasticized films (molar mass 92.09 g/mol) were thinner than S-plasticized films (molar mass 182.17 g/mol), possibly due to the G-plasticized had lower molar mass. Sanyang et al. [43], Aitboulahsen et al. [45], and Ghasemlou et al. [46] stated that S-plasticized films produce higher thickness films than G-plasticized films. The density values of the various plasticizer kinds did not differ much. However, using the same plasticizer concentration, the following is the sequence of density reduction: GS- S- G-plasticized films.

7.3.2 Water absorption (swelling index) SPS/SPNCC incorporated CEO

Figures 7.1 and 7.2 show the water absorption and swelling index test result of the SPS/SPNCC incorporated CEO nanocomposite film. Because of their water-soluble and inherently hygroscopic qualities, increasing plasticizer concentration lowered water absorption in the films. The formation of hydrogen bonds with starch was attributed to hydroxyl groups present within the plasticizer molecules due to an increased tendency of water absorption into plasticized polymers [18]. Hydrophilic compounds are expected to induce film swelling, while hydrophobic ones are expected to reduce it [47]. The flexibility of plasticized sheets with higher plasticizer and moisture content was higher [30].

All nanocomposite films gained weight due to water absorption was linearly increased as duration increased and it becomes constant after 1 h, following the Fickian diffusion process. The film swells as water diffuses into the network chains. Figure 7.2 shows uptake-t curves indicate two separate zones, in which the absorption kinetic was very fast at interval $t < 1$ h, but sluggish at interval $t > 1$ h, resulting in

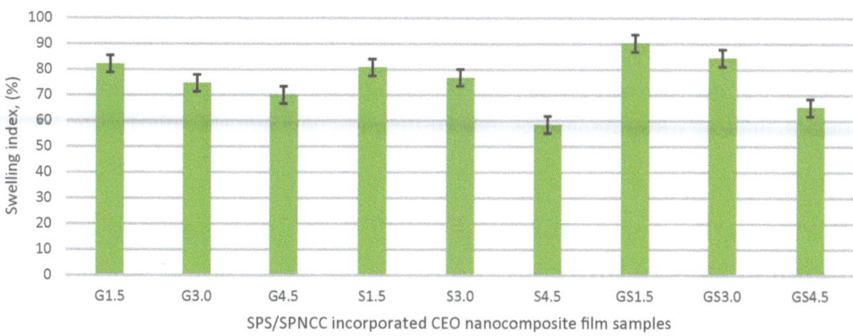

Figure 7.1: Swelling index of SPS/SPNCC incorporated CEO nanocomposite film samples.

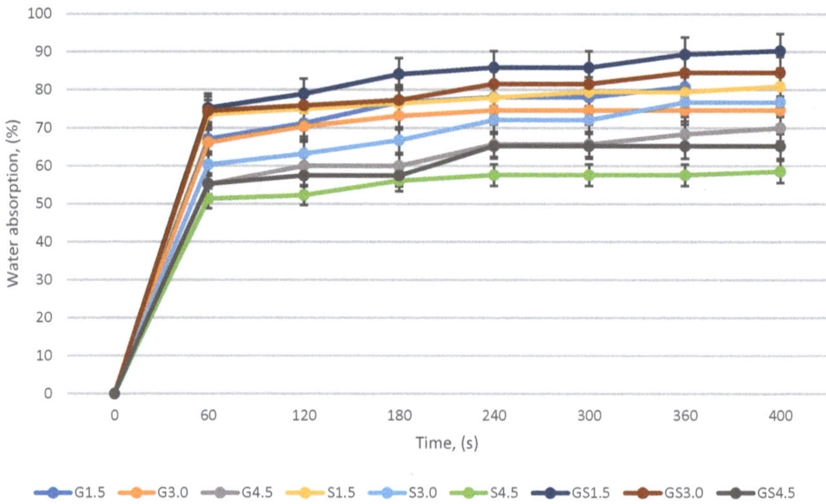

Figure 7.2: Water absorption of SPS/SPNCC incorporated CEO nanocomposite film samples.

equilibrium water uptake. This phenomenon could be caused by the abundance of active hydroxyl groups in films that were unfilled at the start of the absorption process. The active sites, on the other hand, were saturated with time and no longer accommodate any water molecules [43].

Moreover, the interaction between glycerol and starch may promote moisture adsorption by generating a channel that accelerated the accumulation of water molecules due to dimensional changes in the region where the adhesive force between the glycerol-starch surfaces was weak. It may be deduced that as the rate of plasticization increases inside the polymer structure, so does the rate of penetration, which nevertheless produces a barrier effect on water molecule movement at high moisture concentrations [36]. The formation of hydrogen bonds with starch was attributed to hydroxyl groups present within the plasticizer molecules due to an increased tendency of water absorption into plasticized polymers [18, 19]. When plasticizer creates hydrogen bonds with starch, it breaks the existing hydrogen bonds between the hydroxyl groups in the starch molecules. This is because new hydrogen bonds are formed between glycerol and hydroxyl groups [19]. As a result, the starch's plasticizing effect is strengthened, making water molecules more difficult to permeate the plasticized films.

7.3.3 Water vapor permeability of SPS/SPNCC incorporated CEO nanocomposite films

Water vapor permeability (WVP) is one of the most important factors to consider when selecting a food packaging material because it can affect the end product's

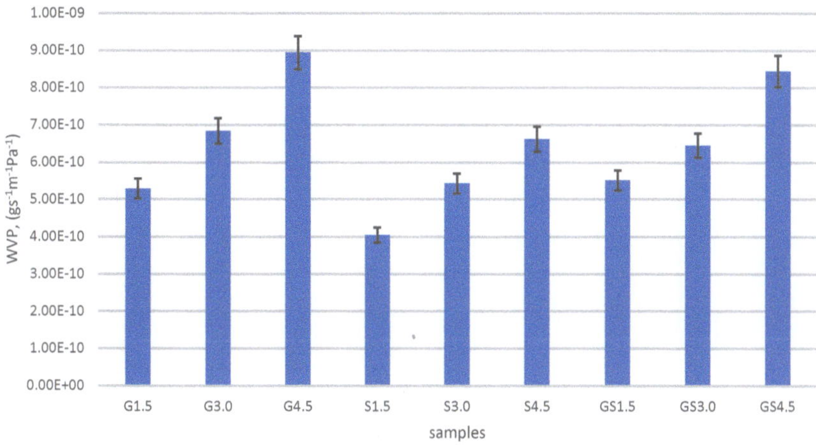

Figure 7.3: WVP of SPS/SPNCC incorporated CEO nanocomposite film samples.

performance, regardless of whether it is a water-sensitive product or not. Furthermore, food packaging applications requires low water vapor permeability included reducing and preventing moisture transfer between the environment and the food [38, 48]. As a result, reducing films' WVP is crucial for their wide range of applications. Figure 7.3 shows the WVP of SPS/SPNCC incorporated CEO nanocomposite film. The highest WVP, as can be observed, was the film that contained the highest content of plasticizer. This is due to their hydrophilic nature, as well as their sensitivity to moisture and humidity, which is a difficult element to control [49]. In addition, the 0.05 wt% of SPNCC decreased tortuous path caused by the presence of rigid structure and dispersion of SPNCC that prolong or hind the path of water molecules [50].

7.3.4 Tear and thermo-seal strength of SPS/SPNCC incorporated CEO nanocomposite films

Tear resistance also another key to issue in the packaging industry, as it provides information on the required strength to open by tearing the package. In general, different materials have varying tearing strengths. When the package was in proper shape, some of them may be quite resistant to tear. So that, pre-cut slit is prepared on the package as they become vulnerable and the strength required to tear was reduced. Figure 7.4a shows the tensile strength required to tear the film for each type of SPS/SPNCC incorporated CEO nanocomposite film. The tensile strength show-decreasing trend according to their concentration of plasticizer. When a cut slit was added to the edge of the films, the tensile strength to tear them was reduced, compared to the actual tensile strength, which is an average of (9–12 MPa) [9]. Figure 7.4b shows the tensile strength of the sealed package SPS/SPNCC incorporated CEO nanocomposite film at the seal area.

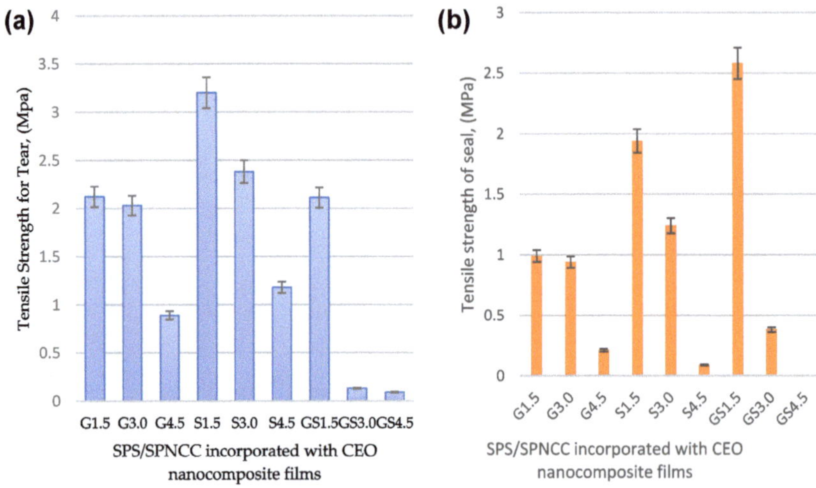

Figure 7.4: Tear and thermo strength of SPS/SPNCC incorporated CEO nanocomposite film. (a) Tensile strength for tear of SPS/SPNCC incorporated CEO nanocomposite film samples. (b) Tensile strength for seal of SPS/SPNCC incorporated CEO nanocomposite film samples.

The strength was decreased according to the increasing contents of the plasticizer. Higher seal strength produced a higher degree of tightness [40]. When the seal has shattered a piece of material corresponding to one of the films was taken out, resulting in cohesive failure mode, which showed for a sample that contain 4.5 wt% of plasticizer. GS4.5 shows no seal strength because the film cannot be sealed as the film is too moist as shown in Figures 7.7 and 7.8. The peeling off method regarding the fact that thermo-sealed films presented a cohesive failure, could easily open packaging bags based on this kind of bionanocomposites.

7.3.5 Solubility of SPS/SPNCC incorporated CEO nanocomposite films

The ability of a film to dissolve in water is an important factor to consider when choosing food packaging plastics. Most of food packaging applications require insolubility to provide water resistance and enhanced the shelf life of food products. Starch has limited applications due to its high water solubility [36]. As a reason, it is envisaged that SPS/SPNCC incorporated with CEO will be able to overcome this constraint. Figure 7.5 shows the films' solubility of SPS/SPNCC incorporated CEO. As seen in Figure 7.5, plasticization of SPS/SPNCC incorporated with CEO bio-nanocomposite films considerably increases film solubility in water. When biofilms were immersed in water, it was discovered that they did not remain intact. Hydrophilic nature of sorbitol and glycerol causes the solubility to increase, regarding their concentration of the

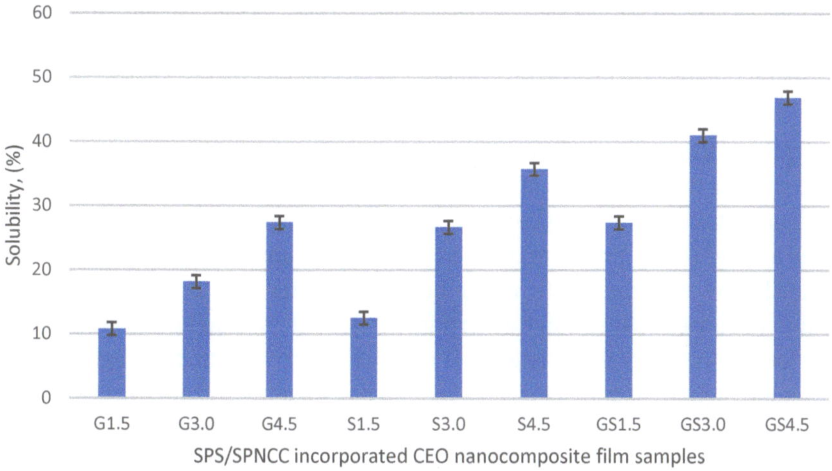

Figure 7.5: Solubility of SPS/SPNCC incorporated CEO nanocomposite film samples.

plasticizer. Glycerol is more hygroscopic than sorbitol, it shows higher solubility and is easier to incorporate into a network of hydrogen bonds. This change, however, could be due to either a variation of types and plasticizer concentrations. An edible film must have low solubility during storage [51]. The result also shows the same findings with Sanyang et al. [43].

7.3.6 Soil burial of SPS/SPNCC incorporated CEO nanocomposite films

The decomposition of materials caused by the activities of fungi, bacteria, and microorganisms, as well as other biological processes, is referred to as degradation rate. The

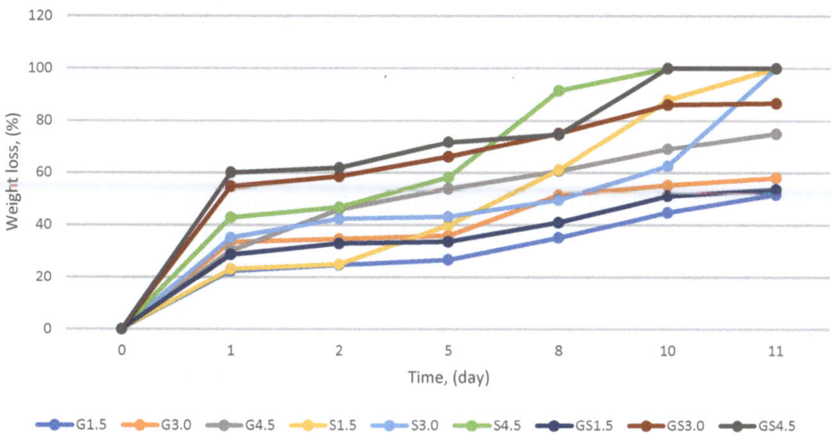

Figure 7.6: Weight loss of SPS/SPNCC incorporated CEO nanocomposite film samples.

degradation of the polymer begins when these microbial organisms get into touch with the biodegradable polymer [36, 52]. These microorganisms break down the polymer by an enzymatic or metabolic process, resulting in smaller compounds with a lower average molecular weight. As a result, material degradation in the environment was favoured. Mineralization process occurred when all the samples had been degraded. Figure 7.6 showed the weight loss of the SPS/SPNCC incorporated CEO nanocomposite after biodegradation testing. A higher concentration of plasticizer causes it to degrade faster than a lower concentration of plasticizer. The water absorption of the SPS/SPNCC incorporated CEO nanocomposite could be regarded for this occurrence. Faster degradation possesses higher water absorption. Figure 7.6, it shows 4.5 wt% of S- and GS-plasticized films totally degrade after 11 days while for S-plasticized degrade 75%. Higher content of plasticizer (4.5%) made the film had high water content and cause it more exposed to microbial organism attack. The presence of water medium causes this microbial organism to attack the SPS [36, 52]. The film loses structural integrity as microbial organisms devour the surrounding starch, lowering mechanical characteristics and allowing the microbial organisms to attack SPS [49]. The findings showed that the SPS/SPNCC incorporated CEO nanocomposite films would not cause any harmful/ ecological impact effect to the environment, implying that the SPS/SPNCC incorporated CEO nanocomposite film will be biodegradable.

7.3.7 Morphological properties of SPS/SPNCC incorporated CEO nanocomposite films

Figures 7.7 and 7.8 showed the camera image and FESEM image of the surface topography of SPS/SPNCC incorporated CEO nanocomposite. The visual appearance of the obtained nanocomposite films was described in the figure. All the films were kind of brownish and transparent. The microstructure of the film surfaces was investigated to assess the impact on morphological surface. Higher content of plasticizer, which was at 4.5 wt% cause the film surface, has agglomerated area (circled in red). In contrast with GS-plasticized film that the films were released the EO on the surface that causes the film moist (circled in blue). There was an oil spot was observed on the GS4.5 films. In fact, plasticizers added with the optimum amount to the starch-based films improve the complete solubilisation of starch granules and improve the smoothness of the film surface. High content of plasticizers produces many agglomerated areas. GS-plasticized films were found to be smoother, uniform, and compact for all plasticizers types. Plasticizers intraction with SPS molecules to weaken intermolecular and intermolecular hydrogen bonds in starch, as shown in FESEM image (Figure 7.8). Nanocomposite film with 4.5 wt% plasticizer was found to be more flexible than the film with lower plasticizer concentration. Plasticizer concentration decreased, which made the flexibility reduced. Plasticizers' small molecular size allowed them to creep into the polymer chains' intermolecular gaps, weakening intermolecular hydrogen bonds, and

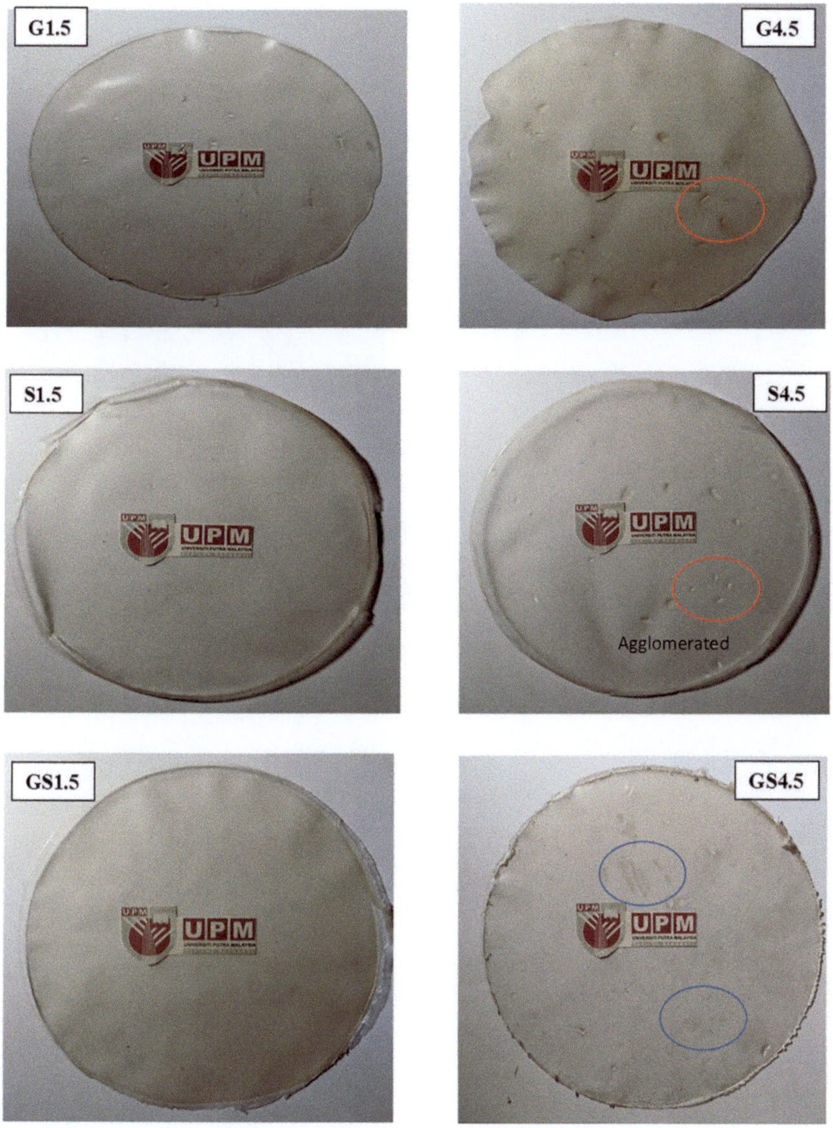

Figure 7.7: Image of SPS/SPNCC incorporated CEO nanocomposite film samples.

enhancing molecular mobility. As the plasticizer concentration went from 1.5 to 4.5 wt%, the intermolecular hydrogen bond of SPS films deteriorated. Nonetheless, changes in plasticizer type had a considerable impact on the flexibility of their resultant films at similar plasticizer concentrations. Thus, GS-plasticized bionanocomposite films revealed higher flexibility than G- and S-plasticized films.

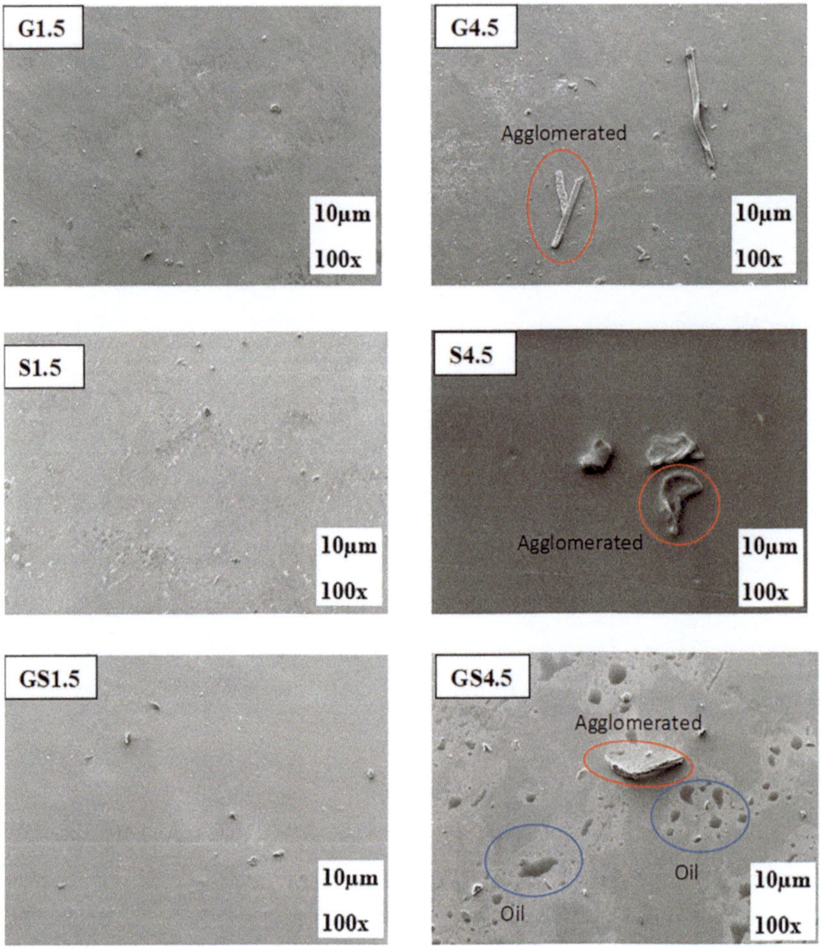

Figure 7.8: FESEM image of SPS/SPNCC incorporated CEO nanocomposite film samples.

7.4 Conclusions

Optimum usage of plasticizer and EO can produce a better bio-nanocomposite films packaging. From this research, different plasticizers type and concentrations incorporated with CEO were produced to investigate the effect of water barrier and biodegradable properties of CEO/sugar palm nanocellulose/starch biopolymer composites. The result shows that the types and concentrations of plasticizers incorporated with CEO influenced water barrier and biodegradable properties. The films with lower concentrations of plasticizer show good barrier properties compared to higher plasticizer concentrations. Generally, GS-plasticized films with the lowest concentration show an optimum formulation that can be further used in food packaging applications. In conclusion, this paper describes a biopolymer that could be used to make biodegradable films.

References

1. Tarique J, Sapuan SM, Khalina A, Sherwani SFK, Yusuf J, Ilyas RA. Recent developments in sustainable arrowroot (Maranta arundinacea Linn) starch biopolymers, fibres, biopolymer composites and their potential industrial applications: a review. J Mater Res Technol 2021;13:1191–219.
2. Jumaidin R, Khiruddin MAA, Asyul Sutan Saidi Z, Salit MS, Ilyas RA. Effect of cogon grass fibre on the thermal, mechanical and biodegradation properties of thermoplastic cassava starch biocomposite. Int J Biol Macromol 2020;146:746–55.
3. Diyana ZN, Jumaidin R, Selamat MZ, Ghazali I, Julmohammad N, Huda N, et al. Physical properties of thermoplastic starch derived from natural resources and its blends: a review. Polymers 2021;13:1–20.
4. Abral H, Basri A, Muhammad F, Fernando Y, Hafizulhaq F, Mahardika M, et al. A simple method for improving the properties of the sago starch films prepared by using ultrasonication treatment. Food Hydrocolloids 2019;93:276–83.
5. Punia Bangar S, Nehra M, Siroha AK, Petrů M, Ilyas RA, Devi U, et al. Development and characterization of physical modified pearl millet starch-based films. Foods 2021;10:1609.
6. Fakhouri FM, Martelli SM, Caon T, Velasco JI, Buontempo RC, Bilck AP, et al. The effect of fatty acids on the physicochemical properties of edible films composed of gelatin and gluten proteins. LWT–Food Sci Technol 2017;87:293–300.
7. Kaya M, Ravikumar P, Ilk S, Mujtaba M, Akyuz L, Labidi J, et al. Production and characterization of chitosan based edible films from Berberis crataegina's fruit extract and seed oil. Innovat Food Sci Emerg Technol 2018;45:287–97.
8. Syafiq RMO, Sapuan SM, Zuhri MRM. Effect of cinnamon essential oil on morphological, flammability and thermal properties of nanocellulose fibre–reinforced starch biopolymer composites. Nanotechnol Rev 2020;9:1147–59.
9. Syafiq R, Sapuan SM, Zuhri MRM. Antimicrobial activity, physical, mechanical and barrier properties of sugar palm based nanocellulose/starch biocomposite films incorporated with cinnamon essential oil. J Mater Res Technol 2021;11:144–57.
10. Syafiq R, Sapuan SM, Zuhri MYM, Ilyas RA, Nazrin A, Sherwani SFK, et al. Antimicrobial activities of starch-based biopolymers and biocomposites incorporated with plant essential oils: a review. Polymers 2020;12: 2403.
11. Othman SH, Amirah S, Edwal M, Risyon NP, Basha RK, Talib RA. Water sorption and water permeability properties of edible film made from potato peel waste. Food Sci Technol 2017;37:63–70.
12. Nordin N, Othman SH, Rashid SA, Basha RK. Effects of glycerol and thymol on physical, mechanical, and thermal properties of corn starch films. Food Hydrocolloids 2020;106:105884.
13. Menzel C. Improvement of starch films for food packaging through a three-principle approach: antioxidants, cross-linking and reinforcement. Carbohydr Polym 2020;250:116828.
14. Nazrin A, Sapuan SM, Zuhri MYM, Tawakkal ISMA, Ilyas RA. Water barrier and mechanical properties of sugar palm crystalline nanocellulose reinforced thermoplastic sugar palm starch (TPS)/poly(lactic acid) (PLA) blend bionanocomposites. Nanotechnol Rev 2021;10:431–42.
15. Nazrin A, Sapuan SM, Zuhri MYM. Mechanical, physical and thermal properties of sugar palm nanocellulose reinforced thermoplastic starch (TPS)/Poly (lactic acid) (PLA) blend bionanocomposites. Polymers 2020;12: 2216.

16. Li J, Ye F, Lei L, Zhao G. Combined effects of octenylsuccination and oregano essential oil on sweet potato starch films with an emphasis on water resistance. Int J Biol Macromol 2018;115:547–53.
17. Wongphan P, Harnkarnsujarit N. Characterization of starch, agar and maltodextrin blends for controlled dissolution of edible films. Int J Biol Macromol 2020;156:80–93.
18. Hazrati KZ, Sapuan SM, Zuhri MYM, Jumaidin R. Effect of plasticizers on physical, thermal, and tensile properties of thermoplastic films based on Dioscorea hispida starch. Int J Biol Macromol 2021;185:219–28.
19. Tarique J, Sapuan SM, Khalina A. Effect of glycerol plasticizer loading on the physical, mechanical, thermal, and barrier properties of arrowroot (Maranta arundinacea) starch biopolymers. Sci Rep 2021;11:1–17.
20. Othman SH, Majid NA, Syafinaz I, Amin M, Basha K, Nordin N, et al. Tapioca starch films reinforced with microcrystalline cellulose for potential food packaging application. Food Sci Technol 2019;39:1–8.
21. Ilyas RA, Sapuan SM, Ishak MR, Zainudin ES. Water transport properties of bio-nanocomposites reinforced by sugar palm (Arenga Pinnata) nanofibrillated cellulose. J Adv Res Fluid Mech Therm Sci J 2018;51:234–46.
22. Nurazzi NM, Asyraf MRM, Rayung M, Norrrahim MNF, Shazleen SS, Rani MSA, et al. Thermogravimetric analysis properties of cellulosic natural fiber polymer composites: a review on influence of chemical treatments. Polymers 2021;13:2710.
23. Ilyas RA, Sapuan SM, Ishak MR, Zainudin ES. Development and characterization of sugar palm nanocrystalline cellulose reinforced sugar palm starch bionanocomposites Development and characterization of sugar palm nanocrystalline cellulose reinforced sugar palm starch bionanocomposites. Carbohydr Polym 2018;202:186–202.
24. Atiqah A, Jawaid M, Sapuan SM, Ishak MR, Ansari MNM, Ilyas RA. Physical and thermal properties of treated sugar palm/glass fibre reinforced thermoplastic polyurethane hybrid composites. J Mater Res Technol 2019;8:3726–32.
25. Suriani M, Sapuan S, Ruzaidi C, Nair D, Ilyas R. Flammability, morphological and mechanical properties of sugar palm fiber/polyester yarn-reinforced epoxy hybrid biocomposites with magnesium hydroxide flame retardant filler. Text Res J 2021;91:2600–11.
26. Hanani ZAN, Mahony JAO, Roos YH, Oliveira PM, Kerry JP. Extrusion of gelatin-based composite films: effects of processing temperature and pH of film forming solution on mechanical and barrier properties of manufactured films. Food Packag Shelf Life 2014;2:91–101.
27. Suderman N, Isa MIN, Sarbon NM. The effect of plasticizers on the functional properties of biodegradable gelatin-based film: a review. Food Biosci 2018;24:111–9.
28. Nobile MAD, Flores S, Mastromatteo M, Conte A, Gerschenson L, Del Nobile M. Influence of glycerol and chitosan on tapioca starch-based edible film properties. J Food Eng 2008;88:159–68.
29. Bakry NF, Isa MIN, Sarbon NM. Effect of sorbitol at different concentrations on the functional properties of gelatin/carboxymethyl cellulose (CMC)/chitosan composite films. Int Food Res J 2017;24:1753–62.
30. Hazrol MD, Sapuan SM, Zainudin ES, Zuhri MYM, Abdul Wahab NI. Corn starch (Zea mays) biopolymer plastic reaction in combination with sorbitol and glycerol. Polymers 2021;13:242.
31. Ribeiro R, Andrade M, Melo NRD, Sanches A. Use of essential oils in active food packaging: recent advances and future trends. Trends Food Sci Technol 2017;61:132–40.
32. Sun H, Li S, Chen S, Wang C, Liu D, Li X. Antibacterial and antioxidant activities of sodium starch octenylsuccinate-based Pickering emulsion films incorporated with cinnamon essential oil. Int J Biol Macromol 2020;159:696–703.
33. Zhou Y, Wu X, Chen J, He J. Effects of cinnamon essential oil on the physical, mechanical, structural and thermal properties of cassava starch-based edible films. Int J Biol Macromol 2021;184:574–83.
34. Han Y, Yu M, Wang L. Physical and antimicrobial properties of sodium alginate/carboxymethyl cellulose films incorporated with cinnamon essential oil. Food Packag Shelf Life 2018;15:35–42.
35. Sahari J, Sapuan SM, Zainudin ES, Maleque MA. Thermo-mechanical behaviors of thermoplastic starch derived from sugar palm tree (Arenga pinnata). Carbohydr Polym 2013;92:1711–6.
36. Ilyas RA, Sapuan SM, Ishak MR, Zainudin ES. Development and characterization of sugar palm nanocrystalline cellulose reinforced sugar palm starch bionanocomposites. Carbohydr Polym 2018;202:186–202.

37. ASTM E96-95. Standard test method for water vapor transmission of materials. West Conshohocken, PA: ASTM International.; 1995.
38. Sanyang M, Sapuan S, Jawaid M, Ishak M, Sahari J. Effect of plasticizer type and concentration on tensile, thermal and barrier properties of biodegradable films based on sugar palm (Arenga pinnata) starch. Polymers 2015;7:1106–24.
39. ASTM D1938 – 14. Standard test method for tear-propagation resistance (trouser tear) of plastic film and thin sheeting by a single-tear method. West Conshohocken, PA: ASTM International.; 2014.
40. López OV, Castillo LA, García MA, Villar MA, Barbosa SE. Food packaging bags based on thermoplastic corn starch reinforced with talc nanoparticles. Food Hydrocolloids 2015;43:18–24.
41. López OV, Lecot CJ, Zaritzky NE, García MA. Biodegradable packages development from starch based heat sealable films. J Food Eng 2011;105:254–63.
42. ASTM F88. Standard test method for seal strength of flexible barrier materials. West Conshohocken, PA: ASTM International.; 2000.
43. Sanyang ML, Sapuan SM, Jawaid M, Ishak MR, Sahari J. Effect of plasticizer type and concentration on physical properties of biodegradable films based on sugar palm (Arenga pinnata) starch for food packaging. J Food Sci Technol 2016;53:326–36.
44. Razavi SMA, Mohammad Amini A, Zahedi Y. Characterisation of a new biodegradable edible film based on sage seed gum: influence of plasticiser type and concentration. Food Hydrocolloids 2015;43:290–8.
45. Aitboulahsen M, El Galiou O, Laglaoui A, Bakkali M, Hassani Zerrouk M. Effect of plasticizer type and essential oils on mechanical, physicochemical, and antimicrobial characteristics of gelatin, starch, and pectin-based films. J Food Process Preserv 2020;44:1–10.
46. Ghasemlou M, Khodaiyan F, Oromiehie A. Physical, mechanical, barrier, and thermal properties of polyol-plasticized biodegradable edible film made from kefiran. Carbohydr Polym 2011;84:477–83.
47. Šuput D, Lazić V, Pezo L, Markov S, Vaštag Ž, Popović L, et al. Characterization of starch edible films with different essential oils addition. Pol J Food Nutr Sci 2016;66:277–85.
48. Dai L, Qiu C, Xiong L, Sun Q. Characterisation of corn starch-based films reinforced with taro starch nanoparticles. Food Chem 2015;174:82–8.
49. Ilyas RA, Sapuan SM, Ishak MR, Zainudin ES. Sugar palm nanocrystalline cellulose reinforced sugar palm starch composite: degradation and water-barrier properties. IOP Conf Ser Mater Sci Eng 2018;368:012006.
50. Sanyang ML, Sapuan SM, Jawaid M, Ishak MR, Sahari J. Effect of sugar palm-derived cellulose reinforcement on the mechanical and water barrier properties of sugar palm starch biocomposite films. Bioresources 2016;11:4134–45.
51. Laohakunjit N, Noomhorm A. Effect of plasticizers on mechanical and barrier properties of rice starch film. Starch-Stärke 2004;56:348–56.
52. Edhirej A, Sapuan SM, Jawaid M, Zahari NI. Preparation and characterization of cassava bagasse reinforced thermoplastic cassava starch. Fibers Polym 2017;18:162–71.

Asmawi Nazrin, Salit Mohd Sapuan*, Mohamed Yusoff Mohd Zuhri, Intan Syafinaz Mohamed Amin Tawakkal and Rushdan Ahmad Ilyas

8 Mechanical degradation of sugar palm crystalline nanocellulose reinforced thermoplastic sugar palm starch (TPS)/poly (lactic acid) (PLA) blend bionanocomposites in aqueous environments

Abstract: The concerning issue regarding petrochemical plastic wastes had prompted scientists and researchers to develop biodegradable plastic in effort to tackle environmental pollution. Alternative bioresources such as poly (lactic acid), sugar palm starch and nanocellulose fibre were utilized in producing cheap, biodegradable and sustainable plastic with satisfactory mechanical properties for food packaging application. In this study, sugar palm crystalline nanocellulose (SPCNC) was priorly dispersed in thermoplastic sugar palm starch (TPS) before melt blended with poly (lactic acid) (PLA) and later compress moulded into a sheet form. Initial biodegradation test of PLA100 and all PLA/TPS blends bionanocomposite samples indicated that PLA60TPS40 has the least variation in weight loss due to the good miscibility between TPS and PLA promoting the reinforcement of SPCNC. Greater weight losses in seawater (17.54%), river water (18.97%) and sewer water (22.27%) result in greater mechanical degradation as observed at the reduction of tensile strength from 12.11 MPa to 2.72 MPa in seawater, 1.48 MPa in river water and 0.40 MPa in sewer water. Similarly, higher weight losses in seawater (22.16%), river water (21.6%) and sewer water (23.09%) correlated with the reduction of flexural strength from 18.37 MPa to

***Corresponding author: Salit Mohd Sapuan**, Advanced Engineering Materials and Composites Research Centre (AEMC), Department of Mechanical and Manufacturing Engineering, Universiti Putra Malaysia, 43400 UPM Serdang, Selangor, Malaysia; and Laboratory of Biocomposite Technology, Institute of Tropical Forestry and Forest Products (INTROP), Universiti Putra Malaysia, 43400 UPM Serdang, Selangor, Malaysia, E-mail: sapuan@upm.edu.my

Asmawi Nazrin, Laboratory of Biocomposite Technology, Institute of Tropical Forestry and Forest Products (INTROP), Universiti Putra Malaysia, 43400 UPM Serdang, Selangor, Malaysia

Mohamed Yusoff Mohd Zuhri, Advanced Engineering Materials and Composites Research Centre (AEMC), Department of Mechanical and Manufacturing Engineering, Universiti Putra Malaysia, 43400 UPM Serdang, Selangor, Malaysia; and Laboratory of Biocomposite Technology, Institute of Tropical Forestry and Forest Products (INTROP), Universiti Putra Malaysia, 43400 UPM Serdang, Selangor, Malaysia. https://orcid.org/0000-0002-1069-7345

Intan Syafinaz Mohamed Amin Tawakkal, Department of Process and Food Engineering, Universiti Putra Malaysia, 43400 UPM Serdang, Selangor, Malaysia

Rushdan Ahmad Ilyas, School of Chemical and Energy Engineering, Faculty of Engineering, Universiti Teknologi Malaysia, 81310 Johor Bahru, Johor, Malaysia. https://orcid.org/0000-0001-6622-2632

As per De Gruyter's policy this article has previously been published in the journal Physical Sciences Reviews. Please cite as: A. Nazrin, S. M. Sapuan, M. Y. Mohd Zuhri, I. S. M. Amin Tawakkal and R. A. Ilyas "Mechanical degradation of sugar palm crystalline nanocellulose reinforced thermoplastic sugar palm starch (TPS)/poly (lactic acid) (PLA) blend bionanocomposites in aqueous environments" *Physical Sciences Reviews* [Online] 2023. DOI: 10.1515/psr-2022-0030 | https://doi.org/10.1515/9783110773606-008

3.5 MPa in seawater, 3.83 MPa in river water and 3.6 MPa in sewer water. The scanning electron microscope (SEM) images of tensile fracture morphology demonstrated clear porous structure due to the removal of starch particles by microbial activity. The homogenous structure of PLA60TPS40 had a steady and consistent degradation, which wholly diminished the interfacial adhesion that led to mechanical properties losses. The mechanical strength reduction clarified that the biodegradation rate within the media used might be able to resolve the excessive non-biodegradable plastic waste in open waters.

Keywords: mechanical degradation; scanning electron microscope (SEM); seawater degradation; sewer water degradation.

8.1 Introduction

The accumulation of petrochemical plastic wastes had been a major concern towards the safety and sustainability of the Mother Nature. Their inability to biodegrade imposed severe impacts to the land, sea and air [1, 2]. In particular, short shelf-life plastics used in food packaging are the major contributor to these issues due to its fast production growth to keep up with the rise of global consumption [3]. The disposal through incineration released harmful substances (dioxins, furans and mercury) into the atmosphere [4] while recycling methods are inefficient because of complexity in polymer separation and high cost to reprocess [5]. Typically, it is discarded carelessly by consumers, which end up in landfill and open waters [6]. This plastic waste crisis had prompted scientists and researchers to develop green plastic based on bioresources in effort to prioritize biodegradability, sustainability and reduce over-reliance on depleting fossil resources [7].

Poly (lactic acid) (PLA) is the most prominent biopolymer as it is a high modulus thermoplastic, which can be synthesized from renewable resources [8–10]. Although PLA is biodegradable, study [11] showed that PLA did not degrade in most liquid environment due to its strong hydrophobic properties. The degradation of PLA occurred through the hydrolysis of the backbone ester groups, which affected by crystallinity, morphology, water diffusion, molecular weight and its distribution [12]. Cheaper biopolymer such as starch is often incorporated into PLA to reduce the high cost of PLA and induce degradation rate [13–15]. The poor interfacial adhesion of starch and PLA weaken the mechanical properties of the new material [16]. However, for application such as food packaging, moderate mechanical properties are required instead of high critical value. Various authors [17] had reported that the bare minimum applicable value of tensile strength is around 12–15 MPa, while elongation at break is around 5–10%. In this study, sugar palm crystalline nanocellulose (SPCNC) reinforced thermoplastic sugar palm starch (TPS) was priorly prepared before blending with PLA. Initial starch gelatinization improve the dispersion of TPS in PLA [18], while SPCNC reinforcement minimize the mechanical strength and water barrier reduction [19]. Unlike other commercial starches (tapioca, potatoes, wheat, rice, cassava and sago), sugar palm starch is considered as a by-product of sugar palm tree, which mainly used in the extraction of sugar palm sap to produce brown sugar blocks [20–23]. Its fibre had been utilized in making seawater-resistant rope for ship cordages [24]. The downsizing of fibre into nano size scale provide greater surface area, improved surface area to volume ratio of 100, outstanding mechanical properties, intense crystallinity, lightweight and low density [19]. Besides, their similar botanical origin are known to have a high affinity of 3D hydrogen bonding networks enhancing the mechanical strength and water barrier properties [25].

In order utilize the reinforcement of SPCNC, a good ratio between TPS and PLA must be attained so that the dispersion of TPS within PLA is uniform and well distributed. All blends bionanocomposite samples were tested for their biodegradation in seawater, river water and sewer water before selecting the compatible ratio for mechanical degradation. The mechanical degradation indicated the defragmentation of a polymer due to microorganism activity that led to mechanical strength deterioration. As the weight loss of a material is often associated with the degradation process, mechanical strength loss is not necessarily occurred even after the weight loss was recorded. To the best of author's knowledge, although frequent research had been carried out on the biodegradability of polymer composites in seawater, only limited research had been done for river water and sewer water. In the present study, the correlation between biodegradation rate of PLA/TPS blend bionanocomposites and mechanical strength deterioration was investigated based on the media used, which were seawater, river water and sewer water. Medium with higher microbial activity might enhance the biodegradation rate as observed through the mechanical strength reduction, which potentially resolve the accumulation of plastic wastes in open waters.

8.2 Materials and methods

8.2.1 Materials

Sugar palm starch was purchased from Hafiz Adha Enterprise, which extracted from the core of sugar palm tree at Kampung Kuala Jempol, Negeri Sembilan, Malaysia. Poly (lactic acid) (PLA) (NatureWork 2003D), glycerol and sorbitol were bought from Mecha Solve Engineering, Petaling Jaya, Malaysia. Sugar palm crystalline nanocellulose (SPCNC) was provided by Laboratory of Biocomposite Technology, Institute of Tropical Forestry and Forest Products (INTROP), Universiti Putra Malaysia (UPM), Selangor, Malaysia. SPCNC was derived from sugar palm fibres, which are entangled at the outer part of sugar palm tree in woven form.

8.2.2 Extraction and preparation of SPS

Unproductive mature sugar palm trees were cut and the inner part was pulverized to obtain sugar palm starch mixed with woody fibres. The mixture was added water and hand kneaded to disturb the mingling of starch and woody fibres. After 1 h, most woody fibres resurfaced at the top while starch accumulated at the bottom. The floating fibres were removed, and finer fibre residues within the starch were filtered using a strainer cloth. Next, the wet starch was spread evenly into a film plastic to remove the excess water. Finally, the starch was dried in an oven at 120 °C for 24 h to obtain starch powder.

8.2.3 Fabrication of SPCNC reinforced TPS/PLA blend bionanocomposites sheet

Prior to the blending of PLA with TPS, TPS was fabricated using solution casting method. Firstly, 1000 mL of distilled water, 0.5 g of SPCNC, 15 g of both glycerol (15%) and sorbitol (15%) were added into a beaker. The dispersion of SPCNC was achieved using sonication to ensure discontinuous particles distribution and uniform dispersion in the solution. Then, the beaker was transferred into a water bath with temperature of 80 °C. Gradually, 100 g of sugar palm starch was added into the beaker with continuous stirring for 30–45 min until a semi-fluid consistency was obtained. The gelatinized starch was poured into glass petri dishes and left to dry in an oven at 60 °C for 24 h. TPS bionanocomposite films were acquired and crushed into granule size to ease the melt-blending process. PLA and TPS were melt blended using Brabender Plastograph (Model 815,651, Brabender GmbH & Co. KG, Duisburg, Germany) at 170 °C for 13 min with a rotor speed of 50 rpm. The composition of PLA and TPS was divided into five different ratios, which were 20:80, 40:60, 60:40, 70:30 and 80:20. Once again, the blend bionanocomposites were crushed into granule size before being fabricated into 150 mm × 150 mm × 3 mm sheet and dumbbell-shaped specimen using compression moulding method (Technovation, Selangor, Malaysia). Figure 8.1 illustrates the overall process of PLA/TPS blend bionanocomposites fabrication. Table 8.1 displays the composition of PLA/TPS blend bionanocomposites.

8.2.4 Biodegradation in aqueous environments

Biodegradation of PLA/TPS blend bionanocomposites were performed under three aqueous media: seawater, river water and sewer waste water. Six specimens (20 × 15 × 3 mm) were placed in a cup filled with 100 mL of seawater, river water and sewer waste water. Prior to testing, specimens were dried at 100 °C for 24 h and weighed to obtain the initial weight, M_i. Four sets of experiments were made for predetermined 1-week interval for a total of 4 weeks. Each experiment was conducted in triplicate and the average value was taken. Finally, specimens were left dried at 100 °C for 24 h and weighed to obtain the final weight, M_f. The weight loss, M_l, can be calculated using Equation (8.1).

$$M_l = \frac{M_i - M_f}{M_i} \times 100\% \tag{8.1}$$

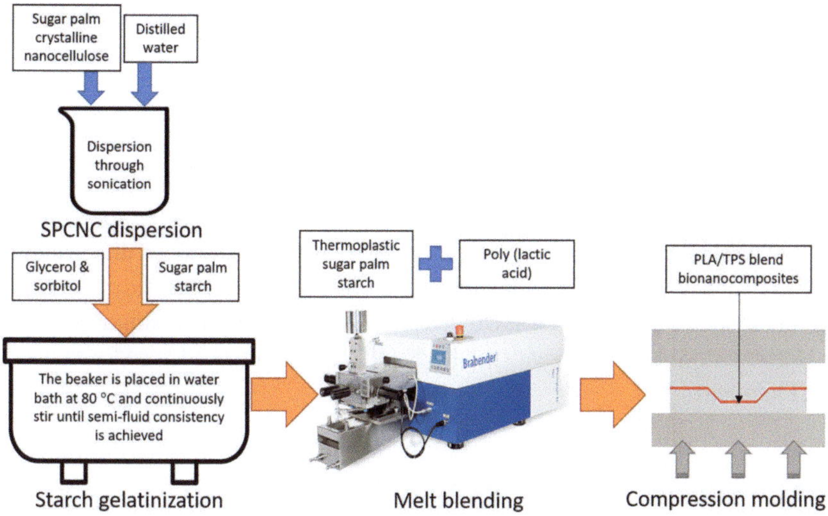

Figure 8.1: The overall process of PLA/TPS blend bionanocomposites fabrication.

PLA60TPS40 was selected to proceed for mechanical degradation test as only minor differences in weight losses in all media were recorded as compared to other blends bionanocmoposite samples. The tensile and flexural specimens of PLA60TPS40 were placed in containers holding seawater, river water and sewer water at room temperature of 23 ± 1 °C for 4 weeks. Every week, the media were renewed to optimize the efficiency of microbial activity for biodegradation process. After 4 weeks, the tensile and flexural specimens were taken out, left dried at 100 °C for 24 h and weighed to obtain the final weight, M_f. Similarly, the weight loss, M_l, can be calculated using Equation (8.1). The method was adopted from Chen et al. [26].

8.2.5 Tensile testing

The tensile test was conducted in accordance with ASTM D-638 at a temperature of 23 ± 1 °C and a relative humidity of 50 ± 5%. The dumbbell-shaped specimens were prepared with a dimension of 115 mm (L) × 19 mm (W) × 3 mm (T). The test was carried out

Table 8.1: The composition of PLA/TPS blend bionanocomposites.

PLA (%)	TPS (%)	Abbreviation
20	80	PLA20TPS80
40	60	PLA40TPS60
60	40	PLA60TPS40
70	30	PLA70TPS30
80	20	PLA80TPS20
100	0	PLA100

on five repetitions using a Universal Testing Machine (INSTRON 5556) with a 5 kN load cell; the crosshead speed was maintained at 1 mm/min.

8.2.6 Flexural testing

The flexural test was conducted in accordance with ASTM D-790 at a temperature of 23 ± 1 °C and a relative humidity of 50 ± 5%. The specimen was cut with a dimension of 130 mm (L) × 13 mm (W) × 3 mm (T). The support span length was set at a ratio of 16:1 to the thickness of the specimen. The test was carried out with five repetitions using a Universal Testing Machine (INSTRON 5556) with a 5 kN load cell; the crosshead speed was maintained at 1 mm/min.

8.2.7 Scanning electron microscope (SEM)

The tensile fractured surface of the specimen was observed under SEM, model JEOL JSM – IT100, operating at an acceleration voltage of 10 kV. All specimens were coated with gold using a Baltec SCD 005 Sputter Coater to avoid unwanted charging.

8.3 Results and discussion

8.3.1 Weight lost in aqueous environments

Figure 8.2 shows the weight loss trend of PLA100 and PLA/TPS blend bionanocomposites for 4 weeks in seawater, river water and sewer water. It was observed that all blends bionanocomposite of samples experienced active biodegradation rate within the first week of immersion. The biodegradation rate seemed to slow down for the following weeks until the fourth week. This indicated that the rapid microbial growth occurred less than a week within all media. In the first week, the specimen with the highest TPS content (PLA20TPS80) experienced weight loss of 49.11% in seawater, 61.86% in river water and 66.02% in sewer water. These results corresponded with the changes in pH value of seawater from 7.47 to 6.75, river water from 7.07 to 5.85 and sewer water from 7.52 to 5.24. Hegyesi et al. [27] mentioned that the fast chain scission of PLA at low pH caused the erosion of PLA and generated carboxylic acids hasten the polymer degradation by further lowering the pH. The significant weight loss was caused by the microbial activity at the dominant TPS content within the sample. The swelled up TPS facilitated entry of water and microorganisms into the sample to reach PLA, thereby facilitating the degradation of PLA matrix itself [28]. As most microorganisms do not strive well in the high salinity of seawater, the degradation rate was slightly lower compared to river water and sewer water. Meanwhile, river water and sewer water with low salinity consist of

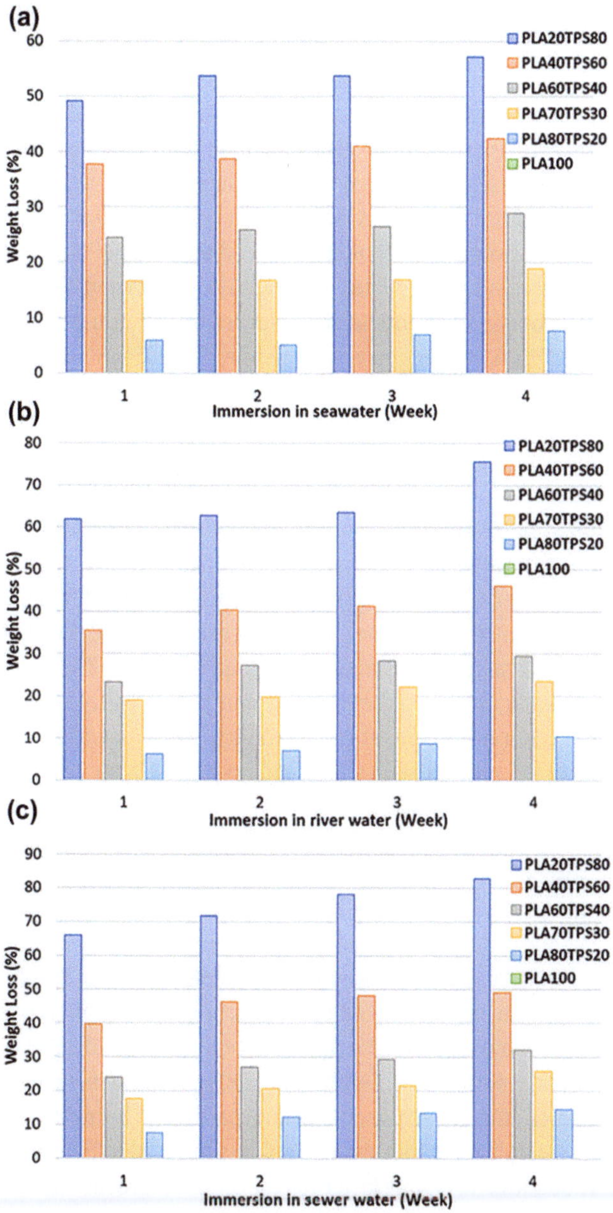

Figure 8.2: The weight loss trends of PLA100 and PLA/TPS blend bionanocomposites in (a) seawater, (b) river water and (c) sewer water.

microorganisms responsible for biological degradation. Other blends bionanocomposite samples also displayed the same weight loss trend as PLA20TPS80 verifying the lack of microbial growth in seawater compared to river water and sewer water. Siracusa [29] reported that PCL had a complete defragmentation in sea water within 56 days but degraded faster in sweetened waters (lake and river). Kliem et al. [30] stated that PCL/TPS achieved complete weight lost in 5 weeks in seawater and 4 weeks in sewer sludge. From the first to fourth week, PLA20TPS80 experienced different weight loss increments in the three media. Seawater had the lowest increment, which was less than 10% while river water and sewer water had increment more than 10%. These findings also supported the minor microorganism concentration in high salinity water if compared to low salinity water.

The reinforcement of SPCNC was observed at PLA40TPS60 and PLA60TPS40 as minor differences in weight loss were recorded for all media. The final weight losses of PLA40TPS60 in seawater, river water and sewer water were 42.41%, 46.17% and 49.22%, respectively. For PLA60TPS40, the final weight loss in seawater, river water and sewer water were 28.93%, 29.52% and 32.21%, respectively. Meanwhile, PLA20TPS80 experienced significant weight loss of 57.13%, 75.61% and 82.8%, respectively. As the dispersion of SPCNC depended on the distribution of TPS within PLA, excessive TPS content within PLA20TPS80 might have agglomerated and provided hot spots for microbial growth. There is also higher concentration of microorganism in river water and sewer water, which facilitate weight loss due to biodegradation. Both PLA40TPS60 and PLA60TPS40 demonstrated minor changes of pH value in seawater, but significant changes were recorded in both river water and sewer water. The minor differences in weight loss for PLA40TPS60 and PLA60TPS40 can be attributed to the good dispersion of SPCNC through TPS uniform distribution within PLA matrix. Ilyas et al. [19] added 0.1% of SPCNC reinforcement into starch and significantly improved the water vapour permeability rate by 6.97%. Gelatinized starch with SPCNC reinforcement formed a tortuous path within the PLA/TPS blends bionanocomposite samples caused by the rigid crystalline structure and the dispersion of SPCNC in TPS. A similar work done by Ayana et al. [18] indicated that priorly dispersed nanoclay in TPS before blending with PLA promoted the formation of tortuous path reducing water absorption rate of the blend composite samples.

The lower TPS content within PLA70TPS30 and PLA80TPS20 did not have significant effect in SPCNC reinforcement was due to the poor compatibility of the compositions. Smaller proportion of TPS was poorly distributed within PLA restricting the reinforcement of SPCNC. The glycerol migration from TPS to PLA during melt-blending process caused large difference in viscosity and lead to the difficulty in mixing [31]. The final weight loss of PLA70TPS30 in seawater, river water and sewer water was 18.99%, 23.45% and 25.81%, respectively meanwhile, PLA80TPS20 recorded values of 7.67%, 10.37% and 14.59%, respectively. The minor TPS content was well covered by the continuous of PLA matrix and less exposed to the water molecules from the media [32]. Thus, the contributing factor for varying biodegradation rates was the microorganism concentration in the medium used.

8.3.2 Mechanical degradation within aqueous environments

The mechanical degradation test was conducted for PLA60TPS40 sample, which was assumed to have well-dispersed SPCNC. PLA60TPS40 was fabricated into tensile and

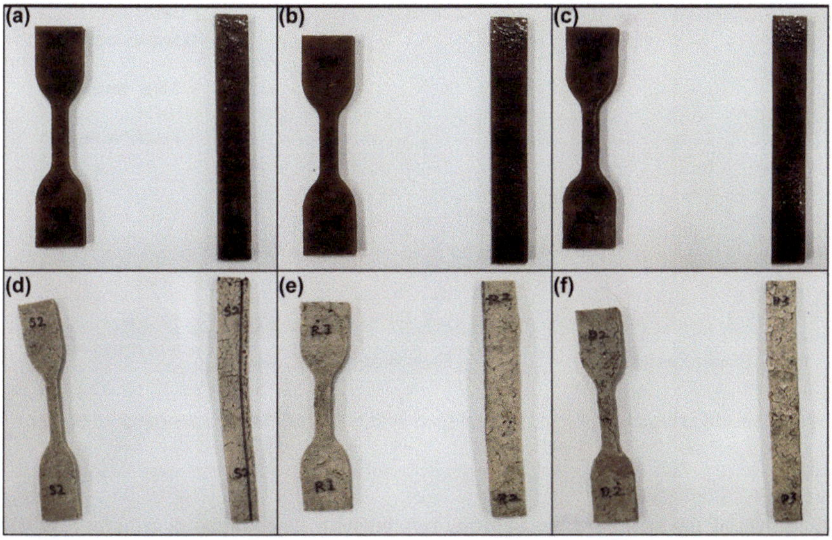

Figure 8.3: The tensile and flexural specimens of PLA60TPS40 bionanocomposites before immersion, (a)–(c) and after 4 weeks of immersion in (d) seawater, (e) river water and (f) sewer water.

flexural specimens in order to determine the effect of SPCNC reinforcement after immersion in various liquid environments. Figure 8.3 shows the physical appearance of tensile and flexural specimens before and after 4 weeks of immersion in seawater, river water and sewer water. All the specimens underwent discolouration with the formation of voids and crack spreading lines. The specimen immersed in the sewer water showed noticeable formation of voids and crack lines as compared to seawater and river water. The effect swelling and shrinking of TPS particles within the specimen was greater within the medium of higher microorganism concentration leaving obvious changes in the surface area. Due to the loss of physical stability, microorganism can easily colonize the surface area [3]. The discolouration was associated with microbial growth due to the biodegradation of the starch. A similar finding was pointed out by Chen et al. [26] that studied on the degradation of starch/PLA composite in seawater.

Figure 8.4 displays the weight loss of tensile and flexural specimens for PLA60TPS40 after 4 weeks of immersion. Tensile specimens experienced the lowest weight loss in seawater (17.54%) followed by river water (18.97%) and sewer water (22.27%). Flexural specimens somehow recorded the lowest weight loss in river water (21.6%) followed by seawater (22.16%) and sewer water (23.09%). The weight loss values in the three media had the similarity with the previous tests in Section 8.3.1 but with slightly lower values due to the difference in specimen geometry [30].

Table 8.2 demonstrates the tensile and flexural strength loss after 4 weeks of immersion in seawater, river water and sewer water. Microorganism invasion breaks down the molecules leaving high mechanical stresses on the macromolecules of the specimens.

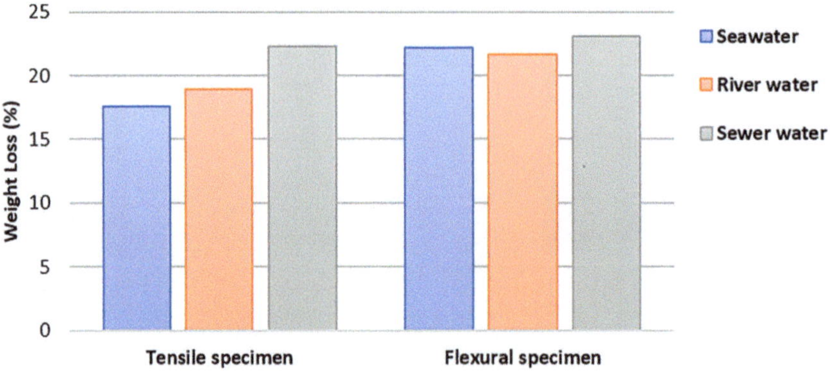

Figure 8.4: The weight losses of tensile and flexural specimens of PLA60TPS40 bionanocomposites after 4 weeks.

The weight loss of the specimen was closely related with mechanical degradation of the specimen. It was observed the higher weight loss in the tensile and flexural specimens caused greater loss in mechanical strength values. Higher microorganism activity promoted the degradation of the material, which breaks down the molecules and made them to be unable to sustain the stress applied. The weight losses in the seawater, river water and sewer water corresponded with the tensile strength values after immersion, which were 2.72 MPa, 1.48 MPa and 0.40 MPa, respectively. The weight loss of flexural specimens also corresponded with the flexural strength values, which were 3.5 MPa, 3.83 MPa and 3.6 MPa, respectively.

The swelling and shrinking of starch molecules progressively diminish the interfacial bonding between PLA and starch causing the deterioration of mechanical properties [32]. The addition of hydrophilic material induced the hydrolysis process of the polymer composites. The event led to the higher weight loss experienced by polymer composites if compared with unmodified polymer such as neat PLA. Lu et al. [33] incorporated crystalline nanocellulose-zinc oxide hybrids (ZnO hybrids) in poly (butylene adipate-co-terephthalate) (PBAT), which result in higher weight loss (5.23%) compared to neat PBAT (1.2%) when immersed in seawater. In quite similar case, Guzman-Sielicka et al. [34] reported that polycaprolactone (PCL) modified with TPS had a significant reduction in tensile strength when immersed in seawater compared to PCL modified with calcium carbonate and unmodified PCL. The addition of TPS into the PLA and PCL enhanced the degradation process correspondingly increased the weight loss, which led to significant mechanical properties reduction.

8.3.3 Morphological properties of the mechanical degradation

Figure 8.5 discloses the tensile fracture morphology of the PLA60TPS40 after being exposed to degradation activity in seawater, river water and sewer water for 4 weeks. The brittle

Table 8.2: The mechanical degradation loss of PLA60TPS40 bionanocomposites in aqueous environments.

Medium	Tensile strength (MPa)			Elongation at break (%)			Flexural strength (MPa)		
	Before immersion	After immersion	Standard deviation	Before immersion	After immersion	Standard deviation	Before immersion	After immersion	Standard deviation
Seawater	12.11	2.72	0.36	2.44	0.63	0.28	18.37	3.5	0.26
River water	12.11	1.48	0.24	2.44	0.22	0.25	18.37	3.83	0.23
Sewer water	12.11	0.40	0.19	2.44	0.24	0.20	18.37	3.6	0.14

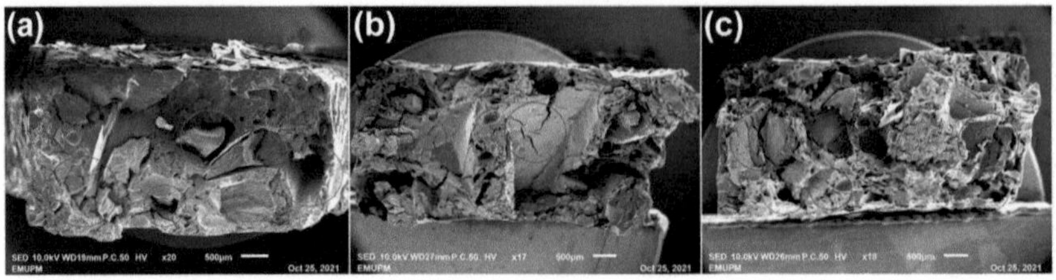

Figure 8.5: The scanning electron microscope (SEM) images of the tensile fracture surfaces of PLA60TPS40 bionanocomposites after 4 weeks of immersion in (a) seawater, (b) river water and (c) sewer water.

fracture of all specimens indicated that the degradation and disappearance of TPS and PLA particles end up leaving a more porous structure. The clear porous structure left behind suggested that the removal of starch particles was due to microbial activity rather than physical removal caused by tensile stress [16]. The large area of degradation gaps presented caused internal hollowness in the specimens and reduced the interfacial adhesion between TPS and PLA phase. Due to this occurrence, there was poor mechanical stress transfer between particles, which led to significant mechanical loss [32]. It was observed that the specimen immersed in seawater has less rough surface as compared to the specimens immersed in river water and sewer water. Numerous voids (red circles) and crack lines (blue circles) were presented on the specimens immersed in river water and sewer water. The dissolution of hydrophilic TPS on the surface and inside promoted channels for the entry of water and microorganism, thereby causing the degradation of the interior of the specimens [28]. The correlation between SEM images and tensile strength loss suggested that higher degradation rate of the specimen further diminished the interfacial bonding of TPS and PLA, which led to brittle fracture as significant reduction in elongation at break values were recorded [35]. These results supported that microbial growths were higher in both media compared to seawater.

8.4 Conclusions

The initial biodegradation test carried out for PLA100 and all PLA/TPS blends bionanocomposite samples verified that the lowest weight loss was attained in high salinity of seawater followed by low salinity of river water and sewer water. PLA60TPS40 bionanocomposites recorded minor variation in the final weight loss owing to the reinforcement of SPCNC through the uniform dispersion of TPS within PLA phase. The greater the weight losses, the higher the tensile and flexural strength losses of the PLA60TPS40 bionanocomposites. The tensile fracture morphology of PLA60TPS40 displayed clear porous structure due to the decomposition of starch particles rather than tensile stress pull out. The well-mixed structure of PLA60TPS40 blend bionanocomposites triggered a uniform and consistent degradation event, which reduced the interfacial adhesion of the entire structure, thus caused major losses in tensile and flexural strength. The mechanical strength reduction proven that the sample was able to biodegrade within the media used, which potentially substituted non-biodegradable plastics that caused environmental pollution. This material could be applied in short shelf-life applications such as disposable food container, agriculture mulching film and drug delivery capsule.

References

1. Azlin MNM, Sapuan SM, Zainudin ES, Zuhri MYM, Ilyas RA. Natural polylactic acid-based fiber composites: a review. In: Al-Oqla FM, Sapuan SM, editors. Advanced processing, properties, and applications of starch and other bio-based polymers, 1st ed. Oxford, UK: Elsevier; 2020:21–34 pp.
2. Ilyas RA, Sapuan SM. The preparation methods and processing of natural fibre bio-polymer composites. Curr Org Synth 2020;16:1068–70.
3. Urbanek AK, Rymowicz W, Mirończuk AM. Degradation of plastics and plastic-degrading bacteria in cold marine habitats. Appl Microbiol Biotechnol 2018;102:7669–78.
4. Verma R, Vinoda KS, Papireddy M, Gowda ANS. Toxic pollutants from plastic waste- a review. Procedia Environ Sci 2016;35:701–8.
5. Wang W, Themelis N, Sun K, Bourtsalas A, Huang Q, Zhang Y, et al. Current influence of China's ban on plastic waste imports. Waste Dispos Sustain Energy 2019;1:67–78.
6. Rhein S, Schmid M. Consumers' awareness of plastic packaging: more than just environmental concerns. Resour Conserv Recycl 2020;162:105063.
7. Moustafa H, Youssef AM, Darwish NA, Abou-Kandil AI. Eco-friendly polymer composites for green packaging: future vision and challenges. Compos B Eng 2019;172:16–25.
8. Ferrarezi MMF, de Oliveira Taipina M, da Silva LCE, Gonçalves MDC. Poly(Ethylene glycol) as a compatibilizer for poly(lactic acid)/thermoplastic starch blends. J Polym Environ 2013;21:151–9.
9. Ilyas RA, Sapuan SM, Harussani MM, Hakimi MYAY, Hakimi MZM, Haziq MZM, et al. Polylactic acid (PLA) biocomposite: processing, additive manufacturing and advanced applications. Polymers 2021;13:1326.
10. Nazrin A, Sapuan SM, Zuhri MYM, Ilyas RA, Syafiq R, Sherwani SFK. Nanocellulose reinforced thermoplastic starch (TPS), polylactic acid (PLA), and Polybutylene succinate (PBS) for food packaging applications. Front Chem 2020;8:1–12.
11. Wang GX, Huang D, Ji JH, Völker C, Wurm FR. Seawater-degradable polymers—fighting the marine plastic pollution. Adv Sci 2021;8:1–26.
12. Farah S, Anderson DG, Langer R. Physical and mechanical properties of PLA, and their functions in widespread applications — a comprehensive review. Adv Drug Delivery Rev 2016;107:367–92.
13. Diyana ZN, Jumaidin R, Selamat MZ, Ghazali I, Julmohammad N, Huda N, et al. Physical properties of thermoplastic starch derived from natural resources and its blends: a review. Polymers 2021;13:1–20.
14. Punia Bangar S, Nehra M, Siroha AK, Petrů M, Ilyas RA, Devi U, et al. Development and characterization of physical modified pearl millet starch-based films. Foods 2021;10:1609.
15. Abral H, Basri A, Muhammad F, Fernando Y, Hafizulhaq F, Mahardika M, et al. A simple method for improving the properties of the sago starch films prepared by using ultrasonication treatment. Food Hydrocolloids 2019;93:276–83.
16. Shogren RL, Doane WM, Garlotta D, Lawton JW, Willett JL. Biodegradation of starch/polylactic acid/poly(hydroxyester-ether) composite bars in soil. Polym Degrad Stabil 2003;79:405–11.
17. Muller J, González-Martínez C, Chiralt A. Combination of poly(lactic) acid and starch for biodegradable food packaging. Materials 2017;10:952.

18. Ayana B, Suin S, Khatua BB. Highly exfoliated eco-friendly thermoplastic starch (TPS)/poly (lactic acid)(PLA)/ clay nanocomposites using unmodified nanoclay. Carbohydr Polym 2014;110:430–9.
19. Ilyas RA, Sapuan SM, Ishak MR, Zainudin ES. Development and characterization of sugar palm nanocrystalline cellulose reinforced sugar palm starch bionanocomposites. Carbohydr Polym 2018;202: 186–202.
20. Nazrin A, Sapuan SM, Zuhri MYM. Mechanical, physical and thermal properties of sugar palm nanocellulose reinforced thermoplastic starch (TPS)/Poly (Lactic Acid) (PLA) blend bionanocomposites. Polymers 2020;12: 2216.
21. Hazrol MD, Sapuan SM, Ilyas RA, Othman ML, Sherwani SFK. Electrical properties of sugar palm nanocrystalline cellulose reinforced sugar palm starch nanocomposites. Polimery 2020;65:363–70.
22. Ilyas RA, Sapuan SM, Atiqah A, Ibrahim R, Abral H, Ishak MR, et al. Sugar palm (Arenga pinnata [Wurmb.] Merr) starch films containing sugar palm nanofibrillated cellulose as reinforcement: water barrier properties. Polym Compos 2020;41:459–67.
23. Ilyas RA, Sapuan SM, Ibrahim R, Abral H, Ishak MR, Zainudin ES, et al. Effect of sugar palm nanofibrillated cellulose concentrations on morphological, mechanical andphysical properties of biodegradable films based on agro-waste sugar palm (Arenga pinnata(Wurmb.) Merr) starch. J Mater Res Technol 2019;8: 4819–30.
24. Ishak MR, Sapuan SM, Leman Z, Rahman MZA, Anwar UMK, Siregar JP. Sugar palm (Arenga pinnata): its fibres, polymers and composites. Carbohydr Polym 2013;91:699–710.
25. Nazrin A, Sapuan SM, Zuhri MYM, Tawakkal ISMA, Ilyas RA. Water barrier and mechanical properties of sugar palm crystalline nanocellulose reinforced thermoplastic sugar palm starch (TPS)/poly(lactic acid) (PLA) blend bionanocomposites. Nanotechnol Rev 2021;10:431–42.
26. Chen X, Wang L, Shi J, Shi H, Liu Y. Environmental degradation of starch/poly(lactic acid) composite in seawater. Polym Polym Compos 2011;19:559–66.
27. Elsawy MA, Kim KH, Park JW, Deep A. Hydrolytic degradation of polylactic acid (PLA) and its composites. Renewable Sustainable Energy Rev 2017;79:1346–52.
28. Huang D, Hu Z, Liu T, Lu B, Zhen Z, Wang G, et al. Seawater degradation of PLA accelerated by water-soluble PVA. E-Polymers 2020;20:759–72.
29. Siracusa V. Microbial degradation of synthetic biopolymers waste. Polymers 2019;11:1066.
30. Kliem S, Kreutzbruck M, Bonten C. Review on the biological degradation of polymers in various environments. Materials 2020;13:1–18.
31. Li H, Huneault MA. Comparison of sorbitol and glycerol as plasticizers for thermoplastic starch in TPS/PLA blends. J Appl Polym Sci 2011;119:2439–48.
32. Acioli-Moura R, Sun XS. Thermal degradation and physical aging of poly(lactic acid) and its blends with starch. Polym Eng Sci 2008;48:829–36.
33. Lu G, Zhu J, Yu H, Jin M, Abdalkarim SYH, Wei Y. Degradation mechanism of green biopolyester nanocomposites with various cellulose nanocrystal based nanohybrids. Cellulose 2021;28:7735–48.
34. Guzman-Sielicka A, Janik H, Sielicki P. Degradation of polycaprolactone modified with TPS or CaCO3 in biotic/abiotic seawater. J Polym Environ 2012;20:353–60.
35. Lv S, Zhang Y, Gu J, Tan H. Physicochemical evolutions of starch/poly (lactic acid) composite biodegraded in real soil. J Environ Manage 2018;228:223–31.

A. Nazrin, A. S. Norfarhana, R. A. Ilyas*, S. M. Sapuan*, A. Khalina, R. M. O. Syafiq, M. R. M. Huzaifah, Ashraf Azmi, P. S. Khoo, Abu Hassan Nordin, H. S. N. Hawanis and S. A. Hassan

9 *Araucaria Araucana* thermoplastic starch nanocomposite films reinforced with nanocellulose

Abstract: *Araucaria Araucana* starch is a raw material with proven potential in obtaining biodegradable polymers. Plasticization improves the flexibility characteristics of starch-based films, however, they still tend to have low tensile strength and high hydrophilicity. The addition of nanocellulose is a technique to improve these characteristics. In this paper, the effects of adding different sources and concentrations of nanocellulose (NC) on the properties of thermoplastic *Araucaria Araucana* films are addressed. One can highlight, the main effects are the increase of tensile strength and transparency of the films along with the reduction of water vapor permeability and water

***Corresponding authors: R. A. Ilyas**, Institute of Tropical Forestry and Forest Products, Universiti Putra Malaysia, Serdang 43400, Selangor, Malaysia; Department of Chemical Engineering, Faculty of Chemical and Energy Engineering, Universiti Teknologi Malaysia, 81310 UTM Johor Bahru, Johor, Malaysia; Centre for Advanced Composite Materials, Universiti Teknologi Malaysia, 81310 UTM Johor Bahru, Johor, Malaysia; and Centre of Excellence for Biomass Utilization, Universiti Malaysia Perlis, 02600, Arau, Perlis, Malaysia, E-mail: ahmadilyas@utm.my. https://orcid.org/0000-0001-6622-2632; and **Salit Mohd Sapuan**, Advanced Engineering Materials and Composites Research Center (AEMC), Universiti Putra Malaysia, Faculty of Engineering, 43400 Seri Kembangan, Selangor, Malaysia, E-mail: sapuan@upm.edu.my

A. Nazrin, Institute of Tropical Forestry and Forest Products, Universiti Putra Malaysia, Serdang 43400, Selangor, Malaysia; Advanced Engineering Materials and Composites Research Center (AEMC), Faculty of Engineering, Universiti Putra Malaysia, 43400 Seri Kembangan, Selangor, Malaysia; and Institute of Power Engineering, Universiti Tenaga Nasional, Jalan IKRAM-UNITEN, Kajang 43000, Selangor, Malaysia

A. S. Norfarhana, Abu Hassan Nordin and H. S. N. Hawanis, Department of Chemical Engineering, Faculty of Chemical and Energy Engineering, Universiti Teknologi Malaysia, 81310 UTM Johor Bahru, Johor, Malaysia. https://orcid.org/0000-0001-7546-3833 (A.H. Nordin)

A. Khalina, Institute of Tropical Forestry and Forest Products, Universiti Putra Malaysia, Serdang 43400, Selangor, Malaysia

R. M. O. Syafiq, Department of Engineering Design and Manufacture, Faculty of Engineering, University of Malaya, 50603 Kuala Lumpur, Malaysia

M. R. M. Huzaifah, Department of Crop Science, Faculty of Agricultural and Forestry Sciences, Universiti Putra Malaysia Bintulu Campus (UPMKB), Bintulu, Sarawak, Malaysia

Ashraf Azmi, School of Chemical Engineering, College of Engineering, Universiti Teknologi MARA, 40450, Shah Alam, Selangor, Malaysia

P. S. Khoo and S. A. Hassan, Centre for Advanced Composite Materials, Universiti Teknologi Malaysia, 81310 UTM Johor Bahru, Johor, Malaysia

As per De Gruyter's policy this article has previously been published in the journal Physical Sciences Reviews. Please cite as: A. Nazrin, A. S. Norfarhana, R. A. Ilyas, S.M. Sapuan, A. Khalina, R. M. O. Syafiq, M.R.M. Huzaifah, Ashraf Azmi, P. S. Khoo, A. H. Nordin, H. S. N. Hawanis and S. A. Hassan. "*Araucaria Araucana* thermoplastic starch nanocomposite films reinforced with nanocellulose" *Physical Sciences Reviews* [Online] 2023. DOI: 10.1515/psr-2022-0021 | https://doi.org/10.1515/9783110773606-009

solubility. Further studies involving the reinforcement of *Araucaria Araucana* starch and nanocellulose should be conducted to overcome the lack of information.

Keywords: *Araucaria Araucana*, biopolymer, celluloses, nanocellulose, nanocomposite, thermoplastic starch

9.1 Introduction

Araucaria is an evergreen coniferous tree from the *Araucariaceae* family. The subfamily of *Araucaria* consists of approximately 19 species, which mostly distributed in the Southern Hemisphere. From prehistoric period until today, the two species seeds namely *Araucaria angustifolia* and *Araucaria Araucana* had been consumed by native of Brazil, Argentina and Chile. The geographical distribution of *A. angustifolia* extended to the south and south east of Brazil and north east of Argentina while *A. Araucana* is restricted to high mountain areas in the south of Argentina and Chile [1]. The large seed of *A. Araucana* measuring 3 cm in length and 1.5 cm in width is locally known as "pehuen" are an important source of carbohydrates, which can be eaten raw, boild or toasted [2]. Figure 9.1 shows the *A. Araucana* tree and seeds. *A. Araucana* embryos have two or rarely three cotyledons, which embedded in the central part of the megagametophyte. Figure 9.2 illustrates the components of *A. Araucana* seed. The starch extracted from the seeds can be utilized to make flour and "mudai" a fermented beverage [3]. The pehuen seeds composed of starch (ca. 80 %) and protein (ca. 6 %). Phenolic compound are typically found in both edible and non-edible plant, which

Figure 9.1: The *Araucaria Araucana* (A) tree and (B) seeds [5].

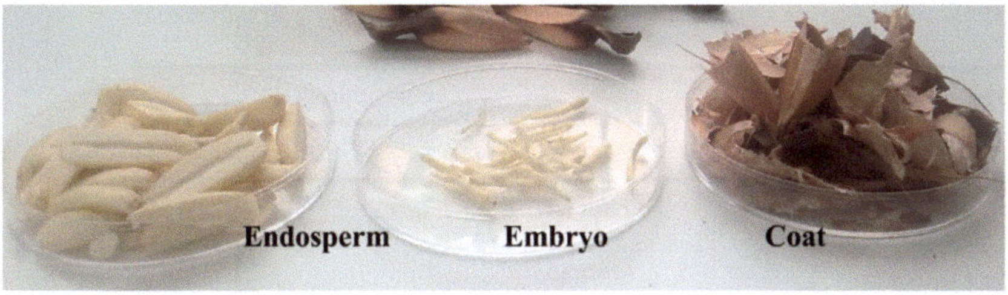

Figure 9.2: The components of *Araucaria Araucana* seed [5].

had been reported to have beneficial biological effect such as antioxidant activity. The phenolic compound varies with different species, cultivator and within different tissues. The seed itself has a very low content of phenolic compound compared to its husk, which are 0.71 ± 0.02 mg of gallic acid equivalents (GAE)/g dry weight (DW) and 66.34 ± 1.89 mg GAE/g DW respectively [4].

9.2 *A. Araucana* starch

The *A. Araucana* starch are made up of low content of ash, crude fibre, lipids and protein. The starch can be easily isolated using water treatment without any additives due to the low content of phenolic compounds. A high starch content and easy extraction process make the pehuen seed to be a promising non-conventional starch source. Its low swelling capacity is potentially used in the noodle making industry while low protein content is suitable for the production of glucose and fructose syrup [6]. The starch granules are small, rounded (ca. 14 µm diameter) and smooth surface with no erosion activity [7]. Several author had reported that there are some inconsistency in the type of crystal structure and amylose content of *A. Araucana* starch. Conforti and Lupano [8] found out that *A. Araucana* starch from Argentina had an A-type structure while Castoña et al. [7] recorded a C-type structure of this starch from Chile. The apparent amylose content of *A. Araucana* starch was reported to be 27.9 %, which was higher than the one extracted from the same species grown in Argentina (17.3 %). Several factors such as cultivation zones, starch extraction method and analytical methods used might be affecting the variation in amylose content. The amylose content greatly influenced the functional properties of starch-based polymer as amylose-rich starches produce harder gel and films. The chromatography analysis of the amylopectin of *A. Araucana* starch indicated a low tendency of long lateral chains to form an inclusion complex with iodine, which are mainly the branch amylopectin chain with degree of polymerization (DP) ≥ 40. The average DP of branch chains of amylopectin is 19.6 %, which consist of mainly a mixture of A-chains (DP 6–12) and B_1-chains (DP 13–24). The distribution shape in the range of DP 6–12 of *A. Araucana* starch is identical to dessert banana (*Musa acuminata L.*) and horse chestnut (*Aesculus hippocastanum*) which are both C-type starches [6]. The proportion of

Table 9.1: The composition of *A. Araucana* seed starch extracted from Argentina [8] and Chile [2].

Component	Composition	
	Argentina	Chile
Amylose	17.3 ± 1.4	27.9
Protein	3.4 ± 0.29	7.81 ± 0.35
Lipid	0.45 ± 0.18	1.11 ± 0.08
Ash	0.42 ± 0.07	2.15 ± 0.01

unit chain of DP 6–12 is associated with the polymorphism type of starch crystals. Table 9.1 indicates the comparison between the composition of *A. Araucana* seed starch extracted from Argentina and Chile regions.

9.2.1 Extraction and preparation of *A. Araucana* starch

The extraction of *A. Araucana* starch involved a simple process by using water treatment without any addition of additives. This starch isolation method was proposed by Henriquez [2]. Firstly, the seeds need to be dehulled and the germs were removed. The peeled pehuen seeds were ground with water in a blender for 5 min. The supernatant need to be filtered through a 60 mesh nytal cloth (particle size of 250 µm) with the supplementary water to ease the starch particles in passing through the mesh. The secondary screening used a 270 mesh nytal cloth (particle size of 53 µm). The yield from the second screening was left at ambient temperature for at least 15 h to decant excess water. The precipitate was suspended into a solution of sodium hydroxide (0.2 w/v of water) for 1 h to solubilize the proteins. The precipitate was proceed to centrifugation process at 5000 rpm for 4 min to separate the starch from the proteins. Concentrated hydrochloric acid solution was used to adjust the pH value around pH 6.0. Later on, the final yield of starch was filtered under vacuum conditions using Whatman No. 4 filter paper and washed with distilled water. Finally, the filtered starch was left to dry in an oven at 60 °C for 1 h.

9.2.2 Properties of *A. Araucana* starch

The starch granules of *A. Araucana* starch are heterogeneous in size with the average size of 8.4 ± 2.9 µm but might differ depending on the botanical source, gene-line variation, growth stage and starch hydrolysis [8]. The starch extracted from *A. Araucana* seed yield 64 % of the dry weight, which is lesser than *A. angustifolia* ranging from 87 % to 72 % of the dry weight. There are several factors that attributing the different starch content such as different ecotypes, growing condition, maturity level and postharvest condition [2]. Figure 9.3 elucidates the starch granules of *Araucaria Araucana* seeds.

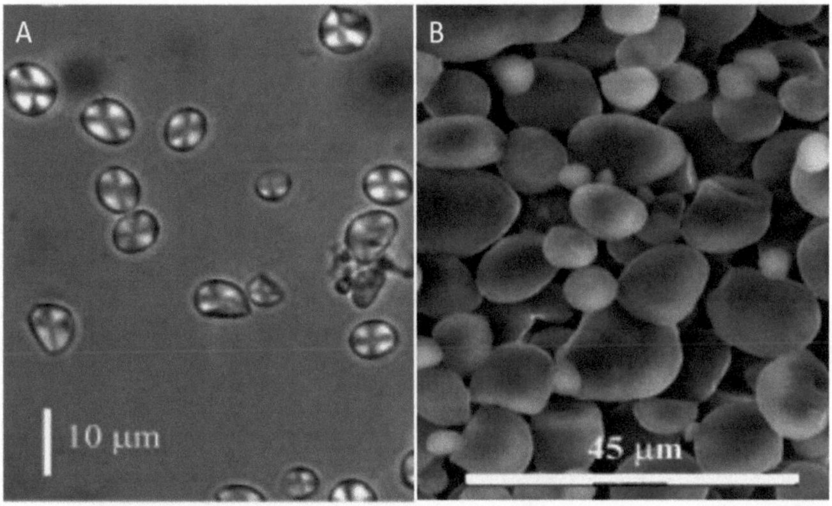

Figure 9.3: *Araucaria Araucana* starch granules. (A) The light micrograph and (B) scanning electron microscope (SEM) images [8].

9.3 Classification of fiber

Biocomposite developed from conventional non-degradable polymer or biopolymer and nonwood biofiber such as flax, hemp, jute, kenaf and husk have gained the interest in recent year for various applications in several sectors including automotive components, aircraft parts, construction, food packaging and sports goods [9]. Advantage of composite materials is strength and stiffness combined with lightness [10, 11]. By choosing an appropriate combination of reinforcement and matrix material, manufacturers can produce properties that exactly fit the requirements for a particular structure for a particular purpose [12, 13]. Major utilization of biofiber as filler is mainly due to their abundancy, low cost, light weight and eco-friendly [14, 15]. Lignocellulosic fiber can be divided into 5 categories based upon the origin of the plant, which are; bast or stem, leaf, fruit or seed, grass and straw fibers [16, 17]. A biofiber is a constitution of cellulose, hemicellulose, lignin and a small content of protein, lipids and starch. The *A. Araucana* seed husk composed of 50 wt% of cellulose, 30 wt% of hemicellulose, 14 wt% of lignin and 0.6 wt% of lipids. Table 9.2 shows the chemical composition of *A. Araucana* husk originated from Chilean forest. A higher hemicellulose and lower lignin content of lignocellulosic agricultural by-products have higher hydrophilic nature compared to wood plants. *A. Araucana* husk has the aspect ratio of 39 with the size distribution of 22.1, 40.3, 21.4, 6.3, 8.0 and 1.9 on the ranges of + 1.0, 0.6–1.0, 0.3–0.6, 0.15–0.3, 0.04–0.3 and less than 0.04 mm, respectively. Figure 9.4 illustrates the external surface of pehuen husk observed through SEM indicated a smooth and homogeneous surface without flaws and damages.

Table 9.2: The chemical composition and physical properties of *A. Araucana* husk from Chilean forest [18].

Properties	Value
Cellulose	50 ± 6
Hemicellulose	31 ± 3
Lignin insoluble	11.7 ± 0.3
Protein	2.1 ± 0.1
Ash	1.4 ± 0.3
Lipids	0.6 ± 0.02
Length	0.89 ± 0.02
Aspect ratio	39
Density	0.305 ± 0.001

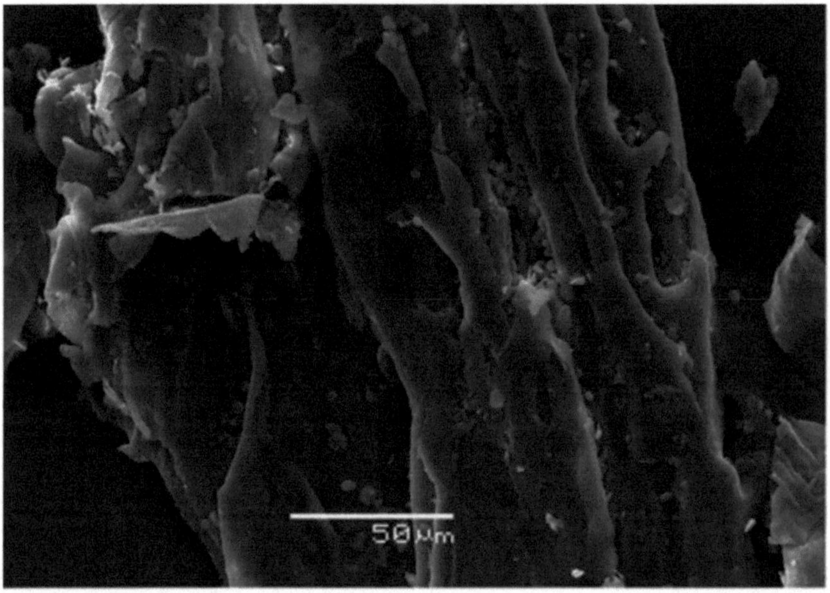

Figure 9.4: The SEM image of pehuen husk external surface.

9.3.1 Cellulose fiber

The husk generated during processing of pehuen seed are used as a raw material or alternative ingredient in human and animal diets. The utilization of *A. Araucana* husk is beneficial to reduce agro-industrial waste as well as to provide high added value in manufacture materials such as composites, drug, paper and food packaging. Cellulose is made up of β-1,4 glycosidic bonds, which constitute around 35–50 % of plant walls. The use of agro-industrial by-products as a mean to extract cellulose is a promising alternative in substituting current forest wood resource in producing paper and cellulose industry. Among other agro-industrial wastes (acai seed, banana pseudostem, soybean husk and passion fruit peel), *A. Araucana* husk indicated higher cellulose content up to 55 %. Both *A. Araucana* husk and extracted cellulose comprise of natural crystalline cellulose phase dispersed in a amorphous matrix, which estimated to have crystallinity values of 19 % for husk and 33 % for extracted cellulose [19]. Hence, it possesses great potential for the purpose of research and application in material science.

9.3.2 Microcrystalline cellulose (MCC)

Abundant lignocellulosic fiber from the by-product of agricultural sector are the main source of renewable cellulose. MCC is a natural substance derived from partially depolymerized pure cellulose. It is traditionally made from processing alpha cellulose using a high concentration of mineral acids [20].The higher content of α-cellulose in A. Araucana husk can be refine to produce MCC. Similarly to MCC from other resources, these biomaterials are applied in food and pharmaceutical industries as a thickening agent [21]. Expend the definition and process to produce MCC.

MCC, also known as microwhiskers or microfibrils, has a high degree of crystallinity, with values ranging from 55 % to 80 % [22]. MCC has a significant potential for usage in specific applications including pharmaceutical, cosmetic, polymer composites, and food industries because of properties such as water insolubility, non-toxicity, strength, stiffness, lightness, fibrous nature, renewability, and biodegradability. It is used in powder form as a filler and binder in medical tablets, food, and most notably as a reinforcing ingredient in the manufacture of polymer composites. MCC is used colloidally in creams and pastes as an emulsifier, suspension stabiliser, viscosity regulator, and water retainer [20].

The properties and application areas of MCCs are greatly influenced by the raw material [23]. MCC on an industrial scale was produced by hydrolyzing cotton and wood with dilute mineral acids. With the development of modern technologies, scientists' interests have shifted towards the extraction of microcrystalline cellulose from a variety of materials, including agro wastes such as bagasse, cotton sliver, rice straw, fruit of *Lageriana siceraria* (water gourd) [24], banana pseudo-stem fibres [25], oil palm biomass [22], jute [26], waste paper [27], fodder grass [28], corn cobs [29], industrial residue such as cotton rags [30], cotton fabrics [31], and bleached pulps from cotton stalks [32].

There are several methods of MCC extraction, and the most basically one is via chemical approach. Extraction of MCC can be classified into two stages, which are pre-treatment of raw material and isolation of MCC. The common chemicals used in the pre-treatment process are sodium hydroxide (NaOH), sodium hypochlorite (NaOCl), sodium chlorite ($NaClO_2$), hydrogen peroxide (H_2O_2) and sodium silicate (Na_2SiO_3). Achor et al. [24] applied NaOH and NaOCl in the pre-treatment and bleaching of the bark of *L. siceraria*'s fruit. Bleached cotton stalks pulp can be prepared by cotton stalks pulping with NaOH and $NaClO_2$ bleaching agent [32]. Kale et al. [23] prepared a basic chemical pre-treatment using NaOH, H_2O_2, and Na_2SiO_3 in one bath to eliminate organic dyes and hydrophobic substances such as waxes, oils, fats, minerals, and so on, in order to improve the whiteness and hydrophilicity of cotton silver [23].

Then, referring to Ohwoavworhua et al. [33] method, MCC can be extracted via acid hydrolysis using hydrochloric acid or sulphuric acid with continual mechanical agitation [23, 24, 32]. The amorphous phase of bleached pulping material is readily hydrolyzed, resulting in shorter, more crystalline pieces with minimal weight loss [20]. Following the end of the acid hydrolysis, the MCC was subsequently rinsed with distilled water and centrifuged at 2500 rpm for 15 min using a centrifuge machine. The concentrated MCC was rinsed with distilled water until it achieved a pH of 6–7 [23].

9.3.3 Nanocellulose (NC)

Through the refining of extracted cellulose, NC can be obtained in the form of a set of nano-sized cellulose chains. NC having the hydroxyl groups,-OH promote compact and ordered network structure to prevent lower molar mass substances (alcohol and sugar) to penetrate it. Expend the definition and process to produce NCC.

The numerous hydroxyl groups enable a broad range of functionalizations by chemical approaches or surface modification, resulting in the generation of diverse materials with superior characteristics. NC has numerous applications, such as biomedical devices, nanocomposite materials, textiles, and so forth. The two major types of NC are nanostructured celluloses (cellulose microfibrils and cellulose microcrystals) and nanofibers (cellulose nanocrystals, cellulose nanofibrils, and bacterial cellulose) [34]. Various techniques and biomass materials can be used to manufacture a variety of nanocellulose [35]. The origin of the cellulose, isolation and processing techniques, and any pre-treatments, all influence the structure, size, and other characteristics of each nanocellulose type [34].

In general, the extraction of nanocellulose from biomass involves two stages: (1) pretreatment of the biomass and (2) nanocellulose isolation from cellulose [36]. Physical, chemical, biological, and combinations of these treatments could be used to pre-treat biomass [37–39]. Pre-treatment of biomass is an important step to clean the fibre, and eliminate non-cellulosic compounds such as pectin, ashes, waxes, extractives, lignin, hemicellulose, and others [39, 40]. Babaee et al. [41] used sodium hydroxide in pre-treatment followed by a three-stage bleaching procedure to convert kenaf bast fibre to pulp fibres. The fibrous material was then mechanically isolated in a mechanical blender equipment and processed utilising an ultra-fine grinder to apply shear stresses to the longitudinal fibre axis. Aside from the mechanical approach, NC was separated from cellulose fibre utilising an assortment of processes, including acid hydrolysis and enzymatic hydrolysis [35].

9.4 Modification of *A. Araucana* starch films

In its pure form, starch is unstable, hard to process and easily deteriorate. It is a common practice when dealing with starch-based polymers, where modification is required to meet the requirement for intended application [42, 43]. The introduction of supplementary elements are often applied to reconstruct the initial network structure of the polymer in establishing a better one. Depending on the which desired properties are needed, various kind of modifications can employed such as plasticization, polymer blending and fiber addition.

9.4.1 Plasticization of *A. Araucana* starch films

Basically, starch plasticization process is the partial or complete destruction of the initial crystalline order. The addition of suitable plasticizers together with mechanical shear stresses and heat, the molecular chains of starch were disturbed and formed hydrogen bonding interactions with the plasticizers. The incorporation of plasticizers into starch-based materials is crucial in order to improve their processability and mechanical strength. There are various types of plasticizers that are regularly employed such as water, polyols (glycerol, sorbitol, xylitol, maltitol), urea and formamide. In particular, lower molecular weight of polyols are more effective in reducing the hydrogen bonding between polymer chains. Meanwhile, water plasticized starch are brittle since water can be easily evaporated at ambient temperature. Castaño et al. [7] reported that thermoplastic pehuen starch (TPS) plasticized by glycerol had its tensile strength decreases while elongation at break increases as glycerol content was increased from 30 % to 40 %. Vice versa, TPS plasticized by sorbitol had its tensile strength increases while elongation at break decreases as sorbitol content was increased. In another aspect, the number and length of branched chains of amylopectin plus the molecular weight of amylose could promote the entanglements of amylose and branched amylopectin chains to restrict their chain mobility subsequently improve its strength. The morphological structure of tensile fracture surfaces of glycerol plasticized TPS displayed a continuos phase with rough and homogenous surface. In the case of sorbitol plasticized TPS, the surfaces are smooth with plenty of voids associated with the evaporation of extra water molecules and partial miscibility between between plasticizer and pehuen starch. The circular voids were distributed uniformly in the fracture surface, and as the sorbitol content was increased, the voids size become smaller. The lower molecular weight of glycerol compared to sorbitol ease the penetration into the amylopectin chains to form amylopectin gel-ball, which promote starch chain mobility extending the elongation at break value. Figure 9.5 shows the SEM micrographs of the fracture surface of TPS using different type and content of plasticizer.

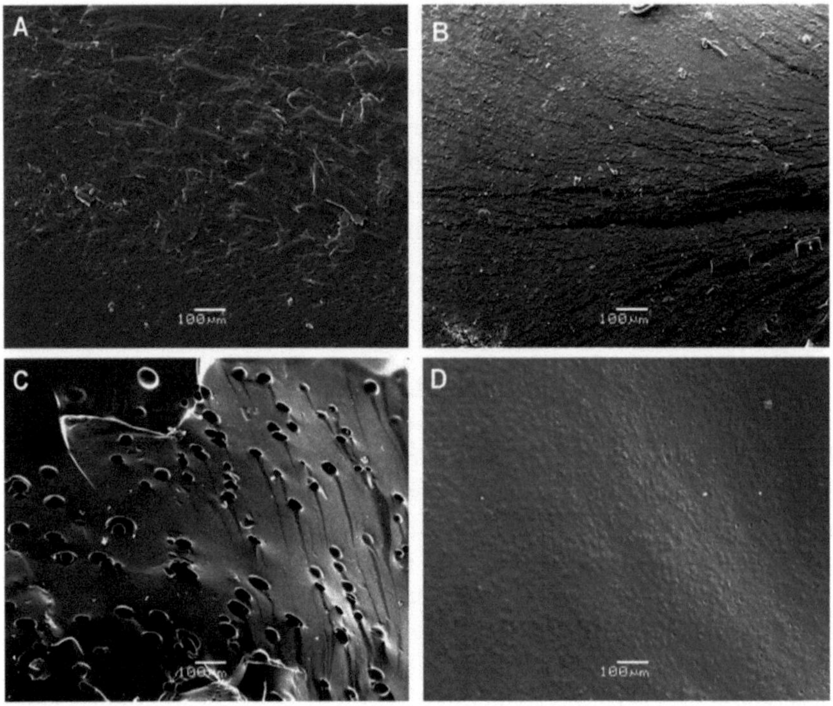

Figure 9.5: The SEM images of the fracture surface of TPS samples plasticized by (A) 30 % of glycerol, (B) 40 % of glycerol, (C) 30 % of sorbitol, (D) 40 % of sorbitol [7].

9.4.2 *A. Araucana* starch blend

Starch-based polymers are well known for their water sensitivity and are often blend with hydrophobic polymer to extend their application that involve moisture contact [44]. Blending TPS with PLA compatibilized using PVA significantly improved their tensile strength approximately 350 %. However, the elongation at break experienced major reduction more than 500 % indicating a rigid material. Similar findings was reported by several author where lower molecular weight of glycerol tends to migrate from TPS into PLA during melt blending process leaving behind sorbitol-rich TPS subsequently reducing the plasticity of the material. The morphological structure of tensile fracture surface of TPS/PLA/PVA blend showed a rough surface and co-continuous phase indicating poor compatibility between TPS and PLA even with the addition of PVA as compatibilizer. Figure 9.6 displays the microstructure image of the fracture surface of TPS/PLA/PVA blend. Meanwhile, TPS/PLA/PVA blend biocomposites revealed microcracks and discontinuities in the interface between the husk and matrix. Figure 9.7 shows the microstructure image of the fracture surface of TPS/PLA/PVA blend composites

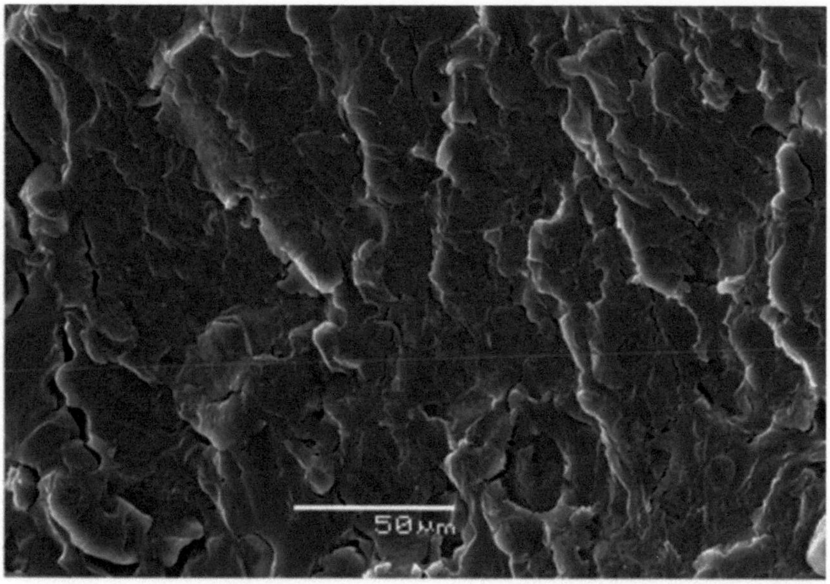

Figure 9.6: The SEM image of the fracture surface of TPS/PLA/PVA blend [18].

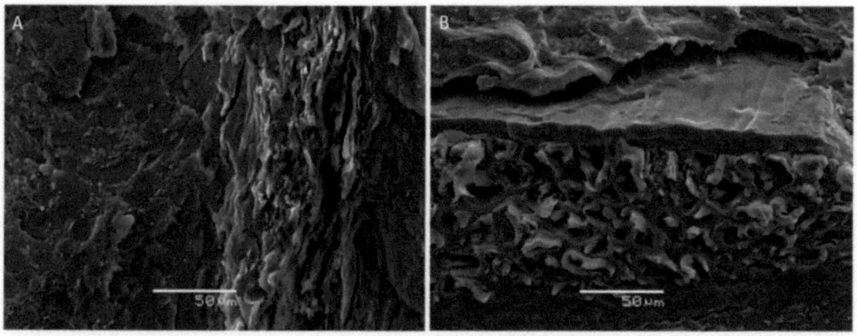

Figure 9.7: The SEM images of (A) TPS5H/PLA/PVA and (B) TPS10H/PLA/PVA [18].

9.4.3 Fiber reinforced *A. Araucana* starch biocomposites

TPS/husk biocomposites established good adhesion bonding as both matrix and filler are made up of hydroxyl group. The SEM images of the samples with 5 % (TPS5H) and 10 % (TPS10H) *A. Araucana* husk loading displayed a random distribution of husk fiber with no obvious gaps between husk/matrix interface. The tensile strength increased from 9 MPa to 9.6 MPa while elongation at break decreased from 40 % to 33 % as the pehuen husk loading was increased from 5 % to 10 %. The reinforcement of fiber established strong fiber and matrix interface bonding, which enables the stress transfer from matrix to fiber but higher loading seemed to diminish the plasticity of the biocomposites. The incorporation of husk fiber into the TPS formed a hydrogen bond with volatile glycerol and restricted the diffusion of oxygen into the polymer, which result in the enhancement of the thermal stability of the biocomposites [45]. At 5 % weight loss, Castaño et al. [7] found out that TPS10H (160 °C) recorded a higher degradation temperature compared to TPS5H (150 °C). Meanwhile, Daudt et al. [46] stated that *A. Angustifolia* starch/husk composite films showed higher water vapor permeability (WVP) compared to the control film. This was associated with the presence of apparent porosity caused by incorporating husk powder that led to the easier penetration of water molecules. The increment in husk loading promote the antioxidant capacity of the composite film as it is rich in phenolic compounds. The natural occurance of antioxidant are beneficial to prolong the packaging film's shelf life and preservation of packed foods. Figure 9.8 reveals the microstructure images of the fracture surface of TPS/husk composite films.

9.4.4 Cellulose reinforced *A. Araucana* starch biocomposites

Cellulose are often employed into starch matrix for the enhancement of the mechanical and water vapor barrier properties in yielding composites with more tear-resistance and more stiff. Furthermore, the good compatibility between starch and cellulose having similar polar hydroxyl groups hinder invasion of water molecules into the composites. In addition of having higher cellulose content, pehuen husk was reported to have greater phenolic compounds as compared to its kernel. This greatly elevate the potential of pehuen cellulose for active packaging. How the cellulose effect the strength of the thermoplastic film?

Cellulose is an essential polysaccharide generating hydrocolloids in the presence of an adequate solvent solution [47]. It has been demonstrated that incorporating cellulose into starch-based films improves biofilm stability and stiffness. Dias et al. [48] discovered that the moisture sorption isotherm of a rice flour composite film reinforced with eucalyptus cellulose had a lower hygroscopicity, a low water solubility coefficient and a

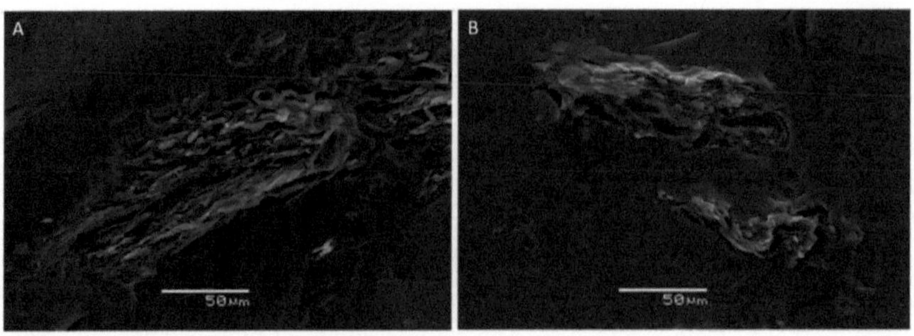

Figure 9.8: The SEM images of the fracture surface of (A) TPS5H, (B) TPS10H [7].

high diffusion coefficient. In the case of similar glycerol content, the reinforcement of cellulose fibres decreased the water vapour permeabilities values by approximately 35 %, whereas for the case of films plasticized with sorbitol, the water vapour permeabilities values decreased by approximately 14 % [48]. This is due to strong linkages between the cellulose fibres and the hydrophilic groups of the starch chain, which replaced for and reduced starch water interactions [48].

Lignin, for example, is thought to be a highly reactive thermoplastic polymer, and it is mostly found on the cellulose surface, forming covalent bonds with cellulose molecules [49]. The hydroxyl groups in lignin can bond with starch and glycerol, serving as an interfacial crosslinking agent between celluloses and starch, resulting in higher tensile strength films [50]. Starch films have higher tensile strength and elongation at break owing to strong interfacial adhesion between hydroxyl groups of starch and cellulose, which can establish hydrogen bonds between interfaces [51].

Tensile strength and elongation at break both increased as cellulose content increased owing to a rise in the overall solid concentration of the film mixture and the high polymerizing capacity of cellulose, which improves the tensile strength and elongation at break of the biopolymer film. Young's modulus, tensile strength, and elongation at break of biopolymer films made from tamarind seed starch reinforced cellulose with optimal concentrations of the film constituents were determined to be 1.55 MPa, 7.68 MPa, and 494.41 %, respectively [52].

9.4.5 Microcrystalline cellulose (MCC) reinforced *A. Araucana* starch

Loaded with abundant cellulose content, the utilization of pehuen husk can benefits the material industries in obtaining raw materials, which are renewable, sustainable and biodegradable for lower price but high functional properties. Under mechanical and chemical treatments, the fibril bundles in the cellulose network are broke down leaving microcrystalline cellulose. By incorporating MCC, the crystallininty of the composites are enhanced, thus, establishing a more compact and complex composite structure. Though, when dealing with hydrophobic polymer matrix, surface modification using stearic acid could be employed to react with the hydroxyl groups of MCC.

9.4.6 Nanocellulose reinforced *A. Araucana* starch

Nano-sized fillers are well-known for its efficiency in elevating the mechanical, chemical, thermal and gas barrier properties of polymer matrix. Due to their remarkable features of high aspect ratio of 100 and surface to volume, the compounded matrix and filler established strong interfacial adhesion bonding. Nanocellulose is yet one of the remarkable nanofiller that had gained the interest of scientists and researchers in material science industry [53, 54]. It is safe, securable and sustainable if compared with current synthetic fiber such as fiberglass and carbon-based fillers. Nanocellulose has a high compatibility with starch-based polymer as both formed hydrogen bonds when compounded. The well dispersed nanocellulose in thermoplastic starch acts as a load-bearing and stress transfer in the nanocomposites result in high resistance against tear and fracture. Furthermore, in active food packaging application, nanocellulose was reported to act as a controlled release agent regulating the concentration of bioactive compound in prolonging the preservation of foods. The innovation of pehun-based nanocellulose reinforced thermoplastic starch to develop bio-based plastic possess a great potential as an alternative to replace current petroleum-based plastic. Please elaborate more on this section.

NC functionalization with active groups capable of incorporating extra functionalities into its structures and materials [55]. Food packaging with antimicrobial properties based on silver nanoparticles (AgNPs), for example, have received a lot of attention [56, 57] because of its broad antibacterial range, high temperature stability, and strong processing capabilities [58]. Dong et al. [56] altered CNFs with carboxylate groups that reacted strongly with AgNPs via TEMPO oxidation method. The inclusion of AgNPs in films inhibited CNF coalescence, most likely due to carboxylate-Ag interaction, which limited water solubilisation and hydrogen bonding with nearby CNFs. Fortunati et al. [57] used melt extrusion to create poly(lactic acid) (PLA) films with CNCs and AgNPs. The antibacterial characteristics of the AgNPs functionalized materials against Gram-positive and Gram-negative bacteria suggest that they could be used in smart applications for food packaging [56, 57]. Yet, when acid phosphate ester of ethoxylated nonylphenol surfactant-treated CNCs were utilised, silver was easily discharged to ethanol 10 %, exceeding the specified migration limitations authorised by legislation [57].

Apart from AgNPs, Salmieri et al. [59] created compression moulded films for sliced ham utilising PLA and CNCs enhanced with nisin. Nisin guaranteed total prevention the growth of *Listeria monocytogenes* beginning on the third day of storage. Gradually, Nisin was removed from the films, reaching 21 % on day 14. The advancement of bio-based food packaging films is impending because of the usage of petroleum-based polymers in food packaging and environmental pollution issues. The incorporation of NC as an effective controlled release agent and stabiliser boosted the antioxidant and antibacterial characteristics of these biodegradable food packaging films greatly.

9.5 Conclusions

The potential of *A. Araucana* based polymer intended for packaging application can exploited vastly as it holds the potential to rival the current synthetic plastic. The high amylose content of pehuen starch is suitable in developing harder gel and film. Additionally, the pehuen husk composed of high cellulose content up to 50 wt% can be utilized as a reinforcement material to substitute cellulose extracted from forest wood. Furthermore, pehuen husk is rich in phenolic compound exhibiting antioxidant properties extending the shelf life of reinforced composite film. In effort to preserve forest plants and fulfilling the need of raw materials, the usage of agricultural biomass such as pehuen husk is necessary.

References

1. Kurtoğlu S, Tan N, Meriçli A, Miski M. Biological activities and chemical constituents of Ferula anatolica Boiss. Planta Med 2013;79:1–25.
2. Henríquez C, Escobar B, Figuerola F, Chiffelle I, Speisky H, Estévez AM. Characterization of piñon seed (Araucaria araucana (Mol) K. Koch) and the isolated starch from the seed. Food Chem 2008;107:592–601.
3. Durán A, Reyes-De-corcuera J, Garay G, Valencia P, Urtubia A. Development of an araucaria araucana beer-like beverage: process and product. Fermentation 2021;7:1–13.
4. Conforti PA, Lupano CE. Selected properties of Araucaria angustifolia and Araucaria araucana seed protein. Int J Food Prop 2011;14:84–91.
5. Gallia MC, Bachmeier E, Ferrari A, Queralt I, Mazzeo MA, Bongiovanni GA. Pehuén (Araucaria araucana) seed residues are a valuable source of natural antioxidants with nutraceutical, chemoprotective and metal corrosion-inhibiting properties. Bioorg Chem 2020;104:104175.
6. Castaño J, Rodríguez-Llamazares S, Bouza R, Franco CML. Chemical composition and thermal properties of Chilean Araucaria araucana starch. Starch/Staerke 2016;68:100–5.
7. Castaño J, Bouza R, Rodríguez-Llamazares S, Carrasco C, Vinicius RVB. Processing and characterization of starch-based materials from pehuen seeds (Araucaria araucana (Mol) K. Koch). Carbohydr Polym 2012;88: 299–307.
8. Conforti PA, Lupano CE. Starch characterisation of Araucaria angustifolia and Araucaria araucana seeds. Starch/Staerke 2007;59:284–9.
9. Suriani MJ, Ilyas RA, Zuhri MYM, Khalina A, Sultan MTH, Sapuan SM, et al. Critical review of natural fiber reinforced hybrid composites: processing, properties, applications and cost. Polymers 2021;13:3514.
10. Chan JX, Wong JF, Petrů M, Hassan A, Nirmal U, Othman N, et al. Effect of nanofillers on Tribological properties of polymer nanocomposites: a review on recent development. Polymers 2021;13:2867.
11. Sharma S, Sudhakara P, Omran AAB, Singh J, Ilyas RA. Recent Trends and developments in conducting polymer nanocomposites for multifunctional applications. Polymers 2021;13:2898.
12. Nurazzi NM, Sabaruddin FA, Harussani MM, Kamarudin SH, Rayung M, Asyraf MRM, et al. Mechanical performance and applications of CNTs reinforced polymer composites—a review. Nanomaterials 2021;11: 2186.
13. Sabaruddin FA, Paridah MT, Sapuan SM, Ilyas RA, Lee SH, Abdan K, et al. The effects of unbleached and bleached nanocellulose on the thermal and flammability of polypropylene-reinforced kenaf core hybrid polymer bionanocomposites. Polymers 2020;13:116.
14. Mohammed AABA, Hasan Z, Omran AAB, Elfaghi AM, Ali YH, Ilyas RA, et al. Effect of sugar palm fibers on the properties of blended wheat starch/polyvinyl alcohol (PVA) -based biocomposite films. J Mater Res Technol 2023;24:1043–55.
15. Nurazzi NM, Khalina A, Chandrasekar M, Aisyah HA, Rafiqah SA, Ilyas RA, et al. Effect of fiber orientation and fiber loading on the mechanical and thermal properties of sugar palm yarn fiber reinforced unsaturated polyester resin composites. Polimery 2020;65:115–24.
16. Asyraf MRM, Ishak MR, Syamsir A, Nurazzi NM, Sabaruddin FA, Shazleen SS, et al. Mechanical properties of oil palm fibre-reinforced polymer composites: a review. J Mater Res Technol 2022;17:33–65.
17. Azman MA, Asyraf MRM, Khalina A, Petrů M, Ruzaidi CM, Sapuan SM, et al. Natural fiber reinforced composite material for product design: a short review. Polymers 2021;13:1917.

18. Castaño J, Rodríguez-Llamazares S, Carrasco C, Bouza R. Physical, chemical and mechanical properties of pehuen cellulosic husk and its pehuen-starch based composites. Carbohydr Polym 2012;90:1550–6.
19. Barros S de S, Pessoa WAG Jr., Júnior AC, Borges ZV, Poffo CM, Regis DM, et al. Value aggregation of pine (Araucaria angustifolia) nuts agro-industrial waste by cellulose extraction. Res Soc Dev 2021;10: e270101018836.
20. Trache D, Hussin MH, Hui Chuin CT, Sabar S, Fazita MRN, Taiwo OFA, et al. Microcrystalline cellulose: isolation, characterization and bio-composites application—a review. Int J Biol Macromol 2016;93:789–804.
21. Ventura-Cruz S, Tecante A. Nanocellulose and microcrystalline cellulose from agricultural waste: review on isolation and application as reinforcement in polymeric matrices. Food Hydrocolloids 2021;118:106771.
22. Haafiz MKM, Eichhorn SJ, Hassan A, Jawaid M. Isolation and characterization of microcrystalline cellulose from oil palm biomass residue. Carbohydr Polym 2013;93:628–34.
23. Kale RD, Bansal PS, Gorade VG. Extraction of microcrystalline cellulose from cotton sliver and its comparison with commercial microcrystalline cellulose. J Polym Environ 2018;26:355–64.
24. Achor M, Oyeniyi YJ, Yahaya A. Extraction and characterization of microcrystalline cellulose obtained from the back of the fruit of Lageriana siceraria (water gourd). J Appl Pharmaceut Sci 2014;4:57–60.
25. Shanmugam N, Nagarkar RD, Khurade M. Microcrystalline cellulose powder from banana pseudostem fibres using biochemical route. Indian J Nat Prod Resour 2015;6:42–50.
26. Jahan MS, Saeed A, He Z, Ni Y. Jute as raw material for the preparation of microcrystalline cellulose. Cellulose 2011;18:451–9.
27. Okwonna O. The effect of pulping concentration treatment on the properties of microcrystalline cellulose powder obtained from waste paper. Carbohydr Polym 2013;98:721–5.
28. Kalita RD, Nath Y, Ochubiojo ME, Buragohain AK. Extraction and characterization of microcrystalline cellulose from fodder grass; Setaria glauca (L) P. Beauv, and its potential as a drug delivery vehicle for isoniazid, a first line antituberculosis drug. Colloids Surf B Biointerfaces 2013;108:85–9.
29. Azubuike CP, Okhamafe AO. Physicochemical, spectroscopic and thermal properties of microcrystalline cellulose derived from corn cobs. Agriculture J Recycl Organic Waste Agriculture 2012;1:1–7.
30. Chauhan YP, Sapkal RS, Sapkal VS, Zamre GS. Microcrystalline cellulose from cotton rags (waste from garment and hosiery industries). Int J Chem Sci 2009;7:681–8.
31. Sun X, Lu C, Liu Y, Zhang W, Zhang X. Melt-processed poly(vinyl alcohol) composites filled with microcrystalline cellulose from waste cotton fabrics. Carbohydr Polym 2014;101:642–9.
32. El-Sakhawy M, Hassan ML. Physical and mechanical properties of microcrystalline cellulose prepared from agricultural residues. Carbohydr Polym 2007;67:1–10.
33. Ohwoavworhua FO, Kunle OO, Ofoefule SI. Extraction and characterization of microcrystalline cellulose derived from Luffa cylindrica plant. Afri J Pharm Res Dev 2004;1:1–6.
34. Trache D, Tarchoun AF, Derradji M, Hamidon TS, Masruchin N, Brosse N, et al. Nanocellulose: from fundamentals to advanced applications. Front Chem 2020;8:392.
35. Phanthong P, Reubroycharoen P, Hao X, Xu G, Abudula A, Guan G. Nanocellulose: extraction and application. Carbon Resour Convers 2018;1:32–43.
36. Autet AC. Production of high value-added nanocellulose from agriculture biomass waste. Barcelona, Spain: Universitat de Barcelona; 2022.
37. Amin FR, Khalid H, Zhang H, Rahman S, Zhang R, Liu G, et al. Pretreatment methods of lignocellulosic biomass for anaerobic digestion. Amb Express 2017;7:72.
38. Anukam A, Berghel J. Biomass pretreatment and characterization: a review. In: Basso TP, Basso TO, Basso LC, editors. Biotechnological applications of biomass, 1st ed. IntechOpen; 2020:19–36 pp.
39. Shrestha S, Kognou ALM, Zhang J, Qin W. Different facets of lignocellulosic biomass including pectin and its perspectives. Waste and Biomass Valorization 2021;12:4805–23.
40. Gupta GK, Shukla P. Lignocellulosic biomass for the synthesis of nanocellulose and its eco-friendly advanced applications. Front Chem 2020;8:1–13.

41. Babaee M, Jonoobi M, Hamzeh Y, Ashori A. Biodegradability and mechanical properties of reinforced starch nanocomposites using cellulose nanofibers. Carbohydr Polym 2015;132:1–8.
42. Diyana ZN, Jumaidin R, Selamat MZ, Ghazali I, Julmohammad N, Huda N, et al. Physical properties of thermoplastic starch derived from natural resources and its blends: a review. Polymers 2021;13:1–20.
43. Tarique J, Sapuan SM, Khalina A, Sherwani SFK, Yusuf J, Ilyas RA. Recent developments in sustainable arrowroot (Maranta arundinacea Linn) starch biopolymers, fibres, biopolymer composites and their potential industrial applications: a review. J Mater Res Technol 2021;13:1191–219.
44. Asrofi M, Sapuan SM, Ilyas RA, Ramesh M. Characteristic of composite bioplastics from tapioca starch and sugarcane bagasse fiber: effect of time duration of ultrasonication (bath-type). Mater Today Proc 2020;46: 1626–30.
45. Sahari J, Sapuan SM, Zainudin ES, Maleque MA. Thermo-mechanical behaviors of thermoplastic starch derived from sugar palm tree (Arenga pinnata). Carbohydr Polym 2013;92:1711–6.
46. Daudt RM, Sinrod AJG, Avena-Bustillos RJ, Külkamp-Guerreiro IC, Marczak LDF, McHugh TH. Development of edible films based on Brazilian pine seed (Araucaria angustifolia) flour reinforced with husk powder. Food Hydrocolloids 2017;71:60–7.
47. Mishra A, Malhotra AV. Tamarind xyloglucan: a polysaccharide with versatile application potential. J Mater Chem 2009;19:8528–36.
48. Dias AB, Müller CMO, Larotonda FDS, Laurindo JB. Mechanical and barrier properties of composite films based on rice flour and cellulose fibers. LWT – Food Sci Technol 2011;44:535–42.
49. Kunanopparat T, Menut P, Morel MH, Guilbert S. Reinforcement of plasticized wheat gluten with natural fibers: from mechanical improvement to deplasticizing effect. Compos Appl Sci Manuf 2008;39:777–85.
50. Mali S, Grossmann MVE, García MA, Martino MN, Zaritzky NE. Microstructural characterization of yam starch films. Carbohydrate Polym 2002;50:379–86.
51. Agustin MB, Ahmmad B, Leon ERPD, Buenaobra JL, Salazar JR, Hirose F. Starch-based biocomposite films reinforced with cellulose nanocrystals from garlic stalks. Polym Compos 2013;34:1325–32.
52. Sudharsan K, Mohan CC, Babu PAS, Archana G, Sabina K, Sivarajan M, et al. Production and characterization of cellulose reinforced starch (CRT) films. Int J Biol Macromol 2016;83:385–95.
53. Kadier A, Ilyas RA, Huzaifah MRM, Harihastuti N, Sapuan SM, Harussani MM, et al. Use of industrial wastes as sustainable nutrient sources for bacterial cellulose (BC) production: mechanism, advances, and future perspectives. Polymers 2021;13:1–51.
54. Norfarhana AS, Ilyas RA, Ngadi N. A review of nanocellulose adsorptive membrane as multifunctional wastewater treatment. Carbohydr Polym 2022;291:119563.
55. Azeredo HMC, Rosa MF, Mattoso LHC. Nanocellulose in bio-based food packaging applications. Ind Crop Prod 2017;97:664–71.
56. Dong H, Snyder JF, Tran DT, Leadore JL. Hydrogel, aerogel and film of cellulose nanofibrils functionalized with silver nanoparticles. Carbohydr Polym 2013;95:760–7.
57. Fortunati E, Armentano I, Zhou Q, Iannoni A, Saino E, Visai L, et al. Multifunctional bionanocomposite films of poly(lactic acid), cellulose nanocrystals and silver nanoparticles. Carbohydr Polym 2012;87:1596–605.
58. Kumar R, Münstedt H. Silver ion release from antimicrobialpolyamide/silver composites. Biomaterials 2005;26:2081–8.
59. Salmieri S, Islam F, Khan RA, Hossain FM, Ibrahim HMM, Miao C, et al. Antimicrobial nanocomposite films made of poly(lactic acid)-cellulose nanocrystals (PLA-CNC) in food applications: part A—effect of nisin release on the inactivation of listeria monocytogenes in ham. Cellulose 2014;21:1837–50.

R. A. Ilyas*, A. Nazrin, M. R. M. Huzaifah, S. M. Sapuan, R. M. O. Syafiq, N. M. Nurazzi, M. R. M. Asyraf, M. N. F. Norrrahim, M. N. A. Uda, K. Z. Hazrati and L. Rajeshkumar

10 Banana starch nanocomposite films reinforced with nanocellulose

Abstract: Owing to the increasing global pollution due to the use of conventional plastic materials, biodegradable films and sheets have been given prominent importance owing to their efficiency in replacing traditional plastic materials. In the recent times, due to globalization and industrialization, the minimization of the use of synthetic petroleum-based materials has been in practice by the global community, and increase in the use of natural materials like biomass waste in place of traditional materials promoting the green and sustainable technology development. Such production of alternative bio-materials development from the lignocellulosic fiber-reinforced composite films and laminates has been seriously investigated by engineers, technologists, and materialists owing to their sustainability, renewability, low toxicity, and abundantly available. Banana is one among the largely consumed tropical crop globally which is due to the

***Corresponding author: R. A. Ilyas**, Department of Chemical Engineering, Faculty of Chemical and Energy Engineering, Universiti Teknologi Malaysia, 81310 UTM Johor Bahru, Johor, Malaysia; Centre for Advanced Composite Materials, Universiti Teknologi Malaysia, 81310 UTM Johor Bahru, Johor, Malaysia; Centre of Excellence for Biomass Utilization, Universiti Malaysia Perlis, 02600, Arau, Perlis, Malaysia; and Institute of Tropical Forestry and Forest Products, Universiti Putra Malaysia, Serdang 43400, Selangor, Malaysia, E-mail: ahmadilyas@utm.my. https://orcid.org/0000-0001-6622-2632

A. Nazrin, Institute of Tropical Forestry and Forest Products, Universiti Putra Malaysia, Serdang 43400, Selangor, Malaysia

M. R. M. Huzaifah, Department of Crop Science, Faculty of Agricultural and Forestry Sciences, Universiti Putra Malaysia Bintulu Campus (UPMKB), Bintulu, Sarawak, Malaysia

S. M. Sapuan, Advanced Engineering Materials and Composites Research Center (AEMC), Faculty of Engineering, Universiti Putra Malaysia, 43400 Seri Kembangan, Selangor, Malaysia

R. M. O. Syafiq, Department of Engineering Design and Manufacture, Faculty of Engineering, University of Malaya, 50603 Kuala Lumpur, Malaysia

N. M. Nurazzi, Bioresource Technology Division, School of Industrial Technology, Universiti Sains Malaysia, 11800, Pulau Pinang, Malaysia

M. R. M. Asyraf, Centre for Advanced Composite Materials, Universiti Teknologi Malaysia, 81310 UTM Johor Bahru, Johor, Malaysia; and Engineering Design Research Group (EDRG), Faculty of Mechanical Engineering, Universiti Teknologi Malaysia, 81310, Johor Bahru, Johor, Malaysia

M. N. F. Norrrahim, Research Centre for Chemical Defence, Universiti Pertahanan Nasional Malaysia (UPNM), Kem Perdana Sungai Besi, 57000, Kuala Lumpur, Malaysia

M. N. A. Uda, Institute of Nano Electronic Engineering, Universiti Malaysia Perlis, Kangar 01000, Malaysia

K. Z. Hazrati, German-Malaysian Institute, Taman Universiti, Jalan Ilmiah, Kajang 43000, Malaysia

L. Rajeshkumar, Centre for Materials and Manufacturing Technology, KPR Institute of Engineering and Technology, Coimbatore 641407, Tamilnadu, India

As per De Gruyter's policy this article has previously been published in the journal Physical Sciences Reviews. Please cite as: R. A. Ilyas, A. Nazrin, M. R. M. Huzaifah, S. M. Sapuan, R. M. O. Syafiq, N. M. Nurazzi, M. R. M. Asyraf, M. N. F. Norrrahim, M. N. A. Uda, K. Z. Hazrati and L. Rajeshkumar "Banana starch nanocomposite films reinforced with nanocellulose" *Physical Sciences Reviews* [Online] 2023. DOI: 10.1515/psr-2022-0034 | https://doi.org/10.1515/9783110773606-010

quantity of cultivation and the availability of calorific values in the fruit. It is also a fruit which is potentially rich in starch content with more than 65 % of starch. This banana starch can be considered as a potential material for the manufacturing of coating material, edible sheet, food preservation and packaging owing to its innocuous, less cost, and biodegradability. The application of banana starch is dictated by the functional, structural and physicochemical properties of the developed material. But an unmodified native banana starch may not be characterized with all the necessary properties for any customized application. In order to widen the application band of the banana starch, it has to be modified by using enzymes, chemical, and physical methods. This review comprehensively deals with the extraction and synthesis of banana starch from various sources, their modification, properties of the banana starch and its characterization. Some aspects regarding the hybridization of the banana starch with nanoparticles has also been given.

Keywords: banana; cellulose; nanocellulose; nanocomposites; packaging; thermoplastic starch.

10.1 Introduction

Petroleum-based plastics have revolutionized our lives by providing us with convenient and affordable products. However, the over-reliance on these non-biodegradable plastics has resulted in environmental problems, such as plastic pollution and greenhouse gas emissions. Bioplastics, made from renewable resources, have emerged as a potential solution to these issues. Some of the problems with current petroleum-based plastics are (1) non-biodegradability: Petroleum-based plastics are non-biodegradable and can persist in the environment for centuries, contributing to plastic pollution, (2) fossil fuel dependency: The production of petroleum-based plastics relies on fossil fuels, a finite resource that has significant environmental impacts, such as greenhouse gas emissions, and (3) recycling challenges: The recycling of petroleum-based plastics is challenging due to their complex composition and the lack of recycling infrastructure, resulting in low recycling rates [1, 2]. Thus, to overcome this issues, petroleum based plastics are recommended to be replaced by bioplastics due to its (1) raw materials: Bioplastics are made from renewable resources, such as corn, sugarcane, and cellulose, reducing our dependence on fossil fuels, (2) biodegradability: Bioplastics can be designed to be biodegradable, breaking down in the environment and reducing plastic pollution, (3) recycling compatibility: Bioplastics can be designed to be compatible with existing recycling infrastructure, improving recycling rates, and (4) functional properties: Bioplastics can be engineered to have the same or better functional properties as petroleum-based plastics, such as strength and durability. There several types of bioplastics such as polylactic acid (PLA), polyhydroxyalkanoates (PHA), polybutylene succinate (PBS), cellulose-based plastics, and thermoplastic starch (TPS).

Thermoplastic starch are renewable and biodegradable materials that have gained considerable attention in recent years due to their environmentally friendly nature. Thermoplastic starch is a biodegradable polymer that is derived from renewable sources. It is an attractive matrix material for biocomposites due to its good mechanical properties, low cost, and biodegradability. Several studies have investigated the use of thermoplastic starch as a matrix material in composites. For example, a study by Yang et al. (2015) found that the addition of thermoplastic starch improved the tensile strength and modulus of the composite. Banana thermoplastic starch (BTPS) is a biodegradable material made from banana peel and thermoplastic starch. The use of BTPS has gained attention as a sustainable alternative to petroleum-based plastics due to its biodegradability and renewability.

While thermoplastic starch (TPS) has many advantages, such as being biodegradable and derived from renewable resources, there are also some disadvantages associated with its use such as low mechanical properties [3]. TPS is sensitive to moisture, which can affect its mechanical properties and lead to degradation over time. Therefore, it requires careful handling and storage. Besides that, TPS can be brittle and prone to cracking, particularly at low temperatures. Thus, in order to overcome this problem, TPS is being reinforced with lignocellulosic fiber to enhance the mechanical properties of TPS composites. Nanocellulose isolated from lignocellulosic fiber, on the other hand, has been studied for its remarkable mechanical and physicochemical properties that make it a promising reinforcement agent for biocomposites. Current review focusses on exploring the latest research on the development of nanocellulose-reinforced banana thermoplastic starch composites.

10.2 Banana

Banana plant (Musa spp.) is a herbaceous perennial plant native to tropical regions of Southeast Asia. It is cultivated worldwide for its fruit, which is a popular food crop consumed by millions of people worldwide. Banana plant belongs to the family Musaceae and the genus Musa. It is a herbaceous plant that grows up to 10–20 feet tall and has large leaves arranged spirally around a central pseudostem, as shown in Figure 10.1 [4]. The fruit of banana plant is a berry, which is long and curved, with a soft and sweet edible pulp. Banana plant is grown in tropical regions with warm and humid climates. It requires well-drained soil, regular watering, and high levels of nutrients, such as potassium and nitrogen. There are two main types of bananas: dessert bananas, which are eaten raw, and plantains, which are cooked and used in savory dishes. Banana plants are propagated through vegetative means, such as suckers, which are small shoots that grow from the base of the parent plant.

Banana fruit is consumed worldwide and is proven to have abundant minerals, vitamins and dietary fibers. It is also used in the preparation of various food products, such as banana chips, banana bread, and smoothies. Banana plant has several medicinal

Figure 10.1: Banana trees [4].

uses. The fruit, leaves, and stem are used in traditional medicine for treating various ailments, such as diarrhea, dysentery, and high blood pressure. The fibers obtained from the stem of the banana plant are used to produce textiles, such as mats, ropes, and paper. Banana plant is also used in agriculture as a source of organic matter for composting and as a natural mulch to enrich the soil and reduce soil erosion.

10.3 Banana starch

10.3.1 Extraction and preparation of banana starch

Banana thermoplastic starch (BTPS) is a biodegradable material made from banana peel. The use of BTPS has gained attention as a sustainable alternative to petroleum-based plastics due to its biodegradability and renewability. Abundant source of starch can be found in the mature green stage of banana, with the green pulp containing around 70–80 % starch on a dry weight basis [5]. Due to the low-level impurities, a wet milling process is ideal for isolating banana starch, and various techniques such as alkaline and non-alkaline extraction have been reported by Pelissari et al. [6]. During the olden days, the starch was squeezed out of the ground short fibers by adding water into it gradually and this separates the starch from the mixture. Then, the mixture of water and starch was sieved to separate them from any possible impurities. After this, the starch in the starch–water mixture was allowed to settle down completely over a period of time and then the water is completely emptied out. Drying of the separated starch is carried out in open air atmosphere first and then using an hot air oven at 120 °C.

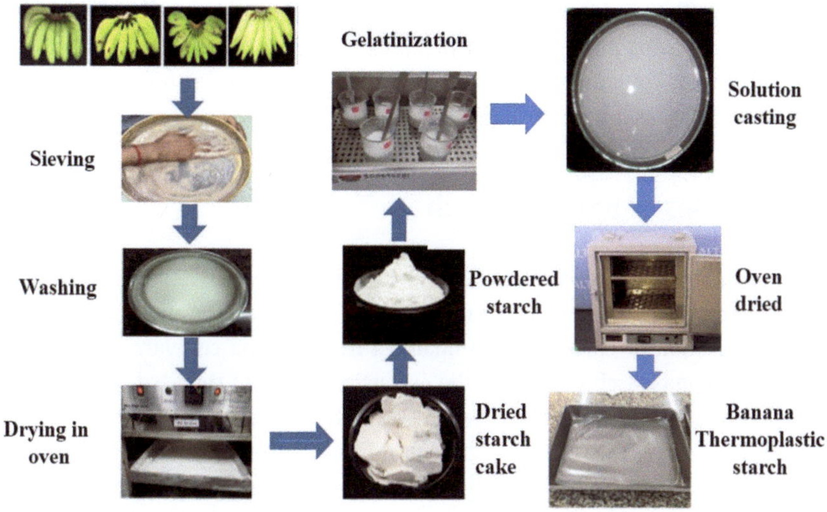

Figure 10.2: Preparation of banana starch and banana thermoplastic starch.

The preparation of banana starch and banana thermoplastic starch is displayed in Figure 10.2. The solution casting method was commonly used to fabricate thermoplastic starch, as described by Ilyas et al. [7]. To make a homogeneous film, a mixture of starch, distilled water, and plasticizers was blended. An amount of 10 g of starch and plasticizers (30 % based on the starch weight) were stirred at 1000 rpm for 20 min at 85 °C in a disperser to gelatinize the starch, ensuring that the granules were uniformly degraded and a homogeneous dispersion was formed. Next, the film-forming suspension was cooled and vacuumed to remove air bubbles before the suspension was cast into molding plate. Finally, the resulting films were allowed to dry at 40 °C overnight before being peeled from the casting surfaces. Evans et al. [8] patented the method of extraction of banana starch with green banana as source. According to his technique, the green banana is soaked in a solution of water with sodium bisulfite salt in it. The resulting liquid suspension is then centrifuged so that the white starch of banana gets separated from the green source material.

Using 95 % ethanol, the segregated banana starch is then washed to make it free from the sodium bisulfite which results in the separation of banana starch. According to Zhang et al. [9] there are 7 steps to make banana peel starch. First, the bananas are to be inspected for their ripening condition to ensure the dryness of the starch after the process. Overriped bananas might not result in a starch which can dry completely. Second, the bananas are to be washed thoroughly and the outer skin must be peeled off. Following that, thin slices of bananas are to be made from peeled bananas. The sliced bananas were then soaked in sodium metabisulfite solution for 1 min. This helps in reducing the pulp in the banana and enhances its drying effectiveness. Then the soaked bananas were mechanically dried in a drier at a temperature of 140–170° until the mixture turn hard. The hardened mixture was then ground to fine powder using a food processing machine using pulverizing setting. Finally, the obtained banana starch was packed in an air-tight container and stored in a dry and cool place so that it could be useful until it was dry, as long as for up to 6 months.

10.3.2 Properties of banana starch

The amount of starch is primarily affected by the ripening stages and banana cultivars, as reported by Zhang et al. [9]. Chávez-Salazar et al. [10] conducted study on isolation and partial characterization of banana starch obtained from banana cultivars in Colombia, shows that the banana starch yield from the extraction has been reported to be 5.79 %–12.71 % with a cumulative content of starch as 75.1 % to 83.9 %. Besides carbohydrates, starch is also comprised of some other microconstituents ash, proteins, and lipids [11, 12]. Starch of the best commercial quality indicates high purity or low non-starch component content.

Table 10.1 presents the starch chemical composition which was obtained from different plantain and bananas. The extracted banana starch contains 6.85 %–14.1 % of moisture and

Table 10.1: Starch chemical composition extracted from different varieties of banana.

Banana variety	Moisture content (%)	Ash content (%)	Lipid content (%)	Protein content (%)	Total starch content (%)	Fiber content (%)	Amylose content (%)	Ref.
Musa—Mysore variety	12.3	0.089	-	0.45	90.1	-	37.9	[13]
White Manzano	-	0.09	0.05	0.23	99.5	-	30.2	[14]
Dwarf Cavendish class	-	0.08	0.08	0.35	99.6	-	26.6	
Nanicao class	14.1	0.13	0.18	0.61	84.94	0.14	-	[11]
Grand Naine	10.84	0.088	0.27	0.37	87.88	0.61	-	
Maca class	8.58	0.27	0.38	0.61	89.68	0.48	-	
Prata-Ana	9.9	0.38	0.46	0.99	87.88	0.38	-	
Fhia 18	10.28	0.34	0.01	1.09	88.08	0.18	-	
Terra variety	7.4	0.12	1.84	1.76	-	-	33.12	[15]
Kluai Khai variety	7.84	0.048	0.032	0.18	-	-	20.33	[16]
Hom Tong variety	7.3	0.05	0.11	0.22	-	-	13.38	
Namwa variety	7.17	0.06	0.13	0.21	-	-	28.05	
Enano variety	7.04	1.42	0.75	0.93	-	-	25.39	
Morado variety	8.75	1.17	0.73	0.83	-	-	21.99	[17]
Valery variety	8.96	1.27	0.78	0.93	-	-	19.33	

Table 10.1: (continued)

Banana variety	Moisture content (%)	Ash content (%)	Lipid content (%)	Protein content (%)	Total starch content (%)	Fiber content (%)	Amylose content (%)	Ref.
Macho variety	7.73	1.12	0.81	0.99	–	–	26.37	
Terra variety	8	0.03	0.02	0.97	94.8	0.28	35	[6]
Karpuravali variety	10.56	1.23	0.21	1.18	87.5	–	27.68	[18]
Poovan variety	9.55	0.77	0.24	0.68	81.72	–	23.2	
Sevvazhai	11.19	1.69	0.25	1.13	89.64	–	32.07	
Thenvazhai variety	8.75	2.09	0.27	0.85	85.64	–	24.65	
Plantain (Fench horn variety)	11.56	0.05	–	–	–	–	29.96	[19]
Plantain (Cadaba variety)	12.86	0.29	–	–	–	–	30.66	
Plantain (Agbaba variety)	13.15	0.45	–	–	–	–	30.91	
Macho variety	9.92	0.53	2.45	2.05	98.15	–	–	[20]
Honduras variety	13.7	0.91	0.73	2.17	69.38	0.48	42.06	[21]
Agbaba variety plantain	11.3	0.61	0.45	2.52	64.0	0.71	38.8	
Kapas variety	8.93	–	0.25	0.98	90.06	–	38.6	[12]
Banana (Kepok variety)	6.83	–	0.06	1.01	82.69	–	40.88	
Banana (Ambon variety)	7.98	–	0.05	1.25	81.64	–	32.56	
Nangka variety	7.99	–	0.10	1.59	81.61	–	37.75	
Plantain (Gros Michel variety)	7.9	0.6	0.9	1.12	74.95	–	22.75	[10]
Plantain (Dominico Hartón variety)	8.3	0.35	0.53	0.92	85	–	31.15	
Plantain (FHIA 20 variety)	7.26	0.27	0.55	0.98	83.4	–	28.6	

0.04 % to 2.11 % of ash content. It also contains 0.011 % to 2.48 % of lipid content and 0.18 % to 2.18 % of protein content and the variations in protein and lipid content was owing to the difference in banana variety and the type of extraction method used. Total starch content varied between 69.41 % and 98.12 %, while the contents of crude fiber and amylose ranged between 0.19 % to 0.48 % and 13.38 %–42.1 %, respectively. For plantains, the amylose content in the starch would typically be in between 22.78 % and 38.8 %.

10.3.3 Properties of banana thermoplastic starch

Banana thermoplastic starch (BTPS) is biodegradable and can be broken down by microorganisms, reducing environmental pollution. BTPS is made from banana peel and starch, both of which are renewable resources, reducing our dependence on non-renewable resources. BTPS has good mechanical characteristics including elongation and tensile behavior, making it suitable for various applications. BTPS has good water resistance, making it suitable for packaging applications.

Alanís-López et al. [22] conducted study on extrusion and characterization of thermoplastic starch sheets from banana (*Musa paradisiacal L.*). Thermoplastic starch was successfully fabricated using conical twin screw extruder. Table 10.2 displays the mechanical characteristics of the sheets. Compared to sheets made from native starch (NS), those created from oxidized starches had a notably higher tensile strength (TS), but there were no significant differences between the TS of sheets prepared out of starch which made to undergo oxidation in sodium hypochlorite solution (2.5 % and 3.5 % denoted as OX25 and OX35, respectively). It was observed from the test results that elastic modulus and elongation at break were higher for the sheets made out of oxidized starch when compared with the sheets made of native starch, but the pattern of their results were similar in both cases. Studies performed earlier suggested that polymeric chain association can be disrupted by the plasticizers increasing sheet flexibility. In particular, glycerol can decrease the rigidity of the network, leading to a less dense sheet matrix and a less ordered sheet structure, which in turn increases the ability of the biopolymer chains to move [23].

Table 10.2 presents the solubility range of starch sheets at 25 °C, which varied from 22 % to approximately 28 %. The sheets made from oxidized starches exhibited

Table 10.2: Mechanical characteristics of banana starch sheets (prepared from native and oxidized starch).

Sample	Tensile strength (MPa)	Elongation at break (%)	Elasticity (MPa)	Glycerol/ amylose ratio	WVP × 10^{-11} (g Pa^{-1} s^{-1} m^{-1})	Solubility (%)
NS	13.62 ± 0.399	9.02 ± 0.359	0.359 ± 0.104	1.32	1.55 ± 0.07	22.55 ± 0.40
OX25	17.31 ± 0.438	13.65 ± 0.306	1.591 ± 0.158	3.39	2.19 ± 0.14	25.14 ± 0.29
OX35	17.48 ± 0.47 b	14.11 ± 0.363	1.619 ± 0.224	5.26	2.74 ± 0.18	27.88 0.39

considerably higher amount of solubility when compared with the sheets made of NS. This solubility increase in samples made of oxidized starch may be attributed to the depolymerization of amylopectin and amylose into more soluble contents with low molecular weight. When carbonyl and carboxyl groups were incorporated into these sheets, they underwent strong interaction with water molecules and a strong network of hydrogen bonds was easily formed within them. Mattehey and Hanna [24] stated that the solubility of the sheets may also be linked to the amylose content and they also stated that when the content of amylose decreased the solubility of the sheets increased. López et al. [23] obtained comparable values for solubility of the sheets made from corn starch which was chemically modified. In term of water vapor permeability (WVP), the value of native banana starch sheets is 1.55 ± 0.07 g Pa^{-1} s^{-1} m^{-1}, as presented in Table 10.2.

Besides that, according to Table 10.2 it has been reported that the water vapor permeability (WVP) values of oxidized starch samples were noticeably higher than those of NS samples, which is consistent with the previously observed trend in solubility [22]. This could be attributed towards the conversion of hydroxyl groups to carbonyl or carboxyl group compounds in the structure of NS during process of chemical modification. The oxidized samples, which are more hydrophilic, exhibited a stronger affinity for water vapor, resulting in higher WVP values. The issue of pollution caused by conventional plastics has led to a renewed interest in biodegradable sheets as a potential replacement for traditional packaging materials. Banana starch is a promising candidate for use in coatings, edible sheets, and food packaging and preservation due to its abundance, harmlessness, biodegradability, and affordability.

10.4 Lignocellulosic fiber: macro to nano-sized banana fiber

The categorization of natural fibers is based on their source: plants, animals, or minerals. Cellulose is the main component of all plant fibers, while the animal fibers are made up of proteins, such as hair, silk, and wool [25–27]. Plant fibers or lignocellulosic fibers can be further divided into bast (also called stem or soft sclerenchyma) fibers, leaf or hard fibers, seed, fruit, wood, cereal straw, and other grass fibers. Figure 10.3 provides a diagram that illustrates the classification of various fibers. Besides that, lignocellulosic fibers are classified based on their source, morphology, and chemical composition. Based on the source, they can be classified as agricultural residues (such as rice straw, sugarcane bagasse, and wheat straw), wood-based fibers (such as softwood and hardwood), and non-woody fibers (such as hemp, flax, sisal, and jute). Based on morphology, they can be classified as long fibers (such as jute and hemp), short fibers (such as wheat straw and rice husk), and particles (such as sawdust and wood chips). Based on chemical composition, they can be classified as cellulose-rich fibers (such as cotton and jute), lignin-rich fibers (such as hardwood and bamboo), and hemicellulose-rich fibers (such as flax and kenaf).

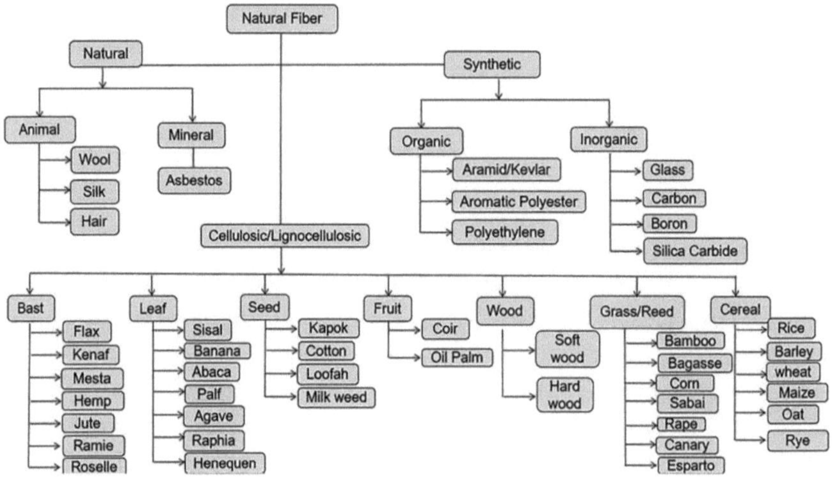

Figure 10.3: Natural fiber classification.

The major constituents of lignocellulosic biomass are cellulose, hemicellulose, and lignin, and their proportion varies according to the source material, as shown in Figure 10.4. Various methods have been proposed in literature to improve the adsorption potential of natural fibers, including the separation of cellulose from other components [28]. Several physicochemical treatments were employed for begetting pure cellulose with augmented hydrophilicity, wider surface area, and active hydroxyl groups on the peripheral surface. Nanocellulose have gained significant attention because of their distinctive characteristics, including high aspect ratio, affordability, and low density. In Figure 10.4, the structure of cellulosic fibers is illustrated, emphasizing the derivation of nanocellulose from cellulosic plant sources, such as cellulose fibers and microfibril bundles that contain both crystalline and amorphous domains.

Banana fibers possess several desirable properties, including lightweight, fire resistance, high strength, low elongation, biodegradability, excellent moisture absorption capabilities, and significant potential for use in various applications. Banana fiber has diameter, density, L/D, microfibrillar angle, initial modulus, ultimate tensile strength (UTS), and elongation of 80–250 µm, 1350 kg/m^3, 150, 10°, 7.7–20.0 GPa, 54–754 MPa, and 10.35 %, respectively, as tabulated in Table 10.3. Another work on mechanical testing of Banana fibers was conducted by Kulkarni et al. [29] revealing that the fibers with diameters ranging from 50 to 250 µm have an ultimate tensile strength (UTS), elastic modulus, and percentage elongation of 711–789 MN m^{-2}, 27–32 GN m^{-2}, and 2.5–3.7 %, respectively, as displayed in Table 10.4. Additionally, when the speed of testing increased from 510 µm/min to 105 mm/min, the tensile strength of banana fiber rose from 729 to 904 MN/m^2 at 105 mm/min. These remarkable mechanical properties of banana fibers are even more impressive when considering that they are derived from the waste of banana cultivation, and can be used to manufacture useful components for various industries.

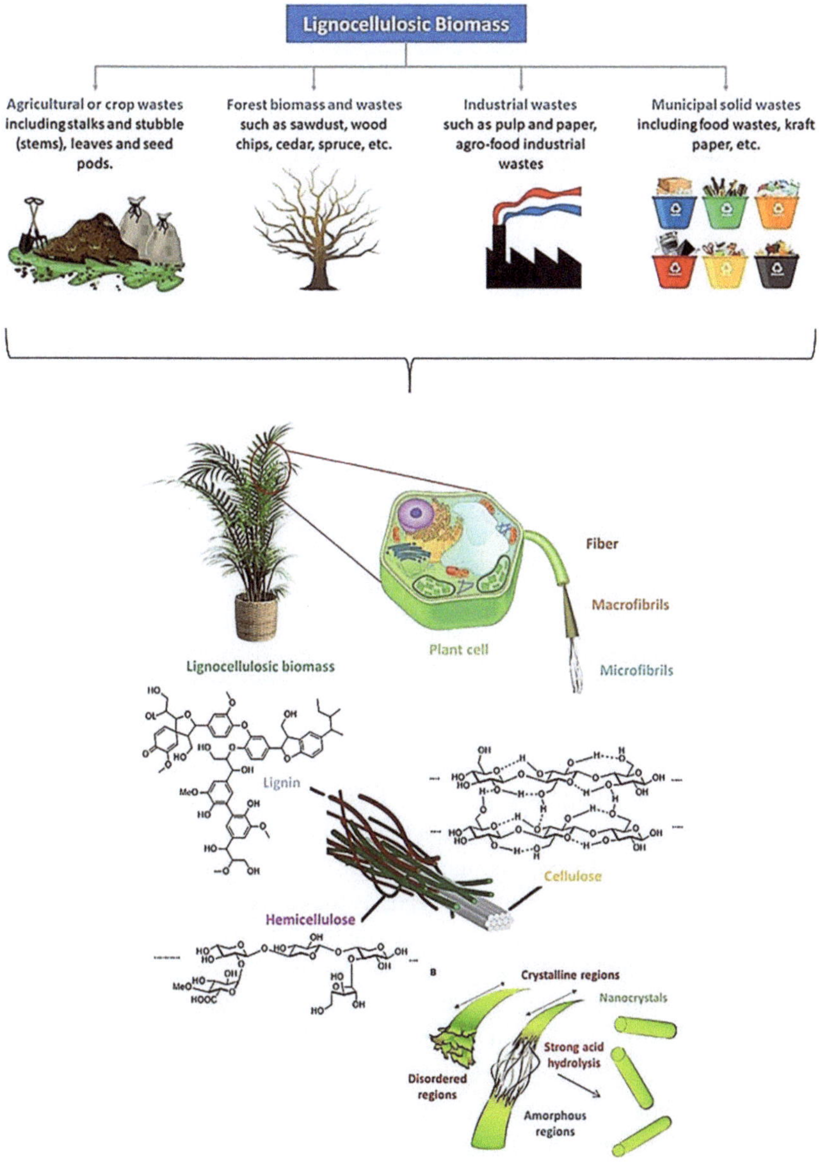

Figure 10.4: Lignocellulosic biomass.

Current day research works gives importance for the development of nanocellulose-reinforced banana thermoplastic starch composites. For example, research work carried out by Wang and his team (2020) stated that addition of nanocellulose enhanced the mechanical characteristics of the starch-based composites. Tensile strength and modulus

Table 10.3: Comparison of physical properties of banana fibers and other fibers Satyanarayan et al. [30].

Name of the fiber	Diameter (μm)	Density (kg/m³)	Aspect ratio	Microfibrillar angle (°)	Initial modulus (GPa)	UTS (MPa)	Elongation at break (%)
Banana	80–250	1350	150	10	7.7–20.0	54–754	10.35
Sisal	50–200	1450	100	10.0–22.0	9.4–15.8	568–640	3.0–7.0
PaLF	20–80	1440	450	8.0–14.0	34.5–82.5	413–1627	0.8–1

Table 10.4: Variation of tensile properties of banana fiber with the banana fiber diameter Kulkarni et al. [29].

Fiber diameter (μm)	Modulus (GPa)	SD initial modulus (GPa)	Fracture strength (MPa)	Strain %
50	32.71	8.21	779.08 ± 210.2	2.76 ± 0.96
100	30.47	4.72	711.68 ± 240.63	2.48 ± 0.81
150	29.75	8.57	773 ± 298.1	3.59 ± 1.12
200	27.71	7.09	789.31 ± 128.57	3.35 ± 0.71
250	29.91	4.06	766.62 ± 165.52	3.26 ± 1.29

of the composite increased with increasing nanocellulose content. Additionally, the thermal stability of the composite was improved with the addition of nanocellulose.

Nanocellulose has garnered attention due to its versatile properties, including sustainability, abundance of –OH groups for modification, biocompatibility, and high aspect ratio. Nanocellulose can be divided into 3 categories that are nanocrystalline cellulose, nanofibrillated cellulose, and bacterial nanocellulose [31–33]. Mishra et al. [34] successfully performed the extraction of nanocrystalline cellulose (NCC) from banana peel powder (BPP) with the aid of bleaching, alkali treatment, and acid hydrolysis methods. NCC was produced from the BPP through acid hydrolysis method while the bleaching and alkali treatment methods were used to remove the hemicellulose and lignin contents. An acid hydrolysis treatment with 200 mL of 55 % dilute sulfuric acid for a period of 60 min at a temperature of 50 °C rendered an NCC yield of 30 %. The zeta potential and the particle size of the above NCC particles were measured to be −44 mV and 290 nm respectively as depicted in Figure 10.5. Removal of hemicellulose and lignin from the BPP after acid hydrolysis, bleaching and alkali treatment was confirmed through attenuated total reflectance Fourier transform infrared spectroscopy (ATR-FTIR). Needle shaped crystals and particles in nanosize were revealed through field emission scanning electron microscopy (FE-SEM) and transmission electron microscopy (TEM). From the results of X-ray diffraction (XRD), it was observed that the cellulose I crystals were present in the extracted NCC and their crystallinity index (CI) was 64.14 %. The extracted NCC was found

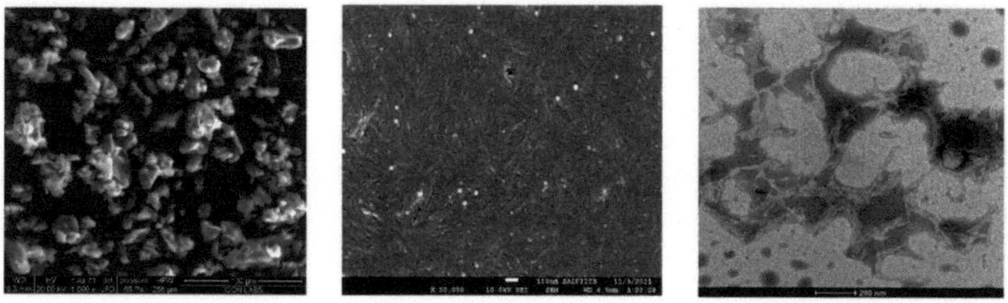

A B C

Figure 10.5: Scanning electron micrographs of (a) BPP, (b) NCC, and (c) TEM micrograph of NCC.

to have wider application potential since their thermal stability was also found to be better through thermogravimetric analysis.

García-Ramón et al. [35] work on banana starch films reinforced with cellulose nanoparticles from plantain rachis. The objective of the research was to study the behavior of plantain rachis derived banana starch films reinforced with nanoparticles for their mechanical and thermal behavior. Sonication and acid hydrolysis methods were used for the synthesis of nanoparticles and they had an average hydraulic diameter of approximately 60 nm. SEM images, as depicted in Figure 10.6, revealed that the thickness of the nanoparticles ranged from 9.8 to 22.3 nm, which is on par with the size (52–145 nm) of commercial or segregated particles derived out of husks. The use of various reagents and intense ultrasonication on banana peel resulted in cellulose nanofibers with diameters ranging from 20 to 35 nm and lengths between 263 and 589 nm, as shown in Figure 10.7. As the ultrasonication power increased, the length of the nanofibers decreased.

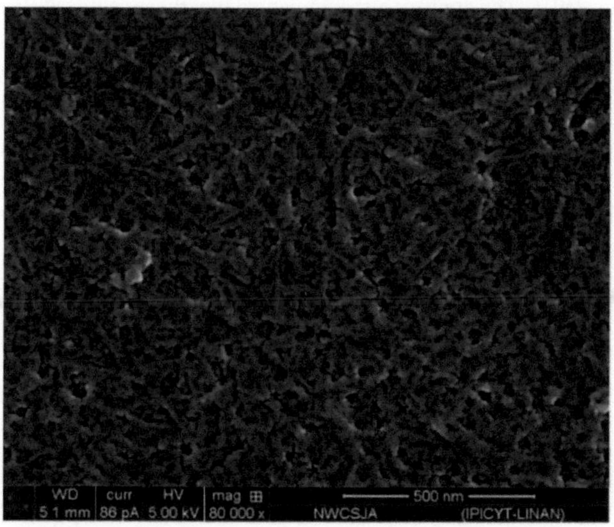

Figure 10.6: SEM micrograph of nanoparticles at 150,000× [35].

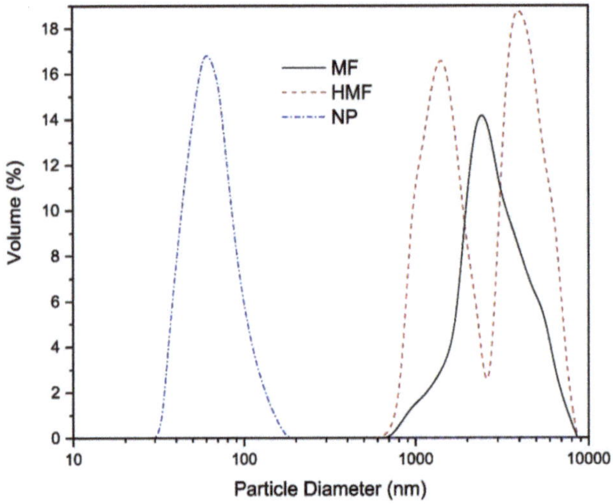

Figure 10.7: Distribution of particle size for MF, HMF, and NP [35].

Significant differences in the nanoparticle pattern were noted, as indicated by a considerable deterioration in the peak at 22.5° and a shift of the peak at 15.5° to approximately 12.5°, as displayed in Figure 10.8. These findings suggest that the nanoparticles possess a Type II crystalline structure for all sized of fiber, microfibers (MF), hydrolyzed microfibers (HMF), and nanoparticles (NP). Furthermore, the CI of the nanoparticles increased by around 76 % comparatively, indicating that the higher amount of crystallinity in the molecular structure of the NPs.

The results of the thermal gravimetric analysis revealed that the nanoparticles displayed excellent thermal stability up to 340 °C. The analysis also highlighted that these cellulose NPs were comprised of considerable amount of OH groups due to which their interaction with the water molecules were better within the same temperature range. The observed peak difference at maximum temperature suggested that acid hydrolysis technique impacted the cellulose structure substantially, thereby making them to get rid of lignin as depicted in Figure 10.9. Notably, the end temperature was similar for all the samples at this particular phase. Furthermore, the nanoparticles exhibited a third phase of weight loss within the temperature range of 430–540 °C, which has been linked to the production of sulfate groups with lower molecular weight, decomposition, and oxidation of carbonized residues.

10.5 Modification of banana starch thermoplastic

The compatibility of starch films are largely attributed to the availability of various starch types and their susceptibility to physical and/or chemical modifications. However, starch's hydrophilic nature poses a significant challenge that constrains starch-based

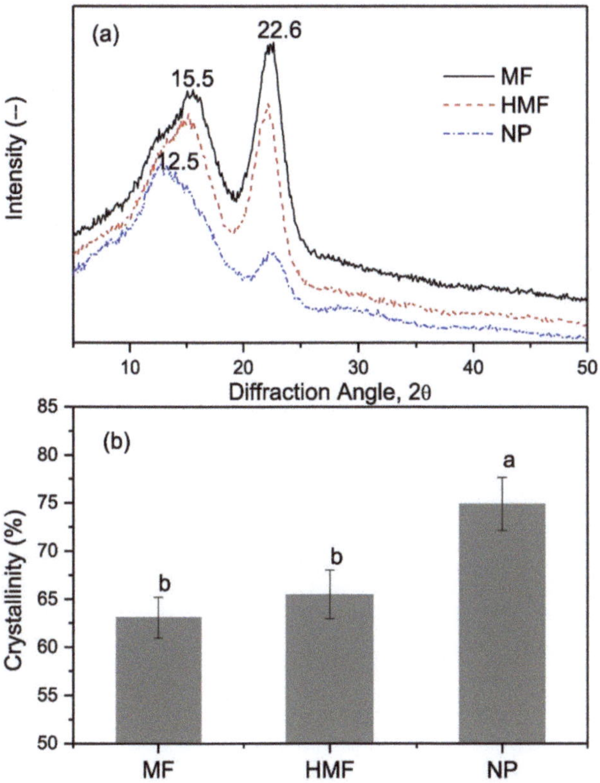

Figure 10.8: X-Ray diffraction analysis. (a) X-ray diffraction (XRD) peaks microfibers (MF), hydrolyzed microfibers (HMF), and nanoparticles (NP), (b) crystallinity degree [35].

materials' development, as their properties are heavily influenced by ambient humidity. To overcome this limitation, modified starches have been explored as an alternative solution. Chemical derivatization, in particular, has been extensively investigated to develop low water-sensitive materials.

López et al. [23] conducted a study on the film-forming capability of chemically modified starches, which demonstrated that the optimal concentration of modified starch was about 5 % weight fraction, resulting in the lowest water vapor permeability value (12.6×10^{-11} g/s m Pa). Incorporating the plasticizer into the film enhanced the mechanical properties of the film. The impact of glycerol-loaded on the performance of plasticized and acetylated starch films was conducted. Unplasticized films were brittle, while those with 1.5 % weight fraction of glycerol were ductile, and the films with higher plasticizer content were highly flexible. However, concentration of plasticizer exceeding 2 % weight fraction resulted in migration of the plasticizer, which resulted as cloudy films with high opacity values. Therefore, 1.5–1.6 g of glycerol per 100 g of film was the preferred

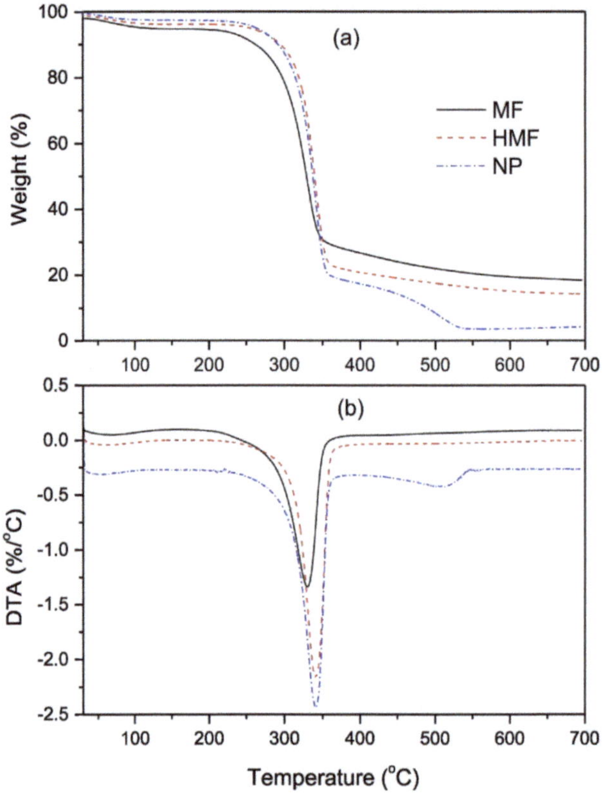

Figure 10.9: Thermal analysis. (a) Weight loss curves, (b) derivative thermogravimetry (DTG) for microfibers (MF), hydrolyzed microfibers (HMF), and nanoparticles (NP) [35].

concentration of the plasticizer, reducing water vapor permeability by 87 % and increasing acetylated starch film flexibility by 34 %.

Matthey and Hanna [24] conducted study on physical and functional properties of twin-screw extruded whey protein concentrate–corn starch blends. They conclude that the addition of whey protein concentrate (WPC) from three distinct sources to the corn starch blends with varying concentration of amylose had an impact on several extrudate functional and physical characteristics. Waxy starch displayed different water solubility index (WSI), water absorption index (WAI), and specific mechanical energy (SME) properties compared to amylose starches. It was also observed from the results that the SME was directly correlated with the concentration of WPC while the WSI and WAI were not highly influenced by the addition of WPC. It is plausible that amylose starch molecules retained the WPC molecules, which results in the formation of novel biopolymer characterized with distinct functional properties. The optimal content of WPC was concluded to be 10 % based on expansion considerations. The apparent content amylose, radial ratio of expansion, and the difference in total color were significantly influenced by the interaction between WPC and the amylose molecules. This finding suggested a physico-chemical interaction between these two molecules, resulting in the creation of a concentrate complex made of these two molecules that decreased the starch molecular degradation.

There is a numerous of research available on the characteristics of composite materials strengthened with banana fiber. Motaleb et al. [36] prepared polymer composites reinforced with nonwoven banana fiber, as shown in Figure 10.10. In the study, different banana tree sections including the outer, middle, and inner bark (OB, MB, and IB), and leaf midrib (MR), were used to produce four nonwoven materials through web formation in wet lays. These nonwovens were then combined with epoxy and polyester matrix materials separately resulting in a total of eight composite variants. Various treatments, such as alkali treatment on raw fibers, and gamma irradiation to the composites were posed to evaluate the effects of those treatments on the composite properties, including water absorption and mechanical characteristics such as elongation at break, tensile strength (TS), and flexural strength (FS). OB composites exhibited the highest water absorbency, TS, and FS, and lowest value of elongation at break when compared with the other composite configurations. On an average, epoxy composites had 16.5 % lower water absorption, 42 % higher TS, and 40 % higher FS when compared with banana-polyester composites. The result also showed that the mercerization minimized the water absorption behavior of the composites by 33 %, and a subsequent water repellent treatment further reduced it by 63 %. The alkali treatment also improved the TS and FS of the composites by 71 % and 87 %, respectively, while gamma radiation increased TS and FS by a further 30 % and 35 %.

10.6 Nanocellulose reinforced banana thermoplastic starch composites

Films were prepared using varying fractions of nanoparticles relative to the total solids, along with a glycerol plasticizer, resulting in improved properties such as reduced water

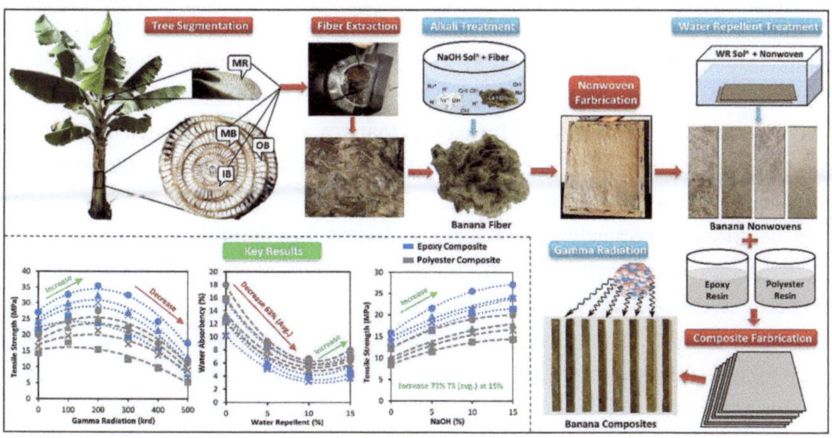

Figure 10.10: Non-woven banana fiber reinforced polymer composites and their properties.

Table 10.5: Properties of biofilms [35].

Sample	Moisture content (%)	TS (MPa)	EB (%)
F0	15.25 ± 0.87	7.68 ± 0.65	7.52 ± 0.55
F2	10.39 ± 0.68	12.53 ± 0.72	12.02 ± 0.63
F3	5.72 ± 0.74	27.46 ± 0.51	17.02 ± 0.60
F5	4.84 ± 0.45	37.22 ± 0.62	19.93 ± 0.50
FNP	2.68 ± 0.65	47.05 ± 0.65	2.03 ± 0.51

vapor permeability, increased tensile strength, and greater flexibility. The molecular structure of the starch network was consolidated through hydrogen bonding effects, serving as crosslinkers, thus improving overall stability.

García-Ramón et al. [35] work on the nanocellulose reinforced banana thermoplastic starch. The incorporation of cellulose nanoparticles into the films resulted in a significant ($p < 0.05$) increase in both tensile strength (TS) and elongation-at-break (EB). Film F0 exhibited TS and EB values of 7.68 MPa and 7.52 %, respectively, which rose to 37.26 MPa and 19.97 % for the F5 film. Film FNP displayed the highest TS value of 47.1 MPa, but a 2.05 % lower EB value was noted lower at 2.03 %, indicating reduced flexibility compared to the other films. This suggests that starch chain and nanoparticle interaction was the primary reason for the enhanced mechanical behavior of the nanocomposites resulting in the formation of network of biopolymers with higher elasticity. Conversely, the absence of starch-nanoparticle interactions in film FNP resulted in weak elasticity. Various studies were reported regarding the enhancement of the mechanical characteristics of biofilms prepared from banana starch and cellulose nanoparticle nanocomposites as listed in Table 10.5.

The inclusion of nanoparticles had a negative impact on the WVP values (shown in Figure 10.11). The F0 film had the highest value of 81.1×10^{-12} g/(m s Pa) and went down to about 2.26 g/(m s Pa) for F5 sample with 0.5 g nanocellulose. As the addition of nanoparticles to the starch induced the higher molecular restructuring, the decline in the WVP values was experienced by the samples. This obstructs the transfer of molecular waves, a phenomenon observed in previous studies by Ma et al. [37]. However, for the FNP film made completely of nanoparticles the WVP value was somewhat higher than that for F5 film. According to Mali et al. [38], the reduction in the WVP of films can be attributed to the CI and film internal structure since the transfer of water molecules happened through the amorphous phases. Thus, the role of nanoparticles here was not only to interact with water molecules and starch chains but also to act as physical fillers. The OH groups present in cellulose nanoparticles can form hydrogen bonds with starch molecules, thereby creating a network with high strength. Additionally, mechanical resistance and flexibility of the hydrogels was facilitated by the glycerol which acted as a plasticizer. The increase in the FTIR ratio indicated that the molecular arrangement of the starch was refined due to the nanoparticles' addition. But the CI of the nanoparticles had a least role to play in this aspect since they value of CI for the starch was as low as 4.1 %.

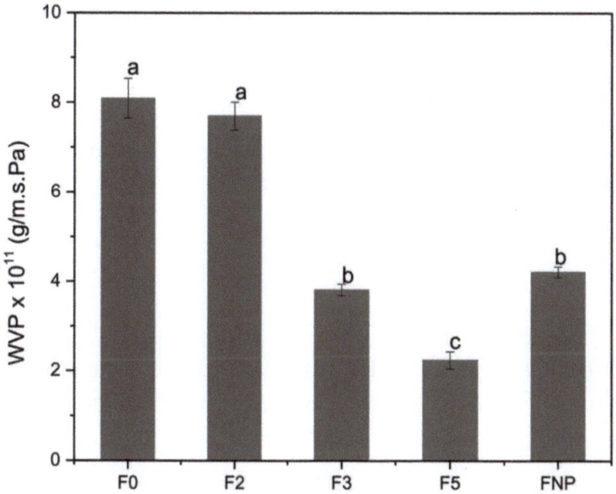

Figure 10.11: Water vapor permeability (WVP) of the different film formulations.

In Figure 10.12, the film labelled FNP displayed a distinct XRD peak at 22.7°, accompanied by two different smaller and broader peaks approximately at 16.8° and 17.2° respectively. This XRD pattern was consistent with type I cellulose [39]. A comparable diffraction pattern was observed by Ilyas et al. [40] for cellulose nanofibers obtained from sugar palm fiber. On the other hand, the starch film made solely from banana starch (F0) exhibited no visible peaks, indicating that the starch's molecular structure was lost at the

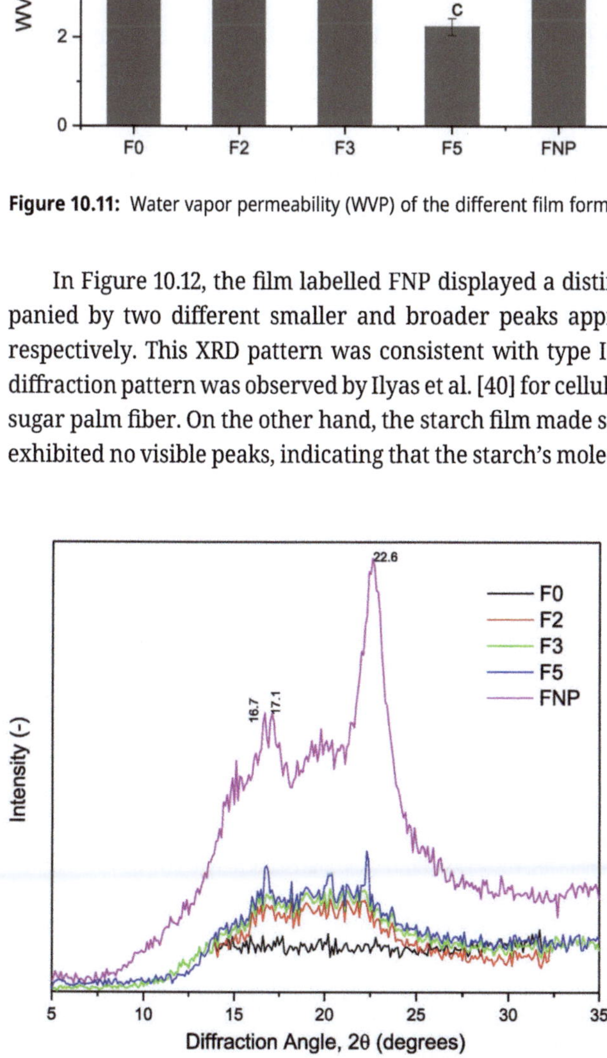

Figure 10.12: Starch film XRD peaks with varying of nanoparticle content.

time of formation of the starch film. Film F2 and film F3 was short of prominent peaks, indicating that the nanoparticle concentrations were insufficient to encourage nanoparticle aggregation. However, the film labeled with 0.5 g nanocellulose (F5) exhibited weak peaks at 16.7° and 17.1°, suggesting the emergence of crystalline phase structures within the film. Ilyas et al. [7] stated a similar finding in nanocomposite starch films created using sugar palm starch and sugar palm nanocellulose.

10.7 Conclusions

Banana thermoplastic starch (BTPS) is a sustainable material with good mechanical properties, water resistance, and biodegradability. The production of BTPS involves the use of renewable resources such as banana peel and starch, reducing our dependence on non-renewable resources. BTPS has potential applications in various industries such as packaging, agriculture, textile, and construction. Further research is needed to optimize the properties of BTPS and to develop new applications for this sustainable material. In conclusion, nanocellulose-reinforced banana thermoplastic starch composites have shown promising mechanical and thermal properties. Banana fibers and thermoplastic starch are renewable and biodegradable materials that make them ideal candidates for sustainable composite materials. Nanocellulose has been found to be an effective reinforcement agent for improving the mechanical and thermal properties of composites. Further research is needed to optimize the properties of these composites and to develop new applications for these sustainable materials.

References

1. Ramesh M, Rajeshkumar LN, Srinivasan N, Kumar DV, Balaji D. Influence of filler material on properties of fiber-reinforced polymer composites: a review. E-Polymers 2022;22:898–916.
2. Akhil UV, Radhika N, Saleh B, Aravind Krishna S, Noble N, Rajeshkumar L. A comprehensive review on plant-based natural fiber reinforced polymer composites: fabrication, properties, and applications. Polym Compos 2023. https://doi.org/10.1002/pc.27274.
3. Rajeshkumar L. Biodegradable polymer blends and composites from renewable resources. In: Mavinkere Rangappa S, Parameswaranpillai J, Siengchin S, editors. Biodegradable polymers, blends and composites, Ramesh Blends and Composites MBT-BP. Elsevier; 2022:527–49 pp.
4. Jumaidin R, Diah NA, Ilyas RA, Alamjuri RH, Yusof FAM. Processing and characterisation of banana leaf fibre reinforced thermoplastic cassava starch composites. Polymers 2021;13:1420.
5. Siracusa V, Rocculi P, Romani S, Rosa MD. Biodegradable polymers for food packaging: a review. Trends Food Sci Technol 2008;19:634–43.
6. Pelissari FM, Andrade-Mahecha MM, Sobral PJ do A, Menegalli FC. Isolation and characterization of the flour and starch of plantain bananas (Musa paradisiaca). Starch - Stärke 2012;64:382–91.
7. Ilyas RA, Sapuan SM, Ishak MR, Zainudin ESD. Development and characterization of sugar palm nanocrystalline cellulose reinforced sugar palm starch bionanocomposites. Carbohydr Polym 2018;202: 186–202.

8. Evans TC, Gavrilovich E, Mihai RC, Isbasescu IEL. Method of isolating a banana starch from green bananas cross. Pat Appl Publ Madson 2015;002:354.
9. Zhang P, Whistler RL, BeMiller JN, Hamaker BR. Banana starch: production, physicochemical properties, and digestibility—a review. Carbohydr Polym 2005;59:443–58.
10. Chávez-Salazar A, Bello-Pérez LA, Agama-Acevedo E, Castellanos-Galeano F, Álvarez-Barreto C, Pacheco-Vargas G. Isolation and partial characterization of starch from banana cultivars grown in Colombia. Int J Biol Macromol 2017;98:240–6.
11. de Barros Mesquita C, Leonel M, Franco CML, Leonel S, Garcia EL, dos Santos TPR. Characterization of banana starches obtained from cultivars grown in Brazil. Int J Biol Macromol 2016;89:632–9.
12. Marta H, Cahyana Y, Djali M, Arcot J, Tensiska T. A comparative study on the physicochemical and pasting properties of starch and flour from different banana (Musa spp.) cultivars grown in Indonesia. Int J Food Prop 2019;22:1562–75.
13. Fontes Sde M, Cavalcanti MT, Candeia RA, Almedia EL. Characterization and study of functional properties of banana starch green variety of Mysore (Musa AAB - Mysore). Food Sci Technol 2017;37:224–31.
14. Hung PV, Cham NTM, Truc PTT. Characterization of Vietnamese banana starch and its resistant starch improvement. Int Food Res J 2013;20:205–11.
15. Babu AS, Mahalakshmi M, Parimalavalli R. Comparative study on properties of banana flour, starch and autoclaved starch. Trends Carbohydr Res 2014;6:38–44.
16. Nimsung P, Thongngam M, Naivikul O. Compositions, morphological and thermal properties of green banana flour and starch. Kasetsart J/Nat Sci 2007;41:324–30.
17. Utrilla-Coello RG, Rodríguez-Huezo ME, Carrillo-Navas H, Hernández-Jaimes C, Vernon-Carter E, Alvarez-Ramirez J. In vitro digestibility, physicochemical, thermal and rheological properties of banana starches. Carbohydr Polym 2014;101:154–62.
18. Reddy CK, Haripriya S, Vidya PV. Morphology, physico-chemical and functional characteristics of starches from different banana cultivars. J Food Sci Technol 2015;52:7289–96.
19. Kiin Kabari DB, Sanipe P, Owuno F. Physico-chemical and pasting properties of starch from three plantain cultivars grown in Nigeria. Afr J Food Sci 2014;8:484–9.
20. Núñez-Santiago MC, Bello-Pérez LA, Tecante A. Swelling-solubility characteristics, granule size distribution and rheological behavior of banana (Musa paradisiaca) starch. Carbohydr Polym 2004;56:65–75.
21. Olatunde GO, Arogundade LK, Orija OI. Chemical, functional and pasting properties of banana and plantain starches modified by pre-gelatinization, oxidation and acetylation. Cogent Food Agric 2017;3:1283079.
22. Alanís-López P, Pérez-González J, Rendón-Villalobos R, Jiménez-Pérez A, Solorza-Feria J. Extrusion and characterization of thermoplastic starch sheets from “Macho” banana. J Food Sci 2011;76:E465–71.
23. López OV, García MA, Zaritzky NE. Film forming capacity of chemically modified corn starches. Carbohydr Polym 2008;73:573–81.
24. Matthey FP, Hanna MA. Physical and functional properties of twin-screw extruded whey protein concentrate–corn starch blends. LWT–Food Sci Technol 1997;30:359–66.
25. Gurunathan T, Mohanty S, Nayak SK. A review of the recent developments in biocomposites based on natural fibres and their application perspectives. Compos Appl Sci Manuf 2015;77:1–25.
26. Mahmud S, Hasan KMF, Jahid MA, Mohiuddin K, Zhang R, Zhu J. Comprehensive review on plant fiber-reinforced polymeric biocomposites. J Mater Sci 2021;56:7231–64.
27. Asyraf MRM, Syamsir A, Supian ABM, Usman F, Ilyas RA, Nurazzi NM, et al. Sugar palm fibre-reinforced polymer composites: influence of chemical treatments on its Mechanical properties. Materials 2022;15: 1–22.
28. Elumalai S, Agarwal B, Runge TM, Sangwan RS. Advances in transformation of lignocellulosic biomass to carbohydrate-derived fuel precursors. In: Biorefining of biomass to biofuels. Cham, Switzerland: Springer; 2018, vol 4:87–116 pp.
29. Kulkarni AG, Satyanarayana KG, Rohatgi PK, Vijayan K. Mechanical properties of banana fibres (Musa sepientum). J Mater Sci 1983;18:2290–6.

30. Satyanarayana KG, Sukumaran K, Mukherjee PS, Pavithran C, Pillai S. Natural fibre-polymer composites. Cem Concr Compos 1990;12:117–36.
31. Ilyas RA, Sapuan SM, Ibrahim R, Abral H, Ishak M, Zainudin E, et al. Effect of sugar palm nanofibrillated celluloseconcentrations on morphological, mechanical andphysical properties of biodegradable films basedon agro-waste sugar palm (Arenga pinnata(Wurmb.) Merr) starch. J Mater Res Technol 2019;8: 4819–30.
32. Kadier A, Ilyas RA, Huzaifah MRM, Harihastuti N, Sapuan SM, Harussani MM, et al. Use of industrial wastes as sustainable nutrient sources for bacterial cellulose (BC) production: Mechanism, advances, and future perspectives. Polymers 2021;13:1–51.
33. Syafri E, Jamaluddin SNH, Mahardika M, Amanda P, Ilyas RA. Isolation and characterization of cellulose nanofibers from Agave gigantea by chemical-mechanical treatment. Int J Biol Macromol 2022;200:25–33.
34. Mishra S, Prabhakar B, Kharkar PS, Pethe AM. Banana peel waste: an emerging cellulosic material to extract nanocrystalline cellulose. ACS Omega 2023;8:1140–5.
35. García-Ramón JA, Carmona-García R, Valera-Zaragoza M, Aparicio-Saguilán A, Bello-Pérez L, Aguirre-Cruz A, et al. Morphological, barrier, and mechanical properties of banana starch films reinforced with cellulose nanoparticles from plantain rachis. Int J Biol Macromol 2021;187:35–42.
36. Motaleb KZMA, Ahad A, Laureckiene G, Milasius R. Innovative banana fiber nonwoven reinforced polymer composites: pre- and post-treatment effects on physical and Mechanical properties. Polymers 2021;13: 3744.
37. Ma X, Chang PR, Yang J, Yu J. Preparation and properties of glycerol plasticized-pea starch/zinc oxide-starch bionanocomposites. Carbohydr Polym 2009;75:472–8.
38. Mali S, Grossmann MVE, García MA, Martino MN, Zaritzky NE. Effects of controlled storage on thermal, mechanical and barrier properties of plasticized films from different starch sources. J Food Eng 2006;75: 453–60.
39. Xu X, Liu F, Jiang L, Zhu JY, Haagenson D, Wiesenborn DP. Cellulose nanocrystals vs. Cellulose nanofibrils: a comparative study on their microstructures and effects as polymer reinforcing agents. ACS Appl Mater Interfaces 2013;5:2999–3009.
40. Ilyas RA, Sapuan SM, Ibrahim R, Abral H, Ishak M, Zainudin E, et al. Sugar palm (Arenga pinnata (Wurmb.) Merr) cellulosic fibre hierarchy: a comprehensive approach from macro to nano scale. J Mater Res Technol 2019;8:2753–66.

Nur Sharmila Sharip, Tengku Arisyah Tengku Yasim-Anuar,
Hazwani Husin and Mohd Nor Faiz Norrrahim*

11 Barley thermoplastic starch nanocomposite films reinforced with nanocellulose

Abstract: Despite being one of the starch producers, barley has yet to be widely studied for thermoplastic starch applications, including nanocellulose thermoplastic composites, due to its uses in the food and beverage industries. However, only 20% of barley is used in the malting industry to produce both alcoholic and non-alcoholic beverages, and 5% is used as an ingredient in a wide variety of foods. As the fourth most important cereal in the world after wheat, corn, and rice, barley can be considered an interesting biomass source to produce biodegradable thermoplastics, stemming from its starch constitution. Therefore, this review attempts to highlight the barley starch properties and its potential utilization for nanocellulose thermoplastic starch composites. Several studies involving barley-based starch in thermoplastic production and nanocellulose reinforcement for properties enhancement are also reviewed, particularly in the attempt to provide various options to reduce and replace the uses of harmful petroleum-based plastic.

Keywords: barley-based starch; nanocellulose; nanocomposite; thermoplastic starch.

11.1 Barley-based starch

Starch, in general, is made up of amylopectin and amylose [1–3]. These two components differ in terms of their monomer unit arrangement and linkages. As shown in Figure 11.1, amylose is linearly arranged and linked through the α-(1–4) linkages [4] while the amylopectin is constructed in branches joined by α-(1–4) linkages and α-(1–6) linkages at their branches [5]. In comparison, the amylopectin unit chains are rather short compared to the amyloses by which about 4–6% of amylopectin linkages are α-(1–6) linkages, making it highly branched [6]. Starch granules can be found in a variety of seeds, roots, and tubers, including barleys [7]. Produced largely in Russia, Spain, Germany, Canada, France, Australia, Türkiye, United Kingdom, Northern Ireland, and Ukraine (Figure 11.2),

***Corresponding author: Mohd Nor Faiz Norrrahim**, Research Center for Chemical Defence, Universiti Pertahanan Nasional Malaysia, Kem Sungai Besi, 57000 Kuala Lumpur, Malaysia, E-mail: faiz@upnm.edu.my. https://orcid.org/0000-0003-2101-5642
Nur Sharmila Sharip, Tengku Arisyah Tengku Yasim-Anuar and Hazwani Husin, Research and Development Department, Nextgreen Pulp & Paper Sdn Bhd, Taman Tun Dr Ismail, 60000 Kuala Lumpur, Malaysia, E-mail: sharmila.sharip@nextgreenglobal.com (N.S. Sharip), tengkuarisyah@nextgreenglobal.com (T.A. Tengku Yasim-Anuar), hazwani.husin@nextgreenglobal.com (H. Husin). https://orcid.org/0000-0003-3322-9592 (N.S. Sharip)

As per De Gruyter's policy this article has previously been published in the journal Physical Sciences Reviews. Please cite as: N. S. Sharip, T. A. T. Yasim-Anuar, H. Husin and M. N. Faiz Norrrahim "Barley thermoplastic starch nanocomposite films reinforced with nanocellulose" *Physical Sciences Reviews* [Online] 2023. DOI: 10.1515/psr-2022-0025 | https://doi.org/10.1515/9783110773606-011

the worldwide production of barley had cumulatively increased by 27% over the decade from 123 to 157 million tons (Figure 11.3). Even though considered the fourth most important cereal in the world after wheat, corn, and rice, it has been mainly used for feed (approximately 70%), followed by its utilization in malting, brewing, and distilling industries which make up 21% while less than 6% are used for human consumption [8]. In such regards, barley can be an interesting biomass source to produce biodegradable thermoplastics, stemming from its starch constitution.

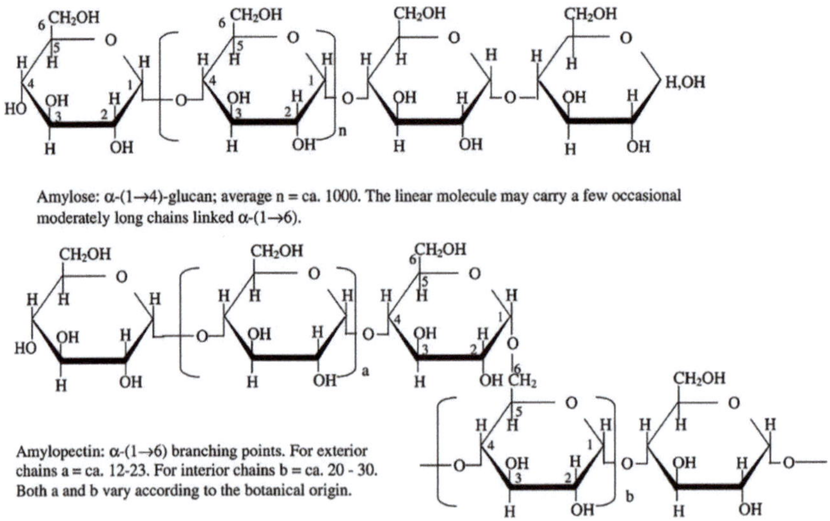

Figure 11.1: The chemical structure of starch components.

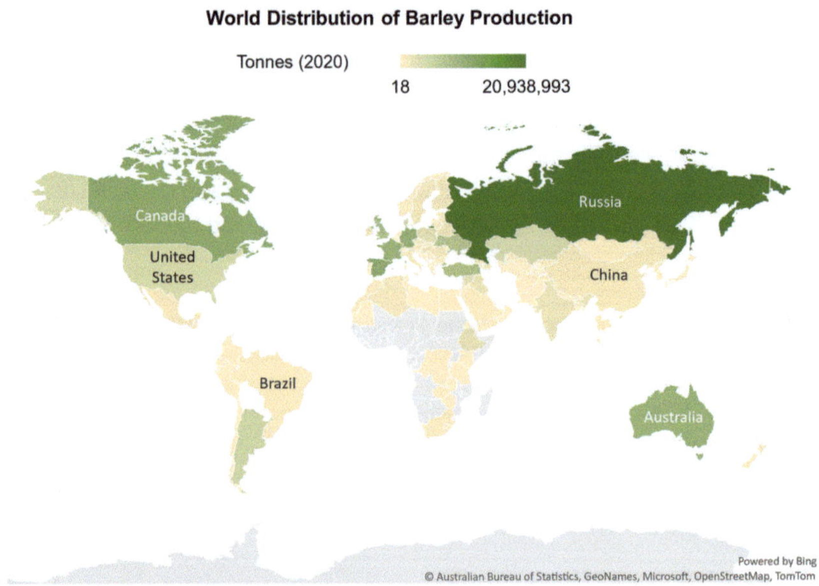

Figure 11.2: The distribution of barley production worldwide [8].

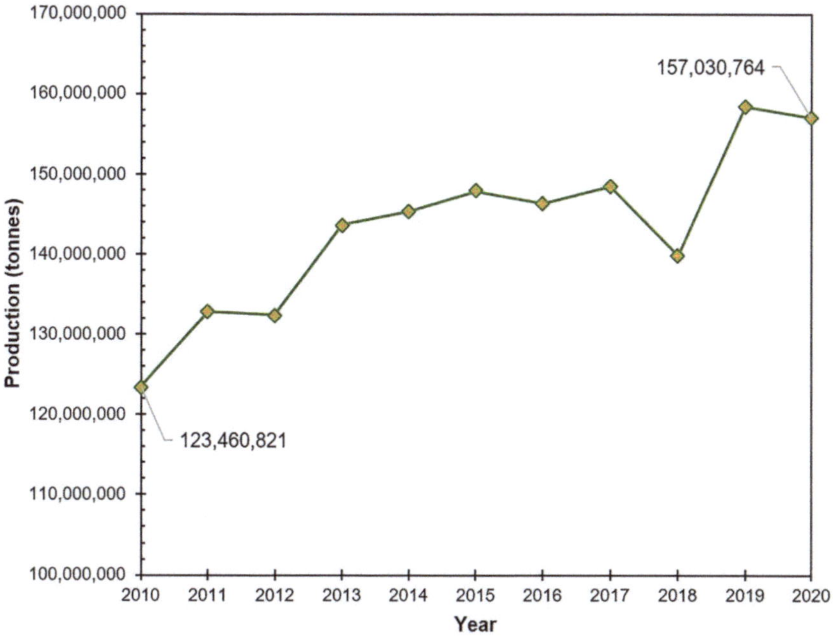

Figure 11.3: The cumulative production of barley from 2010 to 2020 [8].

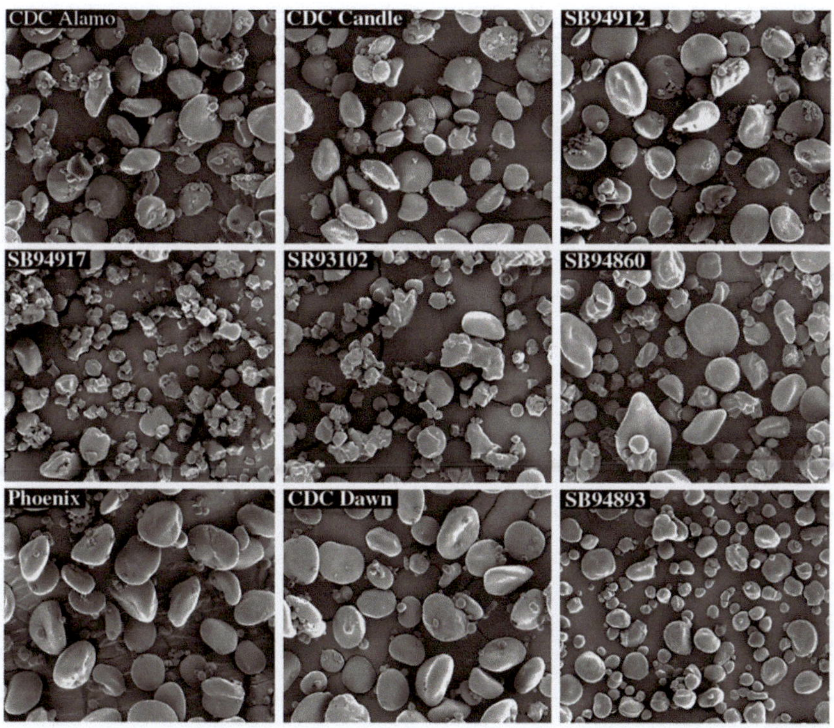

Figure 11.4: The micrograph of barley starches under 450-time magnification by using a scanning electron microscope (figure adopted from Li et al. [11]).

Typically containing amylose and amylopectin, starch constitutes up to 70% of the barley grain dry weight [9]. Some barley starches may have high amylopectin percentage (95–99%) and are referred to as "waxy," whereas others may have a high-amylose concentration (50–70%) [10, 11]. The amount of amylose content in barleys is depending on the growing conditions and granule size [12], which then contributes greatly to the determination of the starch quality [9]. Barley starch is usually made up of a combination of large lenticular granules and smaller, irregularly shaped granules as shown in Figure 11.4.

The granule size distribution of normal and high-amylopectin (waxy) barley starches is bimodal, with big lenticular A-granules and small spherical B-granules, whereby the size ranges from 10–30 μm, and less than 10 μm, respectively. Meanwhile, the starch granule size distribution of high-amylose barley starches is narrower, with smaller A-granules and larger B-granules [13]. Li et al. [11] also stated that the proportion of small and large granules is varying depending on the type of barley. Besides starch, barleys also contain proteins (11.5–14.2%), lipids (4.7–6.8%), and ash (1.8–2.4%). The chemical composition of the different starch compositions in barley is shown in Table 11.1.

Table 11.1: Chemical composition of barley starch.

Genotype	Percentage (%)						References
	Starch	Lipid/Fat	Protein	Ash	β-Glucan	Dietary fibre	
			Normal				
Golf	34.0	2.2	8.7	2.3	4.7	0.3	[13]
AC Hawkeye	61.0	–	12.3	–	3.4	–	[14]
CDC McGwire	72.2	3.0	13.5	–	5.5	11.8	[10]
CDC dawn	64.6	5.2	11.5	1.8	3.7	10.6	[11]
Hull-less	52.9	2.9	14.5	–	3.3	12.9	[15]
			High amylose				
High-amylose glacier	25.0	3.5	10.4	2.5	6.9	2.0	[13]
SB94897	52.8	–	12.7	–	6.9	–	[14]
SH99250	61.1	3.4	14.5	–	9.6	17.4	[10]
SB98497	56.58	4.7	14.0	2.3	7.7	12.6	[11]
Malting	64.4	2.5	12.4	–	3.5	17.9	[15]
			High amylopectin (waxy)				
SW 906129	30.0	3.4	10.5	2.5	6.1	0.4	[13]
CDC candle	56.9	–	11.3	–	6.9	–	[14]
CDC candle	68.5	3.4	13.6	–	6.9	13.6	[10]
CDC candle	64.7	6.1	11.8	1.9	6.4	6.8	[11]

The isolation of starch from barley grains can be done through various laboratory methods including grain steeping, blending followed by screening, deproteinization, and starch recovery by centrifugation [11, 13, 16]. Two main isolation techniques that have been usually used in the starch extraction processes are wet and dry milling. Nevertheless, the industrial method for extracting starch from barley grains is complex because of its various processing phases and time-consuming procedures. Significant water consumption caused by the presence of β-glucans, which absorb water and produce high viscosities, makes the separation of starch by screening and subsequent centrifugation difficult and complex [17, 18]. On average, the normal processing of barley starch yield and extraction efficiency is 44.4% and 70.9% of total grain weight (dry basis), respectively [18]. In the case of waxy barley starch with a high β-glucan concentration, starch yield is 42.1%, while the extraction efficiency is 68.0% [11].

Figure 11.5 shows the example of starch extraction process from barley [18]. The process began with barley dry-milling and dehusking whereby the flour produced and husk are later separated through sifting process. The barley flour is then steeped in cellulolytic enzyme-containing water to liberate the starch granules from other constituents.

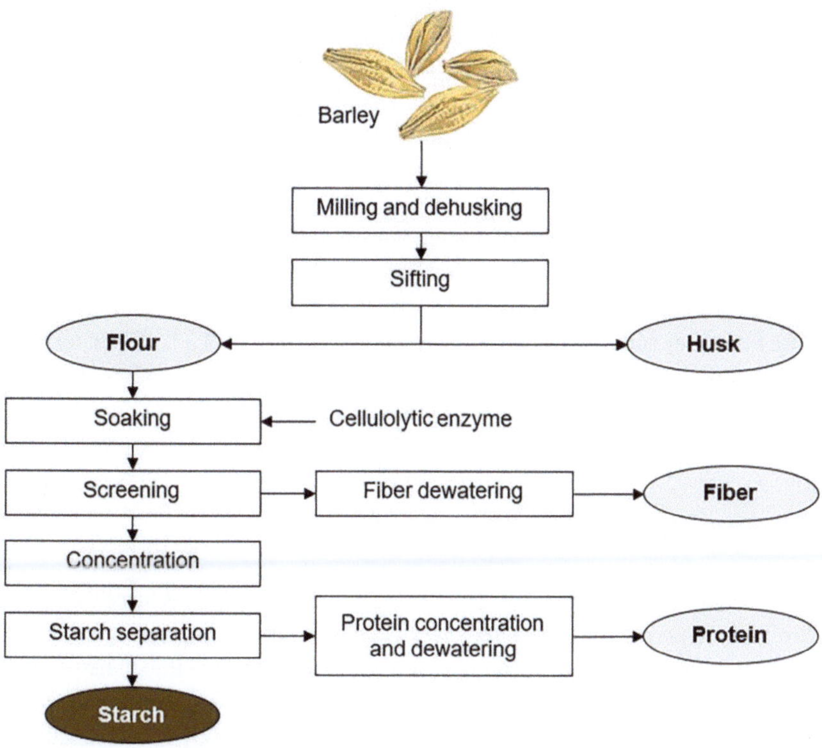

Figure 11.5: The example of barley starch extraction process (figure adapted from Vasanthan and Hoover [18]).

The next step is to filter out any leftover fiber and cell wall components using multistage screens. Starch granules, proteins, and solubles are the main components of the slurry that passes through the screens. Afterwards, centrifugation is utilised to recover the starch.

In terms of properties, a study by Yasui et al. [19] shows that the swelling power of waxy and normal starch correlates with the amylose content of barley starch. Amylose-free starch exhibited the greatest swelling power among the barley starches. Other than that, the key functional attribute of starch which is gelatinization can be influenced by starch composition (amylose to amylopectin ratio, proteins, phosphorus, lipids, and enzymes content) as well as amylopectin molecular structure (main chain length, degree of branching, and molecular mass) [20]. Yasui et al. [19] reported that in comparison to conventional barley starch, waxy barley starch had greater gelatinization temperatures and gelatinization enthalpy. To add, the gelatinization enthalpy of waxy barley starch is directly proportional to the starch granule size, by which smaller granule size contributed to lower gelatinization enthalpy, and vice versa [11]. In general, the gelatinization enthalpy of the starches from barley starch is varied ranging from 6.49 J/g to 9.61 J/g [17, 21, 22], whereby it is also considered a generic measure to predict starch's crystalline structure losses due to the gelatinization process. The highest degree of crystallinity is seen in waxy starch barley, followed by normal and high amylose starches [23].

Another key event in gelatinized starch is retrogradation, which occurs when chains are reconnected after cooling to generate an "ordered" structure. Some starches, particularly those with a high amylose concentration, undergo rapid and high retrogradation, whereas waxy starches undergo gradual and low retrogradation [24]. Amylose produces double helices of 40–70 glucose units during this process, whereas amylopectin crystallises due to re-the association of the outermost short branches. Values of the retrogradation enthalpy are usually in the range of 2.81 J/g to 3.27 J/g [17, 21]. According to Bello–Pérez [22], the amylose level of barley starch has a substantial impact on its functional properties, whereby starch with a high amylose concentration is more susceptible to retrogradation, hence its pasting property is more elastic. The starch pasting temperature for barley starch is usually high, ranging from 86.5 to 91.3 °C [16]. In terms of relative crystallinity (RC) of normal, high amylose and waxy barley starches are in the range of 20–36%, 37–42%, and 33–44%, respectively [25, 26].

11.2 Barley thermoplastic starch

It is well known that starch is a low-cost, globally available renewable resource with great potential to replace petroleum-based polymers when it is plasticized. This interest has resulted in decades of research on the large-scale conversion of starch into thermoplastic starch. Although starch is not a genuine thermoplastic polymer, it can be treated once it has gelatinized by mixing it with enough water and/or plasticizers [27]. Extensive studies have been conducted to identify the potential of various starch-based plants for the development of thermoplastic starch. Maize, wheat, rice, potato, and tapioca are the most frequent sources of starches being used for this purpose around the world [28].

Among early works that focused on barley thermoplastic starch was published by Forssell et al. [29, 30] in the years 1997 and 1999. They reported that barley thermoplastic starch film possessed comparable tensile strength, thermal characteristics, and water vapor rate transmission to other types of thermoplastic starch films, besides effective plasticization of barley starch by glycerol. More recent work involving barley-based bioplastic was reported in 2017 by Sagnelli et al. [31], who produced a cross-linked all-natural barley bioplastic. The engineered barley starch called amylose-only (AO) consisting of 99% amylose was respectively plasticized and cross-linked using glycerol and citrate in a co-rotating inter-meshing twin screw extruder. Their results showed that crosslinked AO exhibited almost three folds higher tensile strength than the crosslinked native barley starch. The plasticization of AO also resulted in better mechanical properties as compared to the plasticized native barley starch. This work was further extended whereby the barley thermoplastic starch was tested as biopolymeric carriers for poly-(glycerol adipate)-based nanoparticles (NP) [32]. The biocompatibility of the derived barley thermoplastic starch-NP material against human intestinal epithelial adenocarcinoma cell line Caco-2 cells provides insight into the potential application of barley thermoplastic starch as coating materials for food or delivery of bioactive molecules.

Despite the promising potential, the utilization and development of barley as thermoplastic starch are rather limited, which could be due to its widespread use in the food sector. As one of the world's staple foods, 75% of barley is used for fodder and feed, 20% is used in the malting industry to make both alcoholic and non-alcoholic beverages, and 5% is being used as an ingredient in a wide variety of foods [33]. This factor may also play a role in why it was not selected for thermoplastic starch development.

11.3 Nanocellulose reinforcement in barley thermoplastic starch

Generally, cellulose including nanocellulose is a carbohydrate polymer made up of repeated β-D-glucopyranose units connected by the β-1,4 glycosidic connections [34]. In terms of nanocellulose, it is a material having at least one of its dimensions in the nanometer range, which is less than 100 nm, and consisted of both amorphous and crystalline structures [35–41]. Plant-derived nanocellulose can be divided into cellulose nanofiber (CNF) and cellulose nanocrystals (CNC) [42, 43], as well as lignocellulose nanofiber (LCNF) [44, 45]. Meanwhile, bacterial-derived nanocellulose is referred to as bacterial nanocellulose (BNC) [46–48]. In addition to being biocompatible and biodegradable, nanocellulose's surface chemistry and functionality can be customized [49–52]. This enables nanocellulose materials to develop the necessary characteristics

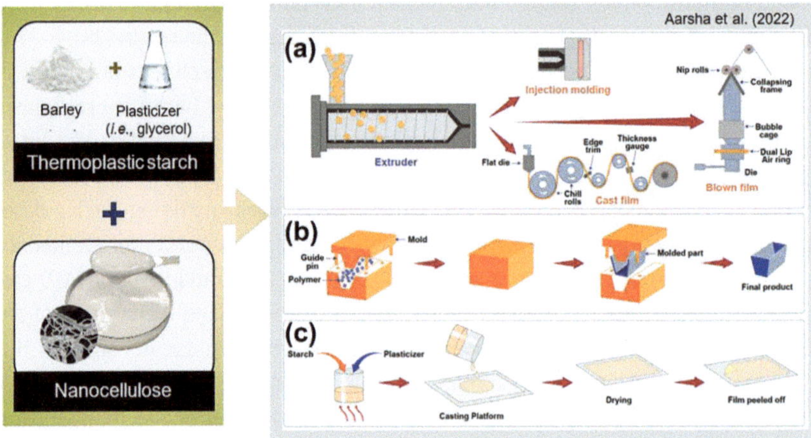

Figure 11.6: The example of thermoplastic starch composite processing method by (a) extrusion, (b) injection molding, and (c) solution casting (figure adopted from Aarsha et al. [56]).

necessary for their intended usage [39, 53–55]. In such ways, the similarity of starch structure with cellulose may enable good compatibility between the two as matrix and filler for better properties of nanocellulose-reinforced thermoplastic starch.

The process of preparing thermoplastic starch and its composites (including nanocomposites) varies, depending on purpose and availability of resources. For instant, utilization of extrusion which is a dependable and continuous process is often chosen for industrial purposes. In this process, melted polymer is pushed through a die with a certain shape to create profiles like sheets, tubes, films, etc. It is a quick and dependable method used to create thermoplastic starch-based goods for industrial uses. Nevertheless, most research uses the solution casting process to create the thermoplastic nanocomposites mixes. This wet procedure is typically used in laboratories to examine the mechanisms by which starch plasticizes, gelatinizes, and retrogrades. The main characteristics of this technology are its capacity for uniform thickness distribution, dimensional stability, low haze, and optical purity. Figure 11.6 shows the example of thermoplastic starch processing with nanocellulose being used as the filler.

The performance evaluation of nanocellulose-reinforced starch composites is usually evaluated in terms of their mechanical, thermal, and water absorption properties [57]. To the best of our knowledge, research reported on the performance of nanocellulose-reinforced barley thermoplastic starch composites is still limited. A study conducted by El Halal et al. [58] revealed that the nanocellulose-reinforced barley thermoplastic starch had better properties as compared to the native starch. Based on the morphological analysis, the oxidized starch was more homogeneous as compared to the film of the native starch. In terms of the mechanical analysis, the addition of nanocellulose in the composites increased the tensile strength and decreased elongation. Interestingly, the composite promoted lower water solubility as compared to native

Table 11.2: The mechanical performance of nanocellulose reinforced thermoplastic starch composites.

Starch source	Type of nanocellulose	Tensile strength (MPa)	Young's modulus (MPa)	Reference
Pea	Cellulose nanocrystal from flax fiber	3.9–11.5	31.9–823.9	[60]
Maize	Cellulose microcrystals (or whiskers) from tunicate (a sea animal)	0.24–20	51–315	[61]
Wheat	Cellulose nanocrystals from cottonseed linter	2.5–7.8	36–301	[62]
Potato	Commercial microcrystalline cellulose (Borregaard Chemcell)	13.7	460	[63]
Sugar palm	Cellulose nanocrystals from sugar palm	11.5	178	[64]
Tapioca	Cellulose nanofiber from water hyacinth	5.8	403	[65]

starch. Authors suggested this can be due to the lower hygroscopicity of the fibers to the starch. Moreover, the nanocellulose interacts with the hydrophilic sites of the starch, replacing the starch with water linkages. Besides that, the water vapor permeability of the film of oxidized starch with 20% of nanocellulose was lower than the native starch. Meanwhile, the thermal properties analysis shows that the addition of nanocellulose increased the complexation of glycerol with cellulose and increased the thermal stability of the composites.

Recently, Xu et al. [59] investigated the morphology, crystallinity, chemical heterogeneity, mechanical, permeability, and contact angle of the nanocellulose-reinforced barley thermoplastic starch composite. Glycerol was added as a plasticizer to increase flexibility and elasticity. The increment of strain-at-break by five folds higher was reported with the addition of 25% nanocellulose. In the presence of high glycerol concentration, this property significantly increased with increases in nanocellulose. Differing from the strain-at-break, increased nanocellulose: starch ratio resulted in higher stress-at-break for the films without glycerol, demonstrating that nanocellulose had a significant strengthening effect on the composites. Moreover, the presence of nanocellulose in the starch matrix significantly enhanced the crystallinity. The water contact angle value was also increased signifying the increase in the surface hydrophobicity. There are several other thermoplastic starch sources such as peas, maize, corn, and sugar palm that had been investigated. Table 11.2 demonstrates the mechanical performance of nanocellulose-reinforced thermoplastic starch composites.

11.4 Potential application of nanocellulose reinforced barley thermoplastic starch composites

Concern over the uses of nonbiodegradable materials or thermoplastic polyolefins in various applications is concerning, hence the substitution of those non-biodegradable materials for more environmentally friendly options including starch has been in the spotlight for the past decades [36, 37, 56, 66–70]. For example, the biodegradability properties of starch could make it a potential material to be used as 'plastic' in agriculture for mulch, low tunnels, greenhouse covers, solarization film, fumigation film, and packaging which were conventionally produced from polyethylene and polypropylene. The potential of starch biodegradable polymer in agriculture has been reviewed elsewhere [68], whereby the potential implementation of starch-based polymer in sustainable agriculture practices, circular economy, waste remediation, and green chemistry was reviewed. Additionally, as starch can be used to make biodegradable films, it can be employed in a variety of agro-food, pharmaceutical, and industrial applications as well as in the production of packaging, vessels, and sheets [71, 72].

Differing from other types of applications, starch has long been a popular material in the biodegradable polymer packaging industry due to its outstanding film-forming, renewable, recyclable, and affordable qualities [67]. Biodegradable plastics with enough starch can be totally degraded while still having a high tensile strength. Additionally, agricultural waste-derived cellulose acetate (CA) offers the possibility of being used as a biodegradable plastic component in the production of packaging, salt containers, fiber, and plastic components as biodegradable materials [73, 74]. Dogan and McHugh, (2016) [74] previously described the preparation of edible starch films that can serve as protective coatings on food. According to Bideau et al. [75], nanocellulose-based composites had increased antimicrobial-resistant against bacteria occasionally found in food: Gram-positive *Bacillus subtilis (B. subtilis)* and Gram-negative bacteria *Escherichia coli (E. coli)*. Besides that, the reinforcement of sugar palm nanocellulose in starch had greatly improved the water vapor barrier properties by 19.94% compared to native film [76], hence may enhance the shelf life of the film-coated products.

In terms of barley thermoplastic starch application, Šubarić et al. [77] highlighted barley starch properties that are relatable to its application, whereby it is found more suitable for applications where high stability is needed during heating and shearing (low breakdown values). Additionally, it may suitable for those applications where high water binding is undesirable as its breakdown viscosities were considerably lower than those of oat-derived starch with lower swelling power and solubility values.

11.5 Conclusions

The utilization of barley starch for thermoplastic applications including nanocellulose thermoplastic composite is yet to be fully explored stemming from its major application in the food and beverages industries. Regardless, it has great potential, especially for food coating and packaging applications due to its similar biocompatibility and biodegradability properties to other starches. Exploration of nanocellulose-incorporated thermoplastic starch using barley will be a beneficial addition to barley industries for example as an alternative use of rejected or low-grade barleys. On the whole, research and development of barley thermoplastic starch reinforced with nanocellulose will widen the barley and the nanocellulose industries, besides offering various options to reduce and replace the uses of harmful petroleum-based plastic.

Acknowledgment: The authors would like to thank Nextgreen Pulp & Paper Sdn. Bhd. for their support and permission to work on this manuscript. Additionally, the authors gratefully acknowledge the technical and financial support from the Universiti Pertahanan Nasional Malaysia (UPNM) in preparation of this review.

References

1. Domene-López D, García-Quesada JC, Martin-Gullon I, Montalbán MG. Influence of starch composition and molecular weight on physicochemical properties of biodegradable films. Polymers 2019;11:1084.
2. Morán JI, Vázquez A, Cyras VP. Bio-nanocomposites based on derivatized potato starch and cellulose, preparation and characterization. J Mater Sci 2013;48:7196–203.
3. Yusuph M, Tester RF, Ansell R, Snape CE. Composition and properties of starches extracted from tubers of different potato varieties grown under the same environmental conditions. Food Chem 2003;82:283–9.
4. Zhu F, Xie Q. Structure and physicochemical properties of starch. In: Physical modifications of starch. SIngapore: Springer Nature; 2018:1–14 pp.
5. Dupuis JH, Liu Q. Potato starch: a review of physicochemical, functional and nutritional properties. Am J Potato Res 2019;96:127–38.
6. Tester RF, Karkalas J, Qi X. Starch – composition, fine structure and architecture. J Cereal Sci 2004;39: 151–65.
7. Xie F, Pollet E, Halley PJ, Avérous L. Starch-based nano-biocomposites. Prog Polym Sci 2013;38:1590–628.
8. Food and Agriculture Organization of the United Nations. FAOSTAT crops and livestock products. Available from: https://www.fao.org/faostat/en/#data/QCL [Accessed 21 Nov 2022].
9. Zhu F. Barley starch: composition, structure, properties, and modifications. Compr Rev Food Sci Food Saf 2017;16:558–79.
10. Asare EK, Jaiswal S, Maley J, Båga M, Sammynaiken R, Rossnagel BG, et al. Barley grain constituents, starch composition, and structure affect starch in vitro enzymatic hydrolysis. J Agric Food Chem 2011;59:4743–54.
11. Li JH, Vasanthan T, Rossnagel B, Hoover R. Starch from hull-less barley: I. Granule morphology, composition and amylopectin structure. Food Chem 2001;74:395–405.
12. Yangcheng H, Gong L, Zhang Y, Jane JL. Pysicochemical properties of Tibetan hull-less barley starch. Carbohydr Polym 2016;137:525–31.
13. Andersson AAM, Andersson R, Åman P. Starch and by-products from a laboratory-scale barley starch isolation procedure. Cereal Chem 2001;78:507–13.
14. Izydorczyk MS, Storsley J, Labossiere D, MacGregor AW, Rossnagel BG. Variation in total and soluble β-glucan content in hulless barley: effects of thermal, physical, and enzymic treatments. J Agric Food Chem 2000;48:982–9.
15. Soares RMD, De Francisco A, Rayas-Duarte P, Soldi V. Brazilian hull-less and malting barley genotypes: I. Chemical composition and partial characterization. J Food Qual 2007;30:357–71.
16. Song Y, Jane J. Characterization of barley starches of waxy, normal, and high amylose varieties. Carbohydr Polym 2000;41:365–77.
17. Pycia K, Gałkowska D, Juszczak L, Fortuna T, Witczak T. Physicochemical, thermal and rheological properties of starches isolated from malting barley varieties. J Food Sci Technol 2015;52:4797–807.

18. Vasanthan T, Hoover R. Barley starch: production, poperties, modification and uses. In: Starch chemistry and technology, 3rd ed. Burlington, MA, USA: Academic Press; 2009:601–28 pp.
19. Yasui T, Seguschi M, Ishikawa N, Fujita M. Starch properties of a waxy mutant line of hull-less barley (*Hordeum vulgare L.*). Starch Staerke 2002;54:179–84.
20. Singh J, Kaur L, McCarthy OJ. Factors influencing the physico-chemical, morphological, thermal and rheological properties of some chemically modified starches for food applications-A review. Food Hydrocolloids 2007;21:1–22.
21. Šubarić D, Babić J, Lalić A, Aćkar D, Kopjar M. Isolation and characterisation of starch from different barley and oat varieties. Czech J Food Sci 2011;29:354–60.
22. Bello-Pérez LA, Sánchez-Rivera MM, Núñez-Santiago C, Rodríguez-Ambriz SL, Román-Gutierrez AD. Effect of the pearled in the isolation and the morphological, physicochemical and rheological characteristics of barley starch. Carbohydr Polym 2010;81:63–9.
23. Naguleswaran S, Vasanthan T, Hoover R, Bressler D. The susceptibility of large and small granules of waxy, normal and high-amylose genotypes of barley and corn starches toward amylolysis at sub-gelatinization temperatures. Food Res Int 2013;51:771–82.
24. Bello-Pérez LA, Agama-Acevedo E. Starch. In: Starch-based materials in food packaging processing, characterization and applications. London, UK: Acadmic Press; 2018:1–18 pp.
25. Tang H, Ando H, Watanabe K, Takeda Y, Mitsunaga T. Physicochemical properties and structure of large, medium and small granule starches in fractions of normal barley endosperm. Carbohydr Res 2001;330: 241–8.
26. Waduge RN, Hoover R, Vasanthan T, Gao J, Li J. Effect of annealing on the structure and physicochemical properties of barley starches of varying amylose content. Food Res Int 2006;39:59–77.
27. Savadekar NR, Mhaske ST. Synthesis of nano cellulose fibers and effect on thermoplastics starch based films. Carbohydr Polym 2012;89:146–51.
28. Xie F, Luckman P, Milne J, McDonald L, Young C, Tu CY, et al. Thermoplastic starch: current development and future trends. J Renew Mater 2014;2:95–106.
29. Forssell PM, Hulleman SHD, Myllärinen PJ, Moates GK, Parker R. Ageing of rubbery thermoplastic barley and oat starches. Carbohydr Polym 1999;39:43–51.
30. Forssell PM, Mikkilä JM, Moates GK, Parker R. Phase and glass transition behaviour of concentrated barley starch-glycerol-water mixtures, a model for thermoplastic starch. Carbohydr Polym 1997;34:275–82.
31. Sagnelli D, Hooshmand K, Kemmer GC, Kirkensgaard JJK, Mortensen K, Giosafatto CVL, et al. Cross-linked amylose bio-plastic: a transgenic-based compostable plastic alternative. Int J Mol Sci 2017;18:2075.
32. Sagnelli D, Cavanagh R, Xu J, Swainson SME, Blennow A, Duncan J, et al. Starch/poly (glycerol-adipate) nanocomposite film as novel biocompatible materials. Coatings 2019;9:482.
33. Fox GP, Watson-Fox L. Barley: current understanding of 'omics data on quality. In: Comprehensive foodomics. Amsterdam, Netherlands: Elsevier; 2021, 1:513–27 pp.
34. Qin C, Soykeabkaew N, Xiuyuan N, Peijs T. The effect of fibre volume fraction and mercerization on the properties of all-cellulose composites. Carbohydr Polym 2008;71:458–67.
35. Sharip NS, Ariffin H. Cellulose nanofibrils for biomaterial applications. Mater Today Proc 2019;16:1959–68.
36. Norrrahim MNF, Kasim NAM, Knight VF, Halim NA, Shah NAA, Noor SAM, et al. Performance evaluation of cellulose nanofiber reinforced polymer composites. Funct Compos Struct 2021;3:024001.
37. Norrrahim MNF, Ariffin H, Yasim-Anuar TAT, Hassan MA, Ibrahim NA, Yunus WMZW, et al. Performance evaluation of cellulose nanofiber with residual hemicellulose as a nanofiller in polypropylene-based nanocomposite. Polymers 2021;13:1064.
38. Thiruganasambanthan T, Ilyas RA, Norrrahim MNF, Kumar TSM, Siengchin S, Misenan MSM, et al. Emerging developments on nanocellulose as liquid crystals: a biomimetic approach. Polymers 2022;14: 1546.

39. Norrrahim MNF, Mohd Kasim NA, Knight VF, Ong KK, Mohd Noor SA, Abdul Halim N, et al. Emerging developments regarding nanocellulose-based membrane filtration material against microbes. Polymers 2021;13:3249.
40. Norrrahim MNF, Farid MAA, Lawal AA, Yasim-Anuar TAT, Samsudin MH, Zulkifli AA. Emerging technologies for value-added use of oil palm biomass. Adv Environ Sci 2022;1:259–75.
41. Norrrahim MNF, Kasim NAM, Knight VF, Ong KK, Noor SAM, Jamal SH, et al. Cationic nanocellulose as promising candidate for filtration material of COVID-19: a perspective. Appl Sci Eng Prog 2021;14:580–7.
42. Klemm D, Cranston ED, Fischer D, Gama M, Kedzior SA, Kralisch D, et al. Nanocellulose as a natural source for groundbreaking applications in materials science: today's state. Mater Today 2018;21:720–48.
43. Dufresne A. Nanocellulose: a new ageless bionanomaterial. Mater Today 2013;16:220–7.
44. Solala I, Iglesias MC, Peresin MS. On the potential of lignin-containing cellulose nanofibrils (LCNFs): a review on properties and applications. Cellulose 2020;27:1853–77.
45. Norrrahim MNF, Kasim NAM, Knight VF, Misenan MSM, Janudin N, Shah NAA, et al. Nanocellulose: a bioadsorbent for chemical contaminant remediation. RSC Adv 2021;11:7347–68.
46. Kargarzadeh H, Mariano M, Gopakumar D, Ahmad I, Thomas S, Dufresne A, et al. Advances in cellulose nanomaterials. Cellulose 2018;25:2151–89.
47. Abitbol T, Rivkin A, Cao Y, Nevo Y, Abraham E, Ben-Shalom T, et al. Nanocellulose, a tiny fiber with huge applications. Curr Opin Biotechnol 2016;39:76–88.
48. Sharma P, Mittal M, Yadav A, Aggarwal NK. Bacterial cellulose: nano-biomaterial for biodegradable face masks-A greener approach towards environment. Environ Nanotechnol Monit Manag 2022;19:100759.
49. Jenol MA, Norrrahim MNF, Nurazzi NM. 17 – nanocellulose nanocomposites in textiles. In: Sapuan SM, Norrrahim MNF, Ilyas RA, Soutis C, editors. Industrial applications of nanocellulose and its nanocomposites. Cambridge, USA: Woodhead Publishing; 2022:397–408 pp.
50. Hakimi MI, Najmuddin SUFS, Yusuff SM, Norrrahim MNF, Janudin N, Yusoff MZM, et al. 18 – nanocellulose as a bioadsorbent for water and wastewater purification. In: Sapuan SM, Norrrahim MNF, Ilyas RA, Soutis C, editors. Industrial applications of nanocellulose and its nanocomposites. Cambridge, USA: Woodhead Publishing; 2022:409–37 pp.
51. Norrrahim MNF, Kasim NAM, Knight VF, Misenan MSM, Janudin N, Shah NAA, et al. 7 – nanocellulose as an adsorbent for heavy metals. In: Sapuan SM, Norrrahim MNF, Ilyas RA, Soutis C, editors. Industrial applications of nanocellulose and Its nanocomposites. Cambridge, USA: Woodhead Publishing; 2022: 197–211 pp.
52. Janudin N, Kasim NAM, Knight VF, Norrrahim MNF, Razak MAIA, Halim NA, et al. Fabrication of a nickel ferrite/nanocellulose-based nanocomposite as an active sensing material for the detection of chlorine gas. Polymers 2022;14:1906.
53. Sharip NS, Yasim-Anuar TAT, Norrrahim MNF, Shazleen SS, Nurazzi NM, Sapuan SM, et al. A review on nanocellulose composites in biomedical application. In: Composites in biomedical applications. Boca Raton, FL, USA: CRC Press; 2020:161–90 pp.
54. Norrrahim MNF, Norizan MN, Jenol MA, Farid MAA, Janudin N, Ujang FA, et al. Emerging development on nanocellulose as antimicrobial material: an overview. Mater Adv 2021;2:3538–51.
55. Norrrahim MNF, Kasim NAM, Knight VF, Ujang FA, Janudin N, Razak MAIA, et al. Nanocellulose: the next super versatile material for the military. Mater Adv 2021;2:1485.
56. Surendren A, Mohanty AK, Liu Q, Misra M. A review of biodegradable thermoplastic starches, their blends and composites: recent developments and opportunities for single-use plastic packaging alternatives. Green Chem 2022;24:8606–36.
57. Ilyas R, Sapuan S, Norrrahim MNF, Yasim-Anuar TAT, Kadier A, Kalil MS, et al. Nanocellulose/starch biopolymer nanocomposites: processing, manufacturing, and applications. In: Advanced processing, properties, and applications of starch and other bio-based polymers. Amsterdam, Netherlands: Elsevier; 2020:65–88 pp.

58. El Halal SLM, Colussi R, Deon VG, Pinto VZ, Villanova FA, Carreño NLV, et al. Films based on oxidized starch and cellulose from barley. Carbohydr Polym 2015;133:644–53.
59. Xu J, Sagnelli D, Faisal M, Perzon A, Taresco V, Mais M, et al. Amylose/cellulose nanofiber composites for all-natural, fully biodegradable and flexible bioplastics. Carbohydr Polym 2021;253:117277.
60. Cao X, Chen Y, Chang PR, Muir AD, Falk G. Starch-based nanocomposites reinforced with flax cellulose nanocrystals. Express Polym Lett 2008;2:502–10.
61. Anglès MN, Dufresne A. Plasticized starch/tunicin whiskers nanocomposite materials. 2. Mechanical behavior. Macromolecules 2001;34:2921–31.
62. Lu Y, Weng L, Cao X. Biocomposites of plasticized starch reinforced with cellulose crystallites from cottonseed linter. Macromol Biosci 2005;5:1101–7.
63. Kvien I, Sugiyama J, Votrubec M, Oksman K. Characterization of starch based nanocomposites. J Mater Sci 2007;42:8163–71.
64. Ilyas RA, Sapuan SM, Ishak MR, Zainudin ES. Development and characterization of sugar palm nanocrystalline cellulose reinforced sugar palm starch bionanocomposites. Carbohydr Polym 2018;202: 186–202.
65. Asrofi M, Abral H, Kasim A, Pratoto A, Mahardika M, Hafizulhaq F. Mechanical properties of a water hyacinth nanofiber cellulose reinforced thermoplastic starch bionanocomposite: effect of ultrasonic vibration during processing. Fibers 2018;6:40.
66. Moshood TD, Nawanir G, Mahmud F, Mohamad F, Ahmad H, Abdulghani A. Sustainability of biodegradable plastics: new problem or solution to solve the global plastic pollution? Curr Res Green Sustainable Chem 2022;5:100273.
67. Bangar SP, Whiteside WS, Ashogbon AO, Kumar M. Recent advances in thermoplastic starches for food packaging: a review. Food Packag Shelf Life 2021;30:100743.
68. Gamage A, Liyanapathiranage A, Manamperi A, Gunathilake C, Mani S, Merah O, et al. Applications of starch biopolymers for a sustainable modern agriculture. Sustainability 2022;14:6085.
69. Norrrahim MNF, Ariffin H, Hassan MA, Ibrahim NA, Yunus WMZW, Nishida H. Utilisation of superheated steam in oil palm biomass pretreatment process for reduced chemical use and enhanced cellulose nanofibre production. Int J Nanotechnol 2019;16:668–79.
70. Ilyas R, Sapuan S, Kadier A, Kalil MS, Ibrahim R, Atikah M, et al. Properties and characterization of PLA, PHA, and other types of biopolymer composites. In: Advanced processing, properties, and applications of starch and other bio-based polymers. Amsterdam, Netherlands: Elsevier; 2020:111–38 pp.
71. Niaounakis M. Introduction. In: Biopolymers: processing and products. Oxford, GB, UK: William Andrew; 2015:1–77 pp.
72. Rujnić-Sokele M, Pilipović A. Challenges and opportunities of biodegradable plastics: a mini review. Waste Manag Res 2017;35:132–40.
73. Moshood TD, Nawanir G, Mahmud F, Mohamad F, Ahmad H, Abdulghani A, et al. Green product innovation: a means towards achieving global sustainable product within biodegradable plastic industry. J Clean Prod 2022;363:132506.
74. Dogan N, McHugh TH. Effects of microcrystalline cellulose on functional properties of hydroxy propyl methyl cellulose microcomposite films. J Food Sci 2007;72:16–22.
75. Bideau B, Bras J, Saini S, Daneault C, Loranger E. Mechanical and antibacterial properties of a nanocellulose-polypyrrole multilayer composite. Mater Sci Eng C 2016;69:977–84.
76. Ilyas RA, Sapuan SM, Ishak MR, Zainudin ES. Sugar palm nanocrystalline cellulose reinforced sugar palm starch composite: degradation and water-barrier properties. In: Proceedings of the IOP conference series: materials science and engineering; 2018, vol 368.
77. Šubarić D, Babić J, LaLić A, Ačkar Đ, Kopjar M. Isolation and characterisation of starch from different barley and oat varieties. Czech J Food Sci 2011;29:354–60.

Nazrin Asmawi, R. A. Ilyas*, Muhammad Huzaifah Mohd Roslim, L. Rajeshkumar, W. Abotbina, Edi Syafri, Ridhwan Jumaidin, R. Syafiq, S. A. Rafiqah, R. Ridwan, Suriani Mat Jusoh and Mohd Zuhri Mohamed Yusoff

12 Cassava starch nanocomposite films reinforced with nanocellulose

Abstract: Recent researchers are keen on developing alternative bioplastic materials from renewable and eco-friendly sources to replace the materials obtained from crude oil and other petroleum-based sources. The measures for these replacements have been made continuously to create a sustainable future for the forthcoming generations. Researchers are focusing on bio-based alternatives due to their numerous

***Corresponding author: R. A. Ilyas,** Institute of Tropical Forest and Forest Products (INTROP), Universiti Putra Malaysia, 43400 UPM Serdang, Selangor, Malaysia; Faculty of Chemical and Energy Engineering, Universiti Teknologi Malaysia, 81310 UTM, Johor, Malaysia; Centre for Advanced Composite Materials, Universiti Teknologi Malaysia (UTM), 81310 Johor Bahru, Malaysia; and Centre of Excellence for Biomass Utilization, Universiti Malaysia Perlis, 02600 Arau, Perlis, Malaysia, E-mail: ahmadilyas@utm.my. https://orcid.org/0000-0001-6622-2632

Nazrin Asmawi, Institute of Tropical Forest and Forest Products (INTROP), Universiti Putra Malaysia, 43400 UPM Serdang, Selangor, Malaysia; and Institute of Power Engineering, Universiti Tenaga Nasional, Jalan IKRAM-UNITEN, 43000 Kajang, Selangor, Malaysia

Muhammad Huzaifah Mohd Roslim, Department of Crop Science, Universiti Putra Malaysia, Serdang 43400, Malaysia

L. Rajeshkumar, Centre for Materials and Manufacturing Technology, KPR Institute of Engineering and Technology, Coimbatore 641407, Tamilnadu, India; and Department of Mechanical Engineering, KPR Institute of Engineering and Technology, Coimbatore 641407, Tamilnadu, India

W. Abotbina, Mechanical and Industrial Engineering Department, Faculty of Engineering, Al-Asmarya Islamic University, Zliten, Libya

Edi Syafri, Departement of Agricultural Technology, Politeknik Pertanian Negeri Payakumbuh, Payakumbuh, West Sumatera, 26271, Indonesia

Ridhwan Jumaidin, Fakulti Teknologi Kejuruteraan Mekanikal dan Pembuatan, Universiti Teknikal Malaysia Melaka, 76100 Durian Tunggal, Melaka, Malaysia

R. Syafiq and S. A. Rafiqah, Institute of Tropical Forest and Forest Products (INTROP), Universiti Putra Malaysia, 43400 UPM Serdang, Selangor, Malaysia

R. Ridwan, Department of Environmental Engineering, Universitas Andalas, Padang, West Sumatera, 25163, Indonesia

Suriani Mat Jusoh, Faculty of Ocean Engineering Technology and Informatics, Universiti Malaysia Terengganu, 21030 Kuala Nerus, Terengganu, Malaysia

Mohd Zuhri Mohamed Yusoff, Institute of Tropical Forest and Forest Products (INTROP), Universiti Putra Malaysia, 43400 UPM Serdang, Selangor, Malaysia; and Department of Mechanical and Manufacturing Engineering, Advanced Engineering Materials and Composites Research Centre (AEMC), Universiti Putra Malaysia, Serdang, Selangor, Malaysia. https://orcid.org/0000-0002-1069-7345

As per De Gruyter's policy this article has previously been published in the journal Physical Sciences Reviews. Please cite as: N. Asmawi, R. A. Ilyas, M. H. Mohd Roslim, L. Rajeshkumar, W. Abotbina, E. Syafri, R. Jumaidin, R. Syafiq, S. A. Rafiqah, R. Ridwan, S. M. Jusoh and M. Z. M. Yusoff "Cassava starch nanocomposite films reinforced with nanocellulose" *Physical Sciences Reviews* [Online] 2023. DOI: 10.1515/psr-2022-0014 | https://doi.org/10.1515/9783110773606-012

benefits, including biodegradability, biocompatibility, nontoxicity, and structural flexibility. The main problem on the current bio-based material such as poly lactic acid, poly butylene succinate and poly L lactide, polyhydroxybuturate, and polyhydroxyalkalonates is the cost of production. Compare with cassava starch, the cost is much cheaper around 0.32 $/kg compare with other bio-based will cost around 1.2–2.4 $/kg. Conversion of biomass into useful materials has been the order of the day, as it reduces the cost of inventory and aims to develop a nature-derived material. The development of nanocomposites from biological sources has progressively experimented with the researchers and the deriving of polysaccharides such as starch, cellulose, and glycogen has aided the development of nanobiocomposites. Corn starch has been the dominant bioplastic material derived out of corn which can handle a variety of reinforcements and render a biocomposite material with better and enhanced properties. Cassava starch is the most economic and cheap polysaccharide derived from the cassava plant and has a greater potential to act as biopolymer material for the development of biocomposites. The development of cassava starch-based biocomposite film was widely used for a wide range of applications mainly for food packaging applications. This review focuses on the extraction, preparation, and properties of cassava starch from cassava plants. The properties of the cassava starch and its composites were also comprehensively dealt with. The development of biocomposite films based on cassava starch for food packaging applications has been reviewed along with the challenges associated with it.

Keywords: biocomposites; cassava starch; cellulose fibre; food packaging.

12.1 Introduction

Highly durable plastic materials have greatly improved human life [1, 2]. Nowadays, it can be traced in almost every corner of the world in all sort of applications [3]. As production keep on growing, the wastes generated are also proportionally increased leaving severe issue for their disposal. Especially in short-term applications such as packaging and agriculture mulching films, their wastes are enormous and mostly non-degradable. The durability of petroleum-based plastics act as a double-edge sword, which prompt scientist to develop environmental-friendly plastic with functional properties rivalling conventional plastics. Currently, natural-based plastics had become a major interest throughout the globe in replacing the synthetic one [4]. The most adapted resources are carbohydrates such as cellulose, starch, chitosan and glycogen. Starch-based polymers are often studied of their potential application in packaging industry [5]. However, hydrophilic nature of starch is the main restriction that must addressed before progressing its application. Particularly, food and beverage packaging plastics must have excellent barrier against water and gas to prevent contamination by microbial organisms. Several approaches were practiced to improve these qualities by adding supplementary compound tailoring the molecular structure of the end product.

Cassava (*Manihot esculenta*) starch is one of the prominent sources utilised for the production of biodegradable polymer. It is a significant source of energy and regard as a staple food in Africa, Asia and Latin America with an estimation cultivation nearly 300 metric tonnes globally. As starch-based polymers are known for their brittleness, poor processability and water sensitive, modifications such as plasticisation and fibre incorporation are typically practiced, which proven to elevate the mechanical, thermal and barrier properties of the composites [6, 7]. Depends on the size and type of fillers imparted, specific functional aspect can be improved significantly [8, 9]. For example, the addition of nanocellulose filler into starch matrix demonstrated good compatibility to elevate the tensile strength, water vapour barrier and thermal degradation temperature of the nanocomposites [10–12]. The uniform dispersion of nanocellulose with high surface area to volume ratio established desirable structure of effective stress transfer, good matrix-filler interfacial interaction and tortuous path [13]. This paper aims to review the potential of cassava-based resources in developing durable, safe and sustainable plastic for packaging application.

12.2 History of cassava plant

Cassava is a perennial woody and tuberous plant, belonging to the family of Euphorbiaceae (spurge family). The roots and plant of Cassava are also termed as mandioca, manioc, and yuca [14, 15]. Origination of the terms cassava was from a term called Casavi and Cazabi in Arawak language (The first indigenous language of people living in Great Antilles) denoting bread [16]. As this plant is grown periodically and harvesting is easy, this plant is available all around the year. Due to these merits, cassava has turned to be a reliable crop for food and other purposes. Though Cassava plant is rich in carbohydrates, thiamine and riboflavin, they are poor in protein content [17]. Cassava crop cultivation regions around the world are dominated by 72 % in Africa, 18 % in Asia, and 10 % each by South America and North America [18]. The yield of cassava plant and crops serves for many applications and industries including the manufacturing of starch, biofuels, bioethanol, animal feeds, and alcohol [19–21]. Portugese commenced the cultivation of cassava crop along the coastal areas of the West Africa and then the cultivation was spread to other regions like East Africa, Central Africa, Malaya, India, Sri Lanka, Madagascar, and Indonesia. Spanish introduced Cassava crop to Asian region at the beginning of 19th Century after their widespread cultivation of cassava in Philippines [22–24]. To date Thailand, Nigeria and Indonesia are the leading producers of Cassava with a rate of production of 30.03, 54.85, and 23.45 million metric tonnes (MMT) in the world as per the data published during 2017 [25]. Figure 12.1 illustrates the yields Cassava from various countries all around the globe.

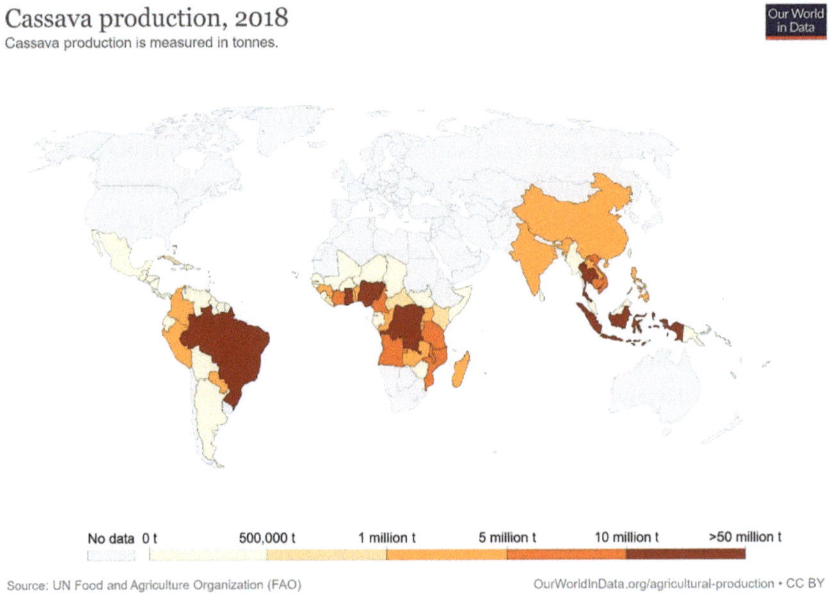

Figure 12.1: Cassava crop production statistics by country in 2018 [26].

12.3 Cassava starch

12.3.1 Extraction and preparation of cassava starch

Cassava starch (CS) can be extracted from the cassava tuber through conventional grating process. The shredded flesh of cassava tuber are mixture of starch and bagasse. Water is added to form a suspension before proceeding to filtration process. Large size of bagasse will be suspended while smaller starch particles will pass through filter cloth. After several repetitions of filtering process, a supernatant consisting primarily starch will be obtained. The supernatant will be left still for 8 h for the starch to sediment at bottom meanwhile bagasse floats. The accumulated starch is collected and spread evenly in pan and left dried in air circulating oven at 80 °C for 24 h.

Cassava starch was produced by the wet milling of fresh cassava extracted from the plants. Figure 12.2 represents the step-by-step process of making cassava starch. Age of the cassava plant and its root quality are the major factors to choose a cassava plant root for the extraction of starch from it. Since the enzymatic cycle of rotting of cassava roots progresses rapidly in one or two days and these roots are very fresh, the treatments of cassava roots like peeling of the skin, rinsing them in water and conversion of the dried roots into finer particles immediately after their harvesting is very important. Following the conversion, the grated pulp are to be sifted to facilitate the starch removal without disturbing the fibres and these fibre contents were washed

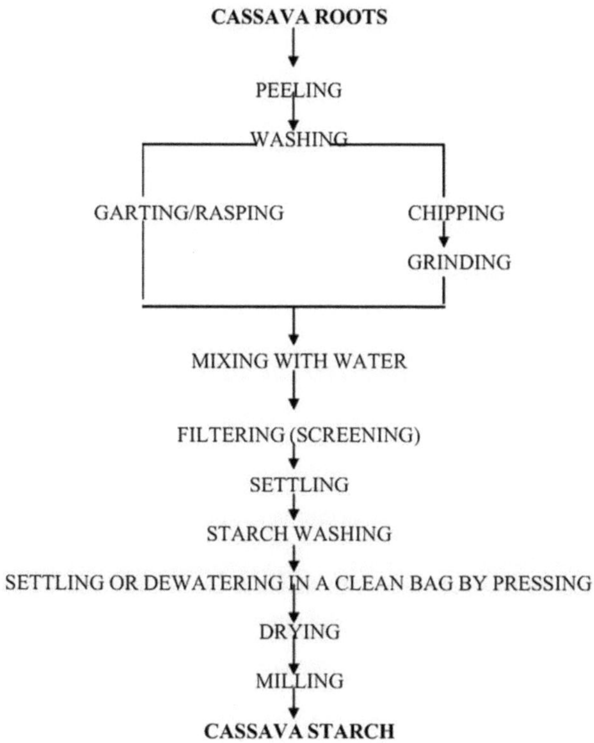

Figure 12.2: Production of cassava starch – steps involved [28].

thrice or four times in a distilled water. After washing, the starch was allowed to be sedimented and the excess fibres were removed. The sedimented starch was washed again with the distilled water to remove the residues of the fibres, if any. The extracted starch was then heated up in a hot air oven at 45 °C to get rid of the moisture content. Finally, the starch was dried in the hot sun for about 4 h and crushed using a dry blender after which they are stored in an airtight container to avoid contamination and moisture prone [27, 28].

12.3.2 Properties of cassava starch (CS)

12.3.2.1 Physical

CS granules are characterised as an oval-shape particle sizing between 5 and 20 µm. They are usually semi-crystalline polymeric materials with A-type crystallinity peaks typically (18.1° and 23.2°) and B-type (15.1°, 17.4° and 26.1°) categorised as C-type crystalline structure [29].

12.3.2.2 Chemical

Basically, the amylose content of CS can be divided into three groups; (1) 0–15 % (waxy starch), (2) 20–35 % (regular starch) and (3) higher than 40 % (high amylose) [30]. The molecular weight of amylose and amylopectin are $2.4–2.7 \times 10^5–10^6$ and $5–40 \times 10^6–10^8$ Da, respectively. Table 12.1 indicates the chemical properties of starch derived from to different location in Sabah, Malaysia.

12.4 Lignocellulosic fibre

12.4.1 Classification of fibre

The constituent of cassava bagasse (CB) made up of cellulose (27 %), hemicellulose (30 %) and lignin (2.7 %) [32]. Cellulose can be further refined to obtain microcrystalline cellulose (MCC) and nanocellulose (NC) through several process of treatments [33–36]. Figure 12.3 displays the segregation of cellulose into MC and three different nanocelluloses. There are vast resources of cellulose, which primarily came from plants and various living organisms such as algae and bacteria.

12.4.2 Cellulose fibre

Cassava bagasse (CB) fibre is the by-product of CS extraction process that made up 30 % of the total solid weight. The composition of CB predominantly made up of starch (40–60 %) and fibre (15–50 %) varies on the cassava's origin and processing method. Figure 12.4 represents the particle size distribution in cassava bagasse fibre. The histogram specifies that the particles with a size between 100 and 200 µm represented the highest percentage (34 %) in cassava bagasse fibre. Overall, the majority of fibres disturbed in the range between 100 and 500 µm made up 83 % of the total particles of

Table 12.1: The chemical composition of cassava starch extracted from Semporna and Tawau [31].

Parameter	Semporna	Tawau
Ash	2.19 ± 0.16	2.43 ± 0.30
Carbohydrate	27.02 ± 1.14	26.39 ± 0.80
Crude fat	0.68 ± 0.01	0.55 ± 0.04
Crude protein	2.69 ± 0.02	2.07 ± 0.11
Moisture content	5.97 ± 1.16	6.33 ± 0.58
Total dietary fibre	2.09 ± 0.01	2.38 ± 0.16

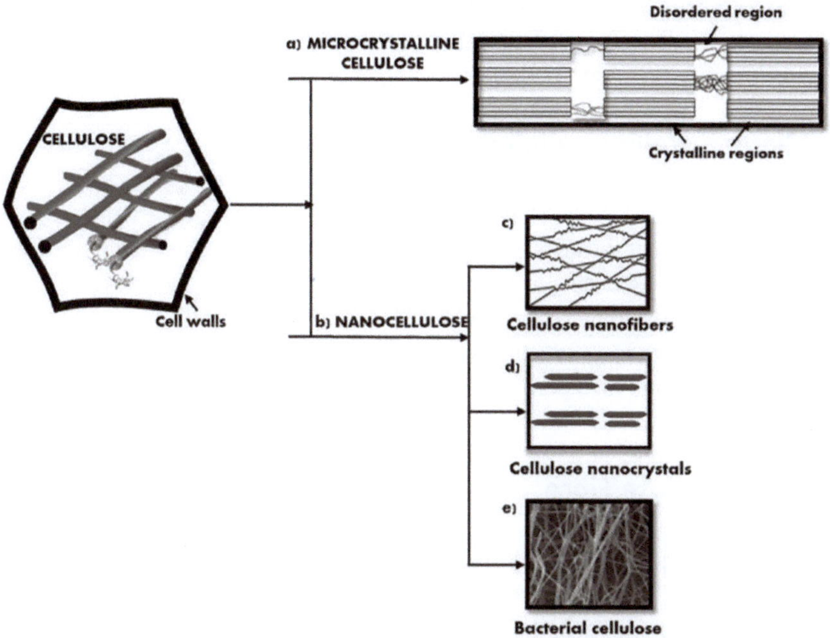

Figure 12.3: Types of plant derived cellulosic fibre, (a) microcrystalline cellulose, (b) nanocellulose, (c) cellulose nanofibres (CNF), (d) cellulose nanocrystals (CNC) and (e) bacterial cellulose (BC) [37].

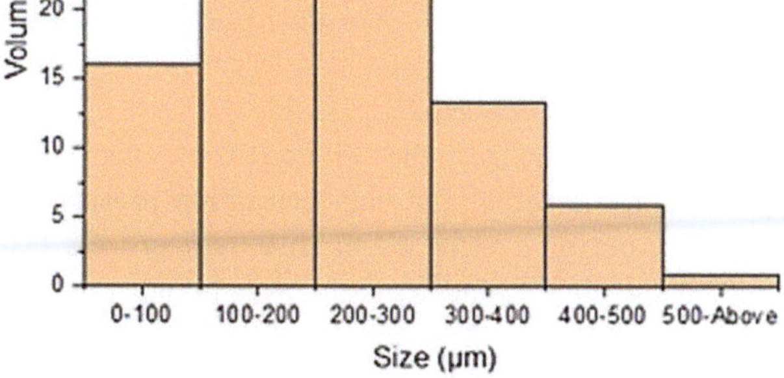

Figure 12.4: Particle size distribution of cassava bagasse [38]

cassava bagasse fibre. The size reduction of fibre particles optimized the composite material's characteristics of biodegradability and tensile.

12.4.3 Microcrystalline cellulose (MCC)

MCC is categorised as pure cellulose having relative high crystallinity. It is extracted from α-cellulose of fibrous plant through acid hydrolysis to remove the associated impurities from the pedicle fibres. According to Ansharullah et al. [39], different concentrations of hydrochloric acid (HCl) influence the yield and size of MCC. Mostly it is utilised as pharmaceutical industry in producing capsules with high hardness, slight friability, quick dissolution time and high percentage of drug release.

12.4.4 Nanocellulose Fibre

Cassava bagasse nanocellulose (CBN) are derived from the cellulose fibres of cassava bagasse (CB). Initially, CB went through acid hydrolysis process under mechanical stirring at 60 °C for 40 min. Then, the retrieved aggregation is washed until it reached pH value of 6–7. Application of non-biodegradable plastics in widespread applications resulted in white pollution which infected the global natural ecosystem [40]. It was quoted in many researches that biodegradable films can be used for manufacturing the food packaging materials from agro industrial waste materials which includes bagasse of cassava, malt, starch and sugar cane [41, 42]. Specifically, cassava bagasse products can be applied in wide range of valuable applications such as in manufacturing organic acid, nanofibres, a-amylase lactic acid, nanoparticles, biofuel, ethanol and in biodegradable packaging materials [43]. Cassava bagasse is composed of 39 % of lignin and cellulose, 38 % of hemicellulose, residual fibres and starch contents [44]. Cassava starch industries produce cassava bagasse and peels in higher quantities. The peel obtained from the cassava root was about 15 % by weight of the root and the remaining portions of the waste root cannot be used for any other purposes while the bagasse can be used as an animal feed. Dried cassava peel contains around 40–55 % of the cellulose and this part can be a rich source of cellulose [45, 46].

Few researchers used cassava peel to extract the cellulose from them using bleaching, mechanical disruption, ultrasonication and homogenization process and prepared cellulose nanofibres (CNF) out of it using alkali treatment process [32]. The extracted CNF were examined for their diameters and it was found that their diameters were 8.2 nm for alkali-treated CNFs and 6.7 nm for acid hydrolysis-treated CNFs after making both the extracts to undergo homogenization and ultrasonication. Figure 12.5 depicts the transmission electron microscopic (TEM) morphology of the CNFs extracted using two different processes. In order to characterize the extracted cellulose and prepared CNFs various characterization techniques such as X-ray

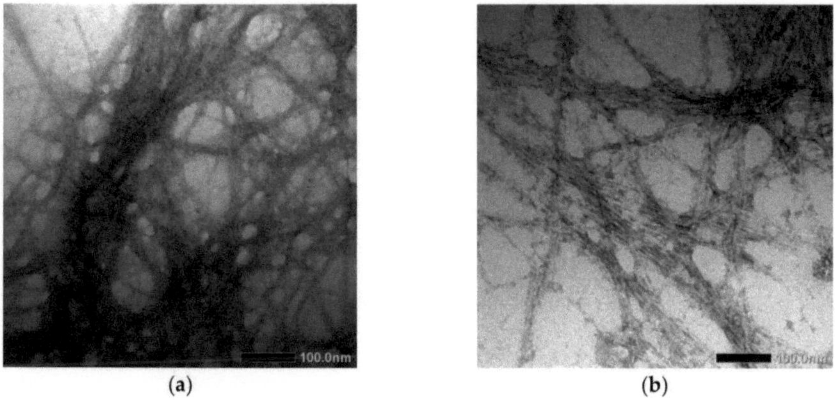

Figure 12.5: TEM micrographs of CNF extracted through (a) acid hydrolysis and (b) alkali treatment [32].

diffraction (XRD) analysis, Fourier Transform Infrared Spectroscopy (FTIR) analysis, Thermogravimetric analysis (TGA), Scanning and Transmission electron microscopy (SEM and TEM) analysis were adopted. Results exhibited clearly that the cellulose and CNF can be successfully prepared by both above methods and materials possessed similar chemical properties and different physical properties than the ones prepared by other methods.

12.5 Modification of cassava starch films

12.5.1 Cassava starch biopolymer reinforced with biofibres

Cassava starch is widely employed as a biopolymer progressively during the recent years owing to its availability, low cost, renewable nature and its biodegradability. Cassava starch is a highly potential materials for the production of sustainable materials. Some experimenters carried out research on the green coir fibre reinforced in cassava starch and examined their tensile behaviour. Biocomposites were prepared using compression moulding technique with the reinforcement concentration in between 5 and 15 % with a lignin content of 35 %. Results portrayed that the tensile strength of the biocompoistes increased, modulus of elasticity decreased with the decrease in fibre diameter and the elongation at break remained constant. It was also deduced from the results that the moisture absorption behaviour of the composites governed the elongation at break and tensile strength of the biocomposites [47]. Table 12.2 enlists the cassava starch-based biofibre reinforced biocomposites.

Table 12.2: Reported works on the natural fibre-reinforced cassava starch-based composites.

Polymer matrix	Fibre	Reference
Cassava starch (CS)	Naive coconut	[47]
CS	Coir	[48]
CS	Cassava bagasse (CB) fibres	[49]
CS	Cellulose CB nanofibrils (CBN)	[50]
CS	CB-Kraft paper	[51]
CS	CNC from kenaf fibres	[52]
Native CS	Cinnamon and clove essential oil/Sucrose ester fatty acids from sugar	[53]
CS	Kapok fibre	[54]
CS	Jute fibre	[54]
CS	Kaolinite-rich clay	[55]
CS	Chitosan (potential material) as compatible malt	[56]
CS	Malt bagasse	[57]
CS	Palm, gluten, Zein, Soy protein and Kraft fibre blends	[58]
Cassava flour (CF)/wheat flour (WF)	Cassava stillage residue (CSR)	[59, 60]
Polyvinyl chloride (PVC)		[61]
Egg albumen: CS and sunflower oil		[62]
CSR	CSR	[45]
CS	Cassava peels (CP)/CB	[57]
CS	Fibrous residue after CS extraction	[63]
CS	Cassava NF	[32]
CS	CS/sugar palm fibres	[64]
CS	Cassava bagasse/CNF	[65]
CS	MCC Avicel PH101	[66]
CS	Rice husk fiber	[67]
CS	Rice husk fiber nanocrystalline cellulose	[67]
CS	Cassava root	[68]
CS	Cassava bagasse	[68]
CS	Lignocellulose nanofibers of CB/nanoclay	[32]
CS	CB/Kraft paper	[69]
CS	Waxy starch nanocrystal	[70]
CS	CB	[71]
CS	Zinc oxide nanofiller	[72]
CS	Acetobacter xylinum BC	[73]
CS	Newspaper pulp fibers (recycled)	[74]
CS	Kenaf CNC	[75]
CS	Montmorillonite clay	[76]
CS	Bamboo nanofibers	[76]
CS	BC	[77]
CS	ZnO/BC	[78]
CS	Carnauba wax/cashew tree gum-based films	[79]
CS	Cotton fibers and natural rubber latex hybrid	[80]
CS	SiO_2 nanoparticles deposited cellulose pulp fibers	[81]

Table 12.2: (continued)

Polymer matrix	Fibre	Reference
CS	Cellulose fibers	[82]
CS	CB	[83]
CS	CP	[83]
CS	Coconut nanocellulose	[84]
CS	Licuri nanocellulose	[84]
CS	Corn stover nanocellulose	[84]
CS	Eucalyptus nanocellulose pulp	[84]
CS	CP	[85]
CS	Kenaf CNCs	[86]
CS	Sugarcane bagasse and Oregano essential oil hybrids	[87]
CS	Nanoclay	[88]
CS	Zein oil	[89]
CS	Gluten oil	[89]
CS	Soy protein oil	[89]
CS	Oil from Kraft fiber	[89]
CS	Palm oil	[89]
CS	Kenaf nanocrystalline cellulose	[75]
CS	Brazilian coir fiber	[90]
CS	Eucalyptus CNCs	[91]
CS	Nanoclay (NC)	[92]
CS	Cassava roots bagasse	[93]
CS	Cellulose fiber/NC	[94]
CS	Montmorillonite clay hybridized with sugarcane bagasse	[95]
CS	Sisal CNF	[96]
CS	Banana fibers	[97]
CS	Pineapple shell fiber	[98]
CS	Soybean hulls fiber	[99]
CS	Soybean hulls microcrystalline cellulose	[99]
CS	Polylactic acid	[100]
CS	Ramie CNF/nano polycrystalline cellulose tapioca starch	[101]
CS	Zeolite	[102]
CS	Beidellite	[102]
CS	Starch nanocrystals (SNCs)	[103]
CS	Nano fiber Straw/ZnO	[104]
CS	Pectin particles	[105]
CS	Cotton fibers	[105]
CS	Ramie cellulose microfibrils	[106]
CS	Coconut fiber nanocellulose	[107]
CS	Bamboo fiber and waste cooking oil (epoxidized)	[108]
CS	Cassava nanofibril	[109]
CS	Carboxymethylcellulose hybridized with lactic acid bacteria	[110]
CS	ZnO nanorods/PVA electrospun mats/rosemary extract	[111]
CS	Polyaniline	[112]
CS	Nanosilica (SiO_2)	[113]
CS	Montmorillonite	[114]

Table 12.2: (continued)

Polymer matrix	Fibre	Reference
CS	Silica	[115]
CS	Sisal fiber	[116]
CS	Cassava bagasse fibers	[117]
CS	Coconut residue fibers	[118]
CS	Cassava cellulose nanocrystals	[119]
CS	Grape stalks	[120]
CS	Carboxymethylcellulose/turmeric oil	[121]
CS	Sisal fiber hybridized with carnauba wax	[122]
CS	Coconut fibers	[123]
CS	Kaolinite	[124]

12.5.2 Plasticization of cassava starch films

The addition of plasticiser into starch matrix is to promote the molecular mobility improving the elongation at break. Smaller molecules of plasticiser will join in the chain network inducing elasticity to the plasticised or thermoplastic starch. According to Edhirej et al. [125], the addition of plasticiser must be maximised at 60 % or else the film will become sticky and difficult to be peel off. Four types of plasticisers were compared namely fructose (F), urea (U), tri-ethylene glycol (TEG) and triethanolamine (TEA) to investigate their variety and concentration toward mechanical, morphological, physical and thermal properties of the CS films. 30 % loading of F-plasticised films indicated a smooth surface of compact structure with highest density (1.74 g/cm^3), relative crystallinity (0.31) tensile strength (4.7 MPa), tensile modulus (69 MPa), lowest water content (10.96 %) and water absorption (110 %). The onset temperature recorded a temperature of 155.3 °C second only to U-plasticised films (160.9 °C).

12.5.3 Cassava starch blend

Polymer blend approach is done to offset the limitation of respective polymer with another polymer. Typically, hydrophilic starch-based polymers are blended with hydrophilic polymers to address water sensitivity and inducing biodegradability. Thermodynamically, these two polymers are not compatible hence grafting method is usually performed to ease the formation intermolecular interaction. Another important matter to achieve homogenous combination of polymer blend is to use the correct composition of polymers. By doing so, a synergistic structure is established resulting functional properties enhancement. In order to improve the polymer blend affinity, Wadaugsorn et al. [126] used propylene oxide of different degree of substitution (DS) to esterify CS before blending with polybutylene adipate terephthalate (PBAT). The increment of DS esterified CS increased the melt flow index (MFI) value. Large hydroxypropyl groups initiated void volume hindering inter- and intra-molecular cohesion force of the polymers, which lower the flow resistance. TPS dispersion was enhanced owing to the compatible interaction of hydrocarbon chain. Relatively, the flexibility of the films improved due to the continuation of plasticised CS particles within PBAT to support the blend structure.

12.5.4 Cassava starch bilayer films

Another method to compensate the drawbacks of starch-based polymer is through layering superior polymers sandwiching the weaker polymers. By doing so, the multi-layered film adapted combined functional properties. This practice is commonly used in single-used packaging application to envelop water-sensitive film with water-resistant. For example, Müller et al. [127] developed an antibacterial bilayer films (BLF) of linear low density polyethylene (LLDPE)/CS reinforced with zinc oxide (ZnO) nano-particles. The BLF consist of 0.2 % of ZnO has greater tensile strength (32.23 MPa) compared to 0.5 % (29.63 MPa) and pure LLDPE (25.72 MPa). Both BLF exhibit inhibition zone for *Lactobacillus sakei*, *Staphylococcus aureus*, and *Salmonella enteritidis*. In this study, LLDPE offset the lack of tensile strength while CS intensify the antimicrobial activity through the fine dispersion of ZnO.

12.5.5 Cellulose reinforced cassava starch biocomposites

Cellulosic fibres are good reinforcement additives in starch matrix. They are abundant, biodegradable, renewable, non-toxic, and have great affinity with starch-based polymer. The composites can be easily fabricated using solution casting method with sonication to promote the dispersion of fibres. Since both consist of hydroxyl groups, they established physical and chemical interaction links when compounded. Edhirej et al. [71] found out that the addition of CB sizing above 300 µm yield stiffer CS biocomposite films. In quite similar study, Versino & García [93] compared three different size ranges of CB for CS reinforcement. It was indicated that the medium size ranging from 53 to 250 µm yield the highest tensile strength meanwhile fine size below 53 µm had the lowest water vapour permeability (WVP) rate. Effective high surface area to volume ratio of fine size seemed to enhanced the starch-fibre interfacial adhesion hindering the diffusion of water molecules. Carvalho [128] found that tensile strength of CB/CS is inversely proportional to elongation at break based on CB loadings. Higher loading yields great tensile strength but diminishes elongation at break due to rigid nature of fibre. Anti-plasticiser effect took place reducing the free volume of polymer chains thus increasing the intermolecular forces and reducing molecular mobility of the composites.

12.5.6 Microcrystalline cellulose (MCC) reinforced cassava starch

The removal of amorphous regions within cellulose producing high refined cellulose. As the name suggested, it is made up of crystalline regions of higher surface area proven to be excellent in strengthen polymeric matrix structure. Désiré et al. [129] verified that water moisture content, solubility rate and tensile strength of MCC/CS composite films improved linearly with the increment of MCC loading up to 30 %. The crystalline cellulose consists of hydroxyl groups interacted with amylose and amylopectin forming extra crystalline zones that stabilise the film structure from dissolving in water. Furthermore, MCC is less hydrophilic compared to starch due to its compact microfibrillar and high crystalline structure making it less sensitive to water.

12.5.7 Nanocellulose reinforced cassava starch

Nano-sized fillers are well-known to be an effective reinforcement owing to their unique feature of high surface area to volume ratio [7, 33, 130]. Significant improvement can be attained by just incorporating minor amount of filler. Travalini et al. [32] reported that 1.3 % loading of CBN into CS significantly reduced the water absorption from 112.48 to 42.15 %. In comparison, the addition of hydrophilic bentonite nanoclay result in reduction to 69.55 %. Huang et al. [41] reported greater enhancement in tensile strength and barrier properties when modified CBN instead of unmodified CBN were introduced into CS. This is associated with the better dispersion establishing strong intermolecular interaction between hydroxyl groups of CS and carboxyl groups of modified CBN.

Basuki et al. [131] experimented on the cassava starch biofoam reinforced with BC and ZnO nanoparticles and evaluated their moisture absorption capability and crystallinity index. Synthesis of cassava starch biofoam composites was carried out with BC and various ZnO contents (varying between 3 and 9 %) using baking process. Experimental results indicated that the moisture absorption capability of the biocomposites reduced by 34 % (to a value of 0.165 gH_2O/g) owing to the addition of maximum ZnO (9 % by weight) when compared with biocomposites without ZnO addition. It was also seen from the XRD results that the crystallinity, intensity of the starch biofoam and diffraction position were improved owing to the addition of ZnO content in the biocomposites.

Lisdayana et al. [132] studied the properties of modified cassava, sago starch and corn/PVA biocomposites film reinforced with the nanocellulose extracted from oil palm empty fruit bunch (OPEFB). The nanocellulose was extracted from the OPEFB through mechanical grinding process using an ultrasonic grinder before the manufacturing of the composites. Biocomposite films were then fabricated using the evaporation casting technique with the nanocellulose content as 0, 1 and 3 %. It was evidently seen form the results that the tensile behaviour of the nanocellulose-starch/ PVA blends increased along with the increase in the nanocellulose content. At the highest content of nanocellulose (3 %) in the biocomposites film, the tensile strength of cassava starch-based biocomposites went up to 3 MPa, while the sago starch and corn starch-based biocompoistes exhibited a tensile strength of 4 and 5 MPa respectively when compared with 0.5 MPa for the unreinforced biocomposite films. This was attributed to the increase in the content of amylose in starch for the biocomposites film.

Qifeng Chen et al. [133] evaluated the mechanical and water absorption behaviour of cassava starch films with microcrystalline cellulose (MCC) and nanocellulose (NCC) as reinforcement agents. Experiments were carried out to assess the water absorption and mechanical properties of the cassava starch biocomposites reinforced with MCC and NCC modified with stearic acid (M-MCC and M-NCC) for biocomposite food packaging applications. MCC was prepared using phosphoric acid hydrolysis (98 % content) and using mechanical stirring process at 600 rpm while the NCC was manufactured using sulfuric acid hydrolysis (65% concentration). Following the preparation of materials, those were modified using stearic acid and the M-MCC/cassava starch and M-NCC/cassava starch biocomposite films were prepared by using casting method. Mechanical, physical, thermal and water absorption behaviour of the biocomposite films were evaluated using universal testing machine, SEM, TGA and contact angle apparatus.

Othman et al. [134] studied on effects of nanocellulose fiber and thymol on mechanical, thermal, and barrier properties of corn starch films. The tensile strength (TS) and Young's modulus (YM) of the CS films were improved with the addition of NCF at 1.5 % weight of starch, while their elongation at break (EAB), oxygen permeability, and water vapour permeability were lowered. Due to the nano-sized cellulose fiber's efficient contact area with the corn starch matrix and subsequent promotion of strong intermolecular interactions, the presence of NCF in the CS films boosted their TS by 18 %.The inclusion of NCF also increased the CS films' thermal stability. The TGA curve was shifted to the right by the addition of NCF in neat CS films, indicating an increase in the films' thermal stability. This resulted from the addition of NCF to the films, which increased their rigidity and crystallinity. Thymol was added to the CS/NCF bio-nanocomposite films, which resulted in a decrease in the TS and YM but an increase in the EAB because of the plasticizing properties of thymol. Thymol was also added, which increased the films' thermal stability but decreased their barrier qualities.

Figure 12.6 depicts the morphology and distribution of diameter of prepared MCC and NCC samples. It was noticed that the MCC morphology was comprised of tiny crystalline particles slightly agglomerated together. From Figure 12.6a, d and c, f specifically, it was observed that the NCC particles were made of very tiny fibril rods linked to form an network arrangement. From the test results, it was deduced by the experimenters that the reinforcement of M-MCC and M-NCC to the cassava starch films enhanced the mechanical and water absorption behaviour of the biocomposites. It was also noted that 0.5 % of M-MCC and 1.5 % of M-NCC was found to have maximum effect on the tensile properties of the biocomposites and an enhancement of tensile strength of the biocomposite films was about 485 % with M-MCC reinforcement and 328 % with M-NCC reinforcement as depicted in Figure 12.7. With regards to the water absorption

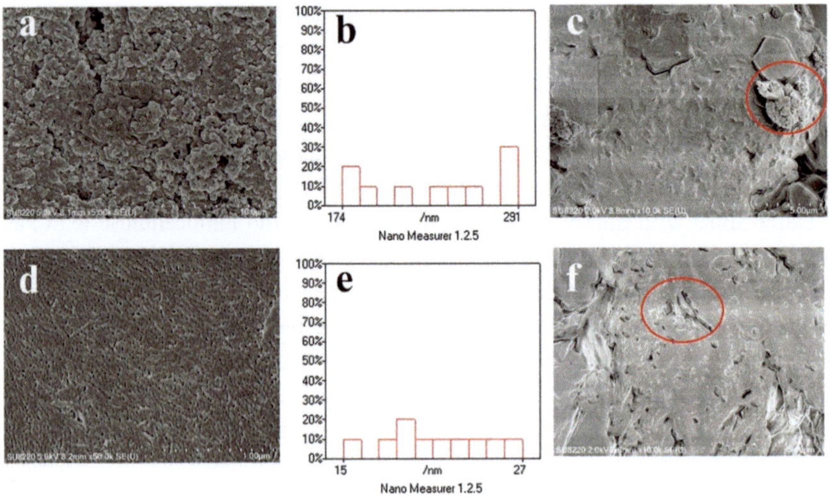

Figure 12.6: SEM micrographs of (a), acid-modified MCC (c), NCC (d), acid-modified NCC (f) and diameter distribution (b), MCC (e) NCC.

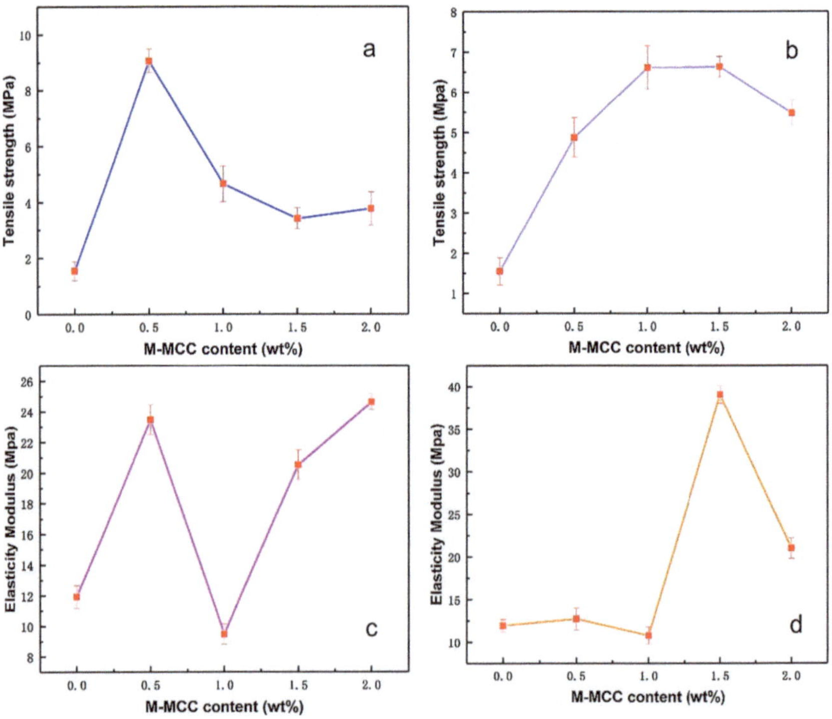

Figure 12.7: Tensile strength of biocomposites with various contents of (a) M-MCC, (b) M-NCC; elasticity modulus of biocomposites with various contents of (c) M-MCC, (d) M-NCC.

property, an increase of 65 % of hydrophobicity was observed when 2 % of M-MCC was reinforced while an increase of 31 % of hydrophobicity was observed for the reinforcement of 0.5 % M-NCC as shown in Figure 12.8. It was stated that the biocomposites with M-MCC reinforcements performed better when compared to the biocomposites with M-NCC reinforcements. However, the thermal stability of the cassava starch biocomposite films decreased with the addition of M-MCC and M-NCC reinforcements.

Owi et al. [135], experimented on the tapioca starch (TS) derived nanocellulose biocomposites crosslinked with *Citrus aurantifolia*. Nanocellulose derived by means of acid hydrolysis from OPEFB were used as a reinforcements in the biocomposite films in varying proportions. These biocomposite films were prepared from TS-based low cost and renewable resources and were used for food packaging applications. From *C. aurantifolia*, two different types of crosslinkers were derived namely lime juice (LJ) and commercial citric acid (CA) and these were mixed with glycerol individually using solvent casting method and comparatively evaluated. These biocomposite films were subjected to physical, mechanical, structural and water vapour permeability (WVP) behaviour tests and the results were comparatively evaluated. FTIR analysis, degree of di-estertification and substitution tests were used to ascertain the crosslinking reaction between the TS and LJ materials. Figure 12.9 depicts the tensile strength and elongation at break values for varying nanocellulose phase concentration for the biocomposite films crosslinked with LJ and CA. It was observed that ten parts per hundred of starch (phs) of LJ increased the elongation at break value of TS-based biocomposites by 106 % and the tensile strength value of this specific biocomposite film sample was noted to be 13.6 MPa. It was also noted that the concentration of nanocellulose reduced the

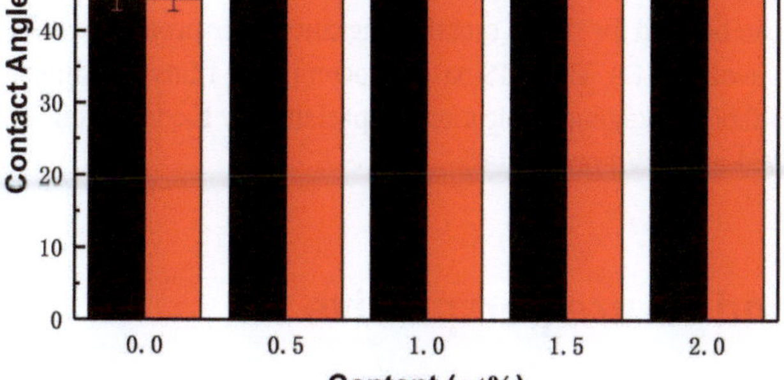

Figure 12.8: Contact angles of biocomposites with varying M-MCC and M-NCC contents.

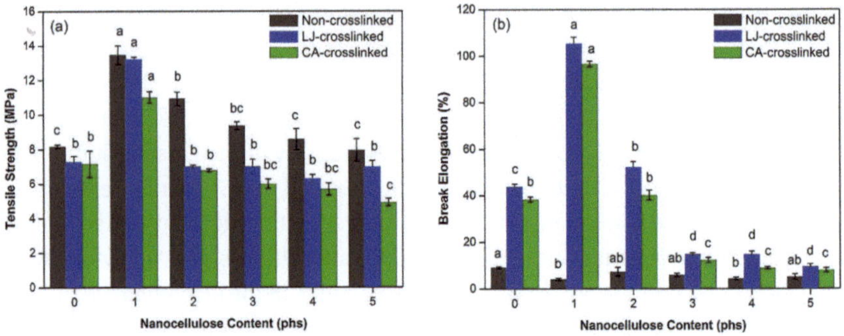

Figure 12.9: (a) Tensile strength and (b) break elongation oil palm empty fruit bunch derived cellulose and its bionanocomposite films [135].

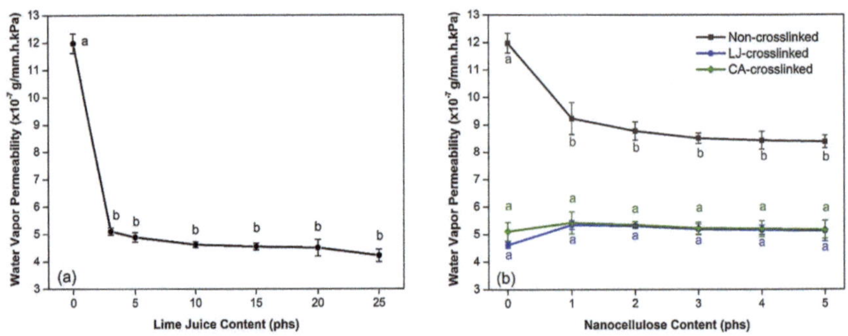

Figure 12.10: Water vapour permeability variation of bionanocomposite films with (a) lime juice content, and (b) nanocellulose content [135].

transparency of the TS-based bionanocomposite samples while the crosslinkers had a very least effect on the optical properties of the nanocomposites. Figure 12.10 denotes the WVP behaviour of all configurations of biocomposites. Results portrayed that the WVP of the crosslinked samples were very low and the WVP of LJ-crosslinked bionanocomposite sample was 55 % lesser than the non-crosslinked biocomposite film. Hence, the experimenters concluded that LJ crosslinker can be used for starch crosslinking since their properties were on par with the commercially available CA. These TS-based biocomposite films crosslinked with LJ and CA can be used in food packaging applications, especially dry foods packaging, owing to their better mechanical and WVP properties.

12.6 Challenges and future perspective

In consideration to widen the versatility of hydrophilic polymer derived from starch and cellulosic fibre, it is essential to modify the hydroxyl groups through surface modification, grafting or esterification. The selection of fibre size also needs to be emphasis, as it affects the molecular bonds established that manipulate the characteristic of end product. Ironically, CS has a broad range of amylose content, which offer particular design and condition to carry out the fabrication process in all kind of applications desired. Removing different substances to improve their performance and recycling issues are still the major concerns associated with nanocellulosic materials. The use of it as a reinforcing material in polymer composites faces numerous obstacles in addition to recent advancements. To produce nanocellulose with the desired properties, more efficient and effective methods must be discovered. To overcome the problems associated to these materials' hydrophilic nature in many applications, further study will be needed before they are widely applied. Nanocellulose reinforcements must be properly dispersed or transformed into a form that can be included into various matrices and strongly bonded to them using techniques and the right chemistry. Bio nanocomposites based on cellulose have several uses in a variety of sectors. The paper and packaging industries have largely accepted nanocellulose and related bio composites for prospective applications. However, the application of cellulose-based nano-composites have gained much attention owing to its nanoscale dimension, high strength and stiffness for use in several industries such as construction, automotive, electronics, pharmacy and cosmetics. To generate nanocellulose, which should be technically possible for large-scale production, novel technologies based on 3D printing must be investigated. These factors will further enable researchers to produce nanocellulose materials and their composites with improved functionalities in the future. Incorporating the production of nanocellulose into the material flow of these biorefineries may be feasible if high value applications can be found, which could potentially boost economics. Nanocellulose-based composites will exhibit superior properties owing to the major breakthroughs in multidisciplinary fields. Additionally, these materials ought to have a minimal impact on the environment. These research gaps will lead towards a new research area of sustainable nanocellulose for high-performance and advanced applications.

12.7 Conclusions

The distinctive traits of cassava-based products can be deployed in producing packaging plastic. Ultimately, as a primary by-product of cassava production, cassava bagasse is proven able to compete with current conventional additives in developing superior packaging plastic. The utilisation of such resource is advantageous in term of availability, material cost and environmental issue. Hence, the reliance on the man-made materials can be reduce and hopefully, in near future, completely replacing the environment-threatening materials.

Acknowledgements: Appreciation is given to the Ministry of Higher Education Malaysia (MOHE) for funding this project through the Fundamental Research Grant Scheme (FRGS/1/2021/TK0/UPM/02/21). Besides that, the authors would like to express gratitude for the financial support received from the Universiti Teknologi Malaysia for the project "The impact of Malaysian bamboos' chemical and fibre characteristics on their pulp and paper properties", grant number PY/2022/02318 – Q.J130000.3851.21H99. The research has been carried out under the programme, Research Excellence Consortium (JPT (BPKI) 1000/016/018/25 (57)), provided by the Ministry of Higher Education Malaysia (MOHE).

References

1. Nurazzi NM, Sabaruddin FA, Harussani MM, Kamarudin SH, Rayung M, Asyraf MRM, et al. Mechanical performance and applications of CNTs reinforced polymer composites—a review. Nanomaterials 2021; 11:2186.
2. Norizan MN, Abdan K, Ilays RA, Zin M, Muthukumar C, Rafiqah S, et al. Effect of fiber orientation and fiber loading on the mechanical and thermal properties of sugar palm yarn fiber reinforced unsaturated polyester resin composites. Polim. -Warsaw 2019;64:34–43.
3. Azammi AMN, Ilyas RA, Sapuan SM, Ibrahim R, Atiqah A, Engineering A, et al. Characterization studies of biopolymeric matrix and cellulose fibres based composites related to functionalized fibre-matrix interface. Sawston, Cambridge: Elsevier; 2020.
4. Diyana ZN, Jumaidin R, Selamat MZ, Ghazali I, Julmohammad N, Huda N, et al. Physical properties of thermoplastic starch derived from natural resources and its blends: a review. Polymers 2021;13:1–20.
5. Asrofi M, Sapuan SM, Ilyas RA, Ramesh M. Characteristic of composite bioplastics from tapioca starch and sugarcane bagasse fiber: effect of time duration of ultrasonication (Bath-Type). Mater Today Proc 2020. https://doi.org/10.1016/j.matpr.2020.07.254.
6. Sari NH, Suteja S, Lokantara IP, Wibowo TG. Influence of pumice particles on the mechanical and morphology properties of polyester-cornhusk fiber composites. J. Fibers Polym. Compos. 2022;1:97–105.
7. Rozilah A, Jaafar CNA, Sapuan SM, Zainol I, Ilyas RA. The effects of silver nanoparticles compositions on the mechanical, physiochemical, antibacterial, and morphology properties of sugar palm starch biocomposites for antibacterial coating. Polymers 2020;12:2605.
8. Azman MA, Asyraf MRM, Khalina A, Petrů M, Ruzaidi CM, Sapuan SM, et al. Natural fiber reinforced composite material for product design: a short review. Polymers 2021;13:1917.
9. Asyraf MRM, Ishak MR, Syamsir A, Nurazzi NM, Sabaruddin FA, Shazleen SS, et al. Mechanical properties of oil palm fibre-reinforced polymer composites: a review. J Mater Res Technol 2022;17:33–65.
10. Norfarhana AS, Ilyas RA, Ngadi N. A review of nanocellulose adsorptive membrane as multifunctional wastewater treatment. Carbohydr Polym 2022;291:119563.
11. Thiruganasambanthan T, Ilyas RA, Norrrahim MNF, Kumar TSM, Siengchin S, Misenan MSM, et al. Emerging developments on nanocellulose as liquid crystals: a biomimetic approach. Polymers 2022; 14:1546.
12. Zulkefli NA, Mustapha R, Jusoh SM, Ruzaidi Ghazali CM, Awang M, Norrrahim MNF, et al. Hybrid nanofiller reinforcement in thermoset and biothermoset applications: a review. Nanotechnol Rev 2023;12: 20220499.
13. Khalili H, Hamid Salim M, Tlemcani SJ, Makhlouf R, Hassani F-ZSA, Ablouh H, et al. Bio-nanocomposite films based on cellulose nanocrystals filled polyvinyl alcohol/alginate polymer blend. J. Fibers Polym. Compos. 2022;1:77–96.

14. Shackelford GE, Haddaway NR, Usieta HO, Pypers P, Petrovan SO, Sutherland WJ. Cassava farming practices and their agricultural and environmental impacts: a systematic map protocol. Environ Evid 2018; 7:1–7.
15. TitleHillocks RJ, Thresh JM, Bellotti A. Cassava: biology, production and utilization. Wallingford, Oxfordshire: CABI Publishing; 2002.
16. Parmar A, Sturm B, Hensel O. Crops that feed the world: production and improvement of cassava for food, feed, and industrial uses. Food Secur 2017;9:907–27.
17. Westby A. Cassava utilization, storage and small-scale processing. Cassava Biol. Prod. Util. 2002;1: 281–300.
18. Byju G, Suja G. Mineral nutrition of cassava, 1st ed. Dewark, Delaware: Elsevier Inc.; 2020, vol. 159.
19. Jarvis A, Ramirez-Villegas J, Campo BVH, Navarro-Racines C. Is cassava the answer to African climate change adaptation? Trop Plant Biol 2012;5:9–29.
20. Eke-Okoro ON, Njoku DN, Madu A, Ezulike TO. Impact of global warming and crop factors on the growth and productivity of four cassava (Manihot esculenta Crantz) cultivars in Nigeria. Sci Res Essays 2009; 4:955–60.
21. Ding Z, Zhang Y, Xiao Y, Liu F, Wang M, Zhu X, et al. Transcriptome response of cassava leaves under natural shade. Sci Rep 2016;6:1–14.
22. TitleKatz SH, Weaver WW. Encyclopedia of food and culture, 1st ed. New York: Scribner; 2003.
23. Aregheore E, Toxicology OA. The toxic effects of cassava (Manihot esculenta grantz) diets on humans: a review. europepmc.org 1991.
24. Olsen KM, Schaal BA. Evidence on the origin of cassava: phylogeography of Manihot esculenta. Proc Natl Acad Sci USA 1999;96:5586–91.
25. Sivamani S, Chandrasekaran AP, Balajii M, Shanmugaprakash M, Hosseini-Bandegharaei A, Baskar R. Evaluation of the potential of cassava-based residues for biofuels production. Rev Environ Sci Biotechnol 2018;17:553–70.
26. Hannah Ritchie Agricultural Production. https://ourworldindata.org/agricultural-production [Accessed 8 Dec 2020].
27. Gunorubon J, Kekpugile K. Journal of engineering modification of cassava starch for industrial uses. Int J Eng Technol 2012;2:913–9.
28. Adetunji A, Isadare D, Akinluwade K, Adewoye O. Waste-to-wealth applications of cassava – a review study of industrial and agricultural applications. Adv. Res. 2015;4:212–29.
29. Teixeira EdM, Curvelo AAS, Corrêa AC, Marconcini JM, Glenn GM, Mattoso LHC. Properties of thermoplastic starch from cassava bagasse and cassava starch and their blends with poly (lactic acid). Ind Crop Prod 2012;37:61–8.
30. Wang Z, Mhaske P, Farahnaky A, Kasapis S, Majzoobi M. Cassava starch: chemical modification and its impact on functional properties and digestibility, a review. Food Hydrocolloids 2022;129:107542.
31. Hasmadi M, Harlina L, Jau-Shya L, Mansoor AH, Jahurul MHA, Zainol MK. Physicochemical and functional properties of cassava flour grown in different locations in Sabah, Malaysia. Food Res 2020;4:991–9.
32. Travalini AP, Lamsal B, Magalhães WLE, Demiate IM. Cassava starch films reinforced with lignocellulose nanofibers from cassava bagasse. Int J Biol Macromol 2019;139:1151–61.
33. Sabaruddin FA, Paridah MT, Sapuan SM, Ilyas RA, Lee SH, Abdan K, et al. The effects of unbleached and bleached nanocellulose on the thermal and flammability of polypropylene-reinforced kenaf core hybrid polymer bionanocomposites. Polymers 2020;13:116.
34. Akli K, Maryam M, Senjawati MI, Ilyas RA. Eco-friendly bioprocessing oil palm empty fruit bunch (Opefb) fibers into nanocrystalline cellulose (Ncc) using white-Rot fungi (Tremetes Versicolor) and cellulase enzyme (Trichoderma Reesei). J. Fibers Polym. Compos. 2022;1:148–63.
35. Rahayu A, Hanum FF, Amrillah NAZ, Lim LW, Salamah S. Cellulose extraction from coconut coir with alkaline delignification process. J Fibers Polym Compos 2022;1:106–16.

36. Arnata IW, Harsojuwono BA, Hartiati A, Anggreni AAMD, Sartika D. Effect of alkaline concentration treatments on the chemical, physical and thermal characteristics of cellulose from tapioca solid waste. J Fibers Polym Compos 2022;1:117–30.
37. Ventura-Cruz S, Tecante A. Nanocellulose and microcrystalline cellulose from agricultural waste: review on isolation and application as reinforcement in polymeric matrices. Food Hydrocolloids 2021;118:106771.
38. Abotbina W, Sapuan SM, Sultan MTH, Alkbir MFM, Ilyas RA. Extraction, characterization, and comparison of properties of cassava bagasse and black seed fibers. J Nat Fibers 2022:1–14. https://doi.org/10.1080/15440478.2022.2068103.
39. Ansharullah A, Saenuddin NMA, Faradilla RF, Asranudin A, Asniar A, Nurdin M. Production of Micro crystalline cellulose from tapioca solid waste: effect of acid concentration on its physico-chemical properties. J Kim Sains dan Apl 2020;23:147–51.
40. Liu EK, He WQ, Yan CR. "White revolution" to "white pollution" – agricultural plastic film mulch in China. Environ Res Lett 2014;9:7–10.
41. Huang L, Zhao H, Yi T, Qi M, Xu H, Mo Q, et al. Reparation and properties of cassava residue cellulose nanofibril/cassava starch composite films. Nanomaterials 2020;10. https://doi.org/10.3390/nano10040755.
42. Ilyas RA, Sapuan SM, Ishak MR, Zainudin ES. Development and characterization of sugar palm nanocrystalline cellulose reinforced sugar palm starch bionanocomposites. Carbohydr Polym 2018;202: 186–202.
43. Dos Santos BH, De Souza Do Prado K, Jacinto AA, Da Silva Spinacé MA. Influence of sugarcane bagasse fiber size on biodegradable composites of thermoplastic starch. J Renew Mater 2018;6:176–82.
44. Shuzhen N, Liang J, Hui Z, Yongchao Z, Guigan F, Huining X, et al. Enhancing hydrophobicity, strength and UV shielding capacity of starch film via novel co-cross-linking in neutral conditions. R Soc Open Sci 2018;5. https://doi.org/10.1098/rsos.181206.
45. Versino F, López OV, García MA. Sustainable use of cassava (Manihot esculenta) roots as raw material for biocomposites development. Ind Crop Prod 2015;65:79–89.
46. Kane SN, Mishra A, Dutta AK. Preparation and characterization of cellulose and nanocellulose from agro-industrial waste – cassava peel. J Phys Conf Ser 2017;755. https://doi.org/10.1088/1742-6596/755/1/011001.
47. Ramírez MGL, Muniz GIBde., Satyanarayana KG, Tanobe V, Iwakiri S. Preparation and characterization of biodegradable composites based on Brazilian cassava starch, corn starch and green coconut fibers. Materia 2010;15:330–7.
48. Kaisangsri N, Kerdchoechuen O, Laohakunjit N. Biodegradable foam tray from cassava starch blended with natural fiber and chitosan. Ind Crop Prod 2012;37:542–6.
49. Vallejos ME, Curvelo AAS, Teixeira EM, Mendes FM, Carvalho AJF, Felissia FE, et al. Composite materials of thermoplastic starch and fibers from the ethanol–water fractionation of bagasse. Ind Crop Prod 2011;33: 739–46.
50. Gutiérrez TJ, Morales NJ, Pérez E, Tapia MS, Famá L. Physico-chemical properties of edible films derived from native and phosphated cush-cush yam and cassava starches. Food Packag Shelf Life 2015;3:1–8.
51. Matsui K. Cassava bagasse-Kraft paper composites: analysis of influence of impregnation with starch acetate on tensile strength and water absorption properties. Carbohydr Polym 2004;55:237–43.
52. Souza AC, Goto GEO, Mainardi JA, Coelho ACV, Tadini CC. Cassava starch composite films incorporated with cinnamon essential oil: antimicrobial activity, microstructure, mechanical and barrier properties. LWT – Food Sci Technol 2013;54:346–52.
53. Belibi PC, Daou TJ, Ndjaka J-MB, Michelin L, Brendlé J, Nsom B, et al. Tensile and water barrier properties of cassava starch composite films reinforced by synthetic zeolite and beidellite. J Food Eng 2013;115:339–46.
54. Prachayawarakorn J, Chaiwatyothin S, Mueangta S, Hanchana A. Effect of jute and kapok fibers on properties of thermoplastic cassava starch composites. Mater Des 2013;47:309–15.

55. Chen G, Liu B, Zhang B. Characterization of composite hydrocolloid film based on sodium cellulose sulfate and cassava starch. J Food Eng 2014;125:105–11.
56. Mello LRPF, Mali S. Use of malt bagasse to produce biodegradable baked foams made from cassava starch. Ind Crop Prod 2014;55:187–93.
57. Versino F, García MA. Cassava (Manihot esculenta) starch films reinforced with natural fibrous filler. Ind Crop Prod 2014;58:305–14.
58. Debiagi F, Kobayashi RKT, Nakazato G, Panagio LA, Mali S. Biodegradable active packaging based on cassava bagasse, polyvinyl alcohol and essential oils. Ind Crop Prod 2014;52:664–70.
59. Shittu TA, Raji AO, Sanni LO. Bread from composite cassava-wheat flour: I. Effect of baking time and temperature on some physical properties of bread loaf. Food Res Int 2007;40:280–90.
60. Zhang Y, Gan T, Li Q, Su J, Lin Y, Wei Y, et al. Mechanical and interfacial properties of poly(vinyl chloride) based composites reinforced by cassava stillage residue with different surface treatments. Appl Surf Sci 2014;314:603–9.
61. Wongsasulak S, Yoovidhya T, Bhumiratana S, Hongsprabhas P, McClements DJ, Weiss J. Thermo-mechanical properties of egg albumen–cassava starch composite films containing sunflower-oil droplets as influenced by moisture content. Food Res Int 2006;39:277–84.
62. Prachayawarakorn J, Pomdage W. Effect of carrageenan on properties of biodegradable thermoplastic cassava starch/low-density polyethylene composites reinforced by cotton fibers. Mater Des 2014. https://doi.org/10.1016/j.matdes.2014.04.051.
63. Belibi PC, Daou TJ, Ndjaka JMB, Nsom B, Michelin L, Durand B. A comparative study of some properties of cassava and tree cassava starch films. Phys Procedia 2014;55:220–6.
64. Edhirej A, Sapuan SM, Jawaid M, Zahari NI. Cassava/sugar palm fiber reinforced cassava starch hybrid composites: physical, thermal and structural properties. Int J Biol Macromol 2017;101:75–83.
65. Teixeira E de M, Pasquini D, Curvelo AAS, Corradini E, Belgacem MN, Dufresne A. Cassava bagasse cellulose nanofibrils reinforced thermoplastic cassava starch. Carbohydr Polym 2009;78:422–31.
66. Maulida, Siagian M, Tarigan P. Production of starch based bioplastic from cassava peel reinforced with microcrystalline celllulose Avicel PH101 using sorbitol as plasticizer. In: Proceedings of the Journal of Physics. Conference Series; 2016.
67. Kargarzadeh H, Johar N, Ahmad I. Starch biocomposite film reinforced by multiscale rice husk fiber. Compos Sci Technol 2017;151:147–55.
68. De Morais Teixeira E, Da Róz AL, De Carvalho AJF, Da Silva Curvelo AA. Preparation and characterisation of thermoplastic starches from cassava starch, cassava root and cassava bagasse. In: Proceedings of the Macromolecular Symposia; 2005.
69. Matsui KN, Larotonda FDS, Paes SS, Luiz DB, Pires ATN, Laurindo JB. Cassava bagasse-Kraft paper composites: analysis of influence of impregnation with starch acetate on tensile strength and water absorption properties. Carbohydr Polym 2004. https://doi.org/10.1016/j.carbpol.2003.07.007.
70. Garcia NL, Ribba L, Dufresne A, Aranguren MI, Goyanes S. Physico-mechanical properties of biodegradable starch nanocomposites. Macromol Mater Eng 2009;294:169–77.
71. Edhirej A, Sapuan SM, Jawaid M, Zahari NI. Preparation and characterization of cassava bagasse reinforced thermoplastic cassava starch. Fibers Polym 2017;18:162–71.
72. Harunsyah, Yunus M, Fauzan R. Mechanical properties of bioplastics cassava starch film with zinc oxide nanofiller as reinforcement. In: Proceedings of the IOP Conference Series: Materials Science and Engineering; 2017.
73. Woehl MA, Canestraro CD, Mikowski A, Sierakowski MR, Ramos LP, Wypych F. Bionanocomposites of thermoplastic starch reinforced with bacterial cellulose nanofibres: effect of enzymatic treatment on mechanical properties. Carbohydr Polym 2010;80:866–73.
74. Sujuthi M, Liew. Properties of bioplastic sheets made from different types of starch incorporated with recycled Newspaper pulp. Trans Sci Technol 2016;3:257–64.

75. Zainuddin SYZ, Ahmad I, Kargarzadeh H. Cassava starch biocomposites reinforced with cellulose nanocrystals from kenaf fibers. Compos Interfac 2013;20:189–99.
76. Llanos JHR, Tadini CC. Preparation and characterization of bio-nanocomposite films based on cassava starch or chitosan, reinforced with montmorillonite or bamboo nanofibers. Int J Biol Macromol 2018;107: 271–382.
77. Da Silva A, Nievola LM, Tischer CA, Mali S, Faria-Tischer PCS. Cassava starch-based foams reinforced with bacterial cellulose. J Appl Polym Sci 2013. https://doi.org/10.1002/app.39526.
78. Basuki MA, Suryanto H, Larasati A, Puspitasari P. Mujiono the effect of ZnO addition against crystallinity and water absorption capacity of biofoam based cassava starch reinforced bacterial cellulose; 2019. p. 050016.
79. Rodrigues DC, Caceres CA, Ribeiro HL, de Abreu RFA, Cunha AP, Azeredo HMC. Influence of cassava starch and carnauba wax on physical properties of cashew tree gum-based films. Food Hydrocolloids 2014;38: 147–51.
80. Sanhawong W, Banhalee P, Boonsang S, Kaewpirom S. Effect of concentrated natural rubber latex on the properties and degradation behavior of cotton-fiber-reinforced cassava starch biofoam. Ind Crop Prod 2017;108:756–66.
81. Raabe J, Fonseca ADS, Bufalino L, Ribeiro C, Martins MA, Marconcini JM, et al. Biocomposite of cassava starch reinforced with cellulose pulp fibers modified with Deposition of silica (SiO_2) nanoparticles. J Nanomater 2015;2015:1–9.
82. Kaisangsri N, Kowalski RJ, Kerdchoechuen O, Laohakunjit N, Ganjyal GM. Cellulose fiber enhances the physical characteristics of extruded biodegradable cassava starch foams. Ind Crop Prod 2019;142:111810.
83. Edhirej A, Sapuan SM, Jawaid M, Zahari NI, Sanyang ML. Effect of cassava peel and cassava bagasse natural fillers on mechanical properties of thermoplastic cassava starch: comparative study. In: Proceedings of the AIP Conference Proceedings 2017.
84. da Silva JBA, Nascimento T, Costa LAS, Pereira FV, Machado BA, Gomes GVP, et al. Effect of source and interaction with nanocellulose cassava starch, glycerol and the properties of films bionanocomposites. Mater Today Proc 2015;2:200–7.
85. Edhirej A, Sapuan SM, Jawaid M, Ismarrubie Zahari N. Preparation and characterization of cassava starch/ peel composite film. Polym Compos 2018. https://doi.org/10.1002/pc.24121.
86. Zainuddin SYZ, Ahmad I, Kargarzadeh H, Abdullah I, Dufresne A. Potential of using multiscale kenaf fibers as reinforcing filler in cassava starch-kenaf biocomposites. Carbohydr Polym 2013;92:2299–305.
87. Ketkaew S, Kasemsiri P, Hiziroglu S, Mongkolthanaruk W, Wannasutta R, Pongsa U, et al. Effect of oregano essential oil content on properties of green biocomposites based on cassava starch and sugarcane bagasse for bioactive packaging. J Polym Environ 2018. https://doi.org/10.1007/s10924-017-0957-x.
88. Wahyuningtiyas NE, Suryanto H. Properties of cassava starch based bioplastic reinforced by nanoclay. J Mech Eng Sci Technol 2018;2:20–6.
89. Kaisangsri N, Kerdchoechuen O, Laohakunjit N. Characterization of cassava starch based foam blended with plant proteins, kraft fiber, and palm oil. Carbohydr Polym 2014;110:70–7.
90. Lomelí Ramírez MG, De Muniz GIB, Satyanarayana KG, Tanobe V, Iwakiri S. Preparation and characterization of biodegradable composites based on Brazilian cassava starch, corn starch and green coconut fibers. In: Proceedings of the 65th ABM International Congress, 18th IFHTSE Congress and 1st TMS/ABM International Materials Congress 2010; 2010.
91. Silva JBA, Pereira FV, Druzian JI. Cassava starch-based films plasticized with sucrose and inverted sugar and reinforced with cellulose nanocrystals. J Food Sci 2012;77:14–9.
92. Suryanto H, Hutomo PT, Wanjaya R, Puspitasari P. Sukarni the stucture of bioplastic from cassava starch with nanoclay reinforcement. In: Proceedings of the AIP Conference Proceedings 2016.
93. Versino F, García MA. Particle size distribution effect on cassava starch and cassava bagasse biocomposites. ACS Sustainable Chem Eng 2019. https://doi.org/10.1021/acssuschemeng.8b04700.

94. El Halal SLM, Bruni GP, do Evangelho JA, Biduski B, Silva FT, Dias ARG, et al. The properties of potato and cassava starch films combined with cellulose fibers and/or nanoclay. Starch/Staerke 2018;70:1700115.
95. Vercelheze AES, Oliveira ALM, Rezende MI, Muller CMO, Yamashita F, Mali S. Physical properties, photo- and bio-degradation of baked foams based on cassava starch, sugarcane bagasse fibers and montmorillonite. J Polym Environ 2013;21:266–74.
96. Santana JS, do Rosário JM, Pola CC, Otoni CG, de Fátima FerreiraSoares N, Camilloto GP, et al. Cassava starch-based nanocomposites reinforced with cellulose nanofibers extracted from sisal. J Appl Polym Sci 2017;134:44637.
97. Mo X, Zhong Y, Liang C, Yu S. Studies on the properties of banana fibers-reinforced thermoplastic cassava starch composites: preliminary results. In: Proceedings of the Advanced Materials Research; 2010.
98. Cabanillas A, Nuñez J, Cruz-Tirado JP, Vejarano R, Tapia-Blácido DR, Arteaga H, et al. Pineapple shell fiber as reinforcement in cassava starch foam trays. Polym Polym Compos 2019. https://doi.org/10.1177/0967391119848187.
99. Merci A, Marim RG, Urbano A, Mali S. Films based on cassava starch reinforced with soybean hulls or microcrystalline cellulose from soybean hulls. Food Packag Shelf Life 2019;20:100321.
100. Ogunrinola TM, Akpan U. Production of cassava starch bioplastic film reinforced with poly-lactic acid (PLA). Int J Eng Res Adv Technol 2018;4:56–62.
101. Syafri E, Kasim A, Abral H, Sudirman, Sulungbudi GT, Sanjay MR, et al. Synthesis and characterization of cellulose nanofibers (CNF) ramie reinforced cassava starch hybrid composites. Int J Biol Macromol 2018; 120:578–86.
102. Belibi PC, Daou TJ, Ndjaka JMB, Michelin L, Brendlé J, Nsomd B, et al. Tensile and water barrier properties of cassava starch composite films reinforced by synthetic zeolite and beidellite. J Food Eng 2013;115:339–46.
103. Dai L, Zhang J, Cheng F. Cross-linked starch-based edible coating reinforced by starch nanocrystals and its preservation effect on graded Huangguan pears. Food Chem 2020;311:125891.
104. Amni C, Marwan M, Mariana M. The making of bioplastic from cassava starch reinforced by nano fiber straw and ZnO. J Litbang Ind 2015;5:91–9.
105. Prachayawarakorn J, Pattanasin, W. Effect of pectin particles and cotton fibers on properties of thermoplastic cassava starch composites. 2016;38:129–36.
106. Syafri E, Kasim A, Asben A, Senthamaraikannan P, Sanjay MR. Studies on Ramie cellulose microfibrils reinforced cassava starch composite: influence of microfibrils loading. J Nat fibers 2020;17:122–31.
107. Machado BAS, Nunes IL, Pereira FV, Druzian JI. Development and evaluation of the effectiveness of biodegradable films of cassava starch with nanocelulose as reinforcement and yerba mate extract as an additive antioxidant. Ciência Rural 2012;42:2085–91.
108. Silviana S, Dzulkarom MC. Synthesis of cassava bagasse starch-based biocomposite reinforced Woven bamboo fibre with lime juice as crosslinker and epoxidized waste cooking oil (EWCO) as bioplasticizer. In: Proceedings of the Journal of Physics. Conference Series; 2019.
109. Huang L, Zhao H, Yi T, Qi M, Xu H, Mo Q, et al. Reparation and properties of cassava residue cellulose nanofibril/cassava starch composite films. Nanomaterials 2020;10:755.
110. Li S, Ma Y, Ji T, Sameen DE, Ahmed S, Qin W, et al. Cassava starch/carboxymethylcellulose edible films embedded with lactic acid bacteria to extend the shelf life of banana. Carbohydr Polym 2020;248:116805.
111. Estevez-Areco S, Guz L, Candal R, Goyanes S. Active bilayer films based on cassava starch incorporating ZnO nanorods and PVA electrospun mats containing rosemary extract. Food Hydrocolloids 2020. https://doi.org/10.1016/j.foodhyd.2020.106054.
112. Arrieta-Almario AA, Mendoza-Fandiño JM, Palencia MS. Composite material elaborated from conducting biopolymer cassava starch and polyaniline. Rev Mex Ing Quim 2020. https://doi.org/10.24275/rmiq/Mat765.
113. Liu Y, Fan L, Pang J, Tan D. Effect of tensile action on retrogradation of thermoplastic cassava starch/nanosilica composite. Iran Polym J (English Ed.) 2020;29:171–83.

114. Huang L, Han X, Chen H, An S, Zhao H, Xu H, et al. Preparation and barrier performance of layer-modified soil-stripping/cassava starchcomposite films. Polymers 2020;12:1611.
115. Liu Y, Fan L, Xiong J, Liang Z. Effect of different particle size of silica on structure, morphology, and properties of thermoplastic cassava starch. Polym Polym Compos 2020;29:863–75.
116. Liu Y, Xiong J, Lan C, Tan D, Liang Z. Properties of thermoplastic cassava starch/sisal fiber composites in retrogradation. Hecheng Shuzhi Ji Suliao/China Synth Resin Plast 2020;37:31–4.
117. Versino F, Urriza M, García MA. Eco-compatible cassava starch films for fertilizer controlled-release. Int J Biol Macromol 2019. https://doi.org/10.1016/j.ijbiomac.2019.05.037.
118. Nansu W, Ross S, Ross G, Mahasaranon S. Effect of crosslinking agent on the physical and mechanical properties of a composite foam based on cassava starch and coconut residue fiber. Proc Mater Today: Proc 2019. https://doi.org/10.1016/j.matpr.2019.06.249.
119. Huang L, Xu H, Zhao H, Xu M, Qi M, Yi T, et al. Properties of thermoplastic starch films reinforced with modified cellulose nanocrystals obtained from cassava residues. New J Chem 2019;43:14883–91.
120. Engel JB, Ambrosi A, Tessaro IC. Development of a cassava starch-based foam incorporated with Grape Stalks using an experimental design. J Polym Environ 2019;27:2853–66.
121. Mustapha FA, Jai J, Sharif ZIM, Yusof NM. Cassava starch/carboxymethylcellulose biocomposite film for food paper packaging incorporated with turmeric oil. In: Proceedings of the IOP Conference Series: Materials Science and Engineering 2019.
122. Rocha Gomes ÁV, De Lima Leite RH, Da Silva Júnior MQ, Gomes Dos Santos FK, Mendes Aroucha EM. Influence of composition on mechanical properties of cassava starch, sisal fiber and carnauba wax biocomposites. Mater Res 2019. https://doi.org/10.1590/1980-5373-MR-2018-0887.
123. Doumbia A, Dable PJMR. Analysis of the thermo-mechanical behavior of composite materials based on plasticized cassava starch reinforced with coconut fibers. Indian J Sci Technol 2019;12:1–10.
124. Mbey JA, Hoppe S, Thomas F. Cassava starch-kaolinite composite film. Effect of clay content and clay modification on film properties. Carbohydr Polym 2012;88:213–22.
125. Edhirej A, Sapuan SM, Jawaid M, Zahari NI. Effect of various plasticizers and concentration on the physical, thermal, mechanical, and structural properties of cassava-starch-based films. Starch/Staerke 2017;69: 1–11.
126. Wadaugsorn K, Panrong T, Wongphan P, Harnkarnsujarit N. Plasticized hydroxypropyl cassava starch blended PBAT for improved clarity blown films: morphology and properties. Ind Crop Prod 2022;176: 114311.
127. Müller L, Zanghelini G, Laroque DA, Laurindo JB, Valencia GA, Costa Cda, et al. Cold atmospheric plasma for producing antibacterial bilayer films of LLDPE/cassava starch added with ZnO-nanoparticles. Food Packag Shelf Life 2022;34:100988.
128. de Carvalho GR, Marques GS, de Matos Jorge LM, Jorge RMM. Cassava bagasse as a reinforcement agent in the polymeric blend of biodegradable films. J Appl Polym Sci 2019;136:1–7.
129. Yao Désiré A, Charlemagne N, Degbeu Claver K, Fabrice Achille T, Marianne S. Starch-based edible films of improved cassava varieties Yavo and TMS reinforced with microcrystalline cellulose. Heliyon 2021;7: e06804.
130. Sharma S, Sudhakara P, Omran AAB, Singh J, Ilyas RA. Recent trends and developments in conducting polymer nanocomposites for multifunctional applications. Polymers 2021;13:2898.
131. Basuki MA, Suryanto H, Larasati A, Puspitasari P, Mujiono. The effect of ZnO addition against crystallinity and water absorption capacity of biofoam based cassava starch reinforced bacterial cellulose. In: International Conference on Biology and Applied Science (ICOBAS), 1st ed. AIP Publishing; 2019, 2120: 050016 p.
132. Lisdayana N, Fahma F, Sunarti TC, Iriani ES. Thermoplastic starch–PVA Nanocomposite films reinforced with nanocellulose from oil palm empty fruit bunches (OPEFBs): effect of starch type. J Nat Fibers 2020;17: 1069–80.

133. Chen Q, Shi Y, Chen G, Cai M. Enhanced mechanical and hydrophobic properties of composite cassava starch films with stearic acid modified MCC (microcrystalline cellulose)/NCC (nanocellulose) as strength agent. Int J Biol Macromol 2020;142:846–54.
134. Othman SH, Nordin N, Azman NAA, Tawakkal ISMA, Basha RK. Effects of nanocellulose fiber and thymol on mechanical, thermal, and barrier properties of corn starch films. Int J Biol Macromol 2021;183:1352–61.
135. Owi WT, Ong HL, Sam ST, Villagracia AR, Tsai C, Akil HM. Unveiling the physicochemical properties of natural Citrus aurantifolia crosslinked tapioca starch/nanocellulose bionanocomposites. Ind Crop Prod 2019;139:111548.

R. M. O. Syafiq, R. A. Ilyas*, L. Rajeshkumar, Faris M. AL-Oqla, Y. Nukman, Mohamed Yusoff Mohd Zuhri, A. Atiqah, S. M. K. Thiagamani, Sneh Punia Bangar, Claudia Barile and Sapuan S.M.

13 Corn starch nanocomposite films reinforced with nanocellulose

Abstract: The need for flexible food packaging has increased substantially in recent years around the world due to the unprecedented growth in food consumption. Due to low cost and ease of processing, petroleum-based synthetic polymer packaging materials are being predominantly used for food packaging these days. To avoid contributing to environmental problems caused by ecological risks, consumers have been searching

***Corresponding author: R. A. Ilyas,** Institute of Tropical Forest and Forest Products (INTROP), Universiti Putra Malaysia, 43400 UPM Serdang, Selangor, Malaysia; Centre for Advanced Composite Materials, Universiti Teknologi Malaysia (UTM), Johor Bahru 81310, Johor, Malaysia; Faculty of Chemical and Energy Engineering, Universiti Teknologi Malaysia, 81310 UTM, Iskandar Puteri, Johor, Malaysia; and Centre of Excellence for Biomass Utilization, Universiti Malaysia Perlis, 02600, Arau, Perlis, Malaysia, E-mail: ahmadilyas@utm.my. https://orcid.org/0000-0001-6622-2632

R. M. O. Syafiq, Institute of Tropical Forest and Forest Products (INTROP), Universiti Putra Malaysia, 43400 UPM Serdang, Selangor, Malaysia

L. Rajeshkumar, Centre for Materials and Manufacturing Technology, KPR Institute of Engineering and Technology, Coimbatore 641407, Tamilnadu, India; and Department of Mechanical Engineering, KPR Institute of Engineering and Technology, Coimbatore 641407, Tamilnadu, India

Faris M. AL-Oqla, Department of Mechanical Engineering, Faculty of Engineering, The Hashemite University, P.O. Box 330127, Zarqa, 13133, Jordan

Y. Nukman, Department of Mechanical Engineering, Faculty of Engineering, University of Malaya, 50603 Kuala Lumpur, Malaysia

Mohamed Yusoff Mohd Zuhri, Institute of Tropical Forest and Forest Products (INTROP), Universiti Putra Malaysia, 43400 UPM Serdang, Selangor, Malaysia; and Department of Mechanical and Manufacturing Engineering, Faculty of Engineering, Universiti Putra Malaysia, 43400 UPM Serdang, Selangor, Malaysia. https://orcid.org/0000-0002-1069-7345

A. Atiqah, PhD, Institute of Microengineering and Nanoelectronics, Universiti Kebangsaan Malaysia, Bangi 43600, Selangor, Malaysia, E-mail: a.atiqah@ukm.edu.my

S. M. K. Thiagamani, Department of Mechanical Engineering, Kalasalingam Academy of Research and Education, Srivilliputhur, India

Sneh Punia Bangar, Department of Food, Nutrition and Packaging Sciences, Clemson University, Clemson, USA

Claudia Barile, Department of Mechanics, Mathematics and Management, Polytechnic University of Bari, Via Orabona n.4, Bari 70125, Italy

Sapuan S.M., Editor, Department of Mechanical and Manufacturing Engineering, Faculty of Engineering, Universiti Putra Malaysia, 43400 UPM Serdang, Selangor, Malaysia, E-mail: sapuan@upm.edu.my. https://orcid.org/0000-0002-4598-6358

As per De Gruyter's policy this article has previously been published in the journal Physical Sciences Reviews. Please cite as: R. M. O. Syafiq, R. A. Ilyas, L. Rajeshkumar, F. M. AL-Oqla, Y. Nukman, M. Y. Mohd Zuhri, A. Atiqah, S. M. K. Thiagamani, S. P. Bangar, C. Barile and Sapuan S. M. "Corn starch nanocomposite films reinforced with nanocellulose" *Physical Sciences Reviews* [Online] 2023.
DOI: 10.1515/psr-2022-0011 | https://doi.org/10.1515/9783110773606-013

for alternatives to plastic food packaging due to the fact that plastics cannot be recycled and are not biodegradable. Therefore, bioplastics made from corn starch, which are biodegradable, can be one solution to these issues. The starch's amylopectin and amylase composition has a significant impact on the polysaccharide's characteristics. Corn starch (CS) is utilized as a food source for humans and other animals, but it also has many other applications in industry. Corn starch, on the other hand, has a few issues that need to be addressed through some modifications. Although starch-based packaging is expensive, they create less waste overall, which helps in the reduction amount of plastic pollution. A variety of natural and synthetic polymers – including polysaccharide-based materials like *Kappaphycus alvarezii* seaweed, collagen, alginate, chitosan, zein, cellulose, gelatin, various amino acids and polymeric materials like acrylic acid, polybutylene (PB), polyacrylic acid, polyethylene (PE), polyvinyl chloride (PVC), polyvinyl alcohol (PVA), and polycaprolactone (PCL) – are used as starch modifiers to produce polymers with a starch base that are completely biodegradable. These biopolymers have the potential to replacing petroleum-based polymers in a variety of commercial and industrial settings. Besides, corn starch nanocomposite films reinforced with nanocellulose have been investigated as a potential eco-friendly packaging material. Consequently, the effects of nanocellulose content on the mechanical, thermal, and barrier properties of corn starch-based films will be discussed in this work.

Keywords: biofilms; corn starch nanocomposites; corn starch composites; food packaging; nanocomposite films; nanocelullose nanocomposites.

13.1 Introduction

Polymers offer remarkable qualities and benefits linked to packing, and they have been utilized for a wide range of uses for a very long time [1]. Due to its low production costs, superior mechanical, and thermal properties, and abundance, it is one of the most widely used materials in the world today [2]. The most extensively utilised polyolefins are polyethylene (PE) and polypropylene (PP). Yet, since PP and PE are not biodegradable, their single usage contributes to the environmental issue when they are disposed of in landfills [3, 4]. Another issue that arises from the use of these plastics is the fact that they are derived from non-renewable resources derived from petroleum [5].

To enhance the mechanical and barrier qualities of starch-based biodegradable packaging sheets, numerous techniques have been developed [6]. These include: (1) replacement of the strong connections between hydroxyl groups with hydrogen bonds formed with starch by the inclusion of additional plasticizers, like urea and formamide to help in the thermoplastic process and boost product flexibility, (2) the addition of synthetic biodegradable polymers results in materials with qualities midway between the two components; the resulting blends can be processed more effectively by extrusion or film blowing and have mechanical and barrier capabilities superior to those of starch alone, and (3) the inclusion of compatibilizers to improve the properties of the blend by reducing the interfacial energy and increasing the miscibility of two incompatible phases. Biodegradable polymers include polyhydroxyalkanoate, cellulosics, polylactic acid, starch- and protein-based plastics, polybutylene succinate, polycaprolactone (PCL), polybutylene adipate-co-terephthalate, polyvinyl alcohol (PVOH), polytrimethylene terephthalate, polyglycolic acid, and polyethylene oxide [5, 7, 8].

Both linear and highly branched amylopectin are polysaccharides that make up starch. Amylose and amylopectin ratios vary depending on the plant source, which has an impact on the starch's physical characteristics and gelatinization patterns. Hydrogen interactions between molecules and amylose chains allow starch to form films [9]. In addition, these interactions may contribute to the nanoparticles achieving a higher level of stability [10–12]. Owing to the combined advantages such as low cost and easy availability, starch which is a natural polymer is regarded as the most potential options [13]. The granular structure of starch must be disrupted entirely or partially in order to create thermoplastic starch (TPS), which is done at high temperatures, pressures, and in the presence of plasticizers [14]. To increase the processability and thermoplastic properties of starch, some plasticizers including sorbitol, water, sugar, and glycerol were utilized. However, due to its hydrophilic nature and high Tg, it has a limited water resistance, which limits its applicability in packing.

The most common source of plant residues is corn, which offers a number of advantages such as high starch levels, outstanding uniformity, performance, ease of availability, and biodegradability [15]. In order to create biodegradable films, corn starch (CS), which is the most abundant feedstock and is inexpensive, has received all of the attention. In addition, due to its high amylose content, CS possesses effective film-forming qualities. Contrary to other polymers, CS-based plastics have a number of disadvantages, such as weak mechanical properties and limited stability brought on by water absorption [1]. In order to improve the mechanical characteristics of the film, researchers have been looking at the possibility of incorporating nanoparticles into the matrix [16]. The usage of montmorillonite, nanocellulose fibers, and starch nanocrystals has been frequent in the development of nanocomposite materials [17–20]. These nanoparticles successfully increased the composite films' mechanical characteristics, barrier properties, and thermal stability [21]. This paper aim to review the nanocellulose reinforced CS.

13.2 Corn plant

Many communities developed their civilization through the farming, living in one place and cultivation of corn, as it was observed from the archaeological data that the cultivation of corn started from 2000 to 2500 B.C.E [22]. From then, corn crop has become the staple or part of food for most humans and animals. As per the statistics, 36 million hectares of corn was cultivated in USA in the year of 2014 and the production reached a quantity of 347.79 million metric tonnes (MMT) in USA during the year 2019 [23]. Across the world, the cultivation of corn has reached a very huge quantity than any other crops. It was shown in the estimates that the cultivation of corn was about 737 MMT between 2005 and 2007, when compared with the rice and wheat production which were about 92 and 122 MMT respectively. This difference in cultivation of corn and other crops is also predicted to continue in future [23]. Corn plants contains various parts including stalk, bottom roots, corn ear, silks, top tassel and husk [24–26]. Corn kernel structure is made of six physical parts, including the hull which is the source of the fiber [27, 28], cap tip, horny starch and gluten, germ and the white starch [29]. The hull, the stalk, the husk, the starch, and the cob are all distinct components of the corn plant as shown in Figure 13.1. Most importantly, corn stalk, cob and corn husk are considered to be the salient sources of fibers when compared to other parts of the corn plant [30–32] (Figure 13.1).

13.3 Corn starch

13.3.1 Extraction and synthesis of corn starch

There are many advantages of using corn as a source of plant leftovers, including its high starch content, outstanding uniformity, performance, ease of availability, and biodegradability [33, 34]. The corn plant is made up of one or more stems that are joined together by a few small roots. At each node of the branch, there are inflorescences. Plant parts are harvested and then made into disposable garbage. Major starch, a type

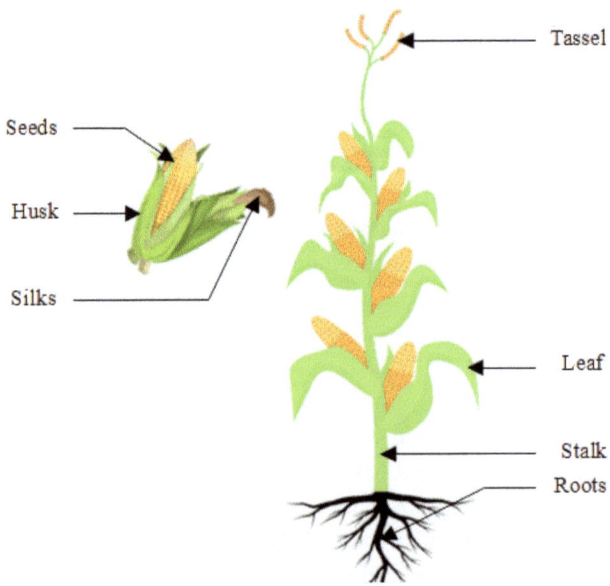

Figure 13.1: Corn plant – various parts.

of alpha-linked glucose – that is frequently employed in food manufacturers as a gelling, water-retention, and bulking agent – is mostly derived from corn. Corn fibers are distinct from other agricultural bio-products in that they save costs by up to 90 % and are widely accessible, in comparison to the availability and accessibility of other natural fibers [19].

The method of extracting corn starch from corn involves doing any one of the following steps: Using sediment separation technique in which the centrifugation of starch has been carried out after steeping the stalk at a temperature of 45 °C for about 24–72 h [35]; Using a simple centrifugation technique, steeping the corn stalks at a temperature of 50 °C for a time period of 20–24 h [36]; Using decantation technique, the steeping of corn stalks were carried out for 30 min duration [37].

13.3.2 Properties of corn starch

13.3.2.1 Physical and chemical structure of CS

Corn starch (CS) is a complex carbohydrate that is widely used in the food industry as a thickener, a stabilizer, and a binder. It is a white, odorless powder that is extracted from the endosperm of corn kernels. The CS has the following chemical composition: 10.45 % moisture, 7.12 % crude lipids, 24.6 % amylose, 74.36 % amylopectin, and 0.61 % char. Table 13.1 shows the composition of physical and chemical properties of CS.

In general, Corn starch comprises of two major components namely amylopectin and amylose. Amylose is a linear chain of glucose molecules that is tightly coiled and forms a helix structure. It makes up about 20–30 % of the total starch content in corn starch. Amylose is responsible for the gelling properties of corn starch and contributes to the viscosity of solutions made with corn starch.

Amylopectin, on the other hand, is a highly branched chain of glucose molecules. It accounts for around 70–80 % of the total starch content in corn starch. Amylopectin not only influences the texture of items created with corn starch but also contributes to the thickening qualities. The physical structure of corn starch can be observed under a microscope. It consists of granules that vary in size from 5 to 100 μm. The granules are

Table 13.1: Composition of chemical and physical characteristics of CS [38].

Description	Quantity in corn starch (CS)
Cellulose content (%)	10.8
Hemicellulose content (%)	60.4
Lignin content (%)	1.99
Ash content (%)	1.96
Density (g/cm^3)	1.42
Moisture content (%)	11.2
Crystallinity index (%)	34.6

Table 13.2: Chemical composition and physical characteristics of CS [42, 45–48].

Characteristic/content	Value	Units
Density of CS	1.41–1.36	g/cm^3
Melting point of CS	257–259	°C
Boiling point of CS	207	°C
Amylose content (per 100 g)	29.5–24.7	g
Amylopectin content (per 100 g)	75.4–72.2	g
Crude fats content (per 100 g)	7.15–0.35	g
Crude proteins content (per 100 g)	7.69–0.39	g
Ash content (per 100 g)	0.61–0.33	g
Phosphor content	0.089	%
Moisture	10.83–10.47	%

mostly oval or spherical in shape and have a distinctive birefringence pattern under polarized light. The size and shape of the granules can affect the functional properties of corn starch, such as its ability to gel, thicken, and bind.

The chemical structure of corn starch is made up of glucose molecules that are linked together by alpha-1,4 alpha-1, and 6 glycosidic bonds. The alpha-1,4 glycosidic bonds form the linear chains of amylose, while the alpha-1,6 glycosidic bonds form the branching points of amylopectin. The arrangement of these bonds determines the physical and chemical properties of corn starch.

Corn starch can be modified chemically to improve its functional properties. Common modifications include cross-linking, hydroxypropylation, and acetylation. Cross-linking can improve the stability of corn starch under high temperatures and acidic conditions. Hydroxypropylation can increase the solubility and viscosity of corn starch, whereas Acetylation can increase the thermal stability and water-holding capacity of corn starch.

According to the previous research, corn grain has a protein content ranging from 8.2 to 11.6 %, a cellulose and hemicellulose content of 10.6 %, a sugar level of 2.1 %, and an ash concentration of 1.6 %. On the other hand, the starch percentage of corn grain ranges from 38 to 75 % [39–41]. During the process of starch isolation, corn hull fiber remains as residual contents and alkalization and some other methods were used for extraction of corn fibers after chemical treatment [42–44]. Table 13.2 enlists the physical characteristics and the chemical composition of CS.

13.4 Lignocellulosic fiber

13.4.1 Classification of fiber: corn

The term “natural fibers" generally refers to a variety of classification of fiber that originated from nature sources such as plants, minerals, and animals [49]. As a result, it is of the utmost importance to clarify that the term "natural fibers" referred to "plant fibers" in the current research [50, 51]. Moreover, "Plant fibers" are also referred to as "lignocellulosic fibers," "cellulosic fibers," or "vegetable fibers." Fabrics, twines, and ropes made from natural fibers have been used extensively in human society ever since they were first introduced. They may be also utilised without harming the environment or depleting resources because they are renewable, biodegradable, and carbon neutral. Natural fibers can be divided into three categories based on their sources: mineral, animal, and plant. Plant fibers can be divided into two primary categories: wood and non-wood bio-fiber which shown in Figure 13.2.

Table 13.3 shows the characteristics of corn cob, husk, hull, and corn stalk. When compared to other prominent sources of natural fibers such as wheat straw, husk and barn as well as rice straw and husk, it is evident that corn husk and cob have a high content of cellulose [53–74].

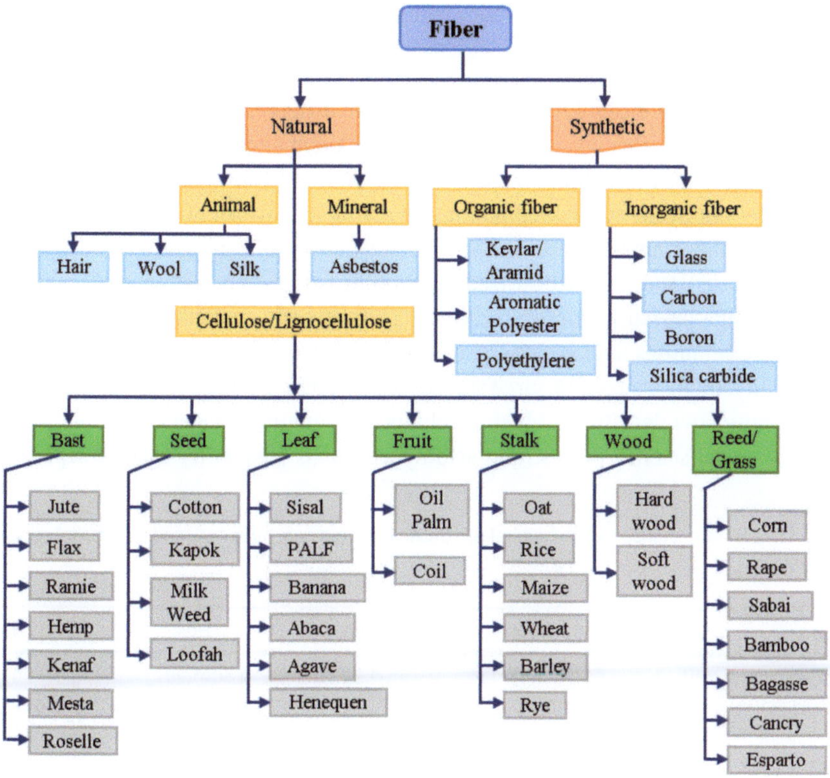

Figure 13.2: Plant fiber categorization [52].

13.4.2 Cellulose fiber

Cellulose, a semicrystalline polymer made of poly(1,4-D-anhydroglucopyranose) units, is the primary component of plant fibers [34]. Strong hydrogen bonds between hydroxyl groups help to generate these units. Lignin and hemicellulose are two additional important constituents of the structure of natural fibers. Hemicellulose is comprised of a wide variety of sugars, including galactose, arabinose, mannose, and xylose, whereas lignin is a highly cross-linked phenolic polymer. Lignin and hemicellulose, however, are both amorphous polymers. The removal of the amorphous region and isolation of NCCs from natural fibers can be accomplished using the well-known method of strong acid hydrolysis. Acid hydrolysis enables removal of the amorphous portions of cellulose fibers under controlled conditions while preserving the crystalline domains as crystalline nanoparticles. In actuality, eliminating the amorphous region affects the fibers' structure, thermal stability, crystallinity, and surface shape.

13.4.3 Nanocellulose fiber

In the past few years, nanocellulose had been produced using cellulose fiber from several sources, such as sugar palm, corn, water hyacinth, oil palm empty fruit bunch (OPEFB), ramie, pineapple leaf, kenaf bast, roselle, cotton, and sugarcane bagasse. Modifying cellulose at the nanoscale is one method for enhancing its performance [34]. There are several ways to make nanocellulose, one of which includes chemical (pulping, bleaching, and acid hydrolysis) and mechanical treatments (ultrasonication, high shear homogenization, and high-pressure homogenization). The attractive inherent properties of nanocrystalline celluloses (NCCs) extracted from plant fibers– including their availability, renewability, biodegradability, large surface area (100 m^2/g^1), high aspect ratio of 100, high crystallinity, low density, and high mechanical strength have generated plenty of interest in material science [84]. Strong acid hydrolysis is frequently used to separate

Table 13.3: Physical characteristics and chemical composition of corn cob, husk, hull, and corn stalk [42, 65, 68–70, 75–83].

Description	Stalk	Hull	Husk	Cob
Density of material (g/cm^3)	1.43	1.33	1.50–1.20	0.81–1.19
Porosity/void (%)	58.52	–	87.1 ± 2.2	67.95
Cellulose content (%)	33–38	15.5	31.4–47.2	40.1–44.2
Hemicellulose content (%)	29.2–42.1	40.5	34.1–43.89	31.2–33.5
Lignin content (%)	5.2–38.2	2.2–2.89	1.52–14.5	16.3–17.8
Ash content (%)	25.2	0.89–1.32	3.32–6.81	2.31–3.22
Moisture content (%)	3.35–6.2	4.3–8.51	7.62–8.73	6.40

NCCs from natural fibers in order to get rid of the amorphous area. Acid hydrolysis allows for the removal of the amorphous components of cellulose fibers while maintaining the crystalline domains as crystalline nanoparticles.

13.5 Modification of corn starch films

13.5.1 Plasticization of corn starch films

If starch is combined with enough non-aqueous plasticizer (generally polyols) or water, it can be transformed into a thermoplastic polymer. Polyols including glycerol, diethylene glycol, ethylene glycol, polyethylene glycol, propylene glycol, sorbitol, and triethylene glycol are the most often utilised hydrophilic plasticizers in starch-based films [85]. Plasticizers are a special category of non-volatile low molecular weight chemicals as well as a special kind of polymer additive. Enhancing the flexibility and fracture resistance of films made of starch is the primary function of plasticizers. Other characteristics were also affected, including crystallinity, permeability to water vapor, and biodegradation. Pure starch films were exceedingly fragile and quickly fractured when handled. Plasticizers are introduced in the right amounts during the film-making process to solve these issues. Plasticizers can diminish the interaction of starch molecules, reduce the stiffness and tensile strength of films, and increase the liquidity of starch-based films [86].

In order to enhance the applicability and flexibility of the starch-based bio composites, plasticizers are generally employed. Since the starch molecules could not be melted in its raw form as hydrogen bonds are present within them, plasticizers were used to impregnate into the starch molecules, breaking the hydrogen bonds at higher pressure, temperature and shear stress [87]. It was reported in a variety of published literature that the most prevalent types of plasticizers for starch were ethanolamine, diethanolamine, ethylene glycol, and triethanolamine [88]. Glycerol is the plasticizer which is prevalently used since it decreases the glass transition temperature of the starch, and thus reduces the strength of the intermolecular hydrogen bonding among the starch molecules [89]. Additionally, glycerol enhances the value of elongation at break and improves the ease of processing of the starch [90]. Nevertheless, the presence of glycerol as a plasticizer in the starch reduces the mechanical properties of the final biocomposites [91, 92].

Plasticizers are usually incorporated into the starch molecules for enhancing their flexibility, plasticity and decreasing the biocomposite material's brittleness. It has been claimed in some research studies that the incorporation of plasticizers into bio composites is required in order to prevent an undesirable decline in the mechanical characteristics of bio composites that do not contain plasticizers. Some researchers strengthened corn starch by adding 8 % corn husk, leading to a 615 MPa in modulus of elasticity and 13 MPa in tensile strength values. Hybridization of the corn starch composites with 6 % of sugar palm fibers enhanced the mechanical properties of the hybrid composites. Tensile strength of the hybrid composites was noted to be 19 MPa and the elasticity modulus reached a value of 1162 MPa. It was also noted that the highest value of elongation at break was obtained for the biocomposites without the sugar palm fibers [93, 94].

In some experiments, corn starch was plasticized with a glycerol mixture and polyol mixtures containing a high molecular weight polyol (HP) including sorbitol, maltitol, and xylitol. The bio composites with a glycerol concentration of 40 parts per hundred (PHR) showed the greatest tensile strength value, at 9.3 MPa, when glycerol was added. Similarly, the high value of elongation at break (EB) was obtained as 118 % at 60 PHR glycerol content. When sorbitol was used as a plasticizer, the highest achieved tensile strength value was 10.4 MPa for a sorbitol content of 50 % in the mixture of polyol, while the EB value reached a highest value of 105 % for 10 % sorbitol content in the polyol mixture. It was also noted that for 20 % glycerol and malitol mixture, a high tensile strength value of 16.27 MPa was recorded [95]. Upon addition of the plasticizer, the breakage of hydrogen bonds occurred which allows the free movement of the molecules, thus increasing the elongation at break. The maximum value of tensile strength was also achieved when starch-plasticizer bonds prevailed over the hydrogen bonds produced between starch molecules [17].

13.5.2 Corn starch blend

CS blends were characterized in terms of their average molecular weight, content of moisture, thermal stability through thermogravimetric analysis (TGA), water solubility, chemical composition through IR spectroscopy (FTIR), chemical characteristics and phase change characteristics through X-ray diffraction analysis (XRD) and differential scanning calorimetry (DSC), antimicrobial, mechanical, and microstructural characteristics, and their solubility in alkaline and acidic mediums [96]. Oliveira et al. [97] developed biodegradable polymer blends based on corn starch and thermoplastic chitosan processed by extrusion. The researchers stated that the blends also showed good thermal stability, and the addition of chitosan resulted in more thermally stable films. Furthermore, SEM images of the blends revealed uniform surfaces, and the criofractured samples exhibited no evidence of chitosan clumping within a totally destructurized starch matrix. The phase transition point in the blending system depends on the blending ratio and solution concentration. The reduction in tensile modulus was discovered to be correlated with the lower transparency point of the mix and phase transition region of hydroxypropyl methylcellulose/hydroxypropyl starch composite from continuous phase to separated phase [98]. The blend system was discovered by the authors to be immiscible, with a phase transition point that depends on the blending ratio and solution concentration. Gutiérrez and Alvarez [96] studied the properties of native and oxidized corn starch/polystyrene (PS) blends with zinc octanoate catalyst prepared using reactive extrusion technique. The findings make it abundantly clear that the catalyst utilized was responsible for the formation of cross-links between the starch and PS. Besides, the oxidative alteration of the starch increased the substance's tendency for chemical activity. A modified system of starch is produced when the starch structures are supplemented with carbonyl and carboxyl groups. This results in the starch having a higher degree of hydrophobicity. Tian et al. [99] experimented on the blends of PVA/corn starch using nanoclay intercalation techniques and observed that the thermal and mechanical characteristics of the bio composites were increased. The strong bond between montmorillonite (MMT) nanolayers and PVA/starch macromolecules may limit the matrix's swelling capacity and hence reduce water absorption. In contrast, the mixture exhibited excellent stability and compatibility when it was combined with electrolyte solutions. Furthermore, the swelling behavior tests revealed that the hydrogel beads had exceptional water absorption capacities [100].

13.5.3 Corn starch bilayer films

Because they are resistant to rapid biodegradation, the newly developed bilayer films have potential for use in studies aimed at improving food packaging [101]. Chen et al. [102] demonstrated that bilayer film has an adhesive interface between the two layers and smooth surface shape with better visibility. Tensile strength and Young's modulus of the bilayer film were significantly increased as compared to the monolayer film. Zuo et al. [103] studied the physical and structural characterization of edible bilayer films made with zein and corn-wheat starch. The addition of zein has an impact on the visual appeal and physical characteristics of films. The authors found that the opacity, WVP, and TS were decreased, whereas the water content, E and antioxidative capacity were increased. The biodegradation test revealed that PLA and TPS biodegrade more quickly than other materials [104]. Scanning electron microscopy revealed robust linkages between the bilayer films and demonstrated the disappearance and reappearance of the network structure inside the films [105]. In comparison to net starch films, a significant improvement was made in tensile and water vapor barrier properties. The films maintained high transparency and oxygen permeability as low as starch films, despite the lower ratio of cast PLA sheet in the bilayer assembly (approximately one third of the film thickness). This, in fact, is the main cause of improving the tensile and water vapor barrier properties [106]. Zhou et al. [107] developed and characterized pea starch and PLA bilayer films for their potential application in the packaging of cherry tomatoes. The authors found that the addition of two layers improved the PS films' water resistance (lower water absorption and solubility), thermal characteristics, and water vapor barrier (lower WVP). However, the laminated structure's synergy effect improved the mechanical and barrier properties of bilayer films, making them a good choice for packaging foods that absorb a lot of moisture. The bilayer films' tensile strength (TS) and elongation at break (EAB) increased by 74.04 % and 348.13 %, respectively [108]. Due to the hydrophobic property of the PLA layer, the integration of the PLA layer greatly decreased the water vapor permeability, water absorption, and solubility of bilayer films [109].

13.5.4 Fiber reinforced corn starch biocomposites

Biocomposites are a promising alternative, based on corn, to traditional composites made from non-renewable materials such as fiberglass and carbon fiber [110–113]. These biocomposites are made by combining corn-based polymers with natural fibers such as corn husks, corn stalks, and corn cobs. Table 13.4 enlists the mechanical characteristics of corn-based biocomposites. In terms of mechanical properties, Corn-based biocomposites have demonstrated high tensile strength, stiffness, and impact resistance. From the table it can be observed that the corn husk (CH) hybridized with sugar palm fiber reinforced corn starch resulted high tensile properties of 19 MPa compared to others reinforcement. The natural fibers in the composites provide reinforcement, while the corn-based polymers act as a matrix to hold the fibers together. This material combination produces a lightweight, robust, and long-lasting composite material. However, the mechanical properties of corn-based bio-composites can vary depending on factors such as the type and proportion of natural fibers used, processing methods, and environmental conditions. Further, more research is needed to optimize the mechanical properties of these bio-composites and ensure their long-term durability and reliability.

The water vapor barrier, mechanical, and optical properties of *Kappaphycus alvarezii* seaweed and corn starch blend bio composite films were investigated, Khalil et al. [116]. The researchers stated that the concentration of seaweed had the biggest impact on the mechanical and optical behavior of the seaweed/starch biocomposite films. However, the water vapor permeability of the composite films improved with increasing amounts of starch and seaweed. Besides that, the FTIR measurement also showed intermolecular interactions between the two compounds, which is in line with the high miscibility degree between the seaweed and the starch in the mixture. The primary peak shift in FTIR to higher wavenumbers and the decrease in crystallinity in XRD are evidences that starch molecules and chitosan may interact via hydrogen bonding [117]. Besides, Ren et al. [117] stated that the film's solubility, tensile properties, elasticity, and water vapor permeability (WVP) of the film were increased with the inclusion of natural fiber. Wang et al. [118] researched on water solubility index and mechanical characteristics of collagen-corn starch biocomposite films, as well as the evaluation of effects of concentration and type of starch. The authors found that the addition of starch to collagen enhanced the composite films' thickness and opacity. Furthermore, the surface of starch-collagen films was rougher than that of pure collagen films, which smoothed out during heating, according to image taken using scanning electron microscopy. SEM image were reported in Figure 13.3.

13.5.5 Cellulose reinforced corn starch biocomposites

Wang et al. [98] reported the relationship between morphologies and mechanical properties of hydroxypropyl methylcellulose (HPMC)/hydroxypropyl starch blends. It

Table 13.4: Mechanical characteristics of corn-based biocomposites.

Reinforcement material	Matrix material	Plasticizer material	Content of plasticizer	Content of reinforcing material (%)	Tensile strength (MPa)	Young's modulus (GPa)	Flexural strength (MPa)	Elongation strength (MPa)	Reference
–	Corn starch	Glycerol	10 %	–	62	63	–	11	[114]
			20 %		31	32		26	
			30 %		18	19		44	
			40 %		10	13		57	
Graphene oxide (GO)	Blend of CS and soy protein isolate (SPI)	Glycerol	30 %	GO 0	6.65 ± 0.9	252 ± 5.2	–	7.25 ± 43	[115]
				GO 0.5	2.84 ± 0.48	236 ± 48		19.73 ± 3.38	
				GO 1	2.32 ± 0.35	332 ± 26		17.9 ± 3.4	
				GO 1.5	3.60 ± 0.6	359 ± 25		10 ± 2.02	
				GO 2	5.9 ± 0.27	449 ± 28		11.83 ± 2.08	
Graphene (G)	Blend of CS and SPI	Glycerol	30 wt%	G 0	6.65 ± 0.9	252 ± 502	–	7.25 ± 0.43	[115]
			30 wt%	G 0.5	16.53 ± 0.26	578 ± 105		2.7 ± 0.12	
			30 wt%	G 1.0	11.38 ± 1.4	380 ± 48		1.2 ± 0.21	
			30 wt%	G 1.5	8.09 ± 1.4	329 ± 34		2.83 ± 1.83	
			30 wt%	G 2.0	6.66 ± 1.3	132 ± 55		6.58 ± 0.98	
Corn husk	Corn starch	Fructose	25 %	2	8.8	320	–	–	[94]
			25 %	4	7.8	190			
			25 %	6	10.9	510			
			25 %	8	12.9	615			
Corn husk (CH) hybridized with sugar palm fiber	Corn starch	Fructose	25 %	8 % CH+ 2–8 % sugar palm fiber	17	1050	–	1.3	[93]
			25 %		17.3	1100		1.15	
			25 %		19	1160		0.9	
			25 %		18	1110		1.1	

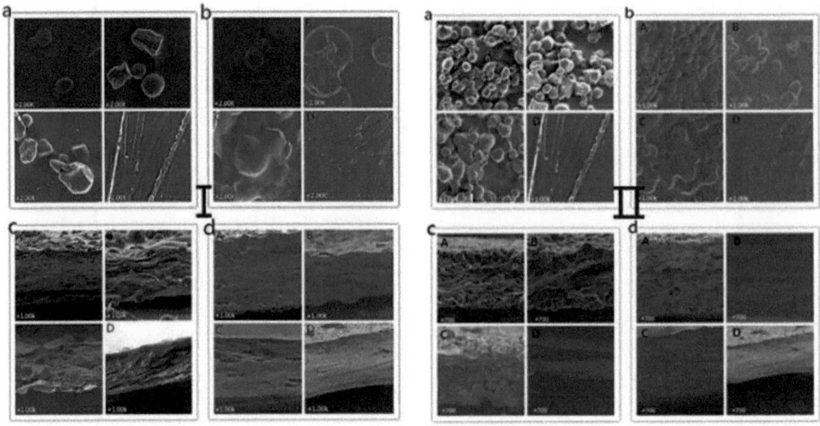

Figure 13.3: SEM micrographs of collagen and collagen-based biocomposite film surface (a) before heating, (b) after heating at a scale of 20 μm; (c) cross section of biocomposite film before heating, and (d) after heating at a scale of 50 μm. Collagen film blended with corn starch (I) 10 wt%, (II) 50 wt% and hybridized with (a) high amylose content, (b) waxy corn starch content, (c) without starch, and (d) neat collagen composites [118].

was revealed by the researchers that as the concentration of the solution increased the modulus and elongation both dropped. This finding is consistent with the observed morphologies, in which it was determined that the HPMC progressively transitioned from a continuous phase to a discrete phase. However, the modulus and elongation both decreased with increasing solution concentration; this is probably due to HPMC's gradual transition from a continuous to a separated phase. Figure 13.4 shows the modulus and elongation at break of HPMC.

13.5.6 Microcrystalline cellulose (MCC) reinforced corn starch

Fabra et al. [119] experimented on the microalgae reinforced in the thermoplastic corn starch films and evaluated their physicochemical and structural characteristics. The kind of microalgae and, more specifically, the degree of cell wall disintegration influence considerably the mechanical characteristics of the biocomposites. With the addition of the various microalgae species, a loss in water permeability of around 54 % was observed.

13.5.7 Nanocellulose reinforced corn starch

One of the novel nanomaterials obtained from the natural lignocellulosic fibers, which could be considered for the potential reinforcement material in the biopolymers to produce green composites, is nanocellulose. When the cellulose polymer is disintegrated

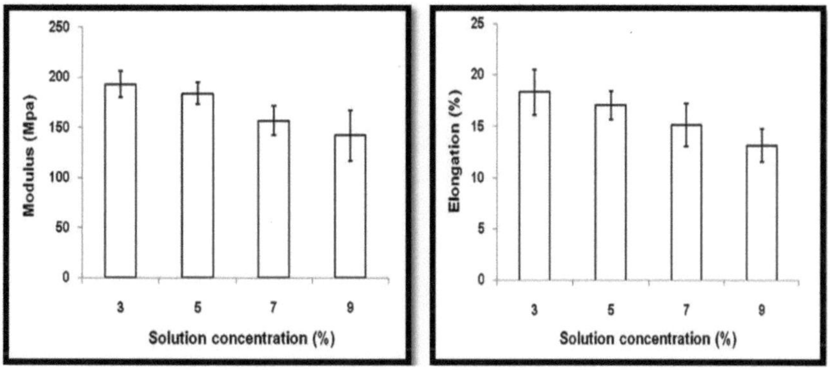

Figure 13.4: Modulus and elongation at break of HPMC [98].

to nanoscale range (10^{-9} m), it is called nanocellulose. Cotton stalks and fibers, coconut fibers, algae, sugarcane bagasse, and other natural fibers are some of the primary agro biomass-based sources to produce cellulose. Good quality nanocellulose can be extracted from the cotton linter which is the by-product of the textile and is considered to the rich source of pure form of cellulose. When compared to other natural fibers, the process of removal of non-cellulosic components like hemicellulose and lignin from the cotton fibers is relatively easier. There are many reasons why NC could be used as a reinforcement material in green polymer matrices, including their high tensile strength (2–10 GPa), higher surface to volume ratio (55–205 m^2/g), excellent thermal stability (225–330 °C), very low density (1.32–1.46 g/cm^3), better rheological properties, hydrophilicity, biodegradability and renewability, high degree of crystallinity, and chemical compatibility. When nanocellulose (NC) was used as reinforcement in biopolymer-based biocomposites, the water barrier and mechanical properties of the biocomposites were found to be enhanced.

Patil et al. [120] studied the corn starch biocomposites films made of nanocellulose reinforcements. Cotton linters were used for the synthesis of NC through chemo-mechanical process which were then used as a reinforcement to manufacture the functionally improved biocomposite films. Chemo-mechanical technique involves the acid hydrolysis of cotton linters after bleaching, and homogenizing them under ultra-high pressure as depicted in Figure 13.5.

Particle size analysis of the NC materials indicated that a particle size of 55.4 nm was there for almost 93.7 % of the particles while the remaining particles had an average particle size of 1103 nm as shown in Figure 13.6a. Colloidal stability of the suspension could be determined by testing the Zeta potential of the material. The absolute value of zeta potential for a stable suspension in any material should be 20 mV. As shown in Figure 13.6b, the average value of zeta potential portrayed by the NC suspension was about 29.06 mV, indicating the production of stable colloidal material of NC from cotton linter using chemo-mechanical technique. It was also observed from the experimental

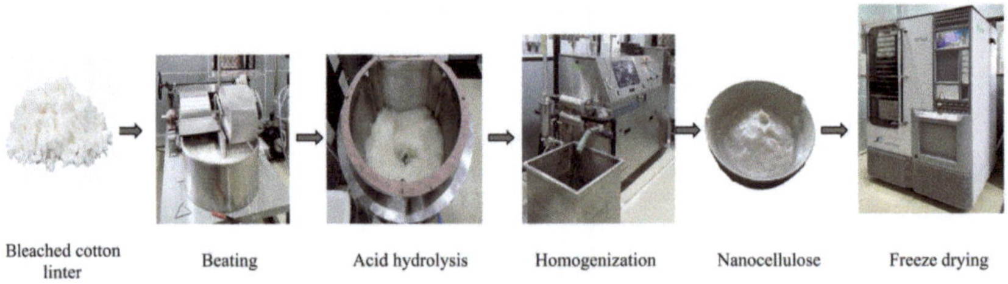

Figure 13.5: Synthesis of nanocellulose.

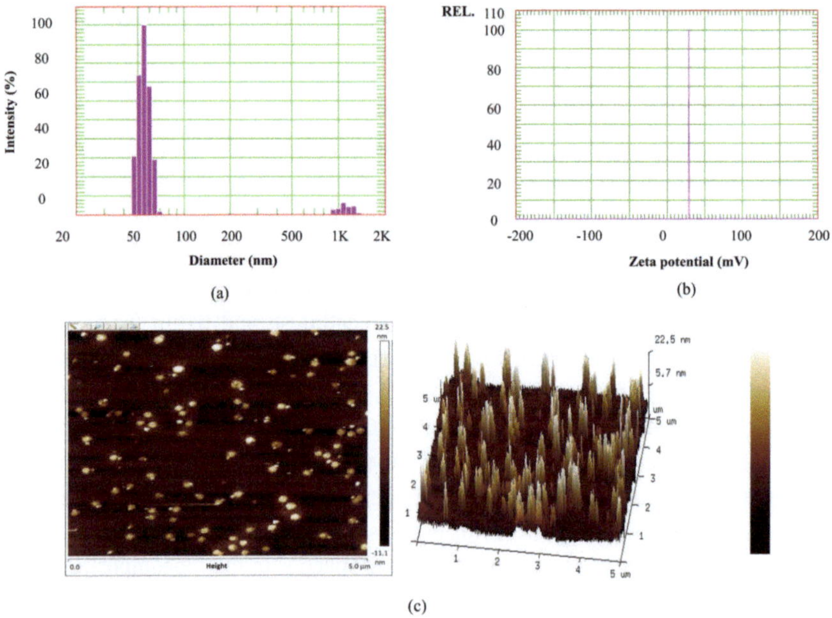

Figure 13.6: Nanocellulose characterization (a) particle size analysis, (b) zeta potential analysis, and (c) 3D atomic force microscopic images.

results that the developed NC material was in nanoscale range. Figure 13.6c illustrates the atomic force microscopic (AFM) analysis of the NC material which further acknowledges the nano-size of NC. After analyzing an AFM image of NC material with the help of Image J software, it has been found that its particles had an average size of 86 nm and a standard deviation of 33 nm.

Few experimenters studied the effect of incorporation of NC at different weight fractions along with different plasticizers in the starch films synthesized using solvent casting method. The tensile strength (TS), elongation at break (EB), water vapor permeability (WVP), and bursting strength (BS) were taken as responses, as well as were optimized using response surface methodology. As per the optimization results, 1.89 % of polyvinyl alcohol (PVOH), 2.54 % of glycerol (GL), and 0.9 % of NC resulted in the optimum formulation of biocomposite film with better properties. As shown in Figures 13.7–13.10, the predicted responses were 8.93 MPa of TS, 41.95 % of EB, 5.57 kPa of BS and 7.09×10^{-10} g/msPa of WVP. *P*-value of the optimization showed that the NC reinforcement influenced the mechanical and other properties significantly with a confidence limit of 0.95. It was also shown in the results that the NC, extracted from the cotton linters, enhanced the mechanical properties of the biocomposite films. Nano size of NC, huge surface area, and strength of the hydrogen bonds formed between starch, were stated to be reasons for the high tensile strength of the biocomposite films. The load bearing capacity of the NC-based composites improved due to the increased mechanical strength and high aspect ratio of the NC. This improved the stress transfer from the matrix of the composites to the NC, which in turn increased the biocomposite film's resistance to breaking [121–125]. It was also portrayed in the experimental results that the effect of NC addition on TS and EB are inversely proportional to each other. Owing to the chemical and structural compatibility between starch and NC, the addition of NC had a negative effect on the EB of the biocomposite films. But this in turn increased the toughness of the

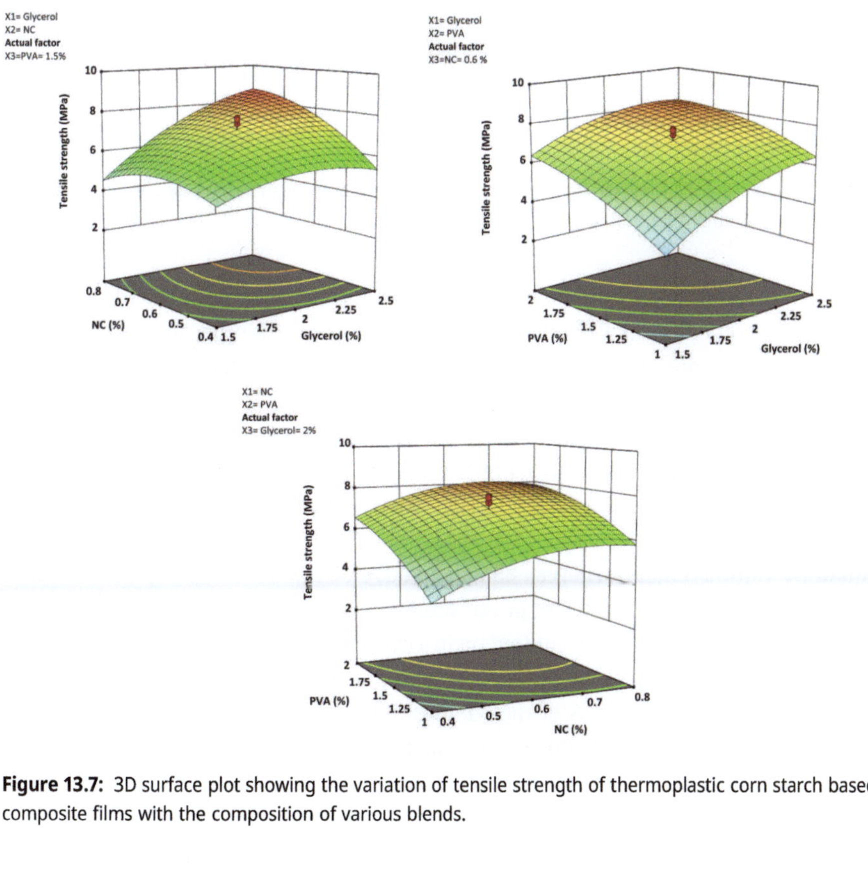

Figure 13.7: 3D surface plot showing the variation of tensile strength of thermoplastic corn starch based composite films with the composition of various blends.

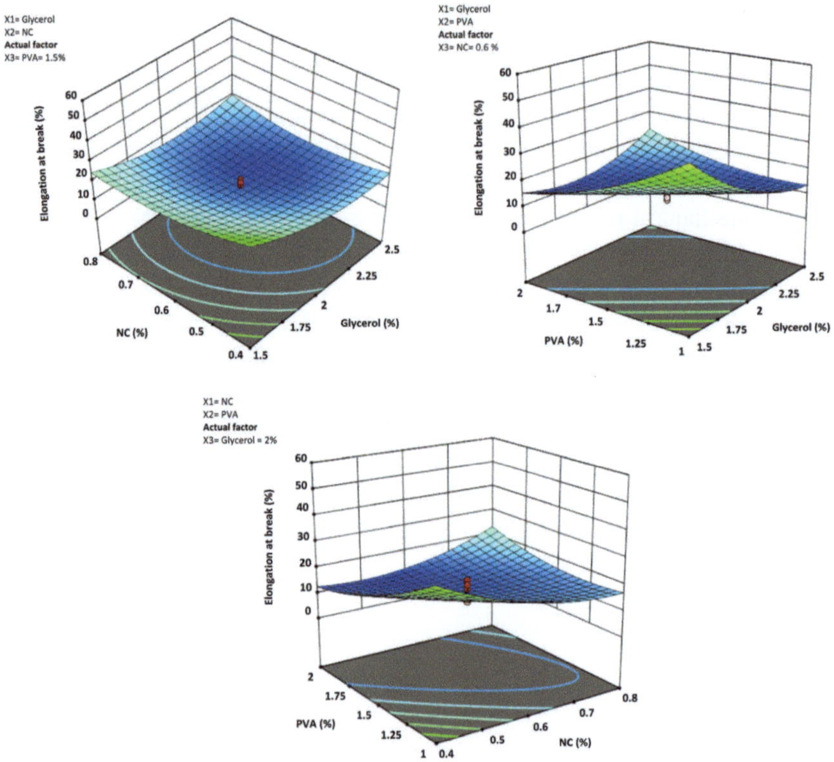

Figure 13.8: 3D surface plot showing the variation of elongation at break of thermoplastic corn starch based composite films with the composition of various blends.

composites through the strengthening of interactions between NC-starch bonds. Few experimenters reported the EB reduced for the starch-based films because of the rigidity of the cellulose nanocrystals [126, 127]. Some experimenters stated that the gradual addition of higher content of PVOH and GL into NC might result in the increase of EB of the composites [120]. Since the GL minimizes the intermolecular forces existing between starch molecules through its plasticizing effect when reinforced in higher contents, the flexibility of the biocomposite films increase considerably. In addition, as depicted in Figure 13.9c, a number of researchers reported that TS and EB had beneficial effects as a result of NC reinforcement, whereas BS had an adverse impact and the quadratic effect had a favorable outcome as a result of reinforcement. The BS value of the biocomposite film dropped with increasing NC content up to a certain point, but after that point, it grew along with NC content, as seen by the convex curved surfaces in Figure 13.9c. Owing to the high surface area of the NC, BS of the composite films increased resulting in a better starch-NC interfacial adhesion which in turn increased the strength of NC films [128–133].

The addition of NC content in the biocomposites restricted the permeation of water molecules inside the films and impacted WVP properties to a larger extent. As seen in

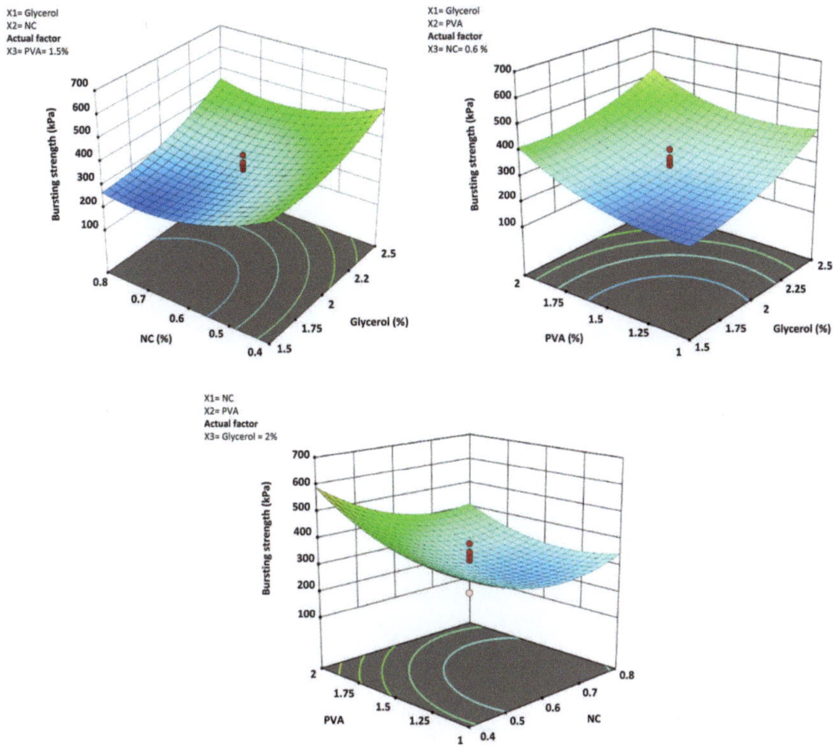

Figure 13.9: 3D surface plot showing the variation of bursting strength of thermoplastic corn starch based composite films with the composition of various blends.

Figure 13.10, the WVP of the biocomposites decreased with the increase in NC contents. The transmission of water vapor through the hydrophilic film is governed by the solubility and diffusivity of the water molecules within the biocomposite film matrix [134]. Water vapor barrier of the composite films were increased and the diffusivity of the composites were decreased by the high aspect ratio of the NC. A rugged pathway was created by the reinforced NC in the polymer matrix which reduces the movement of water by decreasing the diffusion rate of water vapor into the composite [135]. Due to the rigid hydrogen bonding formed due to the high crystallinity of the NC molecules, the pore size of the biocomposites reduced the water diffusion and increased the water vapor barrier properties [136]. The results given by some experimenters suggested that the NC reinforced corn starch-based biocomposite films can be used effectively for storing the edible oil for a longer duration without significantly deteriorating the oil quality in ambient conditions [120]. Due to the high-water vapor permeability properties of these biocomposite films, this application was quite possible. It is a well-known fact that the material used for packaging must possess a low oxygen and water vapor permeability which avoids the oxidation of oil during their storage. Hence, these

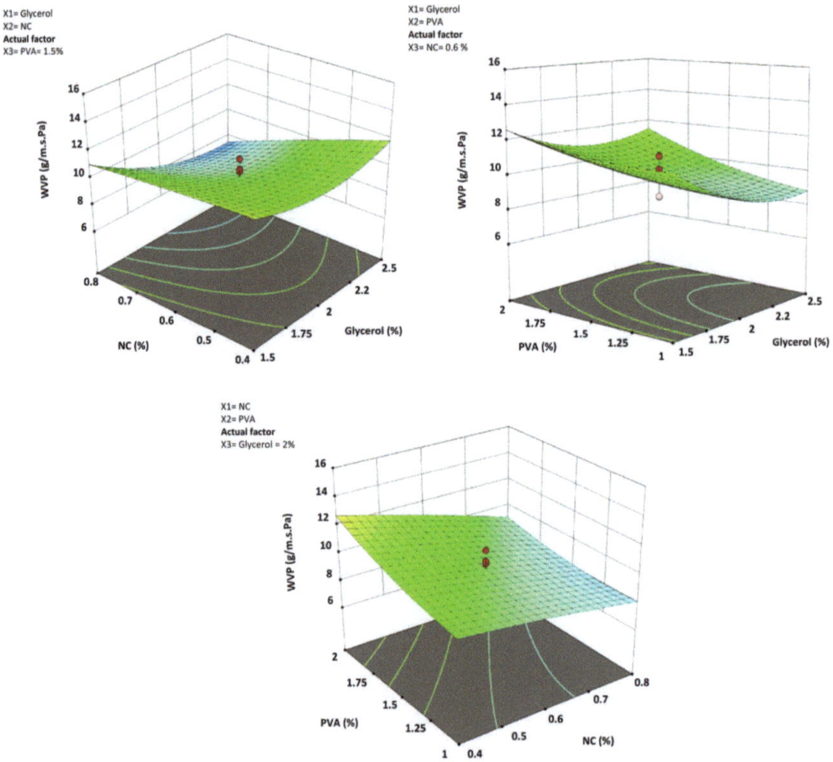

Figure 13.10: 3D surface plot showing the variation of water vapor permeability of thermoplastic corn starch based composite films with the composition of various blends.

NC-based corn starch biocomposite films can be potentially used as a packaging material for edible oil storage.

13.6 Application, challenges and future perspective

One of the major advantages in utilizing the agricultural crop corn as a precursor for biocomposite manufacturing is the renewability of the agricultural-based products. But unless the green cover of the forest is continuously renewed and regenerated, the use of forest plants to produce biocomposites cannot be completely utilized and would be restricted. Various types of papers and paper-boards can be derived from the corn-based straws and fibers due to the presence of high amount of lignin in those crops. Since lignin is characterized by a fuel value and high content of energy, it can be used for burning of fuel. Application of lignin extends to the manufacturing of binders and adhesives despite their low values [137]. Corn-based biocomposites are utilized for manufacturing day-to-day products such as packaging pads, loose fills, foam cushioning, mailer bags, paper padding, and air cushioning.

Applications for nano and micro cellulose, which can be extracted from corn fibers and other fiber sources, are expanding rapidly. As an additive, micro and nano fibrillated cellulose is used in many different industries, including those dealing with cosmetics, paper production, chronic wound healing, pharmaceuticals, medical devices, flexible electronics, coatings, sanitary pads, and high-temperature insulation and thermal barrier applications [138–142]. Owing to various advantageous features like small diameter along the vascular grafts, nanocrystalline cellulose (NCC) was used in medical applications, packaging applications, synthetic plastics or polymers, electronic applications, catalysis, filtration applications, fuel cells, solar cells, tissue engineering, and lithium-ion batteries. Because of their high tensile strength and low thermal expansion, NCC-based biocomposites are used in applications like automobile components and fire extinguishers [143–148]. Microcrystalline cellulose-based biocomposites are also utilized as abrasives in cosmetics, bulking applications, absorbent in water treatment, anti-caking agents, viscosity aqueous raising agents, emulsion stabilizer, binder, texturizer, and slip modifiers [149].

The hydroxyl groups must be altered through surface modification, grafting, or esterification in order to increase the versatility of hydrophilic polymers made from starch and cellulosic fiber. It is important to pay attention to the choice of fiber size since it has an impact on the molecular bonds that are formed and control the characteristics of the final product. Indeed, CS offers a wide variety of amylose content, offering specific design, and conditions to complete the fabrication process in every desired application.

Plastic will eventually be replaced by biodegradable packaging materials for food packaging. Future work will focus on improving the mechanical properties of bioplastics made from renewable bioresources, which are an important possibility. Studies on nanotechnology, in particular, will play a significant role here. For instance, a substance that is inexpensive and has excellent mechanical qualities will enhance the possibility that it will be transformed into such substances as nanotechnology develops. As a result, in the next centuries, the special qualities of bio-nanocomposites made from renewable sources should be the main place to look for new potential and applications for green nanomaterials. Future applications for these nanocomposites with enhanced biocompatibility, biodegradability, and mechanical and thermal qualities include numerous aspects of daily life.

13.7 Conclusions

Corn starch (CS) is a polymer that occurs naturally. CS is widely accessible, affordable, and simple to alter using various physical or chemical techniques. Biopolymers minimize plastic pollution, which is one of the major threats to our environment. They have valuable applications in a variety of fields and can further replace polymers made from petroleum. Starch is combined with natural polymers (cellulose, chitosan, gelatin, alginate, hydroxyapatite, and protein), synthetic polymers (PVA, PVC, PCL, polybutylene, and polyethylene), and inorganic nanoparticles (lithium, silver, and talc) using a variety of chemical and physical techniques. Development of modified CS based bionanocomposites films with better physical and mechanical properties than simple starch-based films– including increased elasticity, break resistance, moisture adsorption capacity, low water vapor permeability value, and higher tensile strength– is one of the most technological advances in the polymer industry. In order to create high-value goods for cutting-edge applications in a variety of fields, modified CS-based polymers has a very bright future.

Acknowledgment: The authors would like to thank the editors R.A. Ilyas, S.M. Sapuan and M.N.F Norrahim for their guidance and review of this article before its publication.

References

1. Jiang B, Li S, Wu Y, Song J, Chen S, Li X, et al. Preparation and characterization of natural corn starch-based composite films reinforced by eggshell powder. CyTA – J Food 2018;16:1045–54.
2. Ghasemlou M, Aliheidari N, Fahmi R, Shojaee-Aliabadi S, Keshavarz B, Cran MJ, et al. Physical, mechanical and barrier properties of corn starch films incorporated with plant essential oils. Carbohydr Polym 2013; 98:1117–26.
3. Harussani MM, Sapuan SM, Rashid U, Khalina A, Ilyas RA. Pyrolysis of polypropylene plastic waste into carbonaceous char: priority of plastic waste management amidst COVID-19 pandemic. Sci Total Environ 2021;803:149911.
4. Harussani MM, Rashid U, Sapuan SM, Abdan K. Low-temperature thermal degradation of disinfected COVID-19 non-woven polypropylene—based isolation gown wastes into carbonaceous char. Polymers 2021;13:3980.
5. Ilyas RA, Sapuan SM, Norrrahim MNF, Yasim-Anuar TAT, Kadier A, Kalil MS, et al. Nanocellulose/starch biopolymer nanocomposites: processing, manufacturing, and applications. In: Al-Oqla F, Sapuan SM, editors. Advanced processing, properties, and applications of starch and other bio-based polymers. Netherlands: Elsevier; 2020:65–88 pp.
6. Rozilah A, Jaafar CNA, Sapuan SM, Zainol I, Ilyas RA. The effects of silver nanoparticles compositions on the mechanical, physiochemical, antibacterial, and morphology properties of sugar palm starch biocomposites for antibacterial coating. Polymers 2020;12:2605.
7. Rajeshkumar L, Ramesh M, Bhuvaneswari V, Balaji D, Deepa C. Synthesis and thermo-mechanical properties of bioplastics and biocomposites: a systematic review. J Mater Chem B 2023;11:3307–37.
8. Rajeshkumar L. Biodegradable polymer blends and composites from renewable resources. In: Rangappa SM, Parameswaranpillai, Siengchin S, Ramesh MBT-BP, editors. Biodegradable polymers, blends and composites. Netherlands: Elsevier; 2022:527–49 pp.
9. Lopez O, Garcia MA, Villar MA, Gentili A, Rodriguez MS, Albertengo L. LWT – food Science and technology thermo-compression of biodegradable thermoplastic corn starch films containing chitin and chitosan. LWT – Food Sci Technol 2014;57:106–15.
10. Florencia O, Alejandra GM, Arce Valeria B, et al. Nanocomposite films with silver nanoparticles synthesized in situ: effect of corn starch content. Food Hydrocoll 2019;97. https://doi.org/10.1016/j.foodhyd.2019.105200.

11. Sharma S, Sudhakara P, Omran AAB, Singh J, Ilyas RA. Recent trends and developments in conducting polymer nanocomposites for multifunctional applications. Polymers 2021;13:2898.
12. Azammi AMN, Ilyas RA, Sapuan SM, Ibrahim R, Atikah MSN, Asrofi M, Atiqah A. Characterization studies of biopolymeric matrix and cellulose fibres based composites related to functionalized fibre-matrix interface. In: Interfaces in Particle and Fibre Reinforced Composites. Netherlands: Elsevier; 2020:29–93 pp.
13. Ibrahim MIJ, Sapuan SM, Zainudin ES, Sapuan SM, Zainudin ES, Zuhri MYM. Physical, thermal, morphological, and tensile properties of cornstarch-based films as affected by different plasticizers. Int J Food Prop 2019;22:925–41.
14. Babaee M, Jonoobi M, Hamzeh Y, Ashori A. Biodegradability and mechanical properties of reinforced starch nanocomposites using cellulose nanofibers. Carbohydr Polym 2015;132:1–8.
15. Harussani MM, Sapuan SM, Firdaus AHM, El-Badry YA, Hussein EE, El-Bahy ZM. Determination of the tensile properties and biodegradability of cornstarch-based biopolymers plasticized with sorbitol and glycerol. Polymers 2021;13:3709.
16. Syafiq R, Sapuan SM, Zuhri MYM, Ilyas RA, Nazrin A, Sherwani SFK, et al. Antimicrobial activities of starch-based biopolymers and biocomposites incorporated with plant essential oils: a review. Polymers 2020;12: 1–26.
17. Tarique J, Sapuan SM, Khalina A. Effect of glycerol plasticizer loading on the physical, mechanical, thermal, and barrier properties of arrowroot (*Maranta arundinacea*) starch biopolymers. Sci Rep 2021;11:1–17.
18. Azeredo HMC, Rosa MF, Mattoso LHC. Nanocellulose in bio-based food packaging applications. Ind Crop Prod 2016;664–71. https://doi.org/10.1016/j.indcrop.2016.03.013.
19. Hazrol MD, Sapuan SM, Zainudin ES, Zuhri MYM, Abdul Wahab NI. Corn starch (*Zea mays*) biopolymer plastic reaction in combination with sorbitol and glycerol. Polymers 2021;13:242.
20. Syafiq R, Sapuan SM, Zuhri MRM. Antimicrobial activity, physical, mechanical and barrier properties of sugar palm based nanocellulose/starch biocomposite films incorporated with cinnamon essential oil. J Mater Res Technol 2021;11:144–57.
21. Syafiq RMO, Sapuan SM, Zuhri MRM. Effect of cinnamon essential oil on morphological, flammability and thermal properties of nanocellulose fibre–reinforced starch biopolymer composites. Nanotechnol Rev 2020;9:1147–59.
22. Serna-Saldivar SO. History of corn and wheat tortillas, 1st ed. Netherlands: Elsevier Ltd.; 2015.
23. Pruitt JD, Pruitt JD. A brief history of corn : looking back to move forward. United States: Lincoln ProQuest Dissertations Publishing; 2016.
24. Hanway J. How a corn plant develops. Spec. Publ. 48. Iowa Agric Home Econ Exp Stn Publ 1966;48:1–18.
25. Cummins DG. Quality and yield of corn plants and component parts when harvested for silage at different maturity stages 1. Agron J 1970;62:781–4.
26. Corn – Living History Farms. Iowa | learning-fields.
27. Mészáros E, Jakab E, Gáspár M, Réczey K, Várhegyi G. Thermal behavior of corn fibers and corn fiber gums prepared in fiber processing to ethanol. J Anal Appl Pyrolysis 2009;85:11–8.
28. Doner LW, Chau HK, Fishman ML, Hicks KB. An improved process for isolation of corn fiber gum. Cereal Chem 1998;75:408–11.
29. Hopkins CG, Smith LH, East EM. The structure of the corn kernel and the composition of its different parts. Seventy Gener Sel Oil Protein Maize 2015, 33–63. https://doi.org/10.2135/1974.seventygenerations.c2.
30. Ganjyal GM, Reddy N, Yang YQ, Hanna MA. Biodegradable packaging foams of starch acetate blended with corn stalk fibers. J Appl Polym Sci 2004;93:2627–33.
31. Saenghirunwattana P, Noomhorm A, Rungsardthong V. Mechanical properties of soy protein based "green" composites reinforced with surface modified cornhusk fiber. Ind Crop Prod 2014;60:144–50.
32. Li M, Cheng YL, Fu N, Li D, Adhikari B, Chen XD. Isolation and characterization of corncob cellulose fibers using microwave-assisted chemical treatments. Int J Food Eng 2014;10:427–36.

33. Sari NH, Pruncu CI, Sapuan SM, Ilyas RA, Catur AD, Suteja S, et al. The effect of water immersion and fibre content on properties of corn husk fibres reinforced thermoset polyester composite. Polym Test 2020;91: 106751.
34. Wróblewska-krepsztul J, Rydzkowski T, Borowski G. Recent progress in biodegradable polymers and nanocomposites based packaging materials for sustainable environment. International Journal of Polymer Analysis and Characterization 2018;5341:383–95.
35. Cinelli P, Chiellini E, Lawton JW, Imam SH. Properties of injection molded composites containing corn fiber and poly(vinyl alcohol). J Polym Res 2006;13:107–13.
36. Taylor PI. Relationships between selected properties of starches from different corn lines. J Food Prop 2007;8:37–41.
37. Singh GD, Bawa AS, Singh S, Saxena DC. Physicochemical, pasting, thermal and morphological characteristics of Indian water chestnut (trapa natans) starch. Starch/Staerke 2009;61:35–42.
38. Ibrahim MIJ, Sapuan SM, Zainudin ES, Zuhri MYM, Edhirej A. Processing and characterization of cornstalk/ sugar palm fiber reinforced cornstarch biopolymer hybrid composites. In: Advanced processing, properties, and applications of starch and other bio-based polymers. Netherlands: Elsevier; 2020:35–46 pp.
39. Mcaloon A, Taylor F, Yee W, Regional E, Ibsen K, Wooley R. Determining the cost of producing ethanol from corn starch and lignocellulosic feedstocks. Technical Report 2000;1–44.
40. Ibrahim MIJ, Sapuan SM, Zainudin ES, Zuhri MYM, Edhirej A, Ilyas RA. Characterization of corn fiber-filled cornstarch biopolymer composites. In: Biofiller-Reinforced Biodegradable Polymer Composites. Boca Raton, USA: CRC Press; 2020:285–301 pp.
41. Chaudhary DP, Kumar S, Langyan S. Maize: nutrition dynamics and novel uses. Maize Nutr Dyn Nov Uses 2013;9788132216:1–161.
42. Ibrahim MIJ, Sapuan SM, Zainudin ES, Zuhri MYM. Extraction, chemical composition, and characterization of potential lignocellulosic biomasses and polymers from corn plant parts. Bioresources 2019. https://doi.org/10.15376/biores.14.3.6485-6500.
43. Yilmaz ND. Effect of chemical extraction parameters on corn husk fibres characteristics. Indian J Fibre Text Res 2013;38:29–34.
44. Kambli N, Basak S, Samanta KK, Deshmukh RR. Extraction of natural cellulosic fibers from cornhusk and its physico-chemical properties. Fibers Polym 2016;17:687–94.
45. Hallauer AR. Specialty corns. CRC press; 2000 Aug 23.
46. Liu H, Yu L, Xie F, Chen L. Gelatinization of cornstarch with different amylose/amylopectin content. Carbohydr Polym 2006;65:357–63.
47. Marichelvam MK, Jawaid M, Asim M. Corn and rice starch-based bio-plastics as alternative packaging materials. Fibers 2019;7.
48. Physical Properties – Heavenly Bath.
49. Sapuan SM, Harussani MM, Syafri E. A short review of recent engineering applications of natural fibres. IOP Conf Ser Earth Environ Sci 2022;1097:012033.
50. Imraan M, Ilyas RA, Norfarhana AS, Bangar SP, Knight VF, Norrrahim MNF. Sugar palm (*Arenga pinnata*) fibers: new emerging natural fibre and its relevant properties, treatments and potential applications. J Mater Res Technol 2023;24:4551–72.
51. Radzi FSM, Suriani MJ, Bakar AA, Khalina A, Ruzaidi CM, Nik WW, et al. Effect of reinforcement of Alkaline-treated sugar palm/bamboo/kenaf and fibreglass/Kevlar with polyester hybrid biocomposites: mechanical, morphological, and water absorption properties. J Mater Res Technol 2023;24:4190–202.
52. Khalid MY, Imran R, Arif ZU, Akram N, Arshad H, Al Rashid A, et al. Developments in chemical treatments, manufacturing techniques and potential applications of natural-fibers-based biodegradable composites. Coatings 2021;11:293.
53. Bledzki AK, Mamun AA, Volk J. Physical, chemical and surface properties of wheat husk, rye husk and soft wood and their polypropylene composites. Compos Appl Sci Manuf 2010;41:480–8.

54. Jagwani D, Joshi P. Deportation of toxic phenol from aqueous system by wheat husk. Int J Plant Anim Environ Sci 2014;4:58–64.
55. Galanakis CM. Food waste recovery: processing technologies and industrial techniques. Netherlands: Elsevier; 2015.
56. Lam PS, Sokhansanj S, Bi X, Mani S, Lim CJ, Womac AR, et al. Physical characterization of wet and dry wheat straw and switchgrass – bulk and specific density. In: *2007 ASABE Annu. Int. Meet. Tech. Pap.* 2007, vol 11.
57. Jiang D, An P, Cui S, Sun S, Zhang J, Tuo T. Effect of modification methods of wheat straw fibers on water absorbency and mechanical properties of wheat straw fiber cement-based composites. Adv Mater Sci Eng 2020;2020:1–14.
58. Bouasker M, Belayachi N, Hoxha D, Al-Mukhtar M. Physical characterization of natural straw fibers as aggregates for construction materials applications. Materials 2014;7:3034–48.
59. Passoth V, Sandgren M. Biofuel production from straw hydrolysates: current achievements and perspectives. Appl Microbiol Biotechnol 2019;103:5105–16.
60. Rehman N, de Miranda MIG, Rosa SML, Pimentel DM, Nachtigall SMB, Bica CID. Cellulose and nanocellulose from maize straw: an insight on the crystal properties. J Polym Environ 2014;22:252–9.
61. Li Y, Yan F, Li T, Zhou Y, Jiang H, Qian M, et al. High-solid anaerobic digestion of corn straw for methane production and pretreatment of bio-briquette. Bioresour Technol 2018;250:741–9.
62. de Carvalho Mendes CA, Ferreira NM, Furtado CR, de Sousa AM. Isolation and characterization of nanocrystalline cellulose from corn husk. Mater Lett 2015;148:26–9.
63. Onipe OO, Beswa D, Jideani AIO. Effect of size reduction on colour, hydration and rheological properties of wheat bran. Food Sci Technol 2017;37:389–96.
64. Mayo B, De Vos WM. The proteolytic system of lactic acid bacteria. Microbiologia 1993;9:90–106.
65. Mendes CADC, Adnet FADO, Leite MCAM, Furtado CRG, Sousa AMF. De Chemical, physical, mechanical, thermal and morphological characterization of corn husk residue. Cellul Chem Technol 2014;49:727–35.
66. Liu L, Song J, Li Y, Li P, Wang H. Robust and cost-saving static solid cultivation method for lipid production using the chlamydospores of Phanerochaete chrysosporium. Biotechnol Biofuels 2019;12:1–14.
67. Matavire TO. Extraction and Modification of hemicellulose from Wheat bran to produce entrapment materials for the controlled release of chemicals and bioactive substances. Doctoral dissertation, Stellenbosch: Stellenbosch University 2018.
68. Pandecha K, Pongtornkulpanich A, Sukchai S. Thermal properties of corn husk fiber as insulation for flat plate solar collector. J. Renew. Energy Smart Grid Technol 2015;10:27–36.
69. Norashikin MZ, Ibrahim MZ. The potential of natural waste (corn husk) for production of environmental friendly biodegradable film for seedling. World Acad. Sci. Eng. Technol. 2009;58:176–80.
70. Table 1 | Physical and Acoustical Properties of Corn Husk Fiber Panels.
71. Nordin R, Ismail H. Properties of rice husk powder/natural rubber composite. In: Solid State Science and Technology, 2nd ed. Malaysia: Solid State Science and Technology; 2007, 15:83–91 pp.
72. Faisal Bukhori MSD. Effect of rice husk waste and rice husk ash composition as filler in plastic bottle drink waste composites on water absorption properties. Int J Sci Res 2015;4:2146–8.
73. Sietske Boschma D, Kees I, Kwant W. Netherlands programmes sustainable biomass. NL Agency Minist Econ Aff 2013:6–30.
74. Kapoor M, Panwar D, Kaira GS. Bioprocesses for enzyme production using agro-industrial wastes: technical challenges and commercialization potential. Netherlands: Elsevier Inc.; 2016.
75. Sugawara M, Suzuki T, Totsuka A, Takeuchi M, Ueki K. Composition of corn hull dietary fiber. Starch/Stärke 1994;46:335–7.
76. Pointner M, Kuttner P, Obrlik T, Jäger A, Kahr H. Composition of corncobs as a substrate for fermentation of biofuels. Agron Res 2014;12:391–6.
77. Fan X, Li M, Zhang J, Tang P, Yuan Q. Optimization of SO2-catalyzed hydrolysis of corncob for xylose and xylitol production. J Chem Technol Biotechnol 2014;89:1720–6.

78. Monette DL. Coating removal techniques in the aerospace industry. Corros Control Aerosp Ind 2009: 225–47. https://doi.org/10.1533/9781845695538.3.225.
79. Composites PH. A review on potentiality of nano filler/natural fiber filled polymer hybrid composites. Polymers 2014;6:2247–73.
80. Daud Z, Hatta MZM, Kassim ASM, Awang H, Aripin AM. Exploring of agro waste (pineapple leaf, corn stalk, and napier grass) by chemical composition and morphological study. Bioresources 2014;9:872–80.
81. Sun YG, Ma YL, Wang LQ, Wang FZ, Wu QQ, Pan GY. Physicochemical properties of corn stalk after treatment using steamexplosion coupled with acid or alkali. Carbohydr Polym 2015;117:486–93.
82. Zhang Y, Ghaly AE, Li B. Physical properties of corn residues. Am J Biochem Biotechnol 2012;8:44–53.
83. Zhang Y, Ghaly AE, Li B. Availability and physical properties of residues from major agricultural crops for energy conversion through thermochemical processes. Am J Agric Biol Sci 2012;7:312–21.
84. Saba N, Safwan A, Sanyang ML, Mohammad F, Pervaiz M, Jawaid M, et al. Thermal and dynamic mechanical properties of cellulose nanofibers reinforced epoxy composites. Int J Biol Macromol 2017;102: 822–8.
85. Gómez-Aldapa C, Velazquez G, Gutierrez MC, Rangel-Vargas E, Castro-Rosas J, Aguirre-Loredo RY. Effect of polyvinyl alcohol on the physicochemical properties of biodegradable starch films. Materials Chemistry and Physics 2020;239:1–7.
86. Wang B, Yu B, Yuan C, Guo L, Liu P, Gao W, et al. An overview on plasticized biodegradable corn starch-based films: the physicochemical properties and gelatinization process. Crit Rev Food Sci Nutr 2021;0:1–11.
87. Mohammadi Nafchi A, Moradpour M, Saeidi M, Alias AK. Thermoplastic starches: properties, challenges, and prospects. Starch/Stärke 2013;65:61–72.
88. Liu L, Yu Y. Improved mechanical and thermal properties of polypropylene blends based on diethanolamine-plasticized corn starch via in situ reactive compatibilization. Washington D.C.: ACS Publication; 2013.
89. Muensri P, Kunanopparat T, Menut P, Siriwattanayotin S. Effect of lignin removal on the properties of coconut coir fiber/wheat gluten biocomposite. Compos A Appl Sci Manuf 2011;42:173–9.
90. Sirviö JA, Visanko M, Ukkola J, Liimatainen H. Effect of plasticizers on the mechanical and thermomechanical properties of cellulose-based biocomposite films. Ind Crop Prod 2018;122:513–21.
91. Fishman ML, Coffin DR, Konstance RP, Onwulata CI. Extrusion of pectin/starch blends plasticized with glycerol. Carbohydr Polym 2000;41:317–25.
92. Chang PR, Jian R, Zheng P, Yu J, Ma X. Preparation and properties of glycerol plasticized-starch (GPS)/ cellulose nanoparticle (CN) composites. Carbohydr Polym 2010;79:301–5.
93. Ibrahim MIJ, Sapuan SM, Zainudin ES, Zuhri MYM. Preparation and characterization of cornhusk/sugar palm fiber reinforced Cornstarch-based hybrid composites. J Mater Res Technol 2020;9:200–11.
94. Ibrahim MIJ, Sapuan SM, Zainudin ES, Zuhri MYM. Potential of using multiscale corn husk fiber as reinforcing filler in cornstarch-based biocomposites. Int J Biol Macromol 2019;139:596–604.
95. Qiao X, Tang Z, Sun K. Plasticization of corn starch by polyol mixtures. Carbohydr Polym 2011;83:659–64.
96. Gutiérrez TJ, Alvarez VA. Properties of native and oxidized corn starch/polystyrene blends under conditions of reactive extrusion using zinc octanoate as a catalyst. React Funct Polym 2017;112:33–44.
97. Pose S, Kirby AR, Paniagua C, Waldron KW, Morris VJ, Quesada MA, Mercado JA, et al. The nanostructural characterization of strawberry pectins in pectate lyase or polygalacturonase silenced fruits elucidates their role in softening. Carbohydr Polym 2015;137:452–8.
98. Wang Y, Zhang L, Liu H, Yu L, Simon GP, Zhang N, et al. Relationship between morphologies and mechanical properties of hydroxypropyl methylcellulose/hydroxypropyl starch blends. Carbohydr Polym 2015;132:134–45.
99. Tian AH, Wang K, Liu D, Yan J, Rajulu AV. Enhanced mechanical and thermal properties of poly (vinyl alcohol)/corn starch blends by nanoclay intercalation. Int J Biol Macromol 2017;101:314–20.
100. Dai R, Woo MW, Chen H, Dang X, Mansouri S, Shan Z. Hydrogel beads based on oxidized corn starch cross-linked with gelatin for tartrazine adsorption from aqueous environments. Nat. Publ. Gr. 2017;49:549–55.

101. Gürler N, Paşa S, Temel H. Silane doped biodegradable starch-PLA bilayer films for food packaging applications: mechanical, thermal, barrier and biodegradability properties. J Taiwan Inst Chem Eng 2021; 000:1–11.
102. Chen X, Cui F, Zi H, Zhou Y, Liu H, Xiao J. Development and characterization of a hydroxypropyl starch/zein bilayer edible film. Int J Biol Macromol 2019;141:1175–82.
103. Zuo G, Song X, Chen F, Shen Z. Physical and structural characterization of edible bilayer films made with zein and corn-wheat starch. J. Saudi Soc. Agric. Sci. 2019;18:324–31.
104. Diaz-Pedraza A, Piñeros-Castro Y, Ortega-Toro R. Bi-layer materials based on thermoplastic corn starch, polylactic acid and modified polypropylene. Rev Mex Ing Quim 2020;19:323–31.
105. Chen F, Chen C, Du Y, Zuo G, Chen F, Zhang L. Effect of storage condition on the physico-chemical properties of corn – wheat starch/zein edible bilayer films. R Soc Open Sci 2020;7:191777.
106. Muller J, González C, Chiralt A. Poly (lactic) acid (PLA) and starch bilayer films, containing cinnamaldehyde, obtained by compression moulding. Eur Polym J 2017;95:56–70.
107. Zhou X, Yang R, Wang B, Chen K. Development and characterization of bilayer films based on pea starch/ polylactic acid and use in the cherry tomatoes packaging. Carbohydr Polym 2019;222:114912.
108. Sun H, Shao X, Jiang R, Shen Z, Ma Z. Mechanical and barrier properties of corn distarch phosphate-zein bilayer films by thermocompression. Int J Biol Macromol 2018. https://doi.org/10.1016/j.ijbiomac.2018.07.069.
109. Sanyang ML, Sapuan SM, Jawaid M, Ishak MR, Sahari J. Development and characterization of sugar palm starch and poly(lactic acid) bilayer films. Carbohydr Polym 2016;146:36–45.
110. Asyraf MRM, Ishak MR, Syamsir A, Nurazzi NM, Sabaruddin FA, Shazleen SS, et al. Mechanical properties of oil palm fibre-reinforced polymer composites: a review. J Mater Res Technol 2022;17:33–65.
111. Asrofi M, Sapuan SM, Ilyas RA, Ramesh M. Characteristic of composite bioplastics from tapioca starch and sugarcane bagasse fiber: effect of time duration of ultrasonication (Bath-Type). Mater Today Proc 2020. https://doi.org/10.1016/j.matpr.2020.07.254.
112. Nurazzi NM, Khalina A, Chandrasekar M, Aisyah HA, Rafiqah SA, Ilyas RA, et al. Effect of fiber orientation and fiber loading on the mechanical and thermal properties of sugar palm yarn fiber reinforced unsaturated polyester resin composites. Polimery 2020;65:115–24.
113. Azman MA, Asyraf MRM, Khalina A, Petrů M, Ruzaidi CM, Sapuan SM, et al. Natural fiber reinforced composite material for product design: a short review. Polymers 2021;13:1917.
114. Mansour G, Zoumaki M, Marinopoulou A, Raphaelides SN, Tzetzis D, Zoumakis N. Investigation on the effects of glycerol and clay contents on the structure and mechanical properties of maize starch nanocomposite films. Starch-Stärke 2020;72:1–37.
115. Mansoori E, Reza A, Esfahani S. Preparation and characterization of corn starch/soy protein biocomposite film reinforced with graphene and graphene oxide nanoplatelets. Polym Adv Technol 2019;30:1–12.
116. Khalil HA, Yap SW, Tye YY, Tahir PM, Rizal S, Fazita MN. Effects of corn starch and *Kappaphycus alvarezii* seaweed blend concentration on the optical, mechanical, and water vapor barrier properties of composite films. Bioresources 2018;13:1157–73.
117. Ren L, Yan X, Zhou J, Tong J, Su X. Influence of chitosan concentration on mechanical and barrier properties of corn starch/chitosan films. Int J Biol Macromol 2017;105:1636–43.
118. Wang K, Wang W, Ye R, Liu A, Xiao J, Liu Y, Zhao Y. Mechanical properties and solubility in water of corn Starch-collagen composite films: effect of starch type and concentrations. Netherlands: Elsivier; 2016.
119. Fabra MJ, Martínez-Sanz M, Gómez-Mascaraque LG, Gavara R, López-Rubio A. Structural and physicochemical characterization of thermoplastic corn starch films containing microalgae. Carbohydr Polym 2018;186:184–91.
120. Patil S, Bharimalla AK, Nadanathangam V, Dhakane-Lad J, Mahapatra A, Jagajanantha P, et al. Nanocellulose reinforced corn starch-based biocomposite films: composite optimization, characterization and storage studies. Food Packag Shelf Life 2022;33:100860.

121. Ilyas RA, Sapuan SM, Atiqah A, Ibrahim R, Abral H, Ishak MR, et al. Sugar palm (*Arenga pinnata* [Wurmb.] Merr) starch films containing sugar palm nanofibrillated cellulose as reinforcement: water barrier properties. Polym Compos 2020;459–67. https://doi.org/10.1002/pc.25379.
122. Ilyas RA, Sapuan SM, Ibrahim R, Abral H, Ishak MR, Zainudin ES, et al. Sugar palm (*Arenga pinnata* (Wurmb.) Merr) cellulosic fibre hierarchy: a comprehensive approach from macro to nano scale. J Mater Res Technol 2019;8:3872–7.
123. Ilyas RA, Sapuan SM, Ibrahim R, Abral H, Ishak MR, Zainudin ES, et al. Thermal, biodegradability and water barrier properties of bio-nanocomposites based on plasticised sugar palm starch and nanofibrillated celluloses from sugar palm fibres. J Biobased Mater Bioenergy 2020;14:234–48. https://doi.org/10.1166/jbmb.2020.1951.
124. Sabaruddin FA, Paridah MT, Sapuan SM, Ilyas RA, Lee SH, Abdan K, et al. The effects of unbleached and bleached nanocellulose on the thermal and flammability of polypropylene-reinforced kenaf core hybrid polymer bionanocomposites. Polymers 2020;13:116.
125. Hazrol MD, Sapuan SM, Ilyas RA, Othman ML, Sherwani SFK. Electrical properties of sugar palm nanocrystalline cellulose reinforced sugar palm starch nanocomposites. Polimery 2020;65:363–70.
126. Chen W, Chen F, Zhang G, Liu X, Kong S, Cai W, et al. Fabrication of cellulose nanocrystal composite filter papers for rapid and highly efficient removal of bacteria from aqueous solutions. Cellulose 2019;26: 7027–35.
127. Slavutsky AM, Bertuzzi MA. Water barrier properties of starch films reinforced with cellulose nanocrystals obtained from sugarcane bagasse. Carbohydr Polym 2014;110:53–61.
128. Norrrahim MNF, Knight VF, Nurazzi NM, Jenol MA, Misenan MSM, Janudin N, et al. The frontiers of functionalized nanocellulose-based composites and their application as chemical sensors. Polymers 2022; 14:4461.
129. Norrrahim MNF, Yasim-Anuar TAT, Sapuan SM, Ilyas RA, Hakimi MI, Najmuddin SUFS, Jenol MA. Nanocellulose reinforced polypropylene and polyethylene composite for packaging application. In: Bio-based packaging. material, environmental and economic aspects. United States: Wiley; 2021.
130. Nazrin A, Sapuan SM, Zuhri MYM, Tawakkal ISMA, Ilyas RA. Flammability and physical stability of sugar palm crystalline nanocellulose reinforced thermoplastic sugar palm starch/poly(lactic acid) blend bionanocomposites. Nanotechnol Rev 2021;11:86–95.
131. Omran AAB, Mohammed AABA, Sapuan SM, Ilyas RA, Asyraf MRM, Koloor SSR, et al. Micro- and nanocellulose in polymer composite materials: a review. Polymers (Basel) 2021;13:231.
132. Norfarhana AS, Ilyas RA, Ngadi N. A review of nanocellulose adsorptive membrane as multifunctional wastewater treatment. Carbohydr Polym 2022;291:119563.
133. Syafiq RMO, Sapuan SM, Zuhri MYM, Othman SH, Ilyas RA. Effect of plasticizers on the properties of sugar palm nanocellulose/cinnamon essential oil reinforced starch bionanocomposite films. Nanotechnol Rev 2022;11:423–37.
134. Almasi H, Ghanbarzadeh B, Entezami AA. Physicochemical properties of starch–CMC–nanoclay biodegradable films. Int J Biol Macromol 2010;46:1–5.
135. Dean KM, Do MD, Petinakis E, Yu L. Key interactions in biodegradable thermoplastic starch/poly(vinyl alcohol)/montmorillonite micro- and nanocomposites. Compos Sci Technol 2008;68:1453–62.
136. Abitbol T, Rivkin A, Cao Y, Nevo Y, Abraham E, Ben-shalom T, et al. ScienceDirect Nanocellulose, a tiny fiber with huge applications. Curr Opin Biotechnol 2016;39:76–88.
137. Bungay HR. Confessions of a bioenergy advocate. Trends Biotechnol 2004;22:67–71.
138. Boufi S, González I, Delgado-Aguilar M, Tarrès Q, Pèlach MÀ, Mutjé P. Nanofibrillated cellulose as an additive in papermaking process: a review. Carbohydr Polym 2016;154:151–66.
139. Grüneberger F, Künniger T, Zimmermann T, Arnold M. Rheology of nanofibrillated cellulose/acrylate systems for coating applications. Cellulose 2014;21:1313–26.
140. Missoum K, Martoïa F, Belgacem MN, Bras J. Effect of chemically modified nanofibrillated cellulose addition on the properties of fiber-based materials. Ind Crop Prod 2013;48:98–105.

141. Gupta P, Singh B, Agrawal AK, Maji PK. Low density and high strength nanofibrillated cellulose aerogel for thermal insulation application. Mater Des 2018;158:224–36.
142. Wang X, Cheng F, Liu J, Smått JH, Gepperth D, Lastusaari M, et al. Biocomposites of copper-containing mesoporous bioactive glass and nanofibrillated cellulose: biocompatibility and angiogenic promotion in chronic wound healing application. Acta Biomater 2016;46:286–98.
143. Asyraf MRM, Rafidah M, Ishak MR, Sapuan SM, Ilyas RA, Razman MR, et al. Integration of TRIZ, Morphological Chart and ANP method for development of FRP composite portable fire extinguisher. Polym Compos 2020:1–6. https://doi.org/10.1002/pc.25587.
144. Brown EE, Hu D, Abu Lail N, Zhang X. Potential of nanocrystalline cellulose-fibrin nanocomposites for artificial vascular graft applications. Biomacromolecules 2013;14:1063–71.
145. George J, Sabapathi SN. Cellulose nanocrystals: synthesis, functional properties, and applications. Nanotechnol Sci Appl 2015;8:45–54.
146. Jawaid M, Boufi S, Abdul Khalil HPS. Cellulose-reinforced nanofibre composites: Production, properties and applications. United Kingdom: Woodhead Publishing; 2017.
147. Yang S, Tang Y, Wang J, Kong F, Zhang J. Surface treatment of cellulosic paper with starch-based composites reinforced with nanocrystalline cellulose. Ind Eng Chem Res 2014;53:13980–8.
148. Lalia BS, Samad YA, Hashaikeh R. Nanocrystalline cellulose-reinforced composite mats for lithium-ion batteries: electrochemical and thermomechanical performance. J Solid State Electrochem 2013;17:575–81.
149. Lynch M. Decorative skin and hair cosmetics containing microcrystalline cellulose as enhancing agent. 2004;1.

Abu Hassan Nordin, Rushdan Ahmad Ilyas*, Norzita Ngadi, Nurul Huda Baharuddin, Muhammad Luqman Nordin and Mohammad Saifulddin Mohd Azami

14 Horse chestnut thermoplastic starch nanocomposite films reinforced with nanocellulose

Abstract: The starch-based materials such as thermoplastic starch film are a promising alternative to non-renewable petroleum-based plastics. The development of an alternative conventional plastic from bio-based materials has gained great interest following its biodegradable, non-hazardous and renewable advantages. Following that, horse chestnut is an exciting source of starch for producing thermoplastic starch film. Nonetheless, the thermoplastic starch film is weak in strength and easily affected by water due to its highly hydrophilic property, thus limiting its practicability. In this regard, the additions of nanocellulose into thermoplastic starch have shown drastic improvement in its mechanical properties and water permeability of the film. This chapter discusses the potential of nanocellulose reinforced plasticized starch from horse chestnut as a replacement for petroleum-based plastic in packaging applications.

Keywords: horse chestnut, nanocellulose, nanocomposites, packaging, thermoplastic starch

***Corresponding author: Rushdan Ahmad Ilyas**, Faculty of Chemical and Energy Engineering, Universiti Teknologi Malaysia (UTM), Skudai 81310, Malaysia; Centre for Advanced Composite Materials (CACM), Universiti Teknologi Malaysia (UTM), Skudai 81310, Malaysia; Institute of Tropical Forest and Forest Products (INTROP), Universiti Putra Malaysia, 43400 UPM, Serdang, Selangor, Malaysia; and Centre of Excellence for Biomass Utilization, Universiti Malaysia Perlis, 02600, Arau, Perlis, Malaysia, E-mail: ahmadilyas@utm.my. https://orcid.org/0000-0001-6622-2632

Abu Hassan Nordin, Faculty of Chemical and Energy Engineering, Universiti Teknologi Malaysia (UTM), Skudai 81310, Malaysia; and Faculty of Applied Sciences, Universiti Teknologi MARA (UiTM), 02600 Arau, Perlis, Malaysia, E-mail: abuhassannordin@gmail.com. https://orcid.org/0000-0001-7546-3833

Norzita Ngadi and Nurul Huda Baharuddin, Faculty of Chemical and Energy Engineering, Universiti Teknologi Malaysia (UTM), Skudai 81310, Malaysia, E-mail: norzita@cheme.utm.my (N. Ngadi), nurul.huda@utm.my (N.H. Baharuddin)

Muhammad Luqman Nordin, Department of Clinical Studies, Faculty of Veterinary Medicine, Universiti Malaysia Kelantan, Pengkalan Chepa, 16100 Kota Bharu, Kelantan, Malaysia, E-mail: luqman.n@umk.edu.my

Mohammad Saifulddin Mohd Azami, Faculty of Applied Sciences, Universiti Teknologi MARA (UiTM), 02600 Arau, Perlis, Malaysia, E-mail: saifulddin@uitm.edu.my

As per De Gruyter's policy this article has previously been published in the journal Physical Sciences Reviews. Please cite as: A. H. Nordin, R. A. Ilyas, N. Ngadi, N. H. Baharuddin, M. L. Nordin and M. S. Mohd Azami "Horse chestnut thermoplastic starch nanocomposite films reinforced with nanocellulose" *Physical Sciences Reviews* [Online] 2023. DOI: 10.1515/psr-2022-0028 | https://doi.org/10.1515/9783110773606-014

14.1 Introduction

The growing volume of plastic trash in landfills and oceans, as well as the strong reliance on fossil petroleum resources for plastic manufacture, has inspired research on bio-based plastic replacements. Starch, a natural renewable polysaccharide produced from a wide range of crops, is one of the most studied and promising raw materials to create biodegradable plastics. Thermoplastic starch (TPS) is a relatively new substance for use as biodegradable plastic, and it is now one of the most researched polymers in this sector [1]. TPS is made by breaking the crystal structure of starch with a plasticizer at high temperatures, both with and without shear stress [2]. There have been multiple examples of starch made for TPS from edible fruits, vegetables, seeds, and other agricultural products [3]. The use of starch extracted from edible sources as a precursor to plastics is controversial, especially in developing nations with limited cultivable land and food shortages. As a result, the usage of starch derived from non-edible materials has gotten a lot of attention [4]. There have been numerous non-edible sources of starch discovered, including the horse chestnut (*Aesculus hippocastanum* L.), which contains more than 35% starch by weight [5]. Many countries, including England, Chile, India and others, have horse chestnut available locally [2, 6]. Horse chestnut seeds are commonly used for decoration, yet the vast majority of them end up in landfills. As a result, TPS derived from horse chestnut starch offers benefits and may be mass-produced in enormous quantities.

Nonetheless, compared to most conventional plastics, TPS has two major drawbacks: it is primarily water-soluble and has low mechanical characteristics. In this regard, natural fibres have been used to strengthen TPS, which has resulted in a significant improvement in mechanical characteristics and water resistance. The fiber described in this study is nanocellulose. Nanocellulose possesses high crystallinity, surface area and mechanical properties [7], thus, incorporating nanocellulose into starch matrix can be an effective way to obtain TPS with higher mechanical properties [8, 9].

14.1.1 Types of packaging material

The original packaging materials were leaves, hollowed-out tree branches, roots, skins, reed baskets and pottery pots [10]. As society progressed, more complex containers were developed to meet specific needs. Plastics, paper, glass and metal are the most often used packaging materials (as shown in Figure 14.1).

Paper is the most commonly used packaging material due to its versatility, low cost, ability to be printed and strong mechanical qualities. Despite this, water and moisture absorption have caused the paper to degrade [11]. Interestingly combining the paper components with a biopolymer covering layer increased the barrier qualities of the paper significantly [12]. Polymers may come in a variety of shapes and sizes, including films. According to studies, nanofiller improved the mechanical and barrier properties of starch papers [13].

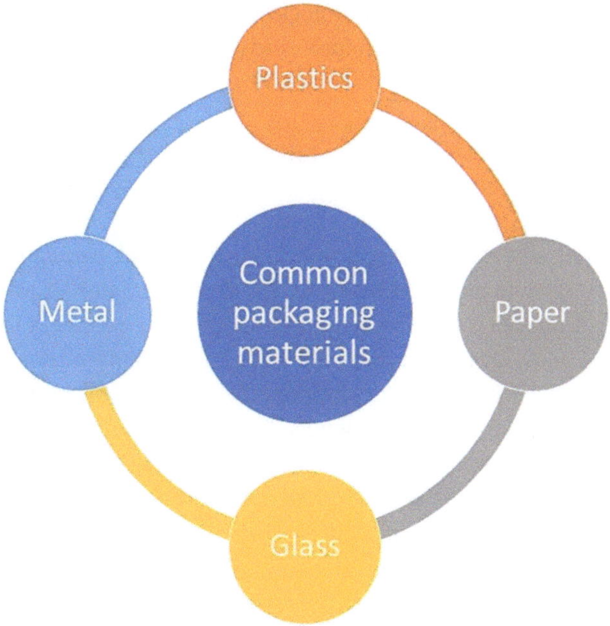

Figure 14.1: Different type of packaging materials.

Around 3000 BC, glass was used as a packaging material for the first time [14]. Glassware was made using the blowing process, and transparent glassware was invented after opaque glass bowls and cups were made using the sand-molding technique. As a food packing material, glass has both benefits and drawbacks. It is vulnerable to physical impact and high pressure damage, yet it has good gas and chemical barrier qualities and is heat processable at higher temperatures. Glass is also rigid and acts as a good insulator in food preservation systems. The majority of glass used in food and beverage storage is both reusable and recyclable.

Meanwhile, metal may protect food products from physical damage or impact, making handling, storing, and transporting them easier. It can also act as moisture, oxygen and other gases and liquids impermeable barrier [15]. On the other hand, metal is unsuitable for packing in a changed environment. Gases and vapour can pass through packing materials by molecular diffusion, macroscopic or microscopic pores, or both.

Although glass and metals create a nearly impenetrable barrier to chemical and other environmental agents, few packages are entirely made of glass or metal because closure devices are included to make filling and emptying easier. Furthermore, the use of the previously stated non-biodegradable and non-renewable materials (plastics, glass and metals) in packaging has raised environmental contamination issues. Every year, a large volume of packaging waste is created for the purpose of use and disposal [16]. Conventional treatment methods for post-consumer plastic waste include burning and landfilling, which are hazardous to human health and the environment [17].

As a result, there is a growing need for items that are ecologically friendly, safe and non-toxic. Also, following the disadvantages of fossil fuels such as risk and resource restrictions, have prompted the economy to turn to alternative green resources.

14.1.2 Bio-based plastic

Plastics are employed in practically every facet of our lives, from food packaging to medical and networking systems to technical uses [18]. For decades, materials scientists have been attempting to develop sustainable and biodegradable polymers to replace or augment traditional synthetic plastics made entirely of petroleum feedstock. Three categories of bioplastics exist: materials predominantly composed of natural polymers, microbially produced polymers, and chemically produced biomaterials generated from renewable source components [19]. Agro-polymers, such as starch, cellulose, pectins and hemicellulose, are natural polymers that combine with proteins (casein, zein, gluten and gelatin) to form semi-rigid three-dimensional polymetric networks in bioplastics.

Natural polymers like polysaccharides (starch, cellulose, pectins and hemicellulose) and proteins (casein, zein, gluten and gelatin) are used to create semi-rigid three-dimensional polymeric networks in bioplastics manufactured from agro-polymers [20]. Instead of synthetic plastics (non-biodegradable and oil-based), bioplastics provide materials from renewable and/or biodegradable sources, raw material availability and a viable solution for the disposal of solid biomass waste. Bioplastics also minimise hazardous gas emissions. A 50% reduction in GHG emissions by 2050 is necessary to prevent a 2 °C rise in global temperature when it comes to environmental concerns like the greenhouse effect, which the Intergovernmental Panel on Climate Change (IPCC) says is a growing global problem. One strategy for reducing global warming is the use of biomaterials like bioplastics and biofuels [21–23].

The disadvantages of traditional plastics, such as sustainability and biodegradability, may be resolved by bio-based polymers. However, due to poor efficiency and variable feedstock qualities, the widespread manufacture of bio-based polymers to replace conventional plastic materials is still difficult (such as location and the time of harvest, high production cost and lack of infrastructure). In the last decade, there has been significant progress in the use of natural building blocks such proteins, lipids, carbohydrates, lignin and other bio-based building blocks derived from agricultural feedstock and agricultural industrial wastes [24].

14.1.3 Biopolymer as packaging materials

Biopolymer manufactured from polysaccharides, proteins and lipids or mixtures of those components are replacing petroleum-based polymers increasingly often as packaging materials [12]. A multistep technique is used to develop bio-based packaging materials. For instance, intermolecular connections are disrupted, followed by the synthesis of a new molecular arrangement and newly generated linkages form a 3D polymeric network [25]. According to Asgher et al. [26], the processing conditions and polymer shape influence the formation of new molecular connections (viz. length, width and ratio). The production of biobased plastic materials has frequently been described using two primary processing methodologies: wet processing and dry processing. Wet processing is concerned with dissolution and the type of solvent used, as well as the pH of the solvent, which can change the polymer conformation. In dry processing, TP characteristics of polymers are addressed, and thermos-mechanical treatment causes the sulfhydryl/disulphide conversion reaction to be initiated. But in wet processing, dissolution, the kind of solvent used, as well as the solvent's pH, which has the potential to alter the conformation of the polymer, are all important considerations [27].

14.1.4 Starch

Starch is made up of two basic components: amylose and amylopectin. The proportions of the two polysaccharides differ depending on the starch source. Lipids and proteins are also found in little concentrations in the starch [28]. Amylose is a linear polysaccharide made up of 1,4 linked D-glucose units. Several plant species have different chain lengths, ranging between 102 and 104 glucose units. In starch (cereal endosperm), the relative quantities of amylose and amylopectin range from 18 to 28 percent and 72–82%, respectively [29]. Starch is semicrystalline, with a crystallinity of 20–40% [30]. The amorphous portion of starch is made up of branching chains of amylose and amylopectin [31].

In comparison to most synthetic plastics, starch is inexpensive and readily available. Starch has also been extensively studied in developing other products such as detergent and insecticide pouches, flushable liners and bags and medical delivery systems and devices following its advantages of being sustainable, green and cost-effective [32]. Starch may be converted to TPS using high temperatures and strong shearing [33]. Moreover, the thermoplastic method relies heavily on water and additional plasticizers. Plasticizers can establish hydrogen bonds with starch molecules, removing the strong interactions between the hydroxyl groups and making starch thermoplastic [34]. This is due to the fact that starch is a polymer with three hydroxyl groups per monomer. Intermolecular and intramolecular hydrogen bonding abound in starch. The typical hydrogen bonds between hydroxyl groups of starch molecules are disrupted when plasticizers create hydrogen bonds with them, causing the starch to show plasticization [35].

14.1.5 Horse chestnut starch

Horse chestnuts have a high starch content (above 35 wt%) and can be used to substitute for other conventional starch sources such as potato, corn, and rice. Previously, horse chestnuts were reported to be used in a variety of food applications, including encapsulating agents, stabilizers, binder, thickeners, and gelling agents [36]. Ahmad and Gani [37] claimed the usage of chestnut starch nanoparticles to protect resveratrol in a nanoencapsulation system from external environmental factors. Resveratrol is more thermally stable after being encapsulated in the starch nanoparticles, based on the higher thermal transition temperatures. Moreover, Hymavathi et al. [38], investigated its usage as a binder for more water retention within the noodles. The results found that the cooking time of noodles developed from composite starch was higher (15–18 min) when compared to the conventional noodles (9.3 min).

14.1.6 Extraction and preparation of horse chestnut starch

The alkaline steeping method was used to extract starch from horse chestnuts, as described by Rafiq et al., [36]. The kernels of dried horse chestnut seeds were chopped into 2 cm^2 pieces, steeped in 0.25% NaOH solution (w/v) in a 1:3 ratio, and stored at 4 °C for 24 h. The soaked kernels and alkali were pulverized in a laboratory grinder, passed through a 100 mesh sieve, and allowed to settle before being washed twice with distilled water. The slurry was centrifuged at 3000 rpm for 15 min after being filtered through a 300 mesh screen. After centrifugation, the aqueous phase was discarded. The non-white layer on top was scraped off, and the white starch layer was resuspended in distilled water and centrifuged 2–3 times. After that, the starch was collected and dried in a hot air oven at 40 °C.

14.1.7 Properties of horse chestnut starch

Previous research had investigated into the physicochemical, pasting, rheological, thermal and morphological properties of starch derived from horse chestnut [36]. The level of amylose was discovered to be 26.10%. The RVA profile revealed a peak viscosity of 4110 cP. Texture profile analyzer was used to measure hardness, cohesiveness, adhesiveness and gumminess. The particle size distribution profile revealed a typical Uni modal size distribution profile with particle sizes ranging from 7.52 to 27.44 m. Starch granules came in a variety of shapes and sizes, including round, irregular, oval, and elliptical with a smooth surface. Horse chestnut starch exhibited a typical C-type pattern with peaks at 5.7, 15.0, 17.3 and 22.3°, according to X-ray diffraction. The temperatures of transition (To, Tp and Tc) and the enthalpy of gelatinization (H) were 53.35, 58.81 and 63.57 °C and 8.76 J/g, respectively. The storage modulus (G/), loss modulus

(G/), and loss factor (tan) at various temperatures were used to determine the rheological characteristics. Peak G/, peak G/ and peak tan were discovered to be 10,400 Pa, 1710 Pa and 0.164, respectively.

14.1.8 Preparation of thermoplastic from horse chestnut

The preparation of thermoplastic from horse chestnut starch was reported by CASTAñO et al. [2]. The authors state that the plasticizers were combined with the starch prior to melt-blending, and that the mixture was then blended for 15 min at 120 °C and 60 rpm. Then, for the creation of TPS, glycerol and malic acid were added to the mixture.

14.1.9 Advantage and disadvantage of biopolymer for packaging

The market for biopolymers is now expanding year over year. One of the key benefits of biopolymers is that they degrade via biodegradation processes into simple molecules already existing in the environment, such as CO2, water or methane [39].

Nevertheless, there are a number of restrictions and technological difficulties related to bioplastics. For instance, if they are not separated at the source, the intermediate products produced during the production of bioplastics might contaminate the recycling stream of conventional plastics [40]. Furthermore, the majority of the bioplastics manufactured today favour first-generation feedstock, such as sugarcane, maize, potato, wheat and cassava. It should be emphasised that the use of first-generation feedstocks, particularly starch-based materials, for creating biofuels, biochemicals and biomaterials is harshly condemned since it competes with the food supply, arable lands, agricultural energy and cost investments [41]. In order to solve this issue, the extraction of natural starch from inedible feedstocks as horse chesnut is proposed.

14.1.10 Lignocellulosic fiber

For increasing the mechanical qualities of bioplastics, natural polymers derived from lignocellulose have been employed [42–44], following their biodegradability and reinforcing capabilities. Lignocellulosic fibres are the world's most abundant source of renewable bioresources. In general, lignocellulosic materials contain 35–55% cellulose, 10–25% lignin, and 20–40% hemicellulose, as well as extracts (such as pectin, resins, waxes and so on), ash, and minerals [45, 46]. Lignocellulosic byproducts from plant biomass are a resource for producing and improving bioplastic properties, such as mechanical properties. Compared to individual polymers, bioplastics made from mixes of multiple polymer types can have better characteristics [47].

14.1.11 Cellulose fiber

Most natural cellulosic fibres include 60–70% cellulose, which is principally made up of C, H, and O_2 and has the basic formula $C_6H_{10}O_5$ [48]. Natural cellulosic fibres are gaining popularity due to their environmental benefits, low cost of production, and higher specific strength/stiffness than their synthetic counterparts. In addition, the procurement procedure for these composites is healthier and provides better working conditions for large-scale industrial applications [49–51]. Natural fibres provide a lot of advantages over synthetic fibres, but they also have a few drawbacks [49, 52]. In some industrial applications, these fibres have disadvantages such as water/moisture absorption sensitivity, polar and hydrophilic nature and poor heat stability [48, 53, 54]. When the advantages of natural cellulosic fibres are evaluated against the disadvantages, the benefits of natural cellulosic fibres outnumber the disadvantages [55].

The addition of cellulose to TPS can improve its water vapor permeability (WVP) properties. For example, Gaspar et al. [56], pointed out that cellulose bioplastics possess lower WVP than pure TPS. Furthermore, literature shows that adding cellulose nanowhiskers to biomaterials improves their mechanical properties [57–60]. Chemical similarity and intermolecular hydrogen bonding between hydroxyl groups of macromolecules enhance interaction between polysaccharides and cellulose as reinforcement [61]. It is important to consider the amount of cellulose utilised in the production of cellulose-based bioplastics, as too much can cause cellulose granule aggregation in the polymeric matrix, therefore, lowering mechanical strength and barrier qualities due to a decrease in hydrogen bond optimization [62, 63].

14.1.12 Microcrystalline cellulose

MCC is a naturally occurring chemical made from partially depolymerized and purified cellulose. It is traditionally made by processing alpha cellulose with a high concentration of mineral acids. At the nanoscale scale, the microfibrils that make up alpha cellulose are made up of paracrystalline and crystalline phases. The paracrystalline zone is made up of an amorphous mass of cellulose chains, whilst the crystalline portions are made up of tightly packed microcrystals arranged in a rigid linear pattern. The crystalline regions are known as cellulose crystallites, and they are generated through van der Waals interactions and hydrogen bonding between cellulose strands. The diameter of these crystallites is comparable to the cellulose microfibrils' diameter [64].

MCC bio-composites are now gaining traction as a viable alternative to traditional polymeric composites [65–67]. MCC has a higher thermal stability than natural fibre, which has led to its acceptance in bio-composites manufacturing. Engineered bio-composites with MCC reinforcement are now being developed to fulfil a wide range of applications for environmentally responsible and cost-effective commodity products that are high-performing, durable, and totally sustainable. The stability of cellulose microcrystals in aqueous solutions is another advantage of MCC over lignocellulosic fillers [68].

14.1.13 Nanocellulose fiber

Nanofibers are useful as filler for composite materials when they demonstrate reinforcing capabilities [69]. Cellulose fibres with one nano/micro scale dimension are appealing as reinforcing fillers in polymeric matrices because of their good mechanical properties, such as high bending strength and stiffness (e.g., Young's modulus of around 150 GPa) [70]. Nanoparticles have a bigger surface area and, as a result, more surface atoms than their microscale counterparts due to their smaller size. Materials with varying electronic characteristics in the nanoscale region might have different optical, catalytic and reactive capabilities [71, 72]. Moreover, well distributed nanoparticles produce an interphase zone with changed mobility around each nanoparticle, resulting in a percolating interphase network in the composite and playing a key role in improving nanocomposite characteristics [73, 74].

According to Jordan et al. [75], a reduction in particle size increases the number of filler particles, bringing them closer together and causing the interface layers from nearby particles to overlap, dramatically affecting the bulk characteristics. To summarise, nanocomposite materials have superior thermal, mechanical and barrier capabilities compared to conventional composites, even at low reinforcement levels [76].

14.1.14 Modification of horse chestnut starch film

Horse chestnut starch packing films were made using casting methods, where the gelatinized starch solutions were dried in the petri dishes to form a film. Native horse chestnut starch produced thicker, more organized and consistent films. The films had the best mechanical resistance, the least water vapour permeability, the most brightness, the least colour, the least oxygen permeability and the least water holding capacity, but they were less extensible and poorly solubilized [77]. Higher crystallinity in native starch films may cause lower permeability of the films [78]. Meanwhile the low thickness of native starch contributed to the low water permeability and oxygen permeability is affected by water solubility, thickness and the amylose content of the starch films. Also, native starch films have poor solubility, high amylose content, more ordered structure and are thus less permeable to oxygen.

14.1.15 Challenges and future perspective

Starch retrogradation causes poor mechanical characteristics and becomes present limitations. Various strategies have been investigated to address this issue, including starch modification via additional biopolymers and functional additives to the films. While some of these techniques have improved film quality, much more work is still required to produce starch-based films that are comparable to synthetic polymer-based films. This is because many parameters influence the functional qualities of starch-based biodegradable products, including starch type, temperature and time during film manufacturing, plasticizers, symbiotic polymers, storage conditions and others.

As a result, more research is needed to understand how these characteristics affect the fundamental functional performance of starch-based materials. To improve the current properties of starch-based polymers, researchers are adding functional components. As a result, starch-based materials will be able to be employed in the future to create active or smart packaging materials. Another issue in the future will be developing quick and straightforward technologies for mass-producing starch-based products. At the moment, laboratory-scale production of starch-based goods with better qualities is the norm. The development of cost-effective processing technologies for the industrial manufacture of starch-based goods will be crucial in the future. Furthermore, the development of a specific composting method for starch-based biodegradable materials should be addressed to ensure proper treatment and recycling; otherwise, pollution may arise. This will not only help to increase the usage of these materials in the food industry, but it will also help to reduce fossil fuel consumption and promote environmental sustainability.

14.2 Conclusions

Today, the usage of plastic materials has been minimised due to growing environmental contamination concerns. Interest in creating alternative packaging materials that are eco-friendly, non-polluting, and biodegradable has increased. In this work, the starch from horse chestnut has been discussed to replace synthetic plastics for packaging materials. However, the addition of nanocellulose to the TPS matrix increased the crystallinity and tensile strength, which in turn improved the quality of the nanocomposite films. Native starch has less mechanical properties than synthetic plastics, but these properties are still higher than those of natural starch.

References

1. Shogren RL, Fanta GF, Doane WM. Development of starch based plastics-a reexamination of selected polymer systems in historical perspective. Starch Staerke 1993;45:276–80.
2. CASTAñO J, Rodríguez-Llamazares S, Contreras K, Carrasco C, Pozo C, Bouza R, et al. Horse chestnut (Aesculus hippocastanum L.) starch: basic physico-chemical characteristics and use as thermoplastic material. Carbohydr Polym 2014;112:677–85.
3. Van Soest J, Benes K, De Wit D, Vliegenthart J. The influence of starch molecular mass on the properties of extruded thermoplastic starch. Polymer 1996;37:3543–52.
4. Bharadwaj D. Use and environment impact of biodegradable plastics-a review. CAAS 2010;2:65–9.

5. Čukanović J, Ninić-Todorović J, Ognjanov V, Mladenović E, Ljubojević M, Kurjakov A. Biochemical composition of the horse chestnut seed (Aesculus hippocastanum L.). Arch Biol Sci 2011;63:345–51.
6. Shah U, Gani A, Ashwar BA, Shah A, Wani IA, Masoodi FA. Effect of infrared and microwave radiations on properties of Indian Horse Chestnut starch. Int J Biol Macromol 2016;84:166–73.
7. Faradilla RF, Lee G, Rawal A, Hutomo T, Stenzel MH, Arcot J. Nanocellulose characteristics from the inner and outer layer of banana pseudo-stem prepared by TEMPO-mediated oxidation. Cellulose 2016;23: 3023–37.
8. Khalil HA, Bhat A, Yusra AI. Green composites from sustainable cellulose nanofibrils: a review. Carbohydr Polym 2012;87:963–79.
9. Montero B, Rico M, Rodríguez-Llamazares S, Barral L, Bouza R. Effect of nanocellulose as a filler on biodegradable thermoplastic starch films from tuber, cereal and legume. Carbohydr Polym 2017;157: 1094–104.
10. Raheem D. Application of plastics and paper as food packaging materials-an overview. Emir J Food Agric 2013:177–88. https://doi.org/10.9755/ejfa.v25i3.11509.
11. Mishra SP, Manent A-S, Chabot B, Daneault C. Production of nanocellulose from native cellulose–various options utilizing ultrasound. Bioresources 2012;7:0422–36.
12. Khwaldia K, Arab-Tehrany E, Desobry S. Biopolymer coatings on paper packaging materials. Compr Rev Food Sci Food Saf 2010;9:82–91.
13. Abdelgawad AM, El-Naggar ME, Hudson SM, Rojas OJ. Fabrication and characterization of bactericidal thiol-chitosan and chitosan iodoacetamide nanofibres. Int J Biol Macromol 2017;94:96–105.
14. Robertson GL. Food packaging: principles and practice. Milan: CRC Press; 2016.
15. Su Y, Kravets V, Wong S, Waters J, Geim AK, Nair RR. Impermeable barrier films and protective coatings based on reduced graphene oxide. Nat Commun 2014;5:1–5.
16. Marsh K, Bugusu B. Food packaging—roles, materials, and environmental issues. J Food Sci 2007;72: R39–R55.
17. Davis G, Song J. Biodegradable packaging based on raw materials from crops and their impact on waste management. Ind Crop Prod 2006;23:147–61.
18. Spierling S, Knüpffer E, Behnsen H, Mudersbach M, Krieg H, Springer S, et al. Bio-based plastics-A review of environmental, social and economic impact assessments. J Clean Prod 2018;185:476–91.
19. Abe MM, Martins JR, Sanvezzo PB, Macedo JV, Branciforti MC, Halley P, et al. Advantages and disadvantages of bioplastics production from starch and lignocellulosic components. Polymers 2021;13:2484.
20. Tharanathan R. Biodegradable films and composite coatings: past, present and future. Trends Food Sci Technol 2003;14:71–8.
21. de Paula FC, de Paula CB, Contiero J. Prospective biodegradable plastics from biomass conversion processes. In: Biofuels-state of development; 2018:245–72 pp.
22. Naik SN, Goud VV, Rout PK, Dalai AK. Production of first and second generation biofuels: a comprehensive review. Renew Sustain Energy Rev 2010;14:578–97.
23. Neuling U, Kaltschmitt M. Review of biofuel production-feedstock, processes and markets. J Oil Palm Res 2017;29:137–67.
24. Mekonnen T, Mussone P, Khalil H, Bressler D. Progress in bio-based plastics and plasticizing modifications. J Mater Chem 2013;1:13379–98.
25. Galić K, Ščetar M, Kurek M. The benefits of processing and packaging. Trends Food Sci Technol 2011;22: 127–37.
26. Asgher M, Qamar SA, Bilal M, Iqbal HM. Bio-based active food packaging materials: sustainable alternative to conventional petrochemical-based packaging materials. Food Res Int 2020;137:109625.
27. Blanco-Pascual N, Fernández-Martín F, Montero M. Effect of different protein extracts from Dosidicus gigas muscle co-products on edible films development. Food Hydrocolloids 2013;33:118–31.
28. Liu H, Xie F, Yu L, Chen L, Li L. Thermal processing of starch-based polymers. Prog Polym Sci 2009;34: 1348–68.

29. Buleon A, Colonna P, Planchot V, Ball S. Starch granules: structure and biosynthesis. Int J Biol Macromol 1998;23:85–112.
30. Whistler RL, BeMiller JN, Paschall EF. Starch: chemistry and technology. Academic Press; 2012.
31. Liu H, Yu L, Simon G, Dean K, Chen L. Effects of annealing on gelatinization and microstructures of corn starches with different amylose/amylopectin ratios. Carbohydr Polym 2009;77:662–9.
32. Fishman M, Coffin D, Konstance R, Onwulata C. Extrusion of pectin/starch blends plasticized with glycerol. Carbohydr Polym 2000;41:317–25.
33. Forssell PM, Mikkilä JM, Moates GK, Parker R. Phase and glass transition behaviour of concentrated barley starch-glycerol-water mixtures, a model for thermoplastic starch. Carbohydr Polym 1997;34:275–82.
34. Hulleman SH, Janssen FH, Feil H. The role of water during plasticization of native starches. Polymer 1998;39: 2043–8.
35. Ma X, Yu J, Kennedy JF. Studies on the properties of natural fibers-reinforced thermoplastic starch composites. Carbohydr Polym 2005;62:19–24.
36. Rafiq SI, Jan K, Singh S, Saxena D. Physicochemical, pasting, rheological, thermal and morphological properties of horse chestnut starch. J Food Sci Technol 2015;52:5651–60.
37. Ahmad M, Gani A. Ultrasonicated resveratrol loaded starch nanocapsules: characterization, bioactivity and release behaviour under in-vitro digestion. Carbohydr Polym 2021;251:117111.
38. Hymavathi T, Thejasri V, Roberts TP. Enhancing cooking, sensory and nutritional quality of finger millet noodles through incorporation of hydrocolloids. Int J Chem Stud 2019;7:877–81.
39. Popescu PA, Popa EE, Mitelut AC, Popa ME. Development of recyclable and biodegradable food packaging materials–opportunities and risks. Curr Trends Nat Sci 2020;9:142–6.
40. Arikan EB, Ozsoy HD. A review: investigation of bioplastics. J Civ Eng Architect 2015;9:188–92.
41. Okolie JA, Rana R, Nanda S, Dalai AK, Kozinski JA. Supercritical water gasification of biomass: a state-of-the-art review of process parameters, reaction mechanisms and catalysis. Sustain Energy Fuels 2019;3:578–98.
42. Ochi S. Development of high strength biodegradable composites using Manila hemp fiber and starch-based biodegradable resin. Compos Appl Sci Manuf 2006;37:1879–83.
43. Stevens E, Klamczynski A, Glenn G. Starch-lignin foams. Express Polym Lett 2010;4:311–20.
44. Yang J, Ching YC, Chuah CH. Applications of lignocellulosic fibers and lignin in bioplastics: a review. Polymers 2019;11:751.
45. Thakur VK, Thakur MK. Recent advances in green hydrogels from lignin: a review. Int J Biol Macromol 2015; 72:834–47.
46. Lennartsson PR, Niklasson C, Taherzadeh MJ. A pilot study on lignocelluloses to ethanol and fish feed using NMMO pretreatment and cultivation with zygomycetes in an air-lift reactor. Bioresour Technol 2011;102: 4425–32.
47. Khan B, Bilal Khan Niazi M, Samin G, Jahan Z. Thermoplastic starch: a possible biodegradable food packaging material—a review. J Food Process Eng 2017;40:e12447.
48. Akil H, Omar M, Mazuki AM, Safiee S, Ishak ZM, Bakar AA. Kenaf fiber reinforced composites: a review. Mater Des 2011;32:4107–21.
49. El-Saied H, El-Diwany AI, Basta AH, Atwa NA, El-Ghwas DE. Production and characterization of economical bacterial cellulose. Bioresources 2008;3:1196–217.
50. Shalwan A, Yousif B. In state of art: mechanical and tribological behaviour of polymeric composites based on natural fibres. Mater Des 2013;48:14–24.
51. Singha A, Thakur VK. Mechanical, morphological, and thermal characterization of compression-molded polymer biocomposites. Int J Polym Anal Char 2010;15:87–97.
52. Singha A, Thakur VK. Fabrication and characterization of S. cilliare fibre reinforced polymer composites. Bull Mater Sci 2009;32:49–58.
53. Bogoeva-Gaceva G, Avella M, Malinconico M, Buzarovska A, Grozdanov A, Gentile G, et al. Natural fiber eco-composites. Polym Compos 2007;28:98–107.

54. Eichhorn SJ, Dufresne A, Aranguren M, Marcovich N, Capadona J, Rowan SJ, et al. Current international research into cellulose nanofibres and nanocomposites. J Mater Sci 2010;45:1–33.
55. Singha A, Thakur VK. Physical, chemical and mechanical properties of Hibiscus sabdariffa fiber/polymer composite. Int J Polym Mater 2009;58:217–28.
56. Gaspar M, Benkő Z, Dogossy G, Reczey K, Czigany T. Reducing water absorption in compostable starch-based plastics. Polym Degrad Stabil 2005;90:563–9.
57. Agustin MB, Ahmmad B, Alonzo SMM, Patriana FM. Bioplastic based on starch and cellulose nanocrystals from rice straw. J Reinforc Plast Compos 2014;33:2205–13.
58. Chen Y, Liu C, Chang PR, Anderson DP, Huneault MA. Pea starch-based composite films with pea hull fibers and pea hull fiber-derived nanowhiskers. Polym Eng Sci 2009;49:369–78.
59. Chen Y, Liu C, Chang PR, Cao X, Anderson DP. Bionanocomposites based on pea starch and cellulose nanowhiskers hydrolyzed from pea hull fibre: effect of hydrolysis time. Carbohydr Polym 2009;76:607–15.
60. Lu Y, Weng L, Cao X. Morphological, thermal and mechanical properties of ramie crystallites—reinforced plasticized starch biocomposites. Carbohydr Polym 2006;63:198–204.
61. Cao X, Chen Y, Chang P, Muir A, Falk G. Starch-based nanocomposites reinforced with flax cellulose nanocrystals. Express Polym Lett 2008;2:502–10.
62. Guimarães IC, dos Reis KC, Menezes EGT, Rodrigues AC, da Silva TF, de Oliveira IRN, et al. Cellulose microfibrillated suspension of carrots obtained by mechanical defibrillation and their application in edible starch films. Ind Crop Prod 2016;89:285–94.
63. do Lago RC, de Oliveira ALM, Dias MC, de Carvalho EEN, Tonoli GHD, Boas EVdBV. Obtaining cellulosic nanofibrils from oat straw for biocomposite reinforcement: mechanical and barrier properties. Ind Crop Prod 2020;148:112264.
64. Leppänen K, Andersson S, Torkkeli M, Knaapila M, Kotelnikova N, Serimaa R. Structure of cellulose and microcrystalline cellulose from various wood species, cotton and flax studied by X-ray scattering. Cellulose 2009;16:999–1015.
65. Zulkifli NI, Samat N, Anuar H, Zainuddin N. Mechanical properties and failure modes of recycled polypropylene/microcrystalline cellulose composites. Mater Des 2015;69:114–23.
66. Sun X, Lu C, Liu Y, Zhang W, Zhang X. Melt-processed poly (vinyl alcohol) composites filled with microcrystalline cellulose from waste cotton fabrics. Carbohydr Polym 2014;101:642–9.
67. Haafiz MM, Hassan A, Zakaria Z, Inuwa IM, Islam MS, Jawaid M. Properties of polylactic acid composites reinforced with oil palm biomass microcrystalline cellulose. Carbohydr Polym 2013;98:139–45.
68. Davoudpour Y, Hossain S, Khalil HA, Haafiz MM, Ishak ZM, Hassan A, et al. Optimization of high pressure homogenization parameters for the isolation of cellulosic nanofibers using response surface methodology. Ind Crop Prod 2015;74:381–7.
69. Islam MT, Alam MM, Zoccola M. Review on modification of nanocellulose for application in composites. Int J Innov Res Sci Eng Technol 2013;2:5444–51.
70. Cao X, Dong H, Li CM. New nanocomposite materials reinforced with flax cellulose nanocrystals in waterborne polyurethane. Biomacromolecules 2007;8:899–904.
71. Boccuni F, Rondinone B, Petyx C, Iavicoli S. Potential occupational exposure to manufactured nanoparticles in Italy. J Clean Prod 2008;16:949–56.
72. Kahn J. Nano's big future: nanotechnology. Natl Geogr 2006;209:98–114.
73. Qiao R, Brinson LC. Simulation of interphase percolation and gradients in polymer nanocomposites. Compos Sci Technol 2009;69:491–9.
74. De Azeredo HM. Nanocomposites for food packaging applications. Food Res Int 2009;42:1240–53.
75. Jordan J, Jacob KI, Tannenbaum R, Sharaf MA, Jasiuk I. Experimental trends in polymer nanocomposites—a review. Mater Sci Eng 2005;393:1–11.
76. Oksman K, Mathew AP, Bondeson D, Kvien I. Manufacturing process of cellulose whiskers/polylactic acid nanocomposites. Compos Sci Technol 2006;66:2776–84.

77. Hassan I, Wani IA, Hussain PR, Dar AH. Physico-chemical properties of Indian horse chestnut (aesculus indica) starch films as affected by γ-irradiation. J Packaging Technol Res 2021;5:175–84.
78. Wani IA, Jabeen M, Geelani H, Masoodi FA, Saba I, Muzaffar S. Effect of gamma irradiation on physicochemical properties of Indian Horse Chestnut (Aesculus indica Colebr.) starch. Food Hydrocolloids 2014;35:253–63.

Nurfatimah Mohd Thani*, Mazween Mohamad Mazlan,
Nur Izzah Nabilah Haris and Mohd Hafizz Wondi

15 Oat thermoplastic starch nanocomposite films reinforced with nanocellulose

Abstract: With the urgent need to curb plastic pollution, research and development in biodegradable plastic have received a great deal of attention. Starch film is the most promising alternative to conventional petroleum-based plastic. It is biodegradable, affordable, and non-toxic to human health and the environment. Over the last decade, much attention has been paid to cellulose nanofiller-reinforcement to the starch-based polymer to produce nanocomposites. While the use of oat as nano-reinforcement is quite well known, there is still limited studies that investigate the use of nanocellulose for oat starch nanocomposite films reinforcement. There is an ample of possibility in improving the properties of oat starch films, and therefore utilizing nanocellulose reinforcement should be further investigated.

Keywords: nanocellulose; nanocomposite; oat; thermoplastic.

15.1 Introduction

Oat (*Avena sativa*), a minor cereal in the Poaceae/Gramineae family, is majorly used for animal feed and human food. Oat is the seventh most significant grain in terms of economic importance [1]. Around 70 species of oats have been identified to date, with *Avena sativa L.* (hulled oat) and *Avena nuda L.* being the most cultivated [2]. Varies based on the species, oat grains contain 12–20% protein, 5–10% oil, and 3–14% crude fibre [3]. Oat is majorly composed of starch, making it a possible candidate as a raw material to produce thermoplastic starch composite [4].

The emergence of bio-based plastic is due to the deleterious impact of conventional petroleum-based plastics. Conventional plastics generate pollution as a result of

***Corresponding author: Nurfatimah Mohd Thani**, Department of Food Sciences, Faculty of Science and Technology, Universiti Kebangsaan Malaysia, UKM Bangi, Selangor, Malaysia; and Innovation Center for Confectionery Technology, Faculty of Science and Technology, Universiti Kebangsaan Malaysia, UKM Bangi, Selangor, Malaysia, E-mail: nurfatimah@ukm.edu.my
Mazween Mohamad Mazlan, Department of Process and Food Engineering, Faculty of Engineering, Universiti Putra, Mukah, Malaysia
Nur Izzah Nabilah Haris, Institute of Sustainable and Renewable Energy, Universiti Malaysia Sarawak, Sarawak, Malaysia
Mohd Hafizz Wondi, Faculty of Plantation and Agrotechnology, Universiti Teknologi MARA, Mukah, Sarawak, Malaysia

As per De Gruyter's policy this article has previously been published in the journal Physical Sciences Reviews. Please cite as: N. M. Thani, M. M. Mazlan, N. I. Nabilah Haris and M. H. Wondi "Oat thermoplastic starch nanocomposite films reinforced with nanocellulose" *Physical Sciences Reviews* [Online] 2023. DOI: 10.1515/psr-2022-0036 | https://doi.org/10.1515/9783110773606-015

accumulations of non-biodegradable plastics which affects both terrestrial and marine ecosystem. To tackle this issue, at the 5th United Nations Environment Assembly in Nairobi on March 2022, 175 nations have pledged to plastic pollution by 2024, and forged an international legally binding agreement [5]. With the urgent need to curb plastic pollution, research and development in biodegradable plastic have received a great deal of attention. Starches are important resources to develop bio-based blends and composites, as they are easily available biopolymers such as oat, corn, cassava, potato, and rice. This chapter discusses thermoplastic starch nanocomposite films that are developed from oat starch.

15.2 Starch

Starch is a plant polysaccharide commonly found in various plants. It is the major energy reserve predominantly in seeds, tubers, roots, fruits, stems, and leaves [6]. There are five hierarchies in the structure of starch: molecules, crystalline and amorphous lamellae, blocklets, growth rings, and granules [7]. Tang et al. [8] mentioned the shell structure of starch granules corresponds to the concentric ring structure encircling the hilum and is also known as the growth ring. A blocklet structure is found between the shell (growth ring) and lamellae hierarchies. Granular starch is occasionally altered to alter its characteristics and functioning. This alteration might be physical or chemical in nature. The variety of starch qualities from various sources already provides a wide range of capabilities, but the degree and kind of modification can expand this range even more [9]. Starch is found in intracellular granules and serves as a storage polysaccharide in plant cells, rather than serving as a structure-building molecule. Starch builds up in these granules over time and is eventually used as a source of energy during germination. One starch granule or a cluster of starch granules can be held by one amyloplast. The morphological properties of starch granules, such as size, shape, and architecture, change based on their botanical origin, making it relatively straightforward to microscopically determine the starch's origins [9]. According to Jane et al. [10], due to starch granules having a higher density than water, gravity sedimentation allows for efficient separation and purification of starch. The semicrystalline form of starch granules helps to keep the granules together and prevents them from dissolving in water at room temperature. Native starch granules can be gelatinized and dispersed by heating starch in the presence of water (as a plasticizer). Starches of various botanical sources have variable gelatinization capabilities, owing to differences in starch molecule structures and the arrangement of double-helical crystalline structures inside starch granules.

Starches are categorized as rapidly digesting starch (RDS), slowly digesting starch (SDS), or resistant starch (RS) depending on their digestibility. RDS can be completely digested in about 20 min, resulting in a rapid rise in postprandial blood glucose levels. SDS can be entirely digested, but it takes longer, around 2 h. The part of the starch that is not hydrolyzed after 2 h is referred to as RS [11]. There are several factors affecting starch

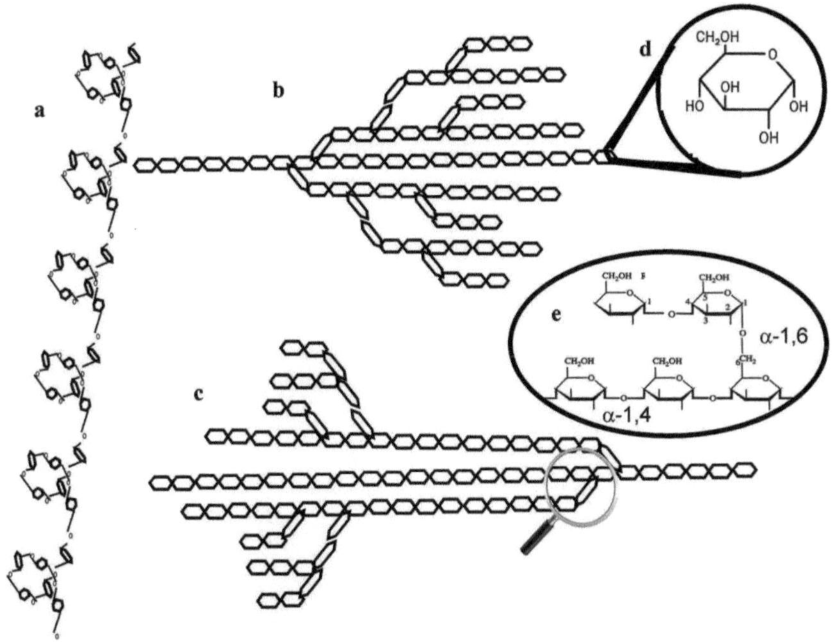

Figure 15.1: Schematic representation of (a) Helical amylose and a possible structure of the amylopectin of (b) complete and (c) substituted triticales reported by [20]. (d) Glucose unit of the structure of both polymers and e) the α-1,4 glycosides links of linear chains and α-1,6 links of the branches. Note: This figure is adapted from [21].

digestion. Some of the external factors are physical treatment [12], recrystallization [13], and chemical treatment [14]. Meanwhile, the general internal factor includes the starch's molecular structure and physicochemical characteristics, the size and shape of the starch particles, the ratio of amylose to amylopectin in the starch, and the starch's crystallinity [15].

Starches from various sources differ in their chemical composition (α-glucans, moisture, lipids, proteins, and phosphorylated residues), as well as the structure of their constituents, all of which will affect the surface, hardness, and crystallinity of the starch granules [16–18]. Starch is chemically composed of two glucan polymers; amylose and amylopectin that are deposited in granules of various sizes and forms, including big A-, medium B- and small C-type granules, as well as disk-like and spherical granules (Figure 15.1) [19].

About 98% of the polymeric carbohydrate content in starch granules is polymeric carbohydrate material. A-D-glucopyranosyl units make up the carbohydrate. Each hemiacetal group, or reducing end, of a starch molecule, is unique. α-1,4 or α-1,6 links are used to connect the glucopyranosyl units [9]. Generally, amylose is in the linear form that accounts for about 18%–33% of the starch carbohydrate weight, while amylopectin which is a highly branched form is responsible for the remaining starch carbohydrate content [22]. The properties and functionality of granule starches are influenced by their physical and chemical characteristics. According to Manners [23], long and short chains are two different types of chains. Because of the overall significantly longer chains in amylose compared to amylopectin, the definitions of long and short chains can be very different. The chains are further broken down into several parts. The section of a chain that extends from the outermost branch point to the nonreducing end is called an external chain. Knowing these features enables the selection of starch with the preferred properties for a certain application, as well as the selection of the starch source and suitable modification process to acquire the functional characteristics required for a specific application [21]. Copeland et al. [24] explain how amylopectin crystallite packing and the structure of the crystalline lamella within starch granules are both influenced by amylose. This is critical for features like swelling and gelatinization that are connected to water intake. Meanwhile, the production of an amylose-lipid helicoidal complex is aided by low molecular weight amylopectin with long branched chains [25].

15.2.1 Amylose

Though amylose is a small component in most granules, it has a significant impact on starch characteristics. Amylose is traditionally defined as a molecule made up of a long, linear chain of (1/4)-linked a-D-glucosyl units [26]. Amylose is a low-molecular-weight polymer ($1.03–4.89 \times 10^5$) made up of linear chains linked by α-1,4 glycosides links that acts like a non-branched molecule [27]. The side branches in amylose are long and have a high molecular weight. Consequently, it behaves and may be thought of as an unbranched linear polymer in practice. Amylose molecules have a helical shape by nature. Iodine, organic alcohols, and fatty acids are all known to interact with amylose. The resulting complexes are known as helical inclusion complexes. If fatty acids are present, the resulting combination is known as an amylose–lipid complex [28]. Ai & Jane [6] further explain that molecules from the families of alcohols, carboxylic acids, ketones, and aldehydes have a tendency to form complexes with amylose, whereas amines and amino acids have minimal potential to form complexes with amylose. Amylose forms a helical complex with free fatty acids, mono- and diglycerides, and phospholipids. Amylose content in starch may vary in different plants (Table 15.1). Some other examples of the starches with high amylose content, such as amylose barley 46.5–48% [29] and maize 62.8–85.6% [25].

The amount of amylose in the starch has an impact on its crystallinity. The absence of amylose has no effect on crystalline polymorphism or granule size, although it does enhance crystallinity [35]. Furthermore, the degree of crystallinity is also influenced by the differences in starch granule types. Because the amylose concentration of A-type starch granules is higher than that of B-type starch granules, the A-type starch granules have a lower % of crystallinity [18]. Putseys et al. [28] briefly explained amylose-inclusion complexes are referred to as 'type I' complexes when they are not organized. 'Type II' amylose inclusion complexes are formed when these amorphous complexes are arranged in lamellae and give rise to the creation of crystallites.

Table 15.1: Amylose content from various starch sources [26].

Starch source	Amylose content (%)	References
Barley	28.3	[30]
Wheat	33.2	[31]
	37.0	[32]
Waxy maize	1.0	[32]
Rice	19.5	[30]
	12.8–24.2	[33]
	13.2	[34]
Sweet potato	19.8	[34]
Tapioca	17.6	[30]
	17.9	[34]
	29.8	[32]
Potato	14.0	[32]
	18.0	[34]

15.2.2 Amylopectin

Amylopectin is the most abundant component of starch, accounting for 65–85% of the matter in starch granules. It is made up of a lot of α-(1,4)-linked D-glucose residues in short chains, and α-(1,6)- linkages are used to connect the chains on their decreasing end side [36]. Aberle et al. [37]; Millard et al. [38]; Buléon et al. [22] mentioned, the accumulation of the chains lead to formation of a massive macromolecule with an average weight (M_w) of 10^7–10^8. As a result, the size is one or two orders of magnitude greater than amylose molecules, which are essentially linear [22, 37]. There are as many nonreducing endings as there are chains in each macromolecule, but only one glucose residue possesses a free reducing end group [36]. According to a study carried out by Takeda et al. [39], among all of the starches they looked at had the most large amylopectin molecules.

In terms of the major chain of amylopectin, Meyer and Bernfield [40] suggested a model implying that external and internal chain segments are two more common types of chain segments. Internal chains are segments that stretch from the outermost branch to the nonreducing end of the chain, whereas external chains are segments that reach from the outermost branch to the nonreducing end of the chain. As a result, A-chains are totally external, whereas each B-chain (and the C-chain) has one external chain segment, and the rest is internal. Internal chain segments in amylopectin are the segments between the branches [36]. Later, a revised model by Gunja et al. [41] showed that amylopectin has a bimodal distribution with two additional groups of major chains, namely short (S) and long (L) chains. L-chains are substantially less than S-chains, and the ratio of S: L-chains in starches from various plants varies greatly. Some amylopectin samples have been shown to include exceedingly long chains, often known as "extra-long" or "superlong" chains, in addition to unit chains with DP up to the order of 100 [39, 42], the same length as ordinary amylose chains since their DP equates to several hundred glucose residues.

15.3 Oat Starch

15.3.1 Extraction and preparation of oat starch

T.B. Osborne [43], a scientist from the United States, created the first approach for classifying plant proteins. To extract, measure, and categorize grain storage proteins, a number of extraction buffers were utilized. There are four solubility fractions of oat: salt-soluble globulin, water-soluble albumin fraction, alcohol-soluble prolamins, and glutelins. Years after that, using Osborne's fractionation, Peterson and Smith [44] characterized proteins from several oats. They later calculated that albumins (water-soluble), globulins (salt-soluble), prolamins, and glutelins represented 14–19%, 52–53%, 7–9%, and 21–27% of total protein, respectively. An observation has been recorded that the percentage of oat protein fractions appears to be affected by buffer content, pH, temperature, and the growth environment [45]. Oat bioactive components are divided into polar and intermediately polar molecules like β-glucan, phenolic acids, and polyphenols, as well as non-polar compounds like tocols. For the extraction of these phytochemicals from these grains, a number of procedures have been tried throughout the years [46]. To minimize unfavorable changes during extraction and concentration, the parameters must be carefully adjusted and monitored. This is true not just in terms of extraction yield and cost efficiency, but also in terms of probable changes in oat β-glucan's physicochemical and physiological characteristics [47].

Oat protein concentrate has a protein concentration ranging from 69 to 90% [48], and can later be extracted via dry and wet fractionation. Wet techniques (alkaline isolation, saline isolation), dry fractionation, and enzymatic extraction may all be used to extract and isolate oat protein, giving it further economic value [49]. Fractionation on the dry side consists of grinding oat grits or grains, then air drying to get a protein-rich fraction with a protein content of 42%, preserving starch and fat, which can help oat protein concentrate retain its physicochemical and technical qualities [50, 51]. Wet fractionation, in contrast to dry techniques, yields isolate with excellent protein purity and yield. Even though it necessitates a large number of solvents and chemicals, it also has the potential to cause the loss of some health-promoting components as well as alterations in protein structure. Furthermore, this wet procedure needs a large amount of water and it consumes a lot of energy, and produces a lot of watery waste, leading to a detrimental effect on the environment [52–54]. The lack of potentially dangerous chemicals during the extraction process is the most significant benefit of dry extraction procedures. However, because numerous cleaning and concentration procedures are required, this approach is quite expensive, even if it produces very low β-glucan concentrations [55].

Among many means of extracting protein from oats, the alkaline approach is the most common method [56]. Alkaline extraction has also been utilized to recover and fractionate arabinoxylans because the hydroxyl ions enhance cellulose swelling and intermolecular hydrogen bond breakdown under these conditions [57]. Furthermore, the carboxyl and acetyl groups of benzyl groups, as well as the uronic acids of lignin, are hydrolyzed by alkaline. The ester bonds between arabinose residues and ferulic acid, resulting in hemicellulose solubilization from the cell's insoluble matrix [58]. The wet extraction method includes extracting proteins in a solvent, centrifuging them, and then isoelectric precipitating them. Protein isolation using the alkaline technique has been reported by a number of publications [53, 59–61]. Prosekov et al. [52] reported, to get oat protein isolates with up to 90% protein content, extraction was done at pH 9.2 to solubilize oat proteins, followed by isoelectric precipitation at pH 5, washing with water, neutralization, and drying. Meanwhile, Liu et al. [53] used 1 M NaOH (pH 9.2–11.5), 20–55 °C, and stirring (100–200 rpm for 60 min), followed by proteins precipitation at lower pH (4–5.5) and centrifugation. Protein recovery from ground oat groats and protein content of oat protein isolates were found to be in the range of 60.4–72.1% and 72.4–92.6%, respectively. In contrast to alkaline extraction, the recovery of oat protein through acid extraction (pH 3) was found to be lower, at 54.6%, compared to 75.5% with alkali extraction (pH 11.5) [59]. Other than alkaline and acid medium, salt extraction has also been used for protein extraction from oat. According to a study by Ma [62] oat protein was isolated using salt (0.5 M $CaCl_2$) and alkaline in two separate types of oat (Hinoat and Sentinel). The isolates contain more than 90% protein; however, the yield was substantially greater (about 60%) in the alkaline-treated isolate than in the salt-treated isolation (25%).

The discovery that oats have serum cholesterol lowering and hypoglycaemic effects has sparked a surge in oat research [63, 64] especially to the soluble fiber fraction, particularly (1–3, 1–4) β-glucan (Figure 15.2). According to [65] oat β-glucans, along with other non-starch polysaccharides, are found in the endosperm cell walls, which encapsulate the grain's starch, matrix protein, and lipid reserves. Endogenous enzymes (β-glucanases and amylases) must normally be inactivated before extracting β-glucans, as leaving these enzymes active might destroy the β-glucan, resulting in a low molecular weight β-glucan product. Maintaining β-glucan's molecular weight (MW) is critical since it dictates its physicochemical qualities, such as viscosity, which in turn impacts its cholesterol-lowering ability. In the extraction of β-glucan, the denaturing impact of high pH may be favorable for protein solubilization, in contrast to acidic extraction [66]. A study by Bhatty [67] reported through a single extraction employing 4 percent of 1 M sodium hydroxide (NaOH) at room temperature yielded the maximum enrichment of β-glucan from oats. 84 percent β-glucan plus pentosans from oat brans were found in the purified high viscosity gums. Moreover, the β-glucan yield, or the percentage of total oat bran recovered, was 61 percent.

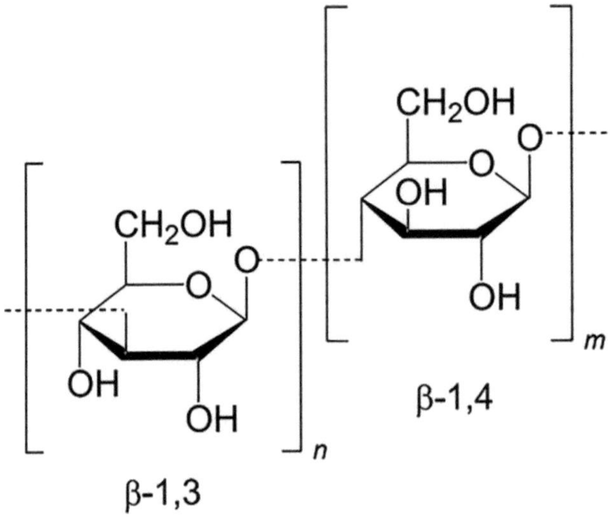

Figure 15.2: β-1,3 and β-1,4 bonds of β-glucan [46].

To improve protein production, pretreatments can be applied to oat grains before extraction or fractionation. According to Lehtinen et al. [68] and Ziegler et al. [69] not only this step can increased the extraction yield, heat treatment of oat grains prevented the development of rancidity in oats during storage by inactivating lipase (by around 60%). Alrahmany et al. [70] reported enzymatic treatment with amyloglucosidase prior to alkaline extraction can improve the release of proteins during extraction by breaking down polysaccharides. By boosting solubility, water retention, and foaming capability, this enzymatic treatment can improve the techno-functionality of proteins [52].

Many studies advised using sequential ultrafiltration systems for the valorization of various grain processing by-products in order to boost selectivity. The benefit of combining membrane systems is the synergy that occurs as a result of their integration, which results in overall considerable improvements in plant compactness, product quality, energy efficiency, and environmental effect [71].

Alternative food processing techniques for extracting nutraceuticals from plants, such as ultrasound-assisted extraction (UAE), supercritical fluid extraction, and accelerated-solvent extraction, have seen a recent spike in popularity. Novel food processing techniques provide several benefits over traditional extraction methods, including being more environmentally friendly, reducing solvent usage, minimizing extraction time, and increasing yield [72].

Some of the recent techniques in extraction of oat are using deep eutectic solvents (DESs) [73], combination of chemical and hydrothermal pretreatments [74], roasting and amylase hydrolysis pretreatments [75], combination of enzyme-aided alkaline (pH 8.0) extraction with ultrafiltration and subsequent diafiltration [76], and subcritical propane and dimethyl ether extraction [77].

15.3.2 Properties of oat starch

Oat has been regarded as a valuable crop since ancient times due to its benefits, such as for hypocholesterolemic effects, glycemic control, and positive impact on gut microbiota, and high protein content [78]. Physical and nutritional features of oat β-glucans, particularly their capacity to lower blood cholesterol, have been extensively researched. However, there is a scarcity of information on the extraction, characterization, and uses of oat protein. In terms of oat grain composition, protein is one of the most important components (12–20%), followed by carbohydrates [79]. According to Zhong et al. [80] the amino acid content and digestibility of cereal grains determine their protein quality. Oat includes a significant level of lysine, asparagine-aspartic acid, and alanine, making it better than other cereal protein sources in terms of acid concentration. Mel and Malalgoda [78] have summarized the techno-functional properties of oat proteins, which are water solubility, foaming ability, emulsifying ability, fat-binding capacity, water hydration capacity, and gelling ability. The impact of alteration on the functional features of oat proteins are shown in Table 15.2.

Oat globulins are less hydrophilic than other plant globulins because they have glutamine-rich fractions at the surface that are exposed to solvent. Because the majority

Table 15.2: Effect of treatments to the functional properties of oat proteins.

Treatment	Outcome	Reference
Enzymatic extraction	Higher oat protein extraction yield with Viscozyme L partial hydrolysis with enzyme (flavourzyme and trypsin)	[56]
	Oat protein isolates with strong water holding capacity have increased gelling qualities (gel strength and texture).	[79]
Acetylation and succinylation and crosslinking	Acetylation and succinylation increased several functional features, such as foaming and emulsification, while excessive crosslinking had a detrimental impact on rheological properties.	[81]
Supercritical carbon dioxide (SC-CO2) extraction	Improved foaming and emulsion activity, as well as increased protein extraction yield.	[82]
Inulin addition to oat protein isolates	Gels with a superior textural character and rheological qualities have been developed.	[79]
Oat protein isolates dextrin conjugation by glycation reaction	Emulsification characteristics have been improved.	[83]
Addition of dextrin and carrageenan polysaccharides	Thermal gelation has improved.	[84]
GDL (glucon-D-lactone) acidification of oat protein isolates (cold set of oat protein isolates)	Cold set gel oat protein with a percolating structure and good mechanical characteristics.	[85]

of the oat protein fractions are globulin, they have a poorer solubility [86]. Meanwhile, the solubility of proteins, pH, and extraction process all affect oat's emulsifying ability. At neutral pH, modifying native protein appears to be a good strategy to increase emulsifying capabilities [49]. The pH at which the protein concentrate was produced determines the solubility, emulsifying capabilities, hydration capacity, and organoleptic qualities of oat proteins [87]. More acidic or alkaline circumstances resulted in an increase in solubility. Oat proteins showed similar declines in nitrogen solubility as ionic strength increased [88].

According to Kumar et al. [49], water holding capacity was improved by chemical changes (acylation and diacylation) and enzymatic hydrolysis (at pH values 3 and 7) of oat protein. Polysaccharides (chitosan, dextrin, and carrageenan) enhanced the water holding capacity of oat proteins, although the effect was pH dependent. In terms of thermal properties of oat, Zhao et al. [89] reported that oat proteins are thermally stable. Oat globulins have a high thermal denaturation temperature of 110 °C.

Other significant physicochemical properties of β-glucan are molecular structure, size, solubility, and viscosity [90]. The molar ratio (DP3/DP4 ratio) directly affects the β-glucan molecule's functional characteristics. The gel forming and high viscosity qualities are mostly due to cellotetraosyl and longer subunits. The cellotriosyl units appear to be responsible for stronger connections between β-glucan chains [91]. The creation of a very viscous solution is thought to be crucial for both technological and physiological reasons. The molecular weight distribution and the concentration in solution, which is restricted by the β-glucan solubility, are the primary determinants of the viscosity of a β-glucan solution. Furthermore, depending on cultivars and climatic circumstances, the viscosity of β-glucan preparations might vary significantly [92–94].

15.4 Modification of oat starch films

15.4.1 Plasticization of oat starch films

Starch film is the most promising alternative to conventional petroleum-based plastic. It is biodegradable, affordable, and non-toxic to human health and the environment. A plethora of investigation was made to utilize various sources of starch such as corn, wheat, rice, oat, potato, and cassava for film application [95, 96]. Stand-alone starch film typically portrays brittle and hydrophilic behaviors which hinders its use for packaging applications [97]. For packaging purposes, particularly for food, a film should provide a good barrier against light and odor and excellent mechanical properties [98]. The downside, as mentioned above, can be overcome by introducing plasticizers such as polyols and glycerol. Plasticizers are capable to improve the film's flexibility and mechanical properties by destroying intramolecular hydrogen bonding of polymer chains which in turns increases the intermolecular spacing and mobility [99]. The reduction of intramolecular force also leads to lower glass transition temperature and degree of crystallinity.

Starch is a hydrophilic material and hence water vapor absorbs easily and affects its structure. The presence of a hydrophobic component in the starch may greatly improve the water vapor sorption. Attempts can be done to improve the properties by adding hydrophobic components such as lipids or fibers in the starch matrix [100]. Unlike typical starches like corn, wheat, and rice, oat starch contains a higher amount of lipid at 1.6%, portraying a potentially better hydrophobicity if used for film application [4].

In 2009, Galdeano and coworkers [101] reported one of the earliest successful development of plasticized oat starch film. Their work analyses the effect of plasticizers on microstructure, moisture sorption, water vapor permeability (WVP) and mechanical properties of oat starch films stored under different relative humidity values. Oat starch films were prepared with glycerol, sorbitol, urea, sucrose, and glycerol-sorbitol mixture as plasticizers. Based on the microscopy images of the films' cross sections, the unplasticized film was made up of a porous and less compact structure. The addition of all plasticizers significantly improves the film structure as the cross section of all films are smooth and dense. Water sorption study revealed that unplasticized films have higher monolayer value (0.052) than sorbitol (0.049), urea (0.033) and sucrose (0.041) films. When plasticizers are added in low amount (<27%), it can attach with starch better, yielding the desired antiplasticizing effect. WVP of plasticized films were significantly lower than unplasticized films, which is non desirable for real application. Mechanical analysis unveiled that unplasticized films have higher stress at break values than all plasticized films. Overall, the pioneering study concluded that the high content of lipids in oat starch act as a water vapor barrier, resulting in a smaller variation of the equilibrium moisture contents in all film formulations under varied relative humidities (RH) and, thus, the plasticizer type would not significantly affect the properties of the final product.

Four years later, Galdeano et al. [102] published another report on the exact same plasticized oat starch films and investigate the influence of the thickness instead of the type of plasticizers on the opacity and mechanical properties. Based on their findings, it was found that the variation of thickness affected the puncture properties. The thicker the film, the higher the force needed to puncture the film except for the sucrose-plasticized oat starch film. Despite that, the thicker films are more rupture resistant, there was no improvement of physical properties shown for all plasticized films. Plasticized films showed worst physical properties than unplasticized films at all RH.

Recently, Zołek-Tryznowska and Kałuźa [103] studied starch films from various origins such as maize, potato, oat, rice, and tapioca with glycerol as a plasticizer. The properties of interest were tensile strength, water vapour transition rate, moisture content, wettability, and their surface free energy. Figure 5 shows the appearance of the developed starch films. It revealed that the oat film had the lowest transparency which appeared yellowish in color and had a rough surface texture that is visible with the naked eye. This appearance may limit its use in the packaging industry as further processes may be needed to obtain the desired surface and color of film. Among all films, oat starch film portrayed the lowest tensile strength, elongation at break, polar component of the surface free energy, and the highest contact angle. Due to the weak properties of oat starch films, it was suggested that oat starch films might be feasible for single-use packaging.

Zamudio-Flores and co-authors [104] attempted a different approach of utilizing starch as a raw material to develop films. They developed oxidized oat starch films as opposed to native starch films as it was stipulated that oxidized starch has film-forming capability and better mechanical characteristic. Oxidized banana starch films were also fabricated in the study. Betalains, a chemical compound typically found in beet and prickly pear, was added to the film formulations due to its known antioxidant capacity and anticancer activity. Based on their results, it was found that oat starch films are more soluble than banana films due to the higher hydrophilicity of oat starch. Oat starch films had higher elongation value and WVP, and lower tensile strength and elastic modulus than banana starch. The data portrays that oat starch films are more flexible, which is critical for food packaging applications. The addition of betalains did not alter the elongation, so these films can additionally produce antioxidant effect when used as covering of products. The summary of plasticized oat starch films studies are depicted in Table 15.3.

Table 15.3: Plasticized oat starch films.

Oat starch type	Plasticizer	Ratio	Result description	References
Native starch	Glycerol Sorbitol Glycerol and sorbitol (1:1) Urea Sucrose	20 g/100 g 25 g/100 g 23 g/100 g 16 g/100 g 55.5 g/100 g	– WVP of plasticized films were lower than unplasticized films. – As RH increase, stress at break and Young's modulus decrease and strain at break increases. – At low RH, sucrose films were the most fragile, but had better properties at high RH. – Urea did not have the expected reinforcing effect. – The thicker the film, the higher the force needed to puncture the film except for the sucrose-plasticized oat starch film. – No improvement of physical properties shown for all plasticized films	[101] [102]
	Glycerol	50 wt%	– Lowest tensile strength and highest contact angle values compared to maize, potato, rice, and tapioca. – Have low wettability and hydrophilicity	[103]
Oxidized starch	Glycerol	50 wt%	– Oat starch film had higher elongation value and WVP, and lower tensile strength and elastic modulus than banana starch – Addition of betalains changed the color and elongation of oat starch films	[104]

15.4.2 Nanocellulose reinforced oat starch nanocomposite films

The emerging demand in biodegradable and eco-friendly materials due to the environmental concern has induced the research and development in the field of material science to develop stronger and better bio-based composite materials. Over the last decade, much attention has been paid to cellulose nanofiller-reinforcement to the starch-based polymer to produce nanocomposites [105]. Nanocellulose has fascinating features, such as large surface area, high crystallinity, low density, and excellent physical properties, making it one of the most important nano-reinforcement in material development. Nanocellulose can be obtained from any cellulose extracted from plants, animals, and microorganisms. Oat is also one of the sources of nanocellulose. Bruni et al. [106] reported the use of cellulose nanocrystals from rice, oat, and eucalyptus rice, as reinforcements for wheat starch films. The incorporation of oat nanocellulose had remarkably increased the tensile strength of the film by 91.3%. Qazanfarzadeh and Kadivar [107] also attempted to use oat-based nano-reinforcement for nanocomposite films. They used oat husk nanocellulose to reinforce whey protein nanocomposite films and found that films prepared with up to 5 wt% oat husk nanocellulose showed the highest tensile strength, Young's modulus, solubility and the lowest elongation at break and moisture content. Diyana et al. [108] wrote a review focuses on various characteristics of biopolymers and their classes, giving an overview of starch as a key ingredient in biopolymers, composites, and prospective uses for starch-based thermoplastics in packaging. While the use of oat as nano-reinforcement is quite well known, there is still limited studies that investigate the use of nanocellulose for oat starch nanocomposite films reinforcement. There is an ample of possibility in improving the properties of oat starch films, and therefore utilizing nanocellulose reinforcement should be further investigated.

15.5 Conclusions

Overall, according to evidence reported in the literature, it can be deduced that adding plasticizers at low concentration may lead to deplasticization effect on oat starch films. Deplasticization is undesired as the final properties are worse than unplasticized starch films. Higher concentration of plasticizers, at 50% ratio to starch, is sufficient to significantly improve the oat starch films properties. Variations in the types of starch used, i.e. native and oxidized, should be further explored as the variation will affect the final characteristic of the oat starch films. Currently, there is a lack of investigations on nanocellulose reinforced oat starch nanocomposite films. Hopefully, this topic will be addressed by scientists and researchers in the near future.

References

1. Boczkowska M, Podyma W, Łapiński B. 4 – oat. In: Singh M, Upadhyaya HD, editors. Genetic and genomic resources for grain cereals improvement. San Diego: Academic Press; 2016:159–225 pp.
2. Jing P, Hu X. Nutraceutical properties and health benefits of oats. In: Cereals and pulses. Iowa: John Wiley & Sons, Ltd; 2012:21–36 pp.
3. Yue J, Gu Z, Zhu Z, Yi J, Ohm JB, Chen B, et al. Impact of defatting treatment and oat varieties on structural, functional properties, and aromatic profile of oat protein. Food Hydrocoll 2021;112:106368.
4. Punia S, Sandhu KS, Dhull SB, Siroha AK, Purewal SS, Kaur M, et al. Oat starch: physico-chemical, morphological, rheological characteristics and its applications - a review. Int J Biol Macromol 2020;154: 493–8.
5. United Nations Environment Program. UN environment assembly concludes with 14 resolutions to curb pollution, protect and restore nature worldwide [Press Release]. Nairobi; 2022. Available from: https://www.unep.org/news-and-stories/press-release/un-environment-assembly-concludes-14-resolutions-curb-pollution#:~:text=Nairobi%2C 02 March 2022 – The,to advance global environmental governance [Accessed 5 Mar 2022].
6. Ai Y, Jane JL. Understanding starch structure and functionality. Duxford: Elsevier Ltd; 2018. Epub ahead of print 2018.
7. Gallant DJ, Bouchet B, Baldwin PM. Microscopy of starch: evidence of a new level of granule organization. Carbohydr Polym 1997;32:177–91.
8. Tang H, Mitsunaga T, Kawamura Y. Molecular arrangement in blocklets and starch granule architecture. Carbohydr Polym 2006;63:555–60.
9. Joye IJ Starch. In: Melton L, Shahidi F, Varelis P, editors, Amsterdam: Encyclopedia of food chemistry. Elsevier; 2008:256–64 pp.
10. Jane J, Chen YY, Lee LF, McPherson AE, Wong KS, Radosavljevic M, et al. Effects of amylopectin branch chain length and amylose content on the gelatinization and pasting properties of starch. Cereal Chem 1999;76:629–37.
11. Englyst HN, Kingman SM, Cummings JH. Classification and measurement of nutritionally important starch fractions. Eur J Clin Nutr 1992;46:S33–50.
12. Tamura M, Singh J, Kaur L, Ogawa Y. Impact of the degree of cooking on starch digestibility of rice – an in vitro study. Food Chem 2016;191:98–104.
13. Frei M, Siddhuraju P, Becker K. Studies on the in vitro starch digestibility and the glycemic index of six different indigenous rice cultivars from the Philippines. Food Chem 2003;83:395–402.
14. Han J-A, BeMiller JN. Preparation and physical characteristics of slowly digesting modified food starches. Carbohydr Polym 2007;67:366–74.
15. Chung H-J, Liu Q, Lee L, Wei D. Relationship between the structure, physicochemical properties and in vitro digestibility of rice starches with different amylose contents. Food Hydrocoll 2011;25:968–75.
16. Tester RF, Karkalas J, Qi X. Starch—composition, fine structure and architecture. J Cereal Sci 2004;39: 151–65.
17. Finnie SM, Jeannotte R, Morris CF, Giroux M, Faubion J. Variation in polar lipids located on the surface of wheat starch. J Cereal Sci 2010;51:73–80.
18. Ao Z, Jane J. Characterization and modeling of the A-and B-granule starches of wheat, triticale, and barley. Carbohydr Polym 2007;67:46–55.
19. Singh S, Singh N, Isono N, Noda T. Relationship of granule size distribution and amylopectin structure with pasting, thermal, and retrogradation properties in wheat starch. J Agric Food Chem 2010;58:1180–8.
20. Cornejo-Ramírez YI, Ramírez-Reyes F, Cinco-Moroyoqui FJ, Rosas-Burgos EC, Martinez-Cruz O, Carvajal-Millan E, et al. Starch debranching enzyme activity and its effects on some starch physicochemical

characteristics in developing substituted and complete triticales (X Triticosecale Wittmack). Cereal Chem 2016;93:64–70.
21. Cornejo-Ramírez YI, Martínez-Cruz O, Del Toro-Sánchez CL, Wong-Corral FJ, Borboa-Flores J, Cinco-Moroyoqui FJ. The structural characteristics of starches and their functional properties. CyTA – J Food 2018;16:1003–17.
22. Buléon A, Colonna P, Planchot V, Ball S. Starch granules: structure and biosynthesis. Int J Biol Macromol 1998;23:85–112.
23. Manners DJ. Recent developments in our understanding of amylopectin structure. Carbohydr Polym 1989; 11:87–112.
24. Copeland L, Blazek J, Salman H, Tang MC. Form and functionality of starch. Food Hydrocoll 2009;23: 1527–34.
25. Li L, Jiang H, Campbell M, Blanco M, Jane J. Characterization of maize amylose-extender (ae) mutant starches. Part I: relationship between resistant starch contents and molecular structures. Carbohydr Polym 2008;74:396–404.
26. Bertoft E. Analyzing starch molecular structure. Duxford: Elsevier; 2018. Epub ahead of print 2018.
27. Mua JP, Jackson DS. Fine structure of corn amylose and amylopectin fractions with various molecular weights. J Agric Food Chem 1997;45:3840–7.
28. Putseys JA, Lamberts L, Delcour JA. Amylose-inclusion complexes: formation, identity and physico-chemical properties. J Cereal Sci 2010;51:238–47.
29. Song Y, Jane J. Characterization of barley starches of waxy, normal, and high amylose varieties. Carbohydr Polym 2000;41:365–77.
30. Bertoft E, Piyachomkwan K, Chatakanonda P, Sriroth K. Internal unit chain composition in amylopectins. Carbohydr Polym 2008;74:527–43.
31. Kalinga DN, Waduge R, Liu Q, Yada RY, Bertoft E, Seetharaman K. On the differences in the granular architecture and starch structure between pericarp and endosperm wheat starches. Starch – Stärke 2013; 65:791–800.
32. Vermeylen R, Goderis B, Reynaers H, Delcour JA. Amylopectin molecular structure reflected in macromolecular organization of granular starch. Biomacromolecules 2004;5:1775–86.
33. Koroteeva DA, Kiseleva VI, Sriroth K, Piyachomkwan K, Bertoft E, Yuryev PV, et al. Structural and thermodynamic properties of rice starches with different genetic background Part 1. Differentiation of amylopectin and amylose defects. Int J Biol Macromol 2007;41:391–403.
34. Srichuwong S, Sunarti TC, Mishima T, Isono N, Hisamatsu M. Starches from different botanical sources I: contribution of amylopectin fine structure to thermal properties and enzyme digestibility. Carbohydr Polym 2005;60:529–38.
35. Yoo S-H, Jane J. Structural and physical characteristics of waxy and other wheat starches. Carbohydr Polym 2002;49:297–305.
36. Nakamura Y. Starch: metabolism and structure; (2015). Japan: Springer; 2015:1–451 pp.
37. Aberle T, Burchard W, Vorwerg W, Radosta S. Conformational contributions of amylose and amylopectin to the structural properties of starches from various sources. Starch – Starke 1994;46:329–35.
38. Millard MM, Dintzis FR, Willett JL, Klavons JA. Light-scattering molecular weights and intrinsic viscosities of processed waxy maize starches in 90% dimethyl sulfoxide and H2O. Cereal Chem 1997;74:687–91.
39. Takeda Y, Shibahara S, Hanashiro I. Examination of the structure of amylopectin molecules by fluorescent labeling. Carbohydr Res 2003;338:471–5.
40. Meyer KH, Bernfeld P. Recherches sur l'amidon V. L'amylopectine. Helv Chim Acta 1940;23:875–85.
41. Gunja-Smith Z, Marshall JJ, Mercier C, Smith E, Whelan W. A revision of the Meyer-Bernfeld model of glycogen and amylopectin. FEBS Lett 1970;12:101–4.
42. Horibata T, Nakamoto M, Fuwa H, Inouchi N. Structural and physicochemical characteristics of endosperm starches of rice cultivars recently bred in Japan. J Appl Glycosci 2004;51:303–13.

43. Osborne TB. The proteins of the wheat kernel/by Thomas B. Osborne. Washington, D.C: Carnegie Institution of Washington; 1907.
44. Peterson DM, Smith D. Changes in nitrogen and carbohydrate fractions in developing oat Groats1. Crop Sci 1976;16:67–71.
45. Peterson DM. Subunit structure and composition of oat seed globulin. Plant Physiol 1978;62:506–9.
46. Gangopadhyay N, Hossain MB, Rai DK, Brunton N. A review of extraction and analysis of bioactives in oat and barley and scope for use of novel food processing technologies. Molecules 2015;20:10884–909.
47. Kaur R, Sharma M, Ji D, Xu M, Agyei D. Structural features, modification, and functionalities of beta-glucan. Fibers 2020;8. https://doi.org/10.3390/fib8010001.
48. Dawkins NL, Nnanna IA. Oat gum and β-glucan extraction from oat bran and rolled oats: temperature and pH effects. J Food Sci 1993;58:562–6.
49. Kumar L, Sehrawat R, Kong Y. Oat proteins: a perspective on functional properties. LWT 2021;152:112307.
50. Sibakov J, Myllymäki O, Holopainen U, Kaukovirta-Norja A, Hietaniemi V, Pihlava J, et al. Lipid removal enhances separation of oat grain cell wall material from starch and protein. J Cereal Sci 2011;54:104–9.
51. Ramadhan K, Foster T. Effects of ball milling on the structural, thermal, and rheological properties of oat bran protein flour. J Food Eng 2018;229:50–6.
52. Prosekov A, Babich O, Kriger O, Ivanova S, Pavsky V, Sukhikh S, et al. Functional properties of the enzyme-modified protein from oat bran. Food Biosci 2018;24:46–9.
53. Liu K. Fractionation of oats into products enriched with protein, beta-glucan, starch, or other carbohydrates. J Cereal Sci 2014;60:317–22.
54. Schutyser MAI, van der Goot AJ. The potential of dry fractionation processes for sustainable plant protein production. Trends Food Sci Technol 2011;22:154–64.
55. Benito-Román O, Alonso E, Lucas S. Optimization of the β-glucan extraction conditions from different waxy barley cultivars. J Cereal Sci 2011;53:271–6.
56. Guan X, Yao H. Optimization of Viscozyme L-assisted extraction of oat bran protein using response surface methodology. Food Chem 2008;106:345–51.
57. Zhang Z, Smith C, Li W. Extraction and modification technology of arabinoxylans from cereal by-products: a critical review. Food Res Int 2014;65:423–36.
58. Galanakis CM. Sustainable applications for the valorization of cereal processing by-products. Foods 2022; 11:1–15.
59. Kriger O, Kashirskikh EV, Babich O, Noskova S. OAT protein concentrate production. Foods Raw Mater 2018;6:47–55.
60. Liu G, Li J, Shi K, Wang S, Chen J, Liu Y, et al. Composition, secondary structure, and self-assembly of oat protein isolate. J Agric Food Chem 2009;57:4552–8.
61. Mirmoghtadaie L, Kadivar M, Shahedi M. Effects of succinylation and deamidation on functional properties of oat protein isolate. Food Chem 2009;114:127–31.
62. Yung Ma C. Preparation, composition and functional properties of oat protein Isolates11contribution no. 509. Can Inst Food Sci Technol J 1983;16:201–5.
63. Anderson JW, Gilinsky NH, Deakins DA, Smith S, O'Neal D, Dillon D, et al. Lipid responses of hypercholesterolemic men to oat-bran and wheat-bran intake. Am J Clin Nutr 1991;54:678–83.
64. åman P. Cholesterol-lowering effects of barley dietary fibre in humans: scientific support for a generic health claim. Scand J Food Nutr 2006;50:173–6.
65. Regand A, Chowdhury Z, Tosh SM, Wolever TM, Wood P. The molecular weight, solubility and viscosity of oat beta-glucan affect human glycemic response by modifying starch digestibility. Food Chem 2011;129: 297–304.
66. Harasym J, Żyła E, Dziendzikowska K, Gromadzka-Ostrowska J. Proteinaceous residue removal from oat β-glucan extracts obtained by alkaline water extraction. Molecules 2019;24. https://doi.org/10.3390/molecules24091729.

67. Saskatoon B, Canada RS University of S. Extraction and enrichment (1 leads to 3), (1 leads to 4)-beta-D-glucan from barley and oat brans. Cereal Chem 1993; 70:73–7.
68. Lehtinen P, Kiiliäinen K, Lehtomäki I, Laakso S. Effect of heat treatment on lipid stability in processed oats. J Cereal Sci 2003;37:215–21.
69. Ziegler V, Ferreira CD, da Silva J, da Rosa Zavareze E, Dias ARG, de Oliveira M, et al. Heat-moisture treatment of oat grains and its effects on lipase activity and starch properties. Starch – Stärke 2018;70: 1700010.
70. Alrahmany R, Avis TJ, Tsopmo A. Treatment of oat bran with carbohydrases increases soluble phenolic acid content and influences antioxidant and antimicrobial activities. Food Res Int 2013;52:568–74.
71. Galanakis CM. Concluding remarks and future perspectives. In: Sustainable recovery and reutilization of cereal processing by-products. Duxford: Elsevier Inc; 2018:319–27 pp.
72. Wang L, Weller CL. Recent advances in extraction of nutraceuticals from plants. Trends Food Sci Technol 2006;17:300–12.
73. Smith EL, Abbott AP, Ryder KS. Deep eutectic solvents (DESs) and their applications. Chem Rev 2014;114: 11060–82.
74. Mitbumrung W, Rungraung N, Muangpracha N, Akanitkul P, Winuprasith T. Approaches for extracting nanofibrillated cellulose from oat bran and its emulsion capacity and stability. Polymers 2022;14. https://doi.org/10.3390/polym14020327.
75. Wang H, Xiang L, Rao P, Ke L, Wu B, Chen S, et al. Effects of pretreatments on structural and functional changes of oat protein isolate. Cereal Chem 2022;99:90–9.
76. Immonen M, Myllyviita J, Sontag-Strohm T, Myllarinen P. Oat protein concentrates with improved solubility produced by an enzyme-aided ultrafiltration extraction method. Foods 2021;10:3050.
77. Li Y, Obadi M, Qi Y, shi J, Liu S, Sun J, et al. Extraction of oat lipids and phospholipids using subcritical propane and dimethyl ether: experimental data and modeling. Eur J Lipid Sci Technol 2021;123:2000092.
78. Mel R, Malalgoda M. Oat protein as a novel protein ingredient: structure, functionality, and factors impacting utilization. Cereal Chem 2022;99:21–36.
79. Nieto-Nieto TV, Wang YX, Ozimek L, Chen L. Inulin at low concentrations significantly improves the gelling properties of oat protein – a molecular mechanism study. Food Hydrocoll 2015;50:116–27.
80. Zhong L, Ma N, Wu Y, Zhao L, Pei F, Hu Q. Characterization and functional evaluation of oat protein isolate-Pleurotus ostreatus β-glucan conjugates formed via Maillard reaction. Food Hydrocolloids 2019;87: 459–69.
81. Mohamed A, Biresaw G, Xu J, Hojilla-Evangelista MP, Rayas-Duarte P. Oats protein isolate: thermal, rheological, surface and functional properties. Food Res Int 2009;42:107–14.
82. Konak Üİ, Ercili-Cura D, Sibakov J, Sontag-Strohm T, Certel M, Loponen J. CO2-defatted oats: solubility, emulsification and foaming properties. J Cereal Sci 2014;60:37–41.
83. Zhang B, Guo X, Zhu K, Peng W, Zhou H. Improvement of emulsifying properties of oat protein isolate–dextran conjugates by glycation. Carbohydr Polym 2015;127:168–75.
84. Nieto TVN, Wang Y, Ozimek L, Chen L. Improved thermal gelation of oat protein with the formation of controlled phase-separated networks using dextrin and carrageenan polysaccharides. Food Res Int 2016; 82:95–103.
85. Yang C, Wang Y, Chen L. Fabrication, characterization and controlled release properties of oat protein gels with percolating structure induced by cold gelation. Food Hydrocoll 2017;62:21–34.
86. Mäkinen OE, Sozer N, Ercili-Cura D, Poutanen K. Chapter 6 - protein from oat: structure, processes, functionality, and nutrition. In: Nadathur SR, Wanasundara JPD, Scanlin LBT-SPS, editors. San Diego: Academic Press; 2017:105–19 pp.
87. Sunilkumar BA, Tareke E. Review of analytical methods for measurement of oat proteins: the need for standardized methods. Crit Rev Food Sci Nutr 2019;59:1467–85.
88. Wu YV, Cluskey JE, Wall JS, GEI. Oat protein concentrates from a wet-milling process: composition and properties. Cereal Chem 1973;50:481–8.

89. Zhao Y, Mine Y, Ma C-Y. Study of thermal aggregation of oat globulin by laser light scattering. J Agric Food Chem 2004;52:3089–96.
90. Schmidt M. Cereal beta-glucans: an underutilized health endorsing food ingredient. Crit Rev Food Sci Nutr 2020:1–20. https://doi.org/10.1080/10408398.2020.1864619.
91. Ryu J-H, Lee S, You S, Shim JH, Yoo SH. Effects of barley and oat β-glucan structures on their rheological and thermal characteristics. Carbohydr Polym 2012;89:1238–43.
92. Lan-Pidhainy X, Brummer Y, Tosh SM, Wolever TM, Wood PJ. Reducing beta-glucan solubility in oat bran muffins by freeze-Thaw treatment attenuates its hypoglycemic effect. Cereal Chem 2007;84:512–7.
93. Moriartey S, Temelli F, Vasanthan T. Effect of formulation and processing treatments on viscosity and solubility of extractable barley $β$-glucan in bread dough evaluated under in vitro conditions. Cereal Chem 2010;87:65–72.
94. Herrera MP, Gao J, Vasanthan T, Temelli F, Henderson K. β-Glucan content, viscosity, and solubility of Canadian grown oat as influenced by cultivar and growing location. Can J Plant Sci 2016;96:183–96.
95. Agarwal S. Major factors affecting the characteristics of starch based biopolymer films. Eur Polym J 2021;160:110788.
96. Lauer MK, Smith RC. Recent advances in starch-based films toward food packaging applications: physicochemical, mechanical, and functional properties. Compr Rev Food Sci Food Saf 2020;19:3031–83.
97. Shah U, Gani A, Ashwar BA, Shah A, Ahmad M, Gani A, et al. A review of the recent advances in starch as active and nanocomposite packaging films. Cogent Food Agric 2015;1:1115640.
98. Coppola G, Gaudio MT, Lopresto CG, Calabro V, Curcio S, Chakraborty S. Bioplastic from renewable biomass: a facile solution for a greener environment. Earth Syst Environ 2021;5:231–51.
99. Özeren HD, Wei X-F, Nilsson F, Olsson RT, Hedenqvist MS. Role of hydrogen bonding in wheat gluten protein systems plasticized with glycerol and water. Polymer 2021;232:124149.
100. Slavutsky AM, Bertuzzi MA. Improvement of water barrier properties of starch films by lipid nanolamination. Food Packag Shelf Life 2016;7:41–6.
101. Galdeano MC, Mali S, Grossmann MVE, Yamashita F, Garcia MA. Effects of plasticizers on the properties of oat starch films. Mater Sci Eng C 2009;29:532–8.
102. Galdeano MC, Wilhelm AE, Mali S, Grossmann MVE. Influence of thickness on properties of plasticized oat starch films. Brazilian Arch Biol Technol 2013;56:637–44.
103. Żołek-Tryznowska Z, Kałuża A. The influence of starch origin on the properties of starch films: packaging performance. Materials 2021;14:1–11.
104. Zamudio-Flores PB, Ochoa-Reyes E, Ornelas-Paz Jde J, Tirado-Gallegos JM, Bello-Pérez LA, Rubio-Ríos A, et al. Physicochemical, mechanical, and structural features of oxidized oat and banana starch films enriched with betalains. Agrociencia 2015;49:483–98.
105. Bangar SP, Whiteside WS. Nano-cellulose reinforced starch bio composite films-a review on green composites. Int J Biol Macromol 2021;185:849–60.
106. Bruni GP, de Oliveira JP, Fonseca LM, Silva FT, Dias ARG, da Rosa Zavareze E. Biocomposite films based on phosphorylated wheat starch and cellulose nanocrystals from rice, oat, and Eucalyptus husks. Starch – Stärke 2020;72:1900051.
107. Qazanfarzadeh Z, Kadivar M. Properties of whey protein isolate nanocomposite films reinforced with nanocellulose isolated from oat husk. Int J Biol Macromol 2016;91:1134–40.
108. Diyana ZN, Jumaidin R, Selamat MZ, Ghazali I, Julmohammad N, Huda N, et al. Physical properties of thermoplastic starch derived from natural resources and its blends: a review. Polymers 2021;13:1396.

Abu Hassan Nordin, Rushdan Ahmad Ilyas*, Norzita Ngadi and Nurul Huda Baharuddin

16 Pea thermoplastic starch nanocomposite films reinforced with nanocellulose

Abstract: Interest in nanocellulose has lately increased as a result of its benefits, such as renewable, biodegradable, high mechanical strength and valuable economically. In this regard, nanocellulose has been frequently employed as reinforcement for the enhancement of mechanical, thermal and biodegradation qualities of nanocomposite materials, such as thermoplastic starch. An overview of the use of pea starch that has been reinforced with nanocellulose for packaging and storage applications is given in this chapter. In comparison to standard sources of starch like maize, wheat and potatoes, it is consequently seen as a comparatively affordable source of starch. These composite polysaccharides (pea thermoplastic starch/nanocellulose) have the potential to replace traditional packaging composed of polymers derived from petroleum.

Keywords: nanocellulose; nanocomposites; packaging; pea; starch.

16.1 Introduction

Plastic packaging has been extensively used to protect and preserve food, cosmetics, pharmaceuticals and chemicals. However, the biodegradability of plastics is low. Plastics are not easily diminished through natural physical, chemical or enzymatic processes because of their strong covalent bonds structure. Consequently, plastic litter is often buried or burnt, causing contamination of the land, water and air, endangering human health and the ecology.

***Corresponding author: Rushdan Ahmad Ilyas,** Faculty of Chemical and Energy Engineering, Universiti Teknologi Malaysia (UTM), Skudai 81310, Malaysia; Centre for Advanced Composite Materials (CACM), Universiti Teknologi Malaysia (UTM), Skudai 81310, Malaysia; Centre of Excellence for Biomass Utilization, Universiti Malaysia Perlis, 02600, Arau, Perlis, Malaysia; and Institute of Tropical Forestry and Forest Products, Universiti Putra Malaysia, Serdang 43400, Selangor, Malaysia, E-mail: ahmadilyas@utm.my. https://orcid.org/0000-0001-6622-2632

Abu Hassan Nordin, Faculty of Chemical and Energy Engineering, Universiti Teknologi Malaysia (UTM), Skudai 81310, Malaysia; and Faculty of Applied Sciences, Universiti Teknologi MARA (UiTM), 02600, Arau, Perlis, Malaysia, E-mail: abuhassannordin@gmail.com. https://orcid.org/0000-0001-7546-3833

Norzita Ngadi and Nurul Huda Baharuddin, Faculty of Chemical and Energy Engineering, Universiti Teknologi Malaysia (UTM), Skudai 81310, Malaysia, E-mail: norzita@cheme.utm.my (N. Ngadi), nurul.huda@utm.my (N.H. Baharuddin)

As per De Gruyter's policy this article has previously been published in the journal Physical Sciences Reviews. Please cite as: A. H. Nordin, R. A. Ilyas, N. Ngadi and N. H. Baharuddin "Pea thermoplastic starch nanocomposite films reinforced with nanocellulose" *Physical Sciences Reviews* [Online] 2023. DOI: 10.1515/psr-2022-0026 | https://doi.org/10.1515/9783110773606-016

Plastics are made mainly by petrochemical engineering, which is largely reliant on non-renewable petroleum resources. As a result, widespread plastic production pollutes the worldwide environment and hastens the decline of petroleum reserves. It has been stated that plastic has spread to many water sources and is responsible for 60–95% of environmental contamination [1]. In this regard, the demand for plant-based goods as an alternative to non-biodegradable plastics is quickly developing in response to the environmental concerns, public pressure and government regulations [2]. The most potential plant-based substitutes are starches, which are easily produced, inexpensive, non-toxic and sustainable, particularly for usage in the packaging sector [3]. Compared to conventional starches, pea starch (PS) is a comparatively inexpensive source of starch found as a secondary product of protein extraction [4].

The procedure of treating starch granules with a plasticizer, such as polyols or amides, in the presence of water, heat and shear conditions results in the production of thermoplastic starch (TPS) [5]. The mechanical properties, moisture absorption rate and heat resistance of starch films are subpar [6, 7]. Therefore, several efforts have been made to mitigate these flaws by including fillers into the starch. Natural fibre has been shown to have considerable potential due to its abundance, remarkable mechanical qualities, versatility and low density [8, 9]. Few studies have reported the incorporation of starch with reinforcing structure from natural fibres (such as kenaf fibres [8], henequen cellulose nanofibers [9] and water hyacinth fibres [10]) in improving the TPS properties. This phenomenon is attributed to the highly crystalline 'hydrophobic' property of the cellulose from natural fibres, as opposed to starch hydrophilic characteristic [11]. In comparison to the macro size of cellulose fibres, incorporating nano-sized cellulose fibre in starch results in better tensile and thermal properties [12]. Therefore, this work discusses the potential development of pea thermoplastic starch nanocomposite films reinforced with nanocellulose for packaging and storage purposes.

16.2 Types of packaging materials

A product's package design and construction significantly impact its shelf life. Adopting the right packing materials and technology keeps product quality and freshness during transportation and storage. Generally, different types of materials were used in packagings, such as glass, metals and plastics.

16.2.1 Glass

Glass provides a lot of advantages for usage in food packaging due to its odorlessness and chemical inertness with practically all food components. Glass containers used for food packaging may be surface coated to prevent line jams, scratching and surface abrasion by lubricating the manufacturing process. Additionally, glass coatings retain and improve the bottle's resilience, minimizing shattering. Due to its resistance to gases and vapours, it keeps food fresh for a long time without compromising taste or flavour. Glass is ideal for heat sterilizing both low-acid and high-acid foods because of its capacity to endure high processing temperatures. Glass is durable, offers excellent insulation and may form into a broad variety of shapes. Glass is transparent, allowing customers to see what they are purchasing, yet because of its colour fluctuations, light-sensitive items may be shielded. Glass packaging is beneficial to the environment since it is recyclable and reusable. The glass industry also uses recycled broken glass (cullet), which makes up to 60% of total raw materials. Every substance contains flaws that are intrinsic, even glass. The heavy weight of the material raises transportation costs despite attempts to utilize thinner glass. Furthermore, there are issues with brittleness and vulnerability to fracture from internal pressure, impact or thermal stress.

16.2.2 Metal

The most flexible material for packaging is metal. It effectively blends physical barriers and protective qualities with formability, decorative potential, recyclability and consumer acceptance. Steel and aluminium are the two metals most often used in packaging.

Aluminium is a thin, silvery-white metal that may be mined from the bauxite rock, where it coexists with oxygen in a compound called alumina. Cans, foil and laminated paper or plastic packaging are often made using it. Aluminium is typically strengthened with magnesium and manganese. Unlike many other metals, aluminium has a natural coating of aluminium oxide that serves as a very efficient defence against the onslaught of air, temperature, moisture and chemicals. This natural coating of aluminium oxide makes aluminium exceedingly resistant to most types of corrosion [13].

Tin-free steel, commonly referred to as electrolytic chromium-coated steel or chrome oxide coated steel, needs an organic material coating to provide full corrosion resistance. It is not appropriate for welding but is perfect for adhering paints, lacquers and inks due to its chrome/chrome oxide composition. Moreover, it is also as strong and formable as tinplate, although being a little less costly. Food cans, can ends, trays, bottle caps, and closures can all be made out of steel that is tin-free. It may also be used to build large containers for bulk sales and storage of unfinished or finished goods.

16.2.3 Plastics

There are many benefits of using plastic for food packaging. Plastics provide manufacturers a variety of options since they are flexible and moldable. Moreover, plastics are affordable, lightweight, and have a variety of physical and cosmetic qualities. In fact, many polymers may be used in manufacturing processes where the package is created, filled, and sealed all on the same line since they are heat sealable, easily printed, and flexible. Polymerization (polycondensation) or addition polymerization is used to add or polymerize monomer units to produce condensation polymers (polyaddition). When a polymer chain extends due to condensation interactions between molecules, polycondensation occurs, which results in the production of low molecular weight byproducts like water and methanol. Alcohol, amine or carboxylic group-containing monomers that have at least two functional groups are polycondensed. When two or more molecules join to produce a larger molecule with no byproducts, this process is known as addition, and it is how polymer chains are created. In polyaddition, double or triple bonds are broken to connect monomer chains, and unsaturated monomers are employed as the starting material.

16.3 Bio-based plastics

There are three main ways to obtain bio-based polymers [14]. One of them entails making some minor modifications to natural polymers while mostly preserving the polymer's mechanical strength [15]. This technique is used to synthesize bio-based polymers and fibres used in non-plastic and non-food applications.

The second method uses a two-stage process for converting biomass, commencing with the creation of bio-based precursors (monomers) by biochemical and/or chemical transformation before polymerizing the monomers in the latter phase (both steps are normally complex and can be divided in several sub-steps). Drop-in monomers are obtained if they are bio-based analogues of conventional monomers (replacements). The advantage of conventional plastics made using drop-in monomers is that they are indistinguishable from their petrochemical equivalents in terms of both structure and performance. In this regard, they are readily penetrated to the current processing and recycling systems. Plastics are produced using bio-based monomers with new structures or those that have not before been employed in commercial applications. These "new" bio-based polymers often have distinctive qualities. They often exhibit increased functioning and, as a result, new markets and applications when compared to the conventional polymers that they potentially replace. Novel bio-based plastics, as opposed to drop-in plastics, need new recycling processes to be developed and put into place. The development of the chemical and biotechnological synthesis of monomers has led to a continuous increase in the relevance of this second method for bio-based plastics.

The third method involves creating a polymeric substance directly in microorganisms or plants, which may then be utilized as plastic without further modifying it. This approach is becoming increasingly viable because of advancements in genetic engineering and biotechnology that make it possible to transfer genes responsible for the creation of a polymer like poly(hydroxyalkanoate) (PHA) from bacteria into crops. Even while direct photosynthesis-based manufacture of bio-based plastics may appear advantageous, there are several significant environmental and regulatory concerns. Despite years of continuous study, there have been no significant amounts of bio-based plastics created in accordance with this technique.

16.4 Starch

Starch is a storage carbohydrate found in the leaf and storage tissues of plants. The shape, size and composition of the granules will change depending on the starch supply and stage of growth [16, 17]. Amylopectin and amylose, both polysaccharides made up of -D-glucose linked by (1 → 4) glycoside bonds, make up the majority of starch. Amylopectin is extensively branched, with (1 → 6) connecting the reducing end side, which accounts for 70–80% of starch. Despite the presence of branches on occasion, amylose is primarily a linear polymer. It is minimally soluble in water and is structured in a helicoidal form in starch granules; it accounts for 20–25% of the starch composition in most cases [18]. Amylose has a crystalline pattern with a straight and lengthy backbone, enabling film production via strong network contact. On the other hand, amylopectin contains short, branched chains that result in a brittle film with poor intermolecular interaction [19].

Tape casting, compression moulding, injection moulding, extrusion, and solvent casting are some of the procedures used to make starch-based plastics. Although solvent casting is an expensive scaling-up approach due to the time-consuming drying stage, it is the primary technique employed at benchtop scale to explore and create innovative film compositions using alternate starch sources [20]. Plasticizers are frequently included in the casting solution since starch does not have thermoplastic characteristics. Plasticizers are low-volatility chemicals that can diminish hydrogen-bonding intermolecular contacts between adjacent portions of polymeric chains, especially in amorphous regions, enhancing flexibility and making the film easier to remove from the supporting mould [21].

16.5 Pea starch (PS)

Pea is farmed in nearly every country. In Asia and South America, the most prevalent use of field pea is for food, where the split or whole seeds are cooked and consumed [4]. In comparison to corn, wheat and potato starches, PS is a comparatively inexpensive source of starch, which is typically found as a secondary product from the protein. Due to its low functional qualities, PS is mostly employed in industrial applications and is rarely used in food. PS contains a high amylose concentration, which varies by species with ranges from 30 to 60%, making them ideal for film formation [22].

16.5.1 Extraction and preparation of PS

It is challenging to separate the starches from peas because they include insoluble flocculent proteins and fine fibre that impede sedimentation and settle with the starch to form a brownish deposit [23]. Pea starches were separated using methods including air classification and pin milling [23, 24]. Among these techniques, the most practical approach is air classification. The technique demands a significant amount of particle size reduction to remove the starch granules from the protein matrix (achieved by pin milling). The most crucial fraction is the low protein starch fraction, which is separated from the fine protein fraction during the air classification procedure. In the starch concentration, starch makes up around 65% of the mixture [25]. All sources of the residual protein connected to the air are categorized. Protein bodies, aggregates, chloroplast membrane fragments (which enclose the starch granule) and a water-soluble substance that seems to have developed from the dehydrated starch are all components of PS granules. Most protein bodies and agglomerates are eliminated by re-milling and reclassifying the starch fraction, and the majority of the remaining associated protein is eliminated by water washing [23]. Protein composition in the washed starch is 0.25% after the purification technique described above. Tyler et al. [26] suggested that the effectiveness of protein separation may be utilized to determine if the starch extraction process of pin-milling and air-classification is feasible. According to the authors, the PS extraction is impacted by seed hardness and the volume of decreased protein that cannot be milled away from the starch granules. Meuser et al. [27] proposed a technique to separate starch from wrinkled peas, where up to 89% of the starch in wrinkled peas could be retrieved, with just around 1% of the protein.

16.5.2 Properties of PS

Ratnayake et al. [24] reported that field pea starches have yields of 35–40% pure starch, 0.52–0.70% protein and 0.01–0.07% ash, respectively. Additionally, the ratios of amylose to amylopectin and the presence of a low-molecular-weight intermediate component in wrinkled pea starches distinguish them from regular pea (smooth pea) starches. The amylose contents of the starches from smooth and wrinkled peas were 33.1–49.6% and 60.5–88%, respectively. Colonna and Mercier [28] found that the branching intermediate fraction of low molecular weight (Mn = 25,100) is present in both the smooth and wrinkled pea starches. The significant differences in amylose content between smooth and wrinkled pea starches may be attributed to the physiological state of the seed and different methods for determining amylose. Additionally, the standard pea starch comes in granule sizes ranging from 2 to 40 μm. The majority of the granules are oval, but there are also those that are spherical, round, elliptical and have irregular shapes.

Gujska et al. [29] found that compared to pinto and navy beans, smooth pea starch had a greater swelling power and solubility. A greater interaction between densely packed amylose chains and/or a higher concentration of amylose–lipid complexes may be the cause of the wrinkled pea starch's decreased swelling ability. The Brabender viscoamylograph (BVA) and rotational viscometers were used to study the rheological characteristics of starches. Compared to cereal starches, pea starches have less information on their rheological behaviour under well-defined flow regimes. The ability of pea starches to paste depends significantly on genetic variables (depending on the type or cultivar) [24, 30].

Starch granules expand irreversibly when heated in excess water over their gelatinization temperature, causing amylose to leak into the solution. If the starch content is high enough, this suspension will form an elastic gel upon chilling. Retrogradation is the term for the chemical reactions that follow cooling (mostly hydrogen bonding between starch chains). We find that the temperature and the passage of time affect these interactions. The degree of smooth pea starch retrogradation (as determined by variations in amylopectin crystallization after storage at 6 °C for 4 days) in comparison to cereal and tuber starches. The degree of retrogradation of smooth pea starch is discovered to be midway between potato and cereal starches (potato > cereal), depending on changes in enthalpy of retrogradation. The differences in retrogradation among the starches were attributed to variations in the average length of the amylopectin chain.

16.5.3 Preparation of thermoplastic pea starch

Granules of starch that are often seen have a crystallinity of 15–45%. Plasticizers must be used in the production of biodegradable thermoplastic starch (TPS) in order to avoid granular structure because they may replace starch's strong intra- and intermolecular hydrogen bonds with hydrogen bonding, which causes plasticization. Polyols, such as glycerol [31], glycol [32], xylitol [33], sorbitol [34] and sugars [35], are often employed as plasticizers for TPS. Amides such as urea [36], formamide [37] and ethylenebisformamide [38] promote the plasticization of starch too. The characteristics of TPS are influenced by the kind of plasticizer. Dai et al. [39] reported that N,N-bis(2-hydroxyethyl)formamide (BHF) was synthesized efficiently and used as a new plasticizer for pea starch to prepare thermoplastic starch (TPS). The Fourier-transform infrared spectra study demonstrated that BHF established robust and stable hydrogen bonds with starch. Starch granules were totally broken up and a homogenous phase was discovered by scanning electron microscopy research. An X-ray diffraction examination showed that BHF plasticized pea starch and caused the starch's crystallinity to vanish.

16.6 Advantages and disadvantages of biopolymer plastics for packaging

There are various benefits of using bio-based plastic for packaging applications, including a reduction in carbon dioxide releases. Bioplastics release between 0.8 and 3.2 metric tonnes less carbon dioxide per tonne than plastics derived from petroleum [40]. According to a Life Cycle Analysis, a typical polypropylene or polystyrene plastic has an approximate carbon footprint of 2.0 kg carbon dioxide equivalents per kilogramme of plastic (from cradle to factory gate). These emissions of carbon dioxide are four times more than those of poly lactic acid (PLA) resin [40]. In addition, biopolymer polymers are a less expensive option since they are more resilient to fluctuations in oil prices. When it comes to waste management, bio-plastics lessen the quantity of harmful runoff produced by the oil-based substitutes. Utilizing biopolymer plastics has positive effects. As nations throughout the globe hunt for oil replacements to save the environment and attain energy security. The main raw element for petroleum-based polymers is fossil fuel. As a result, compared to bioplastics, the creation of plastics using petroleum-based materials like polypropylene and polypropylene requires more energy. There are many end-of-life solutions available, including the possibility of recycling valuable raw materials into new goods to lessen the demand for new virgin materials and to significantly minimize, if not completely eliminate, the harmful environmental effects of "used" plastic items. Table 16.1 shows the comparison between biopolymer and petroleum-based plastics.

16.7 Lignocellulosic fibre

Some of bioplastics have previously been reinforced with lignocellulosic fibres to improve biodegradation and mechanical properties, as this technique resulted in significant increases in the composites' tensile strength [41]. Given their intriguing characteristics, such as their renewability, abundant, non-food agricultural economy, environmental friendly, inexpensive, low density, excellent mechanical properties and

Table 16.1: Comparison between biopolymer and petroleum-based plastics [40].

Factor	Biopolymer plastic	Petroleum-based plastic
Renewability	Renewable	Partially non-biodegradable
Environmental breakdown	Biodegradable and/or compostable	Some degradable by polymer oxidation
Greenhouse gas release	Generally low	Highly significant
Use of fossil fuels	Generally low	Highly significant

modulus and relatively simple processing due to their nonabrasive nature, lignocellulosic seem to be appropriate fillers or reinforcing agents for biodegradable materials [42].

16.7.1 Cellulose fibre

Packaging made of cellulose fibre is critical for storing and transporting goods. Cellulosic fibres often add significantly to the package's strength and structural stability in such applications. The fibres can be utilized as a self-standing thin film, a composite filler or a coating to give excellent barrier qualities [43]. The tensile strength and elastic modulus of the composites were increased as reported in the previous studies using various sources of starch from corn starch [44], tapioca starch [45], rice starch [6] and potato starch [46] mixed with cellulose fibres [47].

Wattanakornsiri et al. [48] investigated the surface morphology of starch–cellulose composites using scanning electron microscopy (SEM). The results revealed that the cellulose fibres appear to be embedded in the matrix and have high adhesion properties, as well as that the TPS matrix stays tightly joined to the fibres even after a cryo-fracture test with no obvious debonding occurrences [49].

16.7.2 Microcrystalline cellulose (MCC)

According to Ma et al. [11], MCC was employed as a reinforcement filler in a glycerol-plasticized TPS matrix. The storage modulus and glass transition temperature of the composites were enhanced after MCC was introduced. At a low water content (13%), MCC improved thermal stability, tensile stress and elongation at break, as well as the water vapour barrier. This was attributed to cellulose derivatives' and starch's good interaction. On the other hand, extra cellulose derivatives caused aggregation in the TPS matrix, thus, lowering their effective content. The addition of MCC to the composites enhanced the storage modulus and glass transition temperature. At low water content (13%), MCC improved thermal stability, tensile stress and elongation at break, as well as the water vapour barrier. The favourable interaction between cellulose derivatives and starch was attributed to this. Surplus cellulose derivatives, on the other hand, caused aggregation in the TPS matrix, lowering their effective contents.

16.7.3 Nanocellulose fibre (NC)

Nanocellulose is a bio-based substance found in abundance in natural plants. Nanosized reinforced composites feature a large surface area with abundant surface hydroxyl groups [50], a high aspect ratio [51], high crystallinity, strong heat resistance [52], good mechanical properties [53] and fewer flaws in the reinforcing portion [54], compared to microsized reinforced composites. In these regards, the use of nanocellulose as a loading-bearing reinforcement in the development of innovative and low-cost biodegradable polymers has grown in popularity.

Montero et al. [5] reported that filler strengthens TPS/NC films by raising the storage modulus. They had a more noticeable effect on starch with a larger amount of amylopectin chains. The addition of NC results in more homogenous surfaces, enhances the amorphous halo and reduces peak intensities in X-ray diffraction (XRD) profiles. It suggests a relationship between nanoparticles and glycerol molecules that hinders the development of an amylose–glycerol complex. This claim is supported by an increase in relaxation temperatures corresponding to glycerol-rich domains found by dynamic mechanical study of TPS/NC films. The presence of NC raises the temperatures of relaxation corresponding to starch-rich domains, which is often ascribed to interactions between the filler and matrix as well as to the development of a hydrogen-bonded cellulose network inside the matrix.

16.8 Modification of starch films

TPS has a number of drawbacks as compared to standard polymers, including a high hydrophilic character and low mechanical qualities, which make it unsuitable for applications such as packaging [47]. As a result, structural modification is required and desirable in the great majority of cases, not just to address these problems but also to introduce new functional capabilities. The modifications have improved its mechanical strength, amphiphilicity, paste clarity, thermal stability, hydrophobicity, freeze–thaw stability and retrogradation resistances.

For example, Ma et al. [55] prepared biodegradable packaging materials for packaging using pea starch–based composites reinforced with citric acid–modified pea starch (CAPS) and citric acid–modified rice starch (CARS). Compared to TPS/CAPS composites, the incorporation of TPS with CARS demonstrated greater storage modulus, tensile strength, elongation at break and gas permeability. Moreover, Zhou et al. [56] recommended to incorporate an octenyl succinic anhydride with pea starch to effectively strengthen the materials' adherence to one another, creating a mechanically improved interpenetrating network. As a result, the films' characteristics were improved, and they had a positive impact on strawberry storage.

16.9 Challenges and future perspective

The development and production of these materials are not cost-effective. It is critical to replace synthetic polymers in packaging applications. However, for customers, the price of packaging stuff outweighs ecological considerations. In order for customers to evaluate the grade, security, lifespan and nutritional value of the materials of the package, additional research is required to add new values to the market. One such value is to make packaging materials simpler while yet being intelligent. A variety of uses in medicine and allied sectors are available for biodegradable materials. Producing biodegradable polymer materials for disposable electrode systems for medical applications and less cost packaging materials for food spoilage detection may be given priority. However, manufacturing lower-cost materials in this industry is critical because the ultimate product cost would be passed on to customers. Therefore, the establishment of these materials necessitates a multidisciplinary approach to study.

16.10 Conclusions

Thermoplastic starch-based polymers have wide applications in industries, including food packaging. Employing starch-based materials for shorter-term usage, such as in the form of mulching films, sacks and biodegradable items, will be fascinating and advantageous. This not only minimizes the possibility of harmful compounds from synthetic polymers being released but also contributes to reduce waste. Since biodegradable TPS and their derivatives have been discovered to have inadequate qualities, there is always room for development of thermal resistance, mechanical capabilities and permeation properties without sacrificing biodegradability. This work discussed the potential of pea thermoplastic starch nanocomposite films reinforced with nanocellulose as a potential biodegradable packaging. The quality of the nanocomposite films improved as a result of the incorporation of nanocellulose to the TPS matrix, which boosted crystallinity and tensile strength.

Acknowledgement: The research has been carried out under the programme, Research Excellence Consortium (JPT (BPKI) 1000/016/018/25 (57)), provided by the Ministry of Higher Education Malaysia (MOHE).

References

1. Scheurer M, Bigalke M. Microplastics in swiss floodplain soils. Environ Sci Technol 2018;52:3591–8.
2. Teixeira EdM, Lotti C, Corrêa AC, Teodoro KB, Marconcini JM, Mattoso LH. Thermoplastic corn starch reinforced with cotton cellulose nanofibers. J Appl Polym Sci 2011;120:2428–33.
3. Mali S, Grossmann MVE, Garcia MA, Martino MN, Zaritzky NE. Microstructural characterization of yam starch films. Carbohydr Polym 2002;50:379–86.
4. Ratnayake WS, Hoover R, Warkentin T. Pea starch: composition, structure and properties—a review. Starch Staerke 2002;54:217–34.
5. Montero B, Rico M, Rodríguez-Llamazares S, Barral L, Bouza R. Effect of nanocellulose as a filler on biodegradable thermoplastic starch films from tuber, cereal and legume. Carbohydr Polym 2017;157: 1094–104.
6. Prachayawarakorn J, Sangnitidej P, Boonpasith P. Properties of thermoplastic rice starch composites reinforced by cotton fiber or low-density polyethylene. Carbohydr Polym 2010;81:425–33.

7. Tomé LC, Fernandes SC, Perez DS, Sadocco P, Silvestre AJ, Neto CP, et al. The role of nanocellulose fibers, starch and chitosan on multipolysaccharide based films. Cellulose 2013;20:1807–18.
8. Zainuddin SYZ, Ahmad I, Kargarzadeh H, Abdullah I, Dufresne A. Potential of using multiscale kenaf fibers as reinforcing filler in cassava starch-kenaf biocomposites. Carbohydr Polym 2013;92:2299–305.
9. Fazeli M, Keley M, Biazar E. Preparation and characterization of starch-based composite films reinforced by cellulose nanofibers. Int J Biol Macromol 2018;116:272–80.
10. Abral H, Dalimunthe MH, Hartono J, Efendi RP, Asrofi M, Sugiarti E, et al. Characterization of tapioca starch biopolymer composites reinforced with micro scale water hyacinth fibers. Starch Staerke 2018;70:1700287.
11. Ma X, Chang PR, Yu J. Properties of biodegradable thermoplastic pea starch/carboxymethyl cellulose and pea starch/microcrystalline cellulose composites. Carbohydr Polym 2008;72:369–75.
12. Kaewtatip K, Thongmee J. Studies on the structure and properties of thermoplastic starch/luffa fiber composites. Mater Des 2012;40:314–8.
13. Marsh K, Bugusu B. Food packaging—roles, materials, and environmental issues. J Food Sci 2007;72: R39–55.
14. Babu RP, O'connor K, Seeram R. Current progress on bio-based polymers and their future trends. Prog Biomater 2013;2:1–16.
15. Aravamudhan A, Ramos DM, Nada AA, Kumbar SG. Natural polymers: polysaccharides and their derivatives for biomedical applications. In: Natural and synthetic biomedical polymers. Elsevier; 2014:67–89 pp.
16. Zabot GL, Silva EK, Emerick LB, Felisberto MHF, Clerici MTPS, Meireles MAA. Physicochemical, morphological, thermal and pasting properties of a novel native starch obtained from annatto seeds. Food Hydrocolloids 2019;89:321–9.
17. Tagliapietra BL, Felisberto MHF, Sanches EA, Campelo PH, Clerici MTPS. Non-conventional starch sources. Curr Opin Food Sci 2021;39:93–102.
18. Molavi H, Behfar S, Shariati MA, Kaviani M, Atarod S. A review on biodegradable starch based film. J Microbiol Biotechnol Food Sci 2021;2021:456–61.
19. Dularia C, Sinhmar A, Thory R, Pathera AK, Nain V. Development of starch nanoparticles based composite films from non-conventional source-water chestnut (Trapa bispinosa). Int J Biol Macromol 2019;136:1161–8.
20. do Val Siqueira L, Arias CILF, Maniglia BC, Tadini CC. Starch-based biodegradable plastics: methods of production, challenges and future perspectives. Curr Opin Food Sci 2021;38:122–30.
21. Nawab A, Alam F, Haq MA, Lutfi Z, Hasnain A. Mango kernel starch-gum composite films: physical, mechanical and barrier properties. Int J Biol Macromol 2017;98:869–76.
22. Hoover R, Hughes T, Chung H, Liu Q. Composition, molecular structure, properties, and modification of pulse starches: a review. Food Res Int 2010;43:399–413.
23. Reichert R, Youngs C. Starch fractions from air-classified field peas. Cereal Chem 1978;55:469–80.
24. Ratnayake W, Hoover R, Shahidi F, Perera C, Jane J. Composition, molecular structure, and physicochemical properties of starches from four field pea (Pisum sativum L.) cultivars. Food Chem 2001;74:189–202.
25. Comer F, Fry M. Purification, modification, and properties of air-classified pea starch. Cereal Chem 1978;55: 818–29.
26. Tyler R, Youngs C, Sosulski F. Air classification of legumes [beans, lentils, peas]. I. Separation efficiency, yield, and composition of the starch and protein fractions. Cereal Chem 1981;58:144–8.
27. Meuser F, Pahne N, Möller M. Extraction of high amylose starch from wrinkled peas. Starch Staerke 1995; 47:56–61.
28. Colonna P, Mercier C. Macromolecular structure of wrinkled-and smooth-pea starch components. Carbohydr Res 1984;126:233–47.
29. Gujska E, Reinhard WD, Khan K. Physicochemical properties of field pea, pinto and navy bean starches. J Food Sci 1994;59:634–6.
30. Davydova N, Leont'Ev S, Genin YV, Sasov AY, Bogracheva TY. Some physico-chemical properties of smooth pea starches. Carbohydr Polym 1995;27:109–15.

31. García NL, Ribba L, Dufresne A, Aranguren M, Goyanes S. Effect of glycerol on the morphology of nanocomposites made from thermoplastic starch and starch nanocrystals. Carbohydr Polym 2011;84: 203–10.
32. Da Róz A, Carvalho A, Gandini A, Curvelo A. The effect of plasticizers on thermoplastic starch compositions obtained by melt processing. Carbohydr Polym 2006;63:417–24.
33. Qiao X, Tang Z, Sun K. Plasticization of corn starch by polyol mixtures. Carbohydr Polym 2011;83:659–64.
34. Li H, Huneault MA. Comparison of sorbitol and glycerol as plasticizers for thermoplastic starch in TPS/PLA blends. J Appl Polym Sci 2011;119:2439–48.
35. Teixeira EdM, Da Roz A, Carvalho A, Curvelo A. The effect of glycerol/sugar/water and sugar/water mixtures on the plasticization of thermoplastic cassava starch. Carbohydr Polym 2007;69:619–24.
36. Ma XF, Yu J, Wan J. Urea and ethanolamine as a mixed plasticizer for thermoplastic starch. Carbohydr Polym 2006;64:267–73.
37. Ma X, Yu J. Formamide as the plasticizer for thermoplastic starch. J Appl Polym Sci 2004;93:1769–73.
38. Yang JH., Yu JG., Ma XF. Preparation of a novel thermoplastic starch (TPS) material using ethylenebisformamide as the plasticizer. Starch Staerke 2006;58:330–7.
39. Dai H, Chang PR, Yu J, Ma X, Zhou P. Preparation and properties of thermoplastic pea starch using N, N-bis (2-hydroxyethyl) formamide as the plasticizer. Polym Eng Sci 2010;50:970–7.
40. Reddy RL, Reddy VS, Gupta GA. Study of bio-plastics as green and sustainable alternative to plastics. Int J Emerg Technol Adv Eng 2013;3:76–81.
41. Tserki V, Matzinos P, Zafeiropoulos N, Panayiotou C. Development of biodegradable composites with treated and compatibilized lignocellulosic fibers. J Appl Polym Sci 2006;100:4703–10.
42. Orts WJ, Shey J, Imam SH, Glenn GM, Guttman ME, Revol J-F. Application of cellulose microfibrils in polymer nanocomposites. J Polym Environ 2005;13:301–6.
43. Ferrer A, Pal L, Hubbe M. Nanocellulose in packaging: advances in barrier layer technologies. Ind Crop Prod 2017;95:574–82.
44. Ma X, Yu J, Kennedy JF. Studies on the properties of natural fibers-reinforced thermoplastic starch composites. Carbohydr Polym 2005;62:19–24.
45. Teixeira EdM, Pasquini D, Curvelo AA, Corradini E, Belgacem MN, Dufresne A. Cassava bagasse cellulose nanofibrils reinforced thermoplastic cassava starch. Carbohydr Polym 2009;78:422–31.
46. Thuwall M, Boldizar A, Rigdahl M. Extrusion processing of high amylose potato starch materials. Carbohydr Polym 2006;65:441–6.
47. Averous L, Boquillon N. Biocomposites based on plasticized starch: thermal and mechanical behaviours. Carbohydr Polym 2004;56:111–22.
48. Wattanakornsiri A, Pachana K, Kaewpirom S, Sawangwong P, Migliaresi C. Green composites of thermoplastic corn starch and recycled paper cellulose fibers. Songklanakarin J Sci Technol 2011;33.
49. Curvelo A, De Carvalho A, Agnelli J. Thermoplastic starch–cellulosic fibers composites: preliminary results. Carbohydr Polym 2001;45:183–8.
50. Majeed K, Jawaid M, Hassan A, Bakar AA, Khalil HA, Salema A, et al. Potential materials for food packaging from nanoclay/natural fibres filled hybrid composites. Mater Des 2013;46:391–410.
51. Tee T-T, Sin LT, Gobinath R, Bee S-T, Hui D, Rahmat A, et al. Investigation of nano-size montmorillonite on enhancing polyvinyl alcohol–starch blends prepared via solution cast approach. Compos B Eng 2013;47: 238–47.
52. Hajaligol M, Waymack B, Kellogg D. Low temperature formation of aromatic hydrocarbon from pyrolysis of cellulosic materials. Fuel 2001;80:1799–807.
53. Minelli M, Baschetti MG, Doghieri F, Ankerfors M, Lindström T, Siró I, et al. Investigation of mass transport properties of microfibrillated cellulose (MFC) films. J Membr Sci 2010;358:67–75.
54. Khalil HA, Davoudpour Y, Islam MN, Mustapha A, Sudesh K, Dungani R, et al. Production and modification of nanofibrillated cellulose using various mechanical processes: a review. Carbohydr Polym 2014;99: 649–65.

55. Ma X, Chang PR, Yu J, Stumborg M. Properties of biodegradable citric acid-modified granular starch/ thermoplastic pea starch composites. Carbohydr Polym 2009;75:1–8.
56. Zhou X, Cheng R, Wang B, Zeng J, Xu J, Li J, et al. Biodegradable sandwich-architectured films derived from pea starch and polylactic acid with enhanced shelf-life for fruit preservation. Carbohydr Polym 2021;251: 117117.

Nur Sharmila Sharip, Tengku Arisyah Tengku Yasim-Anuar,
Farhana Aziz Ujang and Mohd Nor Faiz Norrrahim*

17 Potato thermoplastic starch nanocomposite films reinforced with nanocellulose

Abstract: Potato is a widely available feedstock with biocompatibility and biodegradability properties, making it a strong candidate for producing thermoplastic starch. The application of thermoplastic starch to replace petroleum-based plastic as a sustainable and environmentally friendly approach led to its further improvement through various techniques such as modification and filler reinforcement. Numerous studies have been done addressing the properties enhancement of potato thermoplastic starch through filler reinforcement including nanocellulose. This review focus on the recent and future potential of potato-based starch as one of the feedstocks for producing potato thermoplastic starch composites reinforced with nanocellulose.

Keywords: nanocellulose; potato; thermoplastic starch.

17.1 Introduction

Plastics made from non-renewable petroleum are incredibly useful materials for modern society, yet their waste is undoubtedly increasing in tandem with worsening environmental impacts [1–4]. Thus, extensive research has been conducted to develop renewable plastic resources that can be used as an alternative to nonrenewable plastics on the market, and thermoplastic starch (TPS) is one of the green plastic materials that has gained great attention from researchers and industrial players. TPS is typically composed of starch, glycerol, and water [5]. Yet, despite having better properties than native starch that are poor processability, and lack of flow behavior, TPS is still inferior to the conventional petroleum-based polymer due to its sensitivity to moisture, retrogradation processes, and unsatisfactory mechanical strength or stiffness, hence limiting its use in a variety of applications [6]. Therefore, TPS needs to be incorporated with other materials such as nanocellulose to enhance its physical and chemical properties.

As a versatile green material, nanocellulose has been used as a reinforcement material in various polymers and its effects on these polymers have been widely

***Corresponding author: Mohd Nor Faiz Norrrahim**, Research Center for Chemical Defence, Universiti Pertahanan Nasional Malaysia, Kem Sungai Besi, 57000 Kuala Lumpur, Malaysia, E-mail: faiz@upnm.edu.my
Nur Sharmila Sharip, Tengku Arisyah Tengku Yasim-Anuar and Farhana Aziz Ujang, Research and Development Department, Nextgreen Pulp & Paper Sdn Bhd, Taman Tun Dr. Ismail, 60000 Kuala Lumpur, Malaysia

As per De Gruyter's policy this article has previously been published in the journal Physical Sciences Reviews. Please cite as: N. S. Sharip, T. A. T. Yasim-Anuar, F. A. Ujang and M. N. Faiz Norrrahim "Potato thermoplastic starch nanocomposite films reinforced with nanocellulose" *Physical Sciences Reviews* [Online] 2023. DOI: 10.1515/psr-2022-0015 | https://doi.org/10.1515/9783110773606-017

discussed [7–15]. Its utilization in starch biocomposite in general has also been reviewed elsewhere [16–19]. In such regards, potatoes possess a bright prospective in the area of TPS production and could be a good resource for producing thermoplastic starch nanocellulose composite. Not only due to its huge production worldwide which is 360 million tons per year, but also owing to its high amylose content (>38%) and high starch content (6.9–30.7 wt%) that would enabled good mechanical properties [20–22]. The spherical shape of potato starch granules may also contributes to easier processing [5]. Fahrngruber [20] also highlighted that high amylose and starch content result in strong films or packaging, making it suitable for TPS production.

In this review, the potato-based thermoplastic starch nanocomposite films reinforced with nanocellulose is further elaborated in terms of productions, properties, and potential application.

17.2 Potato based starch

Potato (*Solanum tuberosum* L.) is an extremely alluring crop in agricultural production systems. Having unusually high yield potential with a high nutritional value, potato contains energy, nutrients, proteins, lipids, and vitamins which make it one of the important food sources. It is also used as raw materials for various industrial goods, hence unlike most other crops; potatoes cultivation is also quite appealing as it have a variety of purposes than merely as food [23]. According to the Food and Agriculture Organization (FAO) statistic and worldwide distribution map shown in Figure 17.1, largest producers such as China, India, Ukraine, Russia, and the United States of America (USA) produced about 166 million, 46 million, 22 million, 26 million, and 20 million tons potatoes annually since 2010 [24]. Other countries like Canada, Brazil, and Australia make approximately 5.5 million, 3.7 million, and 1.2 million potatoes in average, every year. With over 150 countries listed as potato producers, the yearly production of potato reached approximately 360 million every year since the last decade (Figure 17.2).

Potato tubers mostly consist of starch, which accounts for 15–20% of their weight [25]. Attributed to the structure of the tissue and the low protein and fat content (values below 4%), the extraction of potato starch is rather straightforward. The only procedures required are milling, decantation, centrifugation, several washing of the starch with distilled water, and drying [26, 27]. The extraction reagent used was only water whereby extraction conditions such as water ratios, samples sizes, mixing times, and sieves sizes can be optimized for better starch yield and purity [28]. Potatoes were mashed and ground before the two solid components were physically separated (starch in the solution and solid fiber) [29]. The potatoes juice was then filtered through a muslin cloth and let to sediment for a few hours. The excess supernatant and the sedimented starch were separated through decantation or centrifugation prior drying off the sedimented starch cake to get the dry potato starch. The separation of sedimented starch from the potatoes

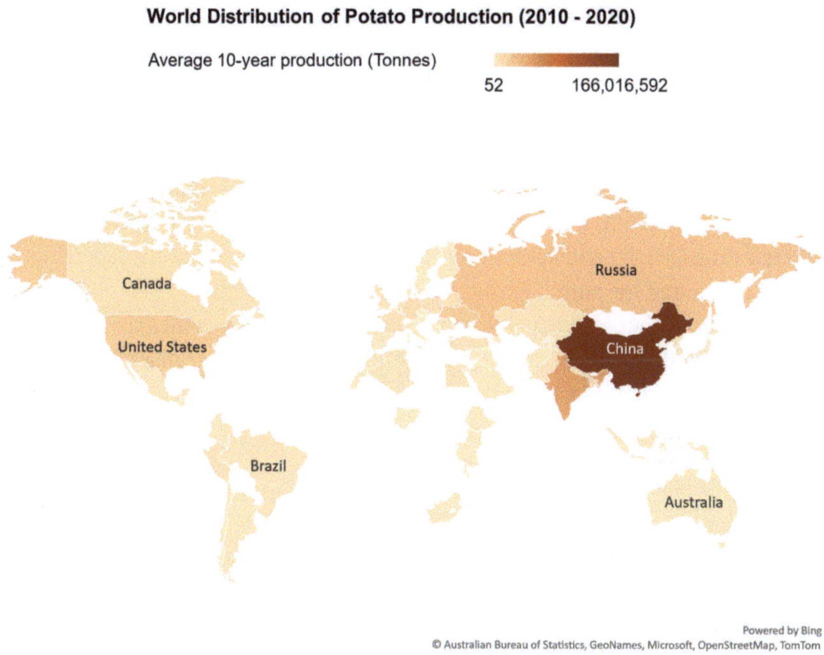

Figure 17.1: The distribution of potato average production worldwide from the year 2010–2020 [24].

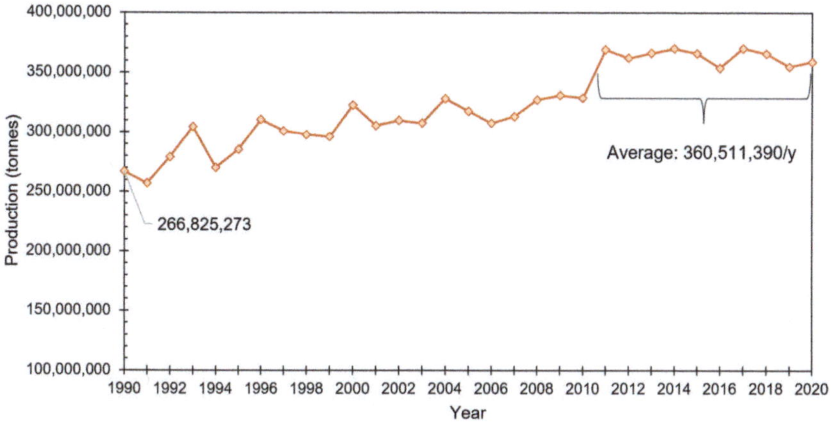

Figure 17.2: The cumulative production of potato from 2010 to 2020 [24].

juice is essentially an important step as the rotation speed and time of centrifugation could influence the potato starch yield [30].

The partially crystalline starch granules are made up of two types of alpha glucan which are amylopectin and amylose as the major and minor components in starches, respectively [31–33]. Figure 17.3 shows that short chains of glucan linked by -(1–4) links and -(1–6) linkages form the highly branched polymer known as amylopectin [34], while glucan chains connected by -(1–4) connections make up the linear polymer amylose with small side branches [35]. According to Tester et al. [36], compared to amylose molecules with a wide distribution profile, amylopectin unit chains are comparatively short, by which approximately 4–6% of amylopectin linkages are α-(1–6) linkages that make it extensively branched. Besides amylose and amylopectin, starch also contains proteins, lipids, and minerals [35], by which these components can vary in quantity depending on the growing season and the starch extraction method or techniques.

In comparison to other starch, potato starch contains higher granules size (24–75 µm) and higher average diameter of the starch granules (46.02 µm) (Table 17.1) [31]. The granules of potato starch are large and smooth, with a high concentration of covalently bonded phosphate, long amylopectin chains, and high-molecular-weight amylose. It typically contains 20–33% amylose on a total starch basis and the balance 80–77% was

Table 17.1: The granules size, diameter of starch granules, amylose content, and particle size dispersion from different type of starch.

Type of starch	Granules size (µm)	Diameter of starch granules (µm)	Amylose content (%)	Particle size dispersion
Potato	24–75	46.02	20.5	1.1
Corn	10–24	16.38	24.8	0.9
Wheat	12–31	20.75	24.5	1.0
Rice	5–52	23.30	16.9	2.9

Table adopted from Domene-López et al. [31].

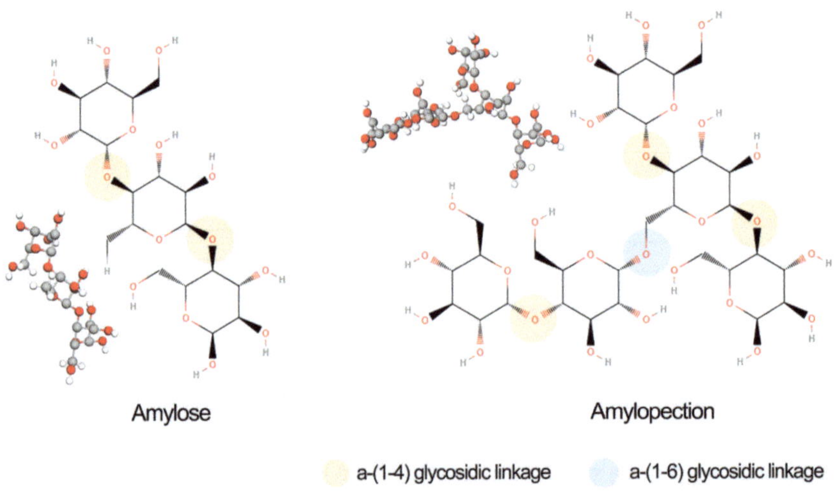

Figure 17.3: Structure of amylose and amylopectin.

amylopectin [34, 37]. It also contains some minerals such as phosphorous, existing in the form of phosphate covalently linked to the amylopectin component. The concentration of this mineral is relatively higher in potato starch as compared to other starch, hence contributing to its higher viscosity, clarity and swelling [36]. Additionally, potato starch is a high hygroscopic material which reached values of around 15% while other starch showed a value around 9–11% [31].

The relatively bigger granules sizes of potato starch make it to be easily sheared hence better swelling than other starches [38, 39]. The potato starch swelling is dependent on both swelling power and water-soluble index where swelling power and water-soluble index tend to increase with a corresponding increase in temperature [40]. The higher swelling power may also be due to the presence of a significant number of phosphate groups on the amylopectin molecule and this increased the solubility of the potato starch [39].

In terms of gelatinization properties, it depends on the molecular and crystalline structure, moisture content, external substances besides starch and water, as well as the processing condition [41]. The breakdown of crystalline structure within the starch granule can cause a variety of functional changes, such as starch swelling, pasting, and birefringence [38]. Differential scanning calorimetry (DSC) can be used to measure the temperatures at which starch gelatinizes and the enthalpy shift that occurs, whereby the parameters include onset temperature (T_o), peak temperature (T_p), conclusion temperatures (T_c), and the gelatinization enthalpy (ΔH) [35, 42]. The gelatinization enthalpy (ΔH) is a general measure of crystallinity that reveals the granule's loss of molecular order whereas peak temperature (T_p) gives a measure of crystalline quality [39]. As such, Liu et al. [41] recorded peak temperature (T_p) and the gelatinization enthalpy (ΔH) for the potato starch at 67–66 °C and 18–17 (J/g) respectively. Other study from shows nearly similar peak temperature (T_p) and gelatinization enthalpy (ΔH) which were at 63 °C and 13 (J/g), respectively [38]. As compared to rice starch and corn starch, potato starch showed higher peak temperature (T_p) and gelatinization enthalpy (ΔH) which may be due to the difference in amounts of longer chains in amylopectins [43].

Pasting entails heat-induced granular swelling, the change of ordered to disordered starch granules in excess water, exudation of molecular components and finally total rupture of the starch granules [44]. A pasting profile is a log of the viscosity change over time and temperature [42]. The pasting temperature (PT) in the pasting profiles is the temperature at which a quantifiable viscosity can be determined. At this temperature, starch starts to thicken, and peak viscosity (PV) is the greatest viscosity level attained during the heating cycle [42]. As reported by Liu et al. [41], the pasting temperature (PT) and the peak viscosity (PV) of potato starch were 61.4 ± 0.9 °C and 5782 ± 331 Cp, respectively. The charged phosphate groups in the starch exhibited a highly significant positive link with peak viscosity and breakdown because potato starch has a greater phosphorous content and therefore raises the viscosity of the starch [34, 45]. The peak viscosity (PV) is reported to be in direct correlation with the granules size. Since potato starch has a bigger granules size, it has higher peak viscosity compared to others starch

[46]. Potato starch also shows lower pasting temperature indicating its high viscosity compared to other starch [44].

17.3 Potato thermoplastic starch

Starch in general naturally exists in the shape of granules, resulting in poor processability and a lack of flow behavior. Many technical challenges during industrial processes are caused by such features. Several efforts have been made to convert native starch to plastic states to overcome these problems, such as subjecting starch granules to shear stress under heating conditions with addition of plasticizer [47]. This results in losses of granules original structure and subsequently forming a new continuous phase in the form of viscous melting substance known as thermoplastic starch (TPS) forms [48]. Additionally, plasticization causes the starch to become less brittle, allowing it to be melted and stiffened repeatedly, hence making it suitable for various plastic fabrication procedures [49].

The process of plasticization lead to conversion of native starch to thermoplastic starch. In general, thermoplastic starch can be defined as deformable thermoplastic materials derived from starch plasticization using plasticizer such as urea, glycerol, sorbitol, glycerine, and water [50]. The processing limitation due to native starch intermolecular forces and hydrogen is overcame through its conversion to thermoplastic starch, thus enable better processability as like conventional thermoplastic polymer which can be shaped by elevated temperature and solidify when cooled, for various intended purposes. Several parameters influence the effectiveness of the native starch to thermoplastic starch conversion process such as the plasticization techniques including kneading, extrusion, compression, and molding, [51–53]; and the type of plasticizers. The latter is rather essential factor determining the characteristics of the resulting thermoplastic starch. Although a variety of plasticizers can be used to generate thermoplastic starch, the most frequent plasticizers utilised in starch are water and glycerol [48]. Some recent works employing potato starch plasticization into thermoplastic starch is as shown in Table 17.2.

In general, plasticization of native starch to thermoplastics is related to starch gelatinization. After its intermolecular bonds are broken down due to shear stress under heating condition, the starch thickens in the presence of water. This allows more water molecules to interact with its hydrogen bonding sites, whereby the plasticizer causes the starch granules to disintegrate permanently in water [62]. Despite the fact that thermoplastic starch appears to be a promising material for replacing petroleum-based polymers, it has numerous drawbacks, including poor mechanical qualities, low heat stability, high moisture sensitivity, and limited dimensional stability [49]. These constraints limit the material's ability to be utilised as a substitute for traditional plastics. As a result, many studies to improve the characteristics of thermoplastic starch have been

Table 17.2: Some related works on potato starch thermoplastic production by using plasticizers.

Plasticizers	Plasticization techniques	Reference
Glycerol (30 wt%) Water (20 wt%)	Melt mixing process in internal mixer at 60 rpm, 160 °C, and 7 min.	[48]
Glycerol + sorbitol (23 wt%) Water (23 wt%)	Blending and drying at 170 °C in a ventilated oven for 40 min.	[54]
Glycerol (10–70 wt% starch) Water (100 mL for 5 g starch)	Solution casting: mixing at 80 °C for 30 min and dried at 45 °C.	–
Sugar alcohol-based deep eutectic solvents (mixture of sorbitol, xylitol, maltitol, choline chloride, and betaine)	Thermocompression at 150 °C, 12 tons (153 bars) for 5 min.	[55]
Glycerol (1.5 wt%) Water (93.5 wt%)	Solution casting: mixing at room temperature for 45 min and dried at 96 °C for 40 min.	[31]
Glycerol and sorbitol (10–50 wt% starch) Water (100 mL for 6 g starch)	Solution casting: mixing at 25 °C for 15 min; heated 95 °C, 500 rpm, for 30 min; dried at 45 °C for 48 h.	[56]
Urea-based deep eutectic solvents and hydroxyethyl ammonium formate ionic liquid (35 wt%)	Thermocompression at 140–150 °C, 12 tons (153 bars) for 5 min.	[57]
Glycerol (35 wt%)	Melt mixing process in a triple screw extruder at 80 °C (introduction zone) and 140 °C (gelatinization zone) using 160 rpm.	[58]
Glycerol and polyglycerol-3 (20–30 wt% starch)	Solution casting: mixing at 80 °C for 1 h and dried at room temperature for 48 h.	[59]
Glycerol (10–40 wt% starch) Water (95 wt%)	Solution casting: mixing (200 rpm) at 100 °C for 30 min and dried at 40 °C for 8 h.	[60]
Glycerol and sorbitol (15–45 wt% starch) Water (100 mL for 8 g starch)	Solution casting: mixing at 95 °C for 30 min and dried at 45 °C.	[61]

conducted including starch modification, blending with other polymers and the use of natural fibers at the macro-, micro-, and nano-scales as thermoplastic reinforcement [63].

17.4 Potato thermoplastic starch composites

Potato thermoplastic starch which was made by gelatinization, casting, and drying has been extensively utilized as a matric or reinforcing material for numerous types of composites and nanofilm owing to its edibility, biodegradability, good stretchability, and high transparency [64]. It is also being considered as one of the most promising degradable bio-packaging materials mainly because of its widespread availability and low cost. Although studies have shown that potato starch has greater mechanical properties than other types of starch, it still falls short of reaching application needs [65].

Throughout the year, numerous studies have been conducted to improve its properties and blending or incorporating with other materials can be considered as one of the easiest and most cost-effective methods. For example, He et al. [65] incorporated PVA and halloysite, a form of nanoclay with potato starch with the objective of maximizing the composite's tensile strength, and they were successful. The tensile strength of starch/PVA/nanoclay composite was 32% higher than the neat starch. A similar finding was reported whereby incorporated montmorillonite (MMT) and titanium dioxide (TiO_2) with potato starch enhanced the structural, barrier, mechanical, and thermal properties of starch [66]. It was reported that tensile strength, glass transition temperature, and melting point of the films were enhanced by 12.55, 106 and 29%, respectively after MMT and TiO_2 incorporation. Since starch also exhibits poor thermal and electrical properties that restrict its widespread use, advanced filler such as graphene was recently been introduced to solve these issues. Gürler and Torğut (2021) reported that by only incorporating 1% graphene into potato starch was able to increase tensile strength and electrical conductivity up to 63 and 116%, respectively. This discovery thus enhances the potential of starch to be used for various applications, including for electronic devices.

Aside from improving mechanical properties, research was also conducted to extend food shelf life by developing starch-based active packaging. As a hydrophilic material, it requires another material to allow it to change or retain the package's atmosphere in a way that extends the shelf life of food, and antioxidant or antibacterial components are great options. Sani and co-workers [26] successfully proved that by incorporating multiflora essential oil (MEO) and zirconium oxide (ZrO2) nanoparticles into starch was able to greatly enhance the antioxidant properties of the film, reduce water vapor permeability and eventually increase the shelf life of quail meat. It was also reported that the addition of these two materials was able to enhance thermal stability of the starch-based film. Similar study was conducted in which Thyme oil microemulsions (ME) and microcapsules (TMS) are combined with potato starch to create antibacterial composite films that can lengthen the shelf life of chilled meat, reduce pollution from nonbiodegradable packaging materials, and enhance the appeal of packaging products to retailers and consumers [64]. This study discovered that the starch/ME/TMS films could gradually release the antibacterial compounds in thyme oil, extending the antibacterial period and effectively preventing the development and reproduction of *E. coli* and *S. aureus*. In fact, the shelf life of chilled meat could be extended to 14 days, which was 8 days longer than the packaging without both ME and TMS.

Briefly, starch is a versatile material that may be used for a variety of applications, ranging from simple foods ingredients to packaging. If its drawbacks, such as hydrophilicity and low mechanical properties, can be resolved, it has a significant commercial value to be used as a primary feedstock in the production of biodegradable plastic packaging to replace synthetic polymers.

17.5 Nanocellulose reinforcement in potato thermoplastic starch

Previous section described the improvement of potato starch through reinforcement with various kind of filler could enable its application in variety of applications, ranging from simple foods ingredients to packaging. As such, utilization of nanocellulose as filler in potato thermoplastic starch could be advantageous attributed to its low cost, high specific surface area, high crystallinity, low density, non-abrasive and combustible nature, non-toxicity, biodegradability, availability, and renewability [67]. In fact, due to the structural similarity between starch [68] and cellulose [69], their affinities can be exploited to improve the mechanical qualities of cellulose thermoplastic starch composites, while preserving the biodegradable quality of the materials. In comparison to microcellulose, the utilization of nanocellulose in starch-based matrix possess a greater interest due to the increased surface area imparted by the nanoscale fibers, hence promoting better cohesive forces between the matrix and the nanocellulose filler [63].

Nanocellulose is polymer membrane having a size of 100 nm (nm) or less, extremely high specific area as well as high pore interconnectivity and porosity [67, 70, 71]. In fact, nanocellulose shows a great effect to enhance the mechanical strength of the material as they have good tensile strength, high permeability, and crystallinity as well as high thermal resistance as compared to neat polymer [72]. The usage of nanocellulose fibers in a variety of products, from consumer goods to high-tech industrial uses including composites, adsorbents, military, biomedical engineering, paints, cosmetics, pharmaceuticals, electronics, meals, and packaging, is made possible by all these favorable qualities [73].

Three commonly known nanocellulose are bacterial nanocellulose (BNC), cellulose nanofiber (CNF) and cellulose nanocrystal (CNC) by which the latter two are generally plant-based materials [74–77]. These three forms of nanocellulose were extracted using various methods, which produced materials with different mechanical, chemical, and crystallographic properties [9, 67, 78–85]. The morphology of the CNC, CNF, and BNC produced via different approaches are shown in Figure 17.4. CNF is typically made by a variety of methods, including mechanical, physical, physiochemical, or by a mix of these, while CNC can only be created by acid pretreatment, which necessitates strong acid hydrolysis. Long, thread-like bundles of cellulose molecules make up the CNF. These bundles are held together laterally by hydrogen bonds formed between the hydroxyl groups of neighboring molecules and their oxygen atoms. The CNC, on the other hand, has a rectangular cross-section and is rod-like, extremely crystalline, and short structured. Their size is a function of the temperature, time, and native cellulose supply material. BNC, on the other hand, is a kind of nanocellulose that is produced by microbes. The BNC is produced as pure cellulose and is distinguished from the CNF and CNC by having average diameters of 20–100 nm and micrometer lengths, which are entangled to form stable network structures. The CNF and CNC may need pretreatment to get rid of lignin and hemicellulose before hydrolysis.

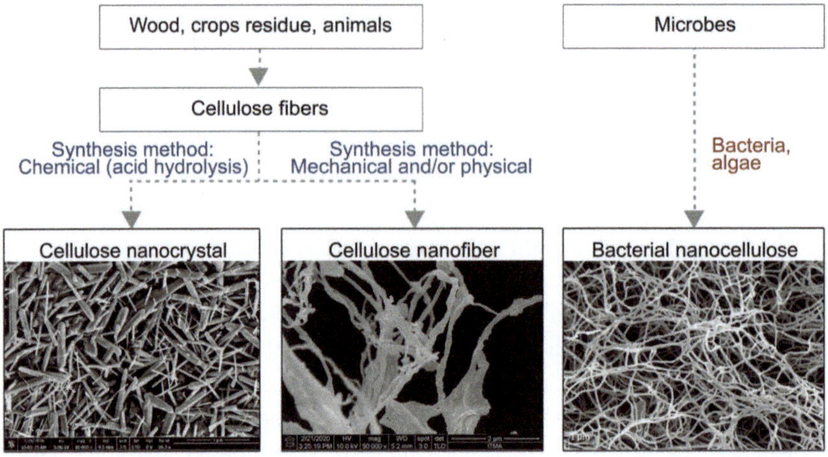

Figure 17.4: Morphology of the cellulose nanocrystal, cellulose nanofiber and bacteria nanocellulose produced via different approaches. This figure was reproduced from Norrrahim et al. [86–88].

In term of manufacturing process, the production of nanocellulose potato thermoplastic starch is similar like other nanocellulose thermoplastic starch which mostly is by solution casting [62]. This is owing to the suitability of this technology at laboratory scale, which only requires solubilization, casting, and drying stages, as well as the compatibility between the starch matrix and nanocellulose filler as well as the plasticizers, which are typically glycerol and water. Not limited to solution casting, nanocellulose thermoplastic starch composites can also be produced through melt mixing with subsequent thermos compression [63]. Figure 17.5 shows the example of potato thermoplastic starch processing with nanocellulose being used as the filler. Focusing on potato thermoplastic starch, several reported studies and enhancement of properties are described in the next section.

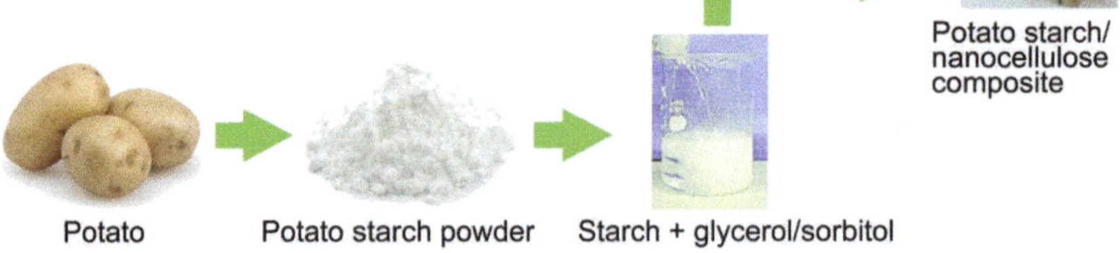

Figure 17.5: The example of thermoplastic starch/nanocellulose composite processing method.

17.5.1 Properties evaluation of nanocellulose reinforced potato thermoplastic starch composites

The performance evaluation of nanocellulose reinforced potato thermoplastic starch composite is usually conducted to evaluate its mechanical, thermal, and water barrier properties. Several studies involving the nanocellulose reinforced potato thermoplastic starch composite have been published in the last decade.

Arun et al. [89] evaluated the performance of a potato starch and reinforced with nanocellulose/ramie fabric composites. In their study, polycaprolactone (PCL) was also coated to the surface of composite to reduce water sensitivity. The composite has a good interface due to chemical similarity between starch and cellulose. The mechanical properties were increased as a result of reinforcement with nanocellulose. The composite has the greatest toughness and more flexible behavior as compared to the native potato starch. This can be seen where the composite has better tensile strength than the native potato starch. Meanwhile, the tensile strength of PCL based composite was lower than the composites without PCL. In term of the elongation at break, the composite has slightly higher value than the native starch. Authors suggested that it can be due to the presence of nanocellulose, which have a very short fiber length that could increase the mobility. The water adsorption analysis found that the composite had lower water adsorption as compared to native starch. This is because the composite has the uniform web-like network of nanocellulose fibers. The PCL based composites show a low water sensitivity as expected. A detailed investigation on the thermal properties of the starch was done by thermogravimetric analysis and flammability studies. It was found that the starch polymer is almost stable up to 300 °C and has better flame-retardant properties.

In a study done by Moran et al. [32] also discovered the effect of nanocellulose towards the performance of potato thermoplastic starch composites. The mechanical properties of composite were evaluated, and they were found it considerably improved with the addition of nanocellulose. Young's Modulus shown a significant increase, which grows steadily in tandem with the rise in nanocellulose concentration. In addition, the inclusion of nanocellulose led to a gradually rising tensile strength. However, as was predicted, elongation at break values had the opposite impact. Their research also revealed that some plasticization was evident even at low reinforcing content levels, which was mostly related to the reinforcing phase's size. Due to starch and nanocellulose's excellent chemical compatibility, the high reinforcing effect has been demonstrated. On the other hand, the addition of nanocellulose results in a decrease in the contact angle value and an increase in polarity because more hydrophilic groups are added.

Recently, Montero et al. [90] also reported on the mechanical, thermal, and water resistance properties of nanocellulose reinforced potato thermoplastic starch. It proved that raw starch with greater amylose chain concentrations produces a rigid amylose network that can extend the plasticization effect and raise the stiffness of composites. The reinforce effect of nanocellulose in composites is proved by the increase of storage modulus. Besides that, a good water resistance and thermal stability were also observed for the composites as compared to the native starch. Based on the thermal stability analysis, the incorporation of nanocellulose increased the degradation temperature. The composites resisted water the most with low diffusivity as compared to the other thermoplastic starch evaluated in their study which are corn and pea. The good interfacial interaction between the nano-sized filler and the starch matrix, regardless of the resources used, caused the increase in nanocellulose loading to reduce the rate of water absorption.

17.5.2 Potential application of nanocellulose-reinforced potato thermoplastic starch composites

This section focuses on the most recent developments in the use of composites made of potato starch and nanocellulose. The discovery on application of nanocellulose reinforced thermoplastic starch composite can be benefited in several fields. Table 17.3 shows starch-based polymer components and its potential applications.

It can be observed, most of the thermoplastic starch composites including potato are useful to be applied as packaging materials. The performance and quality of the products have improved after nanocellulose was added to the starch composites, thus may benefit people and the environment. As previously mentioned, the nonrenewable and nonbiodegradable characteristics of petroleum-based polymers when discarded might lead to environmental issues. Therefore, switching to eco-friendly packaging while keeping food quality and stability is necessary to solve these issues. This reduces the waste materials from nonbiodegradable packaging. Due to their "green" footprint, the nanocellulose

Table 17.3: Nanocellulose based starch thermoplastic composites and its suitable applications.

Source of starch	Potential application	Reference
Plasticized starch	Transparent materials	[91]
Corn	Air permeable, resistant, surface sized paper, and food packaging	[92, 93]
Cassava	Food packaging	[94]
Sugar palm	Food packaging	[95]
Wheat	Food packaging	[96]
Tuber native potato	Packaging	[90]
Cereal corn	Packaging	[90]
Legume pea	Packaging	[90]
Yam bean	Packaging	[97]
Cassava bagasse	Packaging	[94]
Ramie	Packaging	[96]
Bengkuang	Packaging	[98]

reinforced thermoplastic starch had been seen as a promising replacement material, and there are a wide range of products that could be made from these materials, including flushable liners, grocery bags, food and fruit containers or wrapping films, and medical delivery systems.

Although there has been a recent surge in interest in designing cellulose for packaging purposes, nanocellulose fiber reinforced starch-based composites have a wide range of other possible uses, including flexible optoelectronic, transparent materials, and scaffolds in tissue regeneration.

17.6 Conclusions

The use of environmentally friendly nanocellulose in starch thermoplastic production can help to lessen the starch-based bioplastics' two fundamental flaws which are high water affinity and poor mechanical characteristics. Stemmed from its sustainability and environmental friendliness, as well as its availability worldwide, the biodegradable and green potato-based starch thermoplastic composites are a strong contender to replace traditional petroleum-based plastics. In overall, its commercialization is highly plausible whereby nanocellulose filler can increase the value of potato starch-based bioplastics while reducing the accumulation of harmful petroleum-based plastic waste.

Acknowledgment: The authors would like to thank Nextgreen Pulp & Paper Sdn. Bhd. for their support and permission to work on this manuscript. Additionally, the authors gratefully acknowledge the technical and financial support from the Universiti Pertahanan Nasional Malaysia (UPNM) in preparation of this review.

References

1. Norrrahim MNF, Ariffin H, Hassan MA, Ibrahim NA, Nishida H. Performance evaluation and chemical recyclability of a polyethylene/poly(3-hydroxybutyrate-Co-3-Hydroxyvalerate) blend for sustainable packaging. RSC Adv 2013;3:24378–88.
2. Yasim-Anuar TAT, Norrrahim MNF, Sapuan SM, Ilyas RA, Jenol MA, Mamat NA, et al. Polyhydroxyalkanoates for packaging application. In: Bio-based packaging: material, environmental and economic aspects. New York, USA: John Wiley & Sons Ltd; 2021.
3. Ilyas RA, Sapuan SM, Ibrahim R, Atikah MSN, Asyraf MRM, Norrrahim MNF, et al. Environmental advantages and challenges of bio-based packaging materials. In: Bio-based packaging: material, environmental and economic aspects. New York, USA: John Wiley & Sons; 2021.
4. Ariffin H, Norrrahim MNF, Yasim-Anuar TAT, Nishida H, Hassan MA, Ibrahim NA, et al. Oil palm biomass cellulose-fabricated polylactic acid composites for packaging applications. In: Bionanocomposites for packaging applications. Midtown Manhattan, New York City: Springer; 2018:95–105 pp.
5. Offiong EU, Ayodele SL. Preparation and characterization of thermoplastic starch from sweet potato. Int J Sci Eng Res 2016;7:438–43.

6. Cai Z, Čadek D, Šmejkalová P, Kadeřábková A, Nová M, Kuta A. The modification of properties of thermoplastic starch materials: combining potato starch with natural rubber and epoxidized natural rubber. Mater Today Commun 2021;26:1–11.
7. Ting SS, Gie GP, Omar MF, Abdullah MF. Nanocellulose-reinforced biocomposites. In: Binod P, Raveendran S, Pandey A, editors. Biomass, biofuels, biochemicals: biodegradable polymers and composites - process engineering to commercialization. Amsterdam, Netherlands: Elsevier; 2021:461–94 pp.
8. Trache D, Tarchoun AF, Derradji M, Hamidon TS, Masruchin N, Brosse N, et al. Nanocellulose: from fundamentals to advanced applications. Front Chem 2020;8:1–33.
9. Sapuan SM, Norrrahim MNF, Ilyas RA. Industrial applications of nanocellulose and its nanocomposites. Cambridge, United States: Woodhead Publishing; 2022.
10. Norrrahim MNF, Sapuan SM, Yasim-Anuar TAT, Padzil FNM, Sharip NS, Ng LYF, et al. Antimicrobial studies on food packaging materials. In: Food packaging. Boca Raton: CRC Press; 2020:141–70 pp.
11. Norrrahim MNF, Yasim-Anuar TAT, Sapuan SM, Ilyas RA, Hakimi MI, Najmuddin SUFS, et al. Nanocellulose reinforced polypropylene and polyethylene composite for packaging application. In: Bio-based packaging: material, environmental and economic aspects. USA: Wiley Online Library; 2021.
12. Norrrahim MNF, Ariffin H, Yasim-Anuar TAT, Hassan MA, Ibrahim NA, Yunus WMZW, et al. Performance evaluation of cellulose nanofiber with residual hemicellulose as a nanofiller in polypropylene-based nanocomposite. Polymers 2021;13:1064.
13. Norrrahim MNF, Kasim NAM, Knight VF, Halim NA, Shah NAA, Noor SAM, et al. Performance evaluation of cellulose nanofiber reinforced polymer composites. Funct Compos Struct 2021;3. https://doi.org/10.1088/2631-6331/abeef6.
14. Norrrahim MNF, Yasim-Anuar TAT, Jenol MA, Mohd Nurazzi N, Ilyas RA, Sapuan SM. Performance evaluation of cellulose nanofiber reinforced polypropylene biocomposites for automotive applications. In: Biocomposite and synthetic composites for automotive applications. Amsterdam, Netherland: Woodhead Publishing Series; 2020:119–215 pp.
15. Ilyas RA, Azmi A, Nurazzi NM, Atiqah A, Atikah MSN, Ibrahim R, et al. Oxygen permeability properties of nanocellulose reinforced biopolymer nanocomposites. Mater Today Proc 2022;52:2414–9.
16. Bangar SP, Whiteside WS. Nano-cellulose reinforced starch bio composite films- A review on green composites. Int J Biol Macromol 2021;185:849–60.
17. Morán JI, Vázquez A, Cyras VP. Bio-nanocomposites based on derivatized potato starch and cellulose, preparation and characterization. J Mater Sci 2013;48:7196–203.
18. Tibolla H, Czaikoski A, Pelissari FM, Menegalli FC, Cunha RL. Starch-based nanocomposites with cellulose nanofibers obtained from chemical and mechanical treatments. Int J Biol Macromol 2020;161:132–46.
19. Ilyas RA, Sapuan SM, Nazrin A, Syafiq R, Aisyah HA, Norrrahim MNF, et al. Nanocellulose/starch biopolymer nanocomposites: processing, manufacturing, and applications. In: Payne E, editor. Advanced processing, properties, and applications of starch and other bio-based polymers. Amsterdam, Netherlands: Elsevier; 2020:65–88 pp.
20. Fahrngruber B, Eichelter J, Erhäusl S, Seidl B, Wimmer R, Mundigler N. Potato-fiber modified thermoplastic starch: effects of fiber content on material properties and compound characteristics. Eur Polym J 2019;111: 170–7.
21. Bergel BF, Leite Araujo L, dos Santos da Silva AL, Campomanes Santana RM. Effects of silylated starch structure on hydrophobization and mechanical properties of thermoplastic starch foams made from potato starch. Carbohydr Polym 2020;241:116274.
22. Shoja M, Mohammadi-Roshandeh J, Hemmati F, Zandi A, Farizeh T. Plasticized starch-based biocomposites containing modified rice straw fillers with thermoplastic, thermoset-like and thermoset chemical structures. Int J Biol Macromol 2020;157:715–25.
23. Koch M, Naumann M, Pawelzik E, Gransee A, Thiel H. The importance of nutrient management for potato production Part I: plant nutrition and yield. Potato Res 2019;63:97–119.

24. FAO Crops and Livestock Products. Available from: https://www.fao.org/faostat/en/#data/QCL [Accessed 21 Nov 2022].
25. Bertoft E, Blennow A. Structure of Potato Starch. In: Advances in potato chemistry and technology, 2nd ed. Amsterdam, The Netherlands: Elsevier; 2016.
26. Sani IK, Geshlaghi SP, Pirsa S, Asdagh A. Composite film based on potato starch/apple peel pectin/ZrO2 nanoparticles/microencapsulated zataria multiflora essential oil; investigation of physicochemical properties and use in quail meat packaging. Food Hydrocolloids 2021;117:106719.
27. Kringel DH, Halal SLME, Zavareze E, da R, Dias ARG. Methods for the extraction of roots, tubers, pulses, pseudocereals, and other unconventional starches sources: a review. Starch 2020.
28. Torres MD, Fradinho P, Rodríguez P, Falqué E, Santos V, Domínguez H. Biorefinery concept for discarded potatoes: recovery of starch and bioactive compounds. J Food Eng 2020;275:1–10.
29. Singh R, Kaur S, Sachdev PA. A cost effective technology for isolation of potato starch and its utilization in formulation of ready to cook, non cereal, and non glutinous soup mix. J Food Meas Char 2021;15:3168–81. https://doi.org/10.1007/s11694-021-00887-w.
30. Altemimi AB. Extraction and optimization of potato starch and its application as a stabilizer in yogurt manufacturing. Foods 2018;7:1–11.
31. Domene-López D, García-Quesada JC, Martin-Gullon I, Montalbán MG. Influence of starch composition and molecular weight on physicochemical properties of biodegradable films. Polymers 2019;11:1084.
32. Morán JI, Vázquez A, Cyras VP. Bio-nanocomposites based on derivatized potato starch and cellulose, preparation and characterization. J Mater Sci 2013;48:7196–203.
33. Yusuph M, Tester RF, Ansell R, Snape CE. Composition and properties of starches extracted from tubers of different potato varieties grown under the same environmental conditions. Food Chem 2003;82:283–9.
34. Dupuis JH, Liu Q. Potato starch: a review of physicochemical, functional and nutritional properties. Am J Potato Res 2019;96:127–38.
35. Sui Z, Kong X. Physical modifications of starch. Singapore: Springer Singapore; 2018.
36. Tester RF, Karkalas J, Qi X. Starch - composition, fine structure and architecture. J Cereal Sci 2004;39:151–65.
37. Pycia K, Juszczak L, Gałkowska D, Witczak M. Physicochemical properties of starches obtained from polish potato cultivars. Starch/Staerke 2012;64:105–14.
38. Singh N, Chawla D, Singh J. Influence of acetic anhydride on physicochemical, morphological and thermal properties of corn and potato starch. Food Chem 2004;86:601–8.
39. Singh J, Colussi R, McCarthy OJ, Kaur L. Potato starch and its modification. In: Advances in potato chemistry and technology, 2nd ed. Amsterdam, The Netherlands: Elsevier; 2016.
40. Lawal OS, Adebowale KO. Physicochemical characteristics and thermal properties of chemically modified Jack bean (Canavalia ensiformis) starch. Carbohydr Polym 2005;60:331–41.
41. Liu Q, Tarn R, Lynch D, Skjodt NM. Physicochemical properties of dry matter and starch from potatoes grown in Canada. Food Chem 2007;105:897–907.
42. Vamadevan V, Bertoft E. Structure-function relationships of starch components. Starch/Staerke 2015;67: 55–68.
43. Narpinder S, Jaspreet S, Lovedeep K, Navdeep SS, Balmeet SG. Morphological, thermal and rheological properties of starches from different botanical sources. Food Chem 2003;81:1–31.
44. Kumar R, Khatkar BS. Thermal, pasting and morphological properties of starch granules of wheat (Triticum aestivum L.) varieties. J Food Sci Technol 2017;54:2403–10.
45. Karim AA, Toon LC, Lee VPL, Ong WY, Fazilah A, Noda T. Effects of phosphorus contents on the gelatinization and retrogradation of potato starch. J Food Sci 2007;72:132–8.
46. Ye F, Li J, Zhao G. Physicochemical properties of different-sized fractions of sweet potato starch and their contributions to the quality of sweet potato starch. Food Hydrocolloids 2020;108:106023.
47. Sirohi R, Singh S, Tarafdar A, Reddy NBP, Negi T, Gaur VK, et al. Thermoplastic starch. In: Binod P, Raveendran S, Pandey A, editors. Biomass, biofuels, biochemicals. Amsterdam, Netherlands: Elsevier; 2021: 31–49 pp.

48. Altayan MM, Al Darouich T, Karabet F. On the plasticization process of potato starch: preparation and characterization. Food Biophys 2017;12:397–403.
49. Jumaidin R, Zainel SNM, Sapuan SM. Processing of thermoplastic starch. In: Al-Oqla FM, Sapuan SM, editors. Advanced processing, properties, and applications of starch and other bio-based polymers. Amsterdam, Netherland: Elsevier; 2020:11–9 pp.
50. Bangar SP, Whiteside WS, Ashogbon AO, Kumar M. Recent advances in thermoplastic starches for food packaging: a review. Food Packag Shelf Life 2021;30:100743.
51. Mościcki L, Mitrus M, Wójtowicz A, Oniszczuk T, Rejak A, Janssen L. Application of extrusion-cooking for processing of thermoplastic starch (TPS). Food Res Int 2012;47:291–9.
52. González-Seligra P, Guz L, Ochoa-Yepes O, Goyanes S, Famá L. Influence of extrusion process conditions on starch film morphology. LWT (Lebensm-Wiss & Technol) 2017;84:520–8.
53. Kipping T, Rein H. A new method for the continuous production of single dosed controlled release matrix systems based on hot-melt extruded starch: analysis of relevant process parameters and implementation of an in-process control. Eur J Pharm Biopharm 2013;84:156–71.
54. Ren J, Dang KM, Pollet E, Avérous L. Preparation and characterization of thermoplastic potato starch/halloysite nano-biocomposites: effect of plasticizer nature and nanoclay content. Polymers 2018;10:808.
55. Zdanowicz M, Staciwa P, Jedrzejewski R, Spychaj T. Sugar alcohol-based deep eutectic solvents as potato starch plasticizers. Polymers 2019;11:1385.
56. Ballesteros-Mártinez L, Pérez-Cervera C, Andrade-Pizarro R. Effect of glycerol and sorbitol concentrations on mechanical, optical, and barrier properties of sweet potato starch film. NFS J 2020;20:1–9.
57. Zdanowicz M. Starch treatment with deep eutectic solvents, ionic liquids and glycerol. A comparative study. Carbohydr Polym 2020;229:115574.
58. Mazerolles T, Heuzey MC, Soliman M, Martens H, Kleppinger R, Huneault MA. Development of multilayer barrier films of thermoplastic starch and low-density polyethylene. J Polym Res 2020;27:1–15.
59. Merino D, Paul UC, Athanassiou A. Bio-based plastic films prepared from potato peels using mild acid hydrolysis followed by plasticization with a polyglycerol. Food Packag Shelf Life 2021;29:100707.
60. Fu Z, Guo S, Sun Y, Wu H, Huang ZG, Wu M. Effect of glycerol content on the properties of potato flour films. Starch/Staerke 2021;73:1–6.
61. Ng JS, Kiew PL, Lam MK, Yeoh WM, Ho MY. Preliminary evaluation of the properties and biodegradability of glycerol- and sorbitol-plasticized potato-based bioplastics. Int J Environ Sci Technol 2022;19:1545–54.
62. Tan SX, Andriyana A, Ong HC, Lim S, Pang YL, Ngoh GC. A comprehensive review on the emerging roles of nanofillers and plasticizers towards sustainable starch-based bioplastic fabrication. Polymers 2022;14:664.
63. Gutiérrez TJ, Alvarez VA. Cellulosic materials as natural fillers in starch-containing matrix-based films. A review. Polym Bull 2016;74:2401–30.
64. Yuan L, Feng W, Zhang Z, Peng Y, Xiao Y, Chen J. Effect of potato starch-based antibacterial composite films with thyme oil microemulsion or microcapsule on shelf life of chilled meat. LWT (Lebensm-Wiss & Technol) 2021;139:1–7.
65. He Y, Kong W, Wang W, Liu T, Liu Y, Gong Q, et al. Modified natural halloysite/potato starch composite films. Carbohydr Polym 2012;87:2706–11.
66. Oleyaei SA, Almasi H, Ghanbarzadeh B, Moayedi AA. Synergistic reinforcing effect of TiO2 and montmorillonite on potato starch nanocomposite films: thermal, mechanical and barrier properties. Carbohydr Polym 2016;152:253–62.
67. Sharip NS, Yasim-Anuar TAT, Norrrahim MNF, Shazleen SS, Nurazzi NM, Sapuan SM, et al. A review on nanocellulose composites in biomedical application. In: Composites in biomedical applications. Boca Raton, Florida: CRC Press; 2020:161–90 pp.
68. Avérous L. Biodegradable multiphase systems based on plasticized starch: a review. J Macromol Sci, Polym Rev 2004;44:231–74.

69. Shamsudin S, Bahrin EK, Jenol MA, Sharip NS. Characteristics and potential of renewable bioresources. In: Shukor H, Makhtar MMZ, Yaser AZ, editors. Characteristics and potential of renewable bioresources. Singapore: Springer; 2022:21–43 pp.
70. Norrrahim MNF, Mohd Kasim NA, Knight VF, Mohamad Misenan MS, Janudin N, Ahmad Shah NA, et al. Nanocellulose: a bioadsorbent for chemical contaminant remediation. RSC Adv 2021;11:7347–68.
71. Sharip NS, Ariffin H. Cellulose nanofibrils for biomaterial applications. Mater Today Proc 2019;16:1959–68.
72. Kargarzadeh H, Huang J, Lin N, Ahmad I, Mariano M, Dufresne A, et al. Recent developments in nanocellulose-based biodegradable polymers, thermoplastic polymers, and porous nanocomposites. Prog Polym Sci 2018;87:197–227.
73. Norrrahim MNF, Kasim NAM, Knight VF, Halim NA, Shah NAA, Noor SAM, et al. Performance evaluation of cellulose nanofiber reinforced polymer composites. Funct Compos Struct 2021;3:024001.
74. Norrrahim MNF, Norizan MN, Jenol MA, Farid MAA, Janudin N, Ujang FA, et al. Emerging development on nanocellulose as antimicrobial material: an overview. Mater Adv 2021;2:3538–51.
75. Fareez IM, Jasni AH, Norrrahim MNF. Nanofibrillated cellulose based bio-phenolic composites. In: Phenolic polymers based composite materials. Singapore: Springer; 2020:139–51 pp.
76. Thiruganasambanthan T, Ilyas RA, Norrrahim MNF, Kumar TSM, Siengchin S, Misenan MSM, et al. Emerging developments on nanocellulose as liquid crystals: a biomimetic approach. Polymers 2022;14:1–28.
77. Norrrahim MNF, Huzaifah MRM, Farid MAA, Shazleen SS, Misenan MSM, Yasim-Anuar TAT, et al. Greener pretreatment approaches for the valorisation of natural fibre biomass into bioproducts. Polymers 2021;13: 1–23.
78. Norrrahim MNF, Mohd Kasim NA, Knight VF, Ong KK, Mohd Noor SA, Abdul Halim N, et al. Emerging developments regarding nanocellulose-based membrane filtration material against microbes. Polymers 2021;13:3249.
79. Norrrahim MNF, Kasim NAM, Knight VF, Misenan MSM, Janudin N, Shah NAA, et al. Nanocellulose: a bioadsorbent for chemical contaminant remediation. RSC Adv 2021;11:7347–68.
80. Norrrahim MNF, Kasim NAM, Knight VF, Ujang FA, Janudin N, Razak MAIA, et al. Nanocellulose: the next super versatile material for the military. Mater Adv 2021;2:1485–506.
81. Misenan S, Shaffie A, Zulkipli N, Norrrahim F. Nanocellulose in sensors. In: Industrial applications of nanocellulose and its nanocomposites. Sawston, UK: Wood Head Publishing; 2022:213–240 pp.
82. Ariffin H, Tengku Yasim-Anuar TA, Norrrahim MNF, Hassan MA. Synthesis of cellulose nanofiber from oil palm biomass by high pressure homogenization and wet disk milling. In: Nanocellulose: synthesis, structure, properties and applications. Singapore: World Scientific; 2021:51–64 pp.
83. Yasim-anuar TAT, Ariffin H, Norrrahim MNF, Hassan MA. Factors affecting spinnability of oil palm mesocarp fiber cellulose solution for the production of microfiber. Bioresources 2017;12:715–34.
84. Norrrahim MNF, Ariffin H, Yasim-Anuar TAT, Ghaemi F, Hassan MA, Ibrahim NA, et al. Superheated steam pretreatment of cellulose affects its electrospinnability for microfibrillated cellulose production. Cellulose 2018;25:3853–9.
85. Norrrahim MNF, Ariffin H, Hassan MA, Ibrahim NA, Yunus WMZW, Nishida H. Utilisation of superheated steam in oil palm biomass pretreatment process for reduced chemical use and enhanced cellulose nanofibre production. Int J Nanotechnol 2019;16:668–79.
86. Norrrahim MNF, Mohd Kasim NA, Knight VF, Ujang FA, Janudin N, Abdul Razak MAI, et al. Nanocellulose: the next super versatile material for the military. Mater Adv 2021;2:1485–506.
87. Aziz MA, Zubair M, Saleem M. Development and testing of cellulose nanocrystal-based concrete. Case Stud Constr Mater 2021;15:1–21.
88. Marín P, Martirani-Von Abercron SM, Urbina L, Pacheco-Sánchez D, Castañeda-Cataña MA, Retegi A, et al. Bacterial nanocellulose production from naphthalene. Microb Biotechnol 2019;12:662–76.

89. Arun S, Kumar KAA, Sreekala MS. Fully biodegradable potato starch composites: effect of macro and nano fiber reinforcement on mechanical, thermal and water-sorption characteristics. Int J Plast Technol 2012;16: 50–66.
90. Montero B, Rico M, Rodríguez-Llamazares S, Barral L, Bouza R. Effect of nanocellulose as a filler on biodegradable thermoplastic starch films from tuber, cereal and legume. Carbohydr Polym 2017;157: 1094–104.
91. Nasseri R, Mohammadi N. Starch-based nanocomposites: a comparative performance study of cellulose whiskers and starch nanoparticles. Carbohydr Polym 2014;106:432–9.
92. Slavutsky AM, Bertuzzi MA. Water barrier properties of starch films reinforced with cellulose nanocrystals obtained from sugarcane bagasse. Carbohydr Polym 2014;110:53–61.
93. Yang S, Tang Y, Wang J, Kong F, Zhang J. Surface treatment of cellulosic paper with starch-based composites reinforced with nanocrystalline cellulose. Ind Eng Chem Res 2014;53:13980–8.
94. de Teixeira E, Pasquini D, Curvelo AASS, Corradini E, Belgacem MN, Dufresne A. Cassava bagasse cellulose nanofibrils reinforced thermoplastic cassava starch. Carbohydr Polym 2009;78:422–31.
95. Ilyas RA, Sapuan SM, Ishak MR, Zainudin ES. Development and characterization of sugar palm nanocrystalline cellulose reinforced sugar palm starch bionanocomposites. Carbohydr Polym 2018;202: 186–202.
96. Lu Y, Weng L, Cao X. Morphological, thermal and mechanical properties of ramie crystallitesdreinforced plasticized starch biocomposites. Carbohydr Polym 2006;63:198–204.
97. Asrofi M, Abral H, Kasim A, Pratoto A, Mahardika M, Hafizulhaq F. Characterization of the sonicated yam bean starch bionanocomposites reinforced by nanocellulose water hyacinth fiber (WHF): the effect of various fiber loading. J Eng Sci Technol 2018;13:2700–15.
98. Syafri E, Sudirman S, Mashadi, Yulianti E, Deswita, Asrofi M, et al. Effect of sonication time on the thermal stability, moisture absorption, and biodegradation of water hyacinth (Eichhornia crassipes) nanocellulose-filled bengkuang (pachyrhizus erosus) starch biocomposites. J Mater Res Technol 2019;8:6223–31.

S. Silviana* and Febio Dalanta

18 Recent developments in sago starch thermoplastic bio-composites

Abstract: The thermoplastic-based starch has been increasingly demanding in recent years due to its high potential to substitute conventional plastic usage. Sago has been recognized as one of the sources of starch with a large abundance. This chapter mainly emphasized the modification of sago starch using different additives for biodegradable plastic. The other starch modifications are also discussed, including the effect of epoxidized waste cooking oil (EWCO) addition as a plasticizer, lime juice's citric acid as a crosslinker on the physical and chemical characteristics of the modified starch composites. The results mainly showed the advantages of reinforcing sago starch with plasticizers, nanoparticles, nanocellulose, fibers, and cross-linking agents, such as increasing the mechanical strength, higher thermal stability, and higher biodegradability. Therefore, the thermoplastic films from sago starch provide a promising potential to be further developed for practical applications.

Keywords: cellulose fiber; crosslinker agent; sago; starch modification; thermoplastic bio-composite.

18.1 Introduction

Recently, the starch production from new starchy crops has been developed. Among those starchy crops, sago palm (*Metroxylon sagu*) has largely been utilized due to its fast-growing and sustainability [1, 2]. Sago palm is a native crop and distributed across Asia continent, especially in Southeast Asia and Papua New Guinea [3]. This crop can grow in swampy areas, acidic soil, and high salinity condition, where other plants usually tend to fail [3]. It has a robust immunity against extreme rain, flood, wind, and sun exposure. Sago trees can intake a large amount of carbon dioxide (CO_2); thus, it can prevent further drawbacks of global warming and climate change [2]. Hence, it plays an essential role in food security and environmental sustainability [4]. Regarding starch productivity, a single sago palm can produce at least 400 kg of dry starch. On a larger scale, it can produce approximately 25,000 metric tons/ha [5]. This productivity is ten times higher than those of cassava and potato [5]. Until now, sago starch has been utilized in numerous applications in the food

***Corresponding author: S. Silviana**, Department of Chemical Engineering, Faculty of Engineering, Diponegoro University, Semarang, 50275, Indonesia, E-mail: silviana@che.undip.ac.id
Febio Dalanta, Department of Chemical Engineering, Faculty of Engineering, Diponegoro University, Semarang, 50275, Indonesia

As per De Gruyter's policy this article has previously been published in the journal Physical Sciences Reviews. Please cite as: S. Silviana and F. Dalanta "Recent developments in sago starch thermoplastic bio-composites" *Physical Sciences Reviews* [Online] 2023. DOI: 10.1515/psr-2022-0017 | https://doi.org/10.1515/9783110773606-018

industry. Sago starch is classified in the C-type polymorph group [1, 2]. The starch particle morphologies are polymorph or oval [2].

Due to the rapid growth of interest in the utilization of starchy crops as the raw material of biodegradable films, sago has gained massive attention as biodegradable film material [6, 7]. Starch has better characteristics to film manufacturing compared to lipid and protein due to its renewability, availability, biodegradability, and cost-effectiveness [8]. Recently, the various sources of starch can be grouped into two big classes, i.e., grains such as rice, corn, wheat, etc., and tubers such as cassava, potato, sago, etc. [6, 8, 9]. Naturally, starch is composed of two major substances, which are amylose and amylopectin [1]. These two substances play an equally significant role in the properties of the fabricated starch films. However, the native starch does not have thermoplastic properties, which is essential in the biodegradable plastic making [10]. Thermoplastic making from plant starch should be added by some materials such as plasticizer, fiber, cross-linker, and others to produce standardized plastic for practical uses.

In recent years, there have been massive works related to the modification and advancement of sago starch for thermoplastic products. This chapter firstly discusses the fundamental of starch, the physicochemical properties of starch, and the manufacturing process of thermoplastic film from starch. Naturally, starch-based polymers have two significant drawbacks: poor mechanical properties and low water/moisture resistance. Therefore, the current methods for sago starch modification for improving the sago starch-based thermoplastic uses are also reviewed.

18.2 Starch sources and fundamental characteristics

Thermoplastic-based starch bio-composite using various botanical sources has been extensively investigated. These investigations exhibit that starch products from different crops have different characteristics, especially the ratio between amylose and amylopectin [11]. Hence, it provides huge opportunities for developing the thermoplastic bio-composite from starches with different properties. A study reported that starch sources from cassava, corn, and ahipa exhibited good film-forming characteristics, which contained amylose of 15.5%, 23.9%, and 1.6%, respectively [12]. Films that were derived from starch-containing with higher amylose content, such as corn starch, were less strong but were more moisture-resistant than that of the ahipa and cassava. It was also reported that waxy rice, regular rice, and cassava starch have an amylose content of 3.64%, 19.50%, and 26.78%, respectively [13]. Sago starch, which is a type of palm starch, contains the highest amylose content of ~30%. The thermoplastic film made of sago starch exhibited better mechanical strength and moisture resistance. Table 18.1 shows the mechanical and moisture properties of the thermoplastic film derived from different botanical starch sources.

Another study reported that the amylose concentration in the starch controls the film product characteristics [15]. Two types of corn starches with amylose contents of 25 and

Table 18.1: The mechanical and moisture properties of the thermoplastic film are derived from different botanical starch sources.

Starch sources	Amylose (%)	Tensile strength (MPa)	Elongation at break (%)	Water vapor permeability (10^{-11}/Pa g m)	Ref.
Waxy rice	3.64	31.80	4.10	21.03	[13]
Ahipa	11.6	4.70	21.80	16.00	[12]
Normal rice	19.50	28.85	2.01	16.75	[13]
Corn	23.90	31.80	4.10	12	[12]
Cassava	26.78	35.17	2.64	11.22	[13]
Sago	30.00	38.76	10.12	13.43	[14]

80% were studied. It was reported that the thermoplastic films with higher amylose content exhibited higher tensile strength but lower elongation at break compared to the films with lower amylose content [15]. This difference was caused by the fact that amylose has a lower plasticizing effect than amylopectin. The extensibility and flexibility of the films can be improved using starch with higher amylopectin [16]. Furthermore, the thermophysical characteristics of films made of corn, cassava, and ahipa starches have been investigated [12]. They stated that the plasticizers act more effectively in ahipa starch due to the higher content of amylopectin than that of the corn and cassava starches. The higher plasticizer activity is due to the more abundant amount of hydrogen bonding [17]. On the other hand, the amylose was helpful in the moisture-resistant properties of the films. Figure 18.1 depicts the water vapor permeability (WVP) and oxygen permeability of the amylose and amylopectin thermoplastic films at air humidity conditions ranging from 20 to 90%. It can be seen that the amylopectin thermoplastic

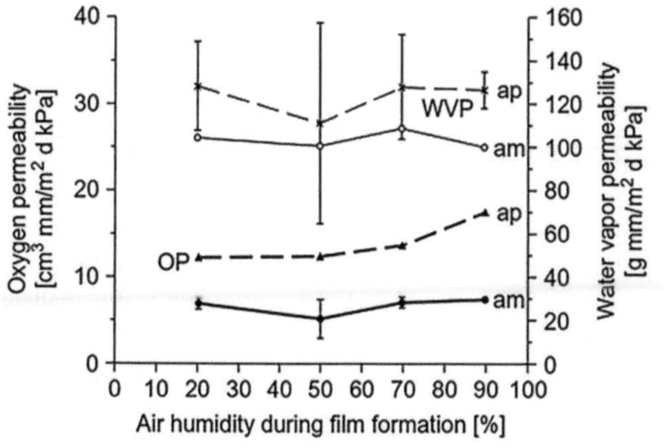

Figure 18.1: The study results of water vapor permeability (WVP) and oxygen permeability (OP) of the amylose (am) and amylopectin (ap) thermoplastic films [11] (reprinted with permission from Elsevier).

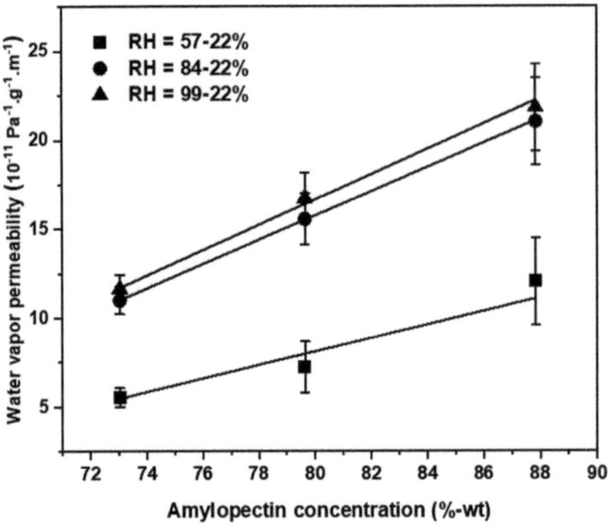

Figure 18.2: Effects of amylopectin contents on the water vapor permeability (WVP) in the thermoplastic films [13] (reprinted with permission from American Chemical Society).

films exhibit more resistance to oxygen and moisture than the amylose thermoplastic films. Corroboratively, Figure 18.2 describes the WVP profiles are relatively proportional to the amylopectin content in the thermoplastic films. They explained that the amylose structures transformed into a B-type structure while the amylopectin structures remained amorphous [13]. The moisture diffusion through the amorphous structure is relatively more accessible than in the crystalline structure. The thermoplastic films made of starch with high content of amylopectin are not recommended for food packaging applications. Therefore, the utilization of sago starch for thermoplastic films has been increasing in recent years due to its acceptable content of amylose and amylopectin, which made it suitable as the starch source in thermoplastic film making.

18.3 Physicochemical properties of sago starch

As one of the essential starch properties, the chemical composition is important to understand the functional properties of a starch source [18]. Sago starch contains amylose ranging from 21.0 to 30.0% [8, 19]. It has been reported that sago starch had a shallow content of other components such as protein (~0.3%), fiber (~0.3%), fat (~0.1%), and ash (~0.4%) [1]. There was also a low content of phosphorous substances (80–96 mg/L) that mainly were attached to C-6 of the glucosyl structure in sago starch [20]. Furthermore, sago starch contains a C-type polymorph structure, which has a granule size ranging from 26.8 to 33.3 μm. The sago starch granules have an oval or polymorphic shape [19, 21]. Figure 18.3 represents the scanning electron microscope (SEM) image of sago starch granules.

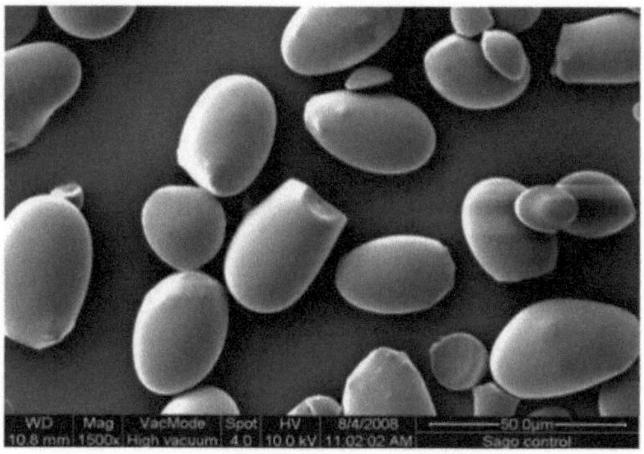

Figure 18.3: Morphological shapes of sago starch by scanning electron microscope (SEM) analysis [21] (reprinted with permission from Elsevier).

Sago amylose has a higher hydrodynamic radius, up to 24.70 nm, compared to other amyloses from potato, cassava, and maize [22]. Additionally, the degree of polymerization (DP) of amylose of sago starch was lower than that of potato and wheat amyloses, but it was higher than that of maize amylose. It was also reported that the molecular weight of sago amylose was 1.40–2.20 MDa [19]. Meanwhile, sago amylopectin had a lower hydrodynamic radius than that of other starches. DP of sago amylopectin was similar to pea amylopectin, lower than that of potato amylopectin, and higher than that of cereal amylopectin [22, 23]. A study reported that the molecular weight of sago amylopectin was 6.70–9.20 MDa [1, 24]. Bertoft and coworkers investigated the internal unit chain composition of various sources of amylopectin, including sago starch [25]. They reported that sago amylopectin had a relatively lower amount of short internal chain structures than another amylopectin from A-type starches, but it has a more extensive content of short internal chain structures than that of the B-type starches. The investigated amylopectin from various botanical sources was classified into four groups according to the internal unit chain composition., as the results are presented in Figure 18.4. Amylopectin from sago starch was categorized into the second group [25]. This classification of amylopectin based on the internal unit chain properties of the sago starch is essential to understanding the physicochemical characteristics of sago starch, especially as the primary material for thermoplastic film applications.

Another essential physicochemical parameter of starch is gelatinization properties. The gelatinization of starch can be analyzed using numerous methods, including rheometer and differential scanning calorimetry (DSC) [26]. A study that conducted the DSC analysis on sago starch revealed that sago starch tends to have a higher or closely similar temperature of gelatinization compared to B-type and C-type starches and has a lower temperature of gelatinization than A-type starches [27]. It was also reported that sago starch has two stages of swelling pattern in the water. The capacity of swelling of sago starch was higher than that of pea and maize starches and was closely similar to that of cassava starch [1]. The thermophysical properties, which are the gelatinization transition temperature and the change of enthalpy (ΔH) of sago starch compared to other sources of starch, are depicted in Figure 18.5 [28]. It can be seen that sago starch has a

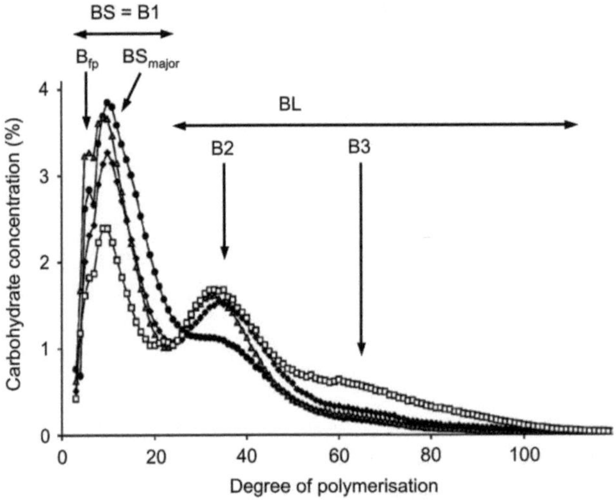

Figure 18.4: Profiles of amylopectin based on the internal short-chain compositions, i.e., 1st group (●), 2nd group (△), 3rd group (◆), and 4th group (▫). Amylopectin from sago starch is classified into the 2nd group. Typical short B-chains (BS or B1) with fingerprint B-chains (B_{fp}), major short B-chains (BS_{major}) as its subgroups. Typical long B-chains (BL) with B2 and B3 chains as their subgroups [25] (reprinted with permission from Elsevier).

relatively high value of enthalpy change compared to other starches. It was similar to cassava and waxy rice starches, higher than standard barley, oat, and medium rice starches, and lower than canna, potato, and lesser yam starches [27]. This enthalpy change difference can be due to the differences in the structure, composition, and crystallinity of starch [17, 28]. As discussed earlier, sago starch has a longer internal unit chain structure of amylopectin than A-type starches. This amylopectin structure created the granules with a double helix structure, which subsequently stabilizes the starch [27].

18.4 Manufacturing process of thermoplastic-based starch films

Thermoplastic films have been commonly made by using casting solution and extrusion methods. In casting solution procedures, starch as raw material, plasticizer, and additives are firstly mixed together in water to form the starch suspension under constant heating to ensure the granules of starch are gelatinized. Then, the suspension is cast onto a smooth surface medium with a certain thickness. The casted solution on the medium is then cooled down and dried until the film can easily be peeled off from the casting medium. The complete schematic illustration of the casting solution method is depicted in Figure 18.6(a). The casting solution composition, heating temperature, and time have

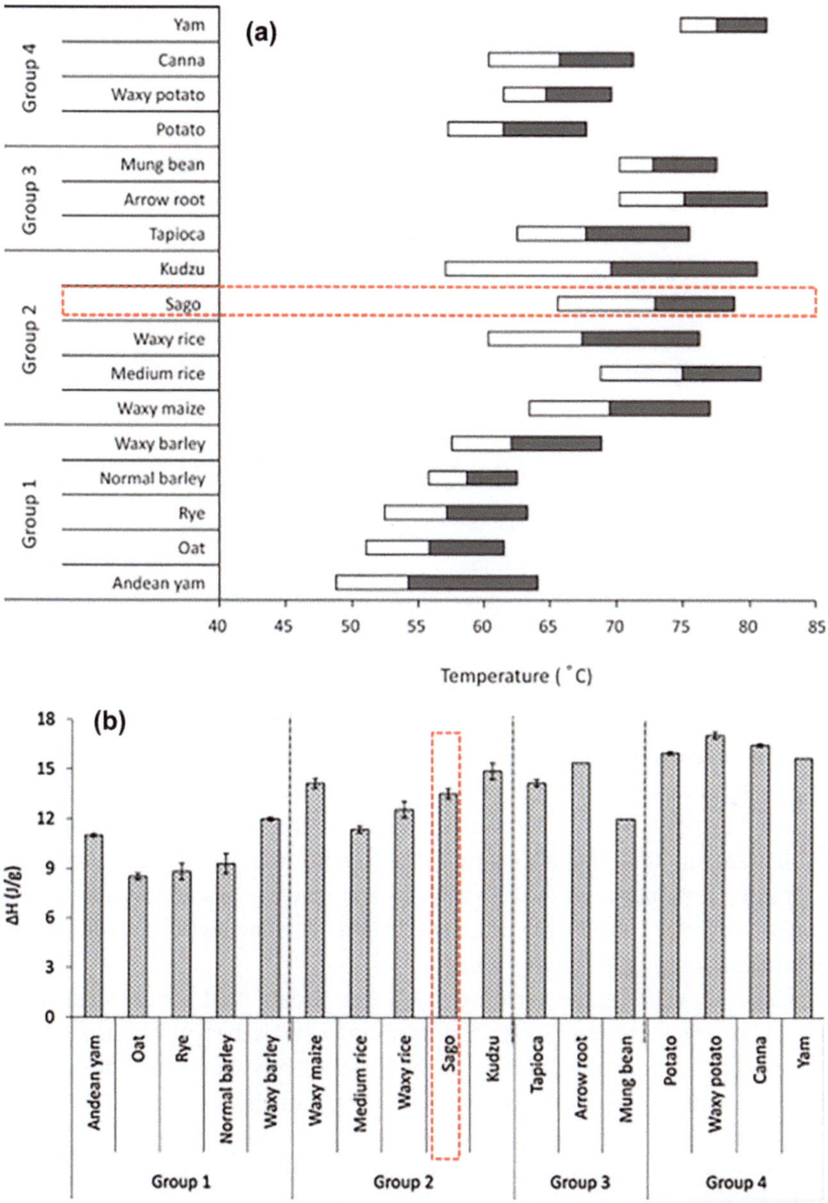

Figure 18.5: (a) Profiles of gelatinization temperature of various starches that are classified into four groups according to the amylopectin structure, and (b) profiles of gelatinization enthalpy change (ΔH) of the classified starches [28] (reprinted with permission from Elsevier).

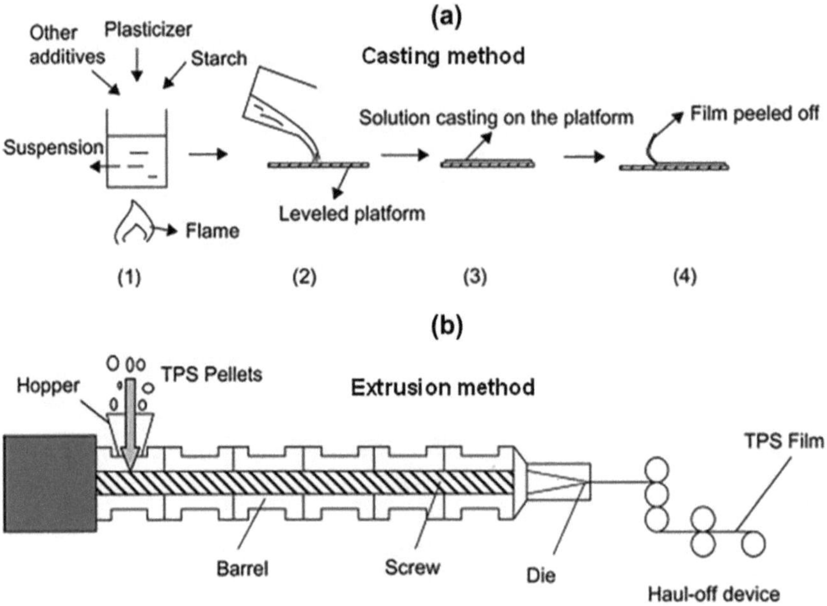

Figure 18.6: Schematic representations of the manufacturing method of thermoplastic-based starch films: (a) Casting method and (b) extrusion method [29] (reprinted with permission from Elsevier).

significantly influenced the characteristics of the thermoplastic films. A study reported that heating solution at 130–150 °C for 5–15 min resulted in degraded starch and fractured thermoplastic films [30]. Heating temperature plays a more dominant effect on thermoplastic film quality than the heating time. This casting method has been commonly applied for laboratory purposes or on a pilot scale to investigate starch modifications, gelatinization, and other thermophysical experiments. This method is not suitable for large-scale manufacturing like in the industry.

For industrial purposes, the extrusion method is widely used. A screw extrusion machine equipped with a hopper, screw feed, barrel, and dies is the typical thermoplastic film production machine in the industry, as the illustration is shown in Figure 18.6(b). Moscicki and coworkers prepared starch pellets by using a single screw extruder [31]. The starch pellet is generated by chopping them with a high-speed die of around 5 mm in size. The thermal process, along with shear stress, weakened the crystalline structure of starch, which resulted in a continuous amorphous viscous solution: the plasticizers, additives, and water molecules attached to starch via hydrogen bonding. Then, the starch becomes melting, fragmenting, and flows along with the barrel chamber. The homogeneous starch mixture is transported to the die of the screw to cut it into pellets. The product quickly solidifies right after out from the die. Usually, before the starch starts to solidify, the extrudates can be blown to form films. Another way the extrudates can be directly molded into desired shapes is by using a molding machine. Furthermore, a study reported that different casting methods resulted in different thermoplastic film product properties [32]. The thermoplastic films prepared using the casting method resulted in an inhomogeneous, translucent, and smooth product, while the films from the extrusion method showed an opaque result. Moreover, the casting method exhibits thermoplastic films that have a lower water vapor permeability and higher mechanical strength than that of the extrusion method [29, 33].

18.5 Sago starch modifications

Similar to other raw starches, raw sago starch cannot be directly used to produce thermoplastic films due to its lack of functionalities. In past decades, numerous modification methods of sago starch have been reported to achieve a better quality of thermoplastic films derived from sago starch. The methods of sago starch modification are discussed in the following section.

18.5.1 Plasticizer addition

Plasticizers are substances with high boiling points that are usually added to the polymer matrix in order to increase the mechanical strength of the polymer. Plasticizers should be nonvolatile and should not cause a phase separation with polymers [11]. Native starch without plasticizer addition cannot be classified into the thermoplastic polymer [34]. Plasticizer plays as the agent to reduce the inherent brittleness of starch by decreasing the glass transition temperature, decreasing the intermolecular forces between the starch chains, increasing the polymer's flexibility, and reducing the melting temperature of the native starches below their decomposition temperatures, which is commonly at 230 °C [35–40]. Therefore, the plasticizer is an essential part of thermoplastic-based starch product preparations.

Several plasticizers have been commonly used in thermoplastic-based starch preparations, such as water, sorbitol, glycerol, amides, fructose, sucrose, urea, and stearic acid until the epoxidized waste cooking oil (EWCO) [36, 41–44] Due to its ease of incorporation, small molecule size, and ease of positioning within the starch network, water is regarded as one of the most effective plasticizers for native starch processing [44]. A study was conducted to investigate the effects of different plasticizers, namely glycerol, sorbitol, and monosaccharide, on the thermophysical properties of the prepared thermoplastic films [44]. It was reported that film with glycerol addition has the lowest glass transition temperature compared to that of sorbitol and monosaccharide, with an approximate value of −70 °C. It indicates that glycerol provided a good plasticization effect for the thermoplastic-based starch films [41]. However, a major drawback of glycerol as a plasticizer is that the glycerol-plasticized thermoplastic films adsorbed moisture faster than the sorbitol-plasticized thermoplastic films. It is explainable due to the higher dielectric constant of glycerol (42.5) than sorbitol (33.5), which makes glycerol-plasticized thermoplastic films adsorbed more moisture [41].

The comparison effects between sorbitol and glycerol on heat sealability of the thermoplastic-based starch films showed that sorbitol has a better heat sealability than the glycerol-plasticized films. Application of monosaccharides such as fructose, glucose, and mannose as the plasticizers exhibited that the monosaccharide-plasticized films have higher tensile strength and elongation at break and lower water vapor permeability [17]. It indicates that the addition of monosaccharide as a plasticizer resulted in better thermoplastic films based on the mechanical properties. It has also been reported that the incorporation of glycerol as a plasticizer exhibited a slightly lower degradation temperature, DTG peak temperature, and glass transition temperature compared to sorbitol-plasticized films [45]. It indicates that sorbitol provides higher thermal stability than that of glycerol [45]. Table 18.2 summarizes the thermophysical properties of thermoplastic-based starch films using different types of plasticizers. Urea, amines, and amides substances have also been tested as the plasticizer for thermoplastic-based starch films. Using the functionalized –CO– and –NH– groups to form a massive hydrogen bonding with starch structure [46–48]. It was reported that the mixture of urea and formamide as plasticizers resulted in robust and more homogeneous thermoplastic films compared to that of glycerol [48]. Ma and coworkers studied the strength of hydrogen bonding for several starch plasticizers, including urea, glycerol, formamide, and acetamide [49]. It showed that the strength of hydrogen bonding with the starch follows this order: urea > formamide > acetamide > glycerol. It can be concluded that the amines and amides compounds showed stronger hydrogen bonding compared to glycerol [47, 49]. Nevertheless, for food safety purposes, plasticizers that contain amide groups are not suitable for edible composite and food packaging. Silviana and coworkers synthesized the EWCO as the plasticizer by using the epoxidation process of waste cooking oil [42, 50, 51]. It was reported that the EWCO contains several oxygenated groups, such as C–O, C=O, and COO, that can be promising to be used as a plasticizer [50].

18.5.2 Nanoparticle addition

Nanoparticles in the form of nanoclay have been dispersed as agent with thermoplastic starch film to improve the mechanical, thermal, and water resistance properties [52–55]. The large contact area between nanoparticles and starch, which resulted in large interaction sites, provides the characteristic improvement of the thermoplastic films [54, 55]. As an example, in perfectly homogenized dispersion, the montmorillonite nanoclay (MMT) can provide a specific surface area of 600–800 m^2/g [29, 53]. Three types of thermoplastic composites incorporated with nanoclay, namely phase-separated, intercalated, and exfoliated structures. Among those three types, only intercalated and exfoliated structures can be considered as the uniform nanoparticle distribution, which can possibly enhance the physical properties of the thermoplastic film composite. Hou and

Table 18.2: Effects of plasticizer additional to the thermophysical properties of the starch-based films [45] (Reprinted with permission from Elsevier).

Samples	Plasticizer type	Concentration (%-wt)	Degradation temperature (°C)			DTG peak temperature (°C)	T_g (°C)	Mass residue (%)
			$T_{10\%}$	$T_{25\%}$	$T_{50\%}$			
Native starch	–	0	173.60	299.80	325.00	301.30	76.88	20.77
G30	Glycerol	30	135.10	280.10	322.10	303.50	70.37	15.21
G45	Glycerol	45	130.30	276.30	319.80	304.60	66.78	11.67
G60	Glycerol	60	140.70	275.60	315.50	306.60	66.44	12.18
S30	Sorbitol	30	150.20	300.50	326.20	311.50	70.47	16.38
S45	Sorbitol	45	140.50	281.50	324.10	311.60	68.66	13.35
S60	Sorbitol	60	151.60	311.00	332.50	338.60	63.79	15.88
SG30	Sorbitol-Glycerol	30	148.50	294.50	319.50	296.80	70.00	10.52
SG45	Sorbitol-Glycerol	45	127.20	282.50	324.80	311.70	65.66	13.19
SG60	Sorbitol-Glycerol	60	124.50	275.30	320.50	314.20	65.61	12.43

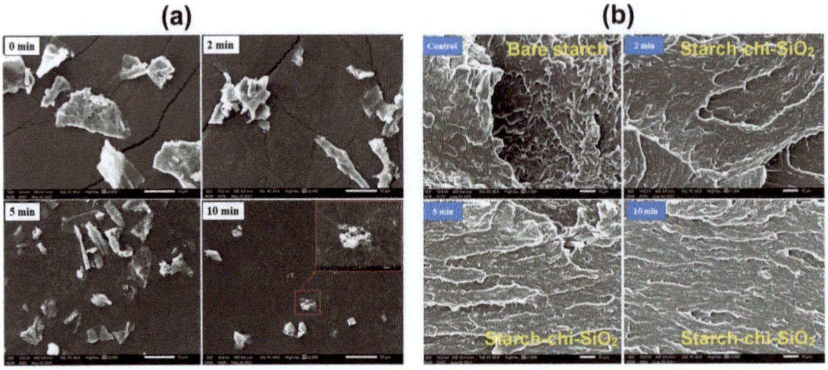

Figure 18.7: (a) Morphologies of chitosan-SiO_2 nanoparticles that prepared for starch's filler under different durations of ultrasonication, and (b) SEM images of starch-chitosan-SiO_2 nanocomposites with different ultrasonic treatment durations [56] (reprinted with permission from Elsevier).

Yan investigated the effects of chitosan-SiO_2 hybrid nanoparticles addition in corn starch-based thermoplastic films [56]. They have reported that the addition of chitosan-SiO_2 that prepared with 10 min of ultrasonic treatment resulted in the improved tensile strength of the fabricated thermoplastic films with good dispersion of chitosan-SiO_2 nanoparticles. Figure 18.7 shows the SEM images of the prepared chitosan-SiO_2 hybrid nanoparticles and the fabricated starch-chitosan-SiO_2 thermoplastic films. They also found that the incorporation of chitosan-SiO_2 improved the water activity due to the abundant hydrogen bonding in the prepared bio-composites [56]. Additionally, to investigate the presence of nanoparticles in the film samples, transmission electron microscopy (TEM) and X-ray diffraction (XRD) analyses are commonly used [53] (Figure 18.8). Also, to analyze the thermal properties of the thermoplastic composite, DTG analysis can be used. Figure 18.9

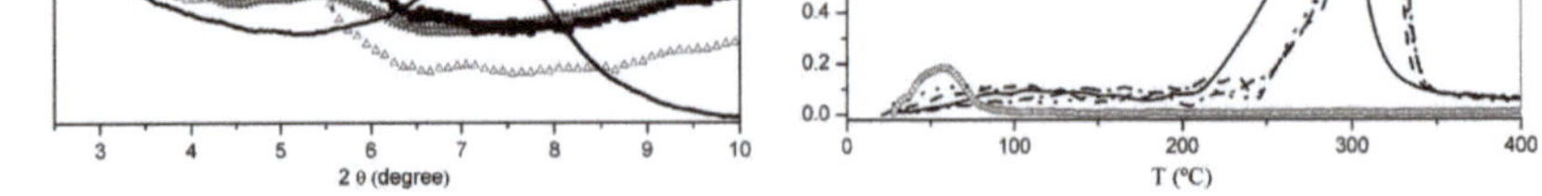

Figure 18.8: (a) XRD and (b) DTG patterns of native starch, MMT nanoparticles, and their thermoplastic film composites [53] (reprinted with permission from Elsevier).

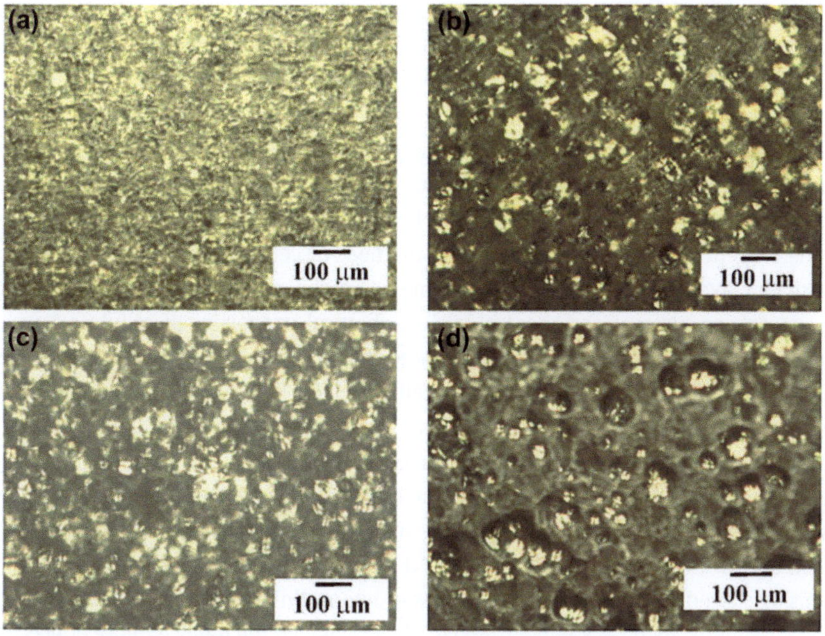

Figure 18.9: TEM micrographs of the films of starch (a) and the composites with 2 wt% (b), 3 wt% (c), and 5 wt % (d) of MMT nanoparticles [53] (reprinted with permission from Elsevier).

shows the reported XRD and DTG results of thermoplastic starch incorporated with different compositions of MMT.

18.5.3 Nanocellulose addition

Nanocellulose can be classified into three main groups; nanofibrillated cellulose (NFC), nanocrystalline cellulose (NCC), and bacterial nanocellulose (BNC) [57]. NFC and NCC can be prepared from the extraction and isolation of agricultural products using whether physical, chemical, and enzymatic methods. NCC was revealed to reduce the WVP of the prepared starch-based bio-composites by increasing the tortuosity of the bio-composite films [58]. NCC can be used as a reinforcing agent for starch-based bio-composite products. It is due to its abundant hydroxyl groups on its surface, which provides a strong hydrogen bonding between the hydrophilic polymer matrix (rich -OH groups) and non-polar matrix. This effect resulted in improved aspect ratio, mechanical properties, and created an affordable starch-based bio-composite products [59]. NCC was reported to have a high compatibility with starch matrix, which enhances its water activity and makes an outstanding feature for bio-packaging uses [60]. On the other hand, BNC can be produced by bacteria fermentation on microfibrils as the cultivation medium. BNCs are usually generated using microbe fermentations such as *Acetobacter xylinum*, *Achromobacter*, *Acanthamoeba*, and *Zooglea*

[61–63]. Implementation of BNC in bio-composite products have some benefits due to its high mechanical strength, nanofibril structure, high purity, good crystallinity, and biocompatible [64]. Some applications of BNCs have been reported in various fields such as biomedicine, electronic papers, audio membrane elements, food packaging, sound insulation, bio-plastics, and self-degradable microbeads [65–67].

Numerous review papers that focused on the processing methods and applications of nanocellulose-reinforced bio-composites have been reported recently. Oksman et al. discussed the current technologies of processing methods of cellulose-based bio-composites with particular focus on melt processing, casting technique, and resin incorporation [68]. Ng and coworkers summarized thermal and mechanical properties of cellulose-starch-based bio-composites [69]. Klemm et al. reviewed the structural details, production line, and physical-chemical characteristics of nanocellulose [70]. They also discussed about the potential implementations of nanocellulose on various application including on medical fields as implants, cosmetic additives, drug carriers, biomedical and tissue engineering, food packaging, and wound dressing. Dufresne and groups reviewed the pros and cons aspects of nanocellulose-based bio-composite products' processing methods especially using melt and wet method for both NCC and NFC based bio-composites [71]. Kagarzadeh and teams discussed an overview on the effects of nanocellulose on thin film composites, piles, and coating for food packaging uses [72]. They reviewed the nanocellulose effects on oxygen barrier, WVP properties, and the response on humidity. Numerous potential strategies to prepare eco-friendly packaging materials from nanocellulose are also summarized. Lastly, Trache et al. reported a review on the recent developments and challenges in the processing method of NCC from some natural resources of cellulose [73].

All three types of nanocellulose i.e., NCC, NFC, and BNC have been utilized to improve the characteristic of starch-based bio-composites. Generally, the nanocellulose introduction into starch-based composites results in flexible bonding for polymeric connection in the composite with the starch chains [74]. This interaction between nanocellulose and starch enhances the mechanical properties of the composite such as the Young's modulus, tensile strength, and elongation at break [75]. The existence of nanocellulose sites in the starch-based composites also improves the WVP and ductility of the bio-composites as the effect of possible network-chain development between the starch and nanocellulose particles [76]. Table 18.3 contained a summary of the recent reported works on the introduction of nanocellulose materials into starch-based bio-composites. Lu and coworkers reported the effects of cotton NCC into starch-based bio-composite that resulted in improved Young's modulus (700%) and tensile strength (200%) properties as well as an enhanced water resistance [76].

Recently, Johansson et al. investigated a new method to create CNF-reinforced starch bio-composite using the layer-by-layer formation (LbL) method [84]. They reported a significant improvement of Young's moduli of the fabricated CNF-reinforced starch-based bio-composite [84]. Plasticizers are not necessary in this method. Prakobna and groups investigated the interaction effects between nanocellulose fillers and starch matrix on mechanical properties [89]. They observed the starch-based nanocomposite that reinforced

Table 18.3: Effects of types and loadings of nanocellulose on the properties' improvements of the starch-based bio-composites (Properties: YM = Young's modulus; TS = tensile strength; WR = water resistance; SM = storage modulus; EB = elongation at break; FE = folding endurance; AP = air permeability; BD = biodegradability).

Type of nanocellulose	Loading (%wt)	Plasticizer	Preparation technique	Improved properties	Ref.
NCC	2.5–30	Glycerol	Solution-casting	YM, TS, WR	[76]
	10	Glycerol	Solution-casting	YM, SM, EB, WR	[75]
	1–15	Glycerol	Ball-milling/melt-compounding	YM, TS, SM	[77]
	2–10	Glycerol/sorbitol	Solution-casting	YM, TS, SM	[78]
	10–20	Glycerol	Solution-casting	SM, TS, FE, AP	[79]
NFC	5–20	Glycerol/sorbitol	Compression	YM, TS, SM, EB	[80]
	5–15	Glycerol	Solution-casting	YM, TS, SM, AP	[81]
	0.1–1.0	Acetic acid/ glycerol	Solution-casting	TS, SM, AP	[82]
	5–20	Sorbitol	Melt extrusion	YM, TS, AP	[83]
	2–10	Glycerol	Solution-casting	AP	[82]
	–	–	Layer-by-layer formation	YM	[84]
	10	Glycerol	Solution-casting	YM, TS, SM, AP, BD	[83]
	0–20	Xylitol/glycerol	Solution-casting	SM, CB	[85]
BNC	2.5	Glycerol	Solution-casting	YM, TS	[86]
	7.8–22	Glycerol	Solution-casting	YM, TS, AP	[87]
	1–5	Glycerol	Solution-casting	YM, TS, SM, AP	[88]

with NFCs using two methods namely mixed and core–shell methods. In the mixed method, the cellulose nanofibril-reinforced amylopectin (CNF/AP) bio-composite was fabricated by direct mixing of CNF and AP, with direct filtration and drying, so that the CNF/AP bio-composite was produced. Under this method, almost all AP structure was scavenged in the CNF structure without significant interaction between the matrix and nanofillers. In core–shell method, the remaining AP in the mixture was removed from the suspension of CNF/AP mixture by several times of washing/centrifugation. The removal of "spent" AC in the mixture resulted in formation of adsorbed AP shell on the core surface of CNF, thus a core–shell CNF/AP was formed. The comprehensive comparison of fabrication methods, and the developed performances are schematically shown in Figure 18.10. Generally, it can be seen that the CNF/AP bio-composite prepared by core–shell method possess stronger mechanical strength that of the mixed method.

18.5.4 Fiber as filler agent

Fiber additions into thermoplastic films have been found to improve some thermoplastic properties such as mechanical, thermal, and water resistance [90–92]. The enhanced mechanical properties were caused by the stronger bonding of the attached fiber and the starch,

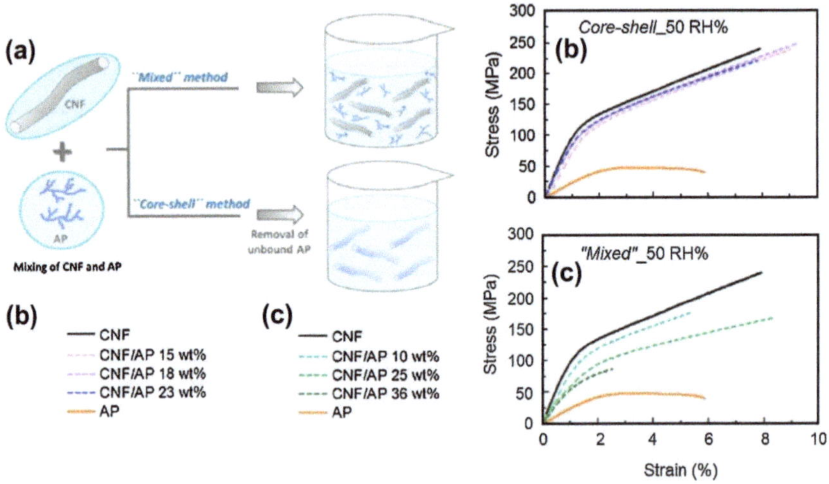

Figure 18.10: Comparison of fabrication methods, and the developed performances of core-shell and mixed methods in the preparation of CNF/AP bio-composites. (a) Schematic illustration. Variations of CNF/AP bio-composites and mechanical strength performance of the CNF/AP bio-composites using (b) Core-shell method and (c) Mixed method [89] (Reprinted with permission from American Chemical Society).

which resulted in better mechanical strength. It was reported that the addition of 5% fiber resulted in improvements in elongation modulus by 24% and tensile strength by 16%; nevertheless, the additional higher than 5% decreased the mechanical strength due to the agglomeration formation of the fiber [90]. The mechanical strength of thermoplastic films was also improved by the addition of fiber content 0 to 20%-w from 2.5 to 15.2 MPa, and it was slightly improved the moisture resistance ability of the fabricated thermoplastic films [49]. Several studies have been utilized the microfibrillated cellulose (MFC) from bamboo as the fiber in the thermoplastic film from sago and other starchy sources [51, 93, 94]. They found that adding 5% MFC to sago starch significantly improved the tensile strength of the thermoplastic film from 4.90 to 21.90 MPa, as shown in Figure 18.11 [95]. Regarding the preparation technique, the mechanical strength of thermoplastic sago starch film can be achieved by using an ultrasonic homogenizer. Recently, there have been several widely used reinforcing fibers such as commercial cellulose, cellulose nano-crystallite, and microfibrillated cellulose [51, 92, 94, 96, 97]. The reinforced thermoplastic with cellulose-based materials showed a significant improvement in mechanical, thermal, and wettability properties.

18.5.5 Polymer blending and cross-linking agents

Another method to improve the mechanical and wettability properties of the thermoplastic starch is polymer blending. Polymer blending can be performed using non-biodegradable polymers such as polyethylene and polypropylene or biodegradable

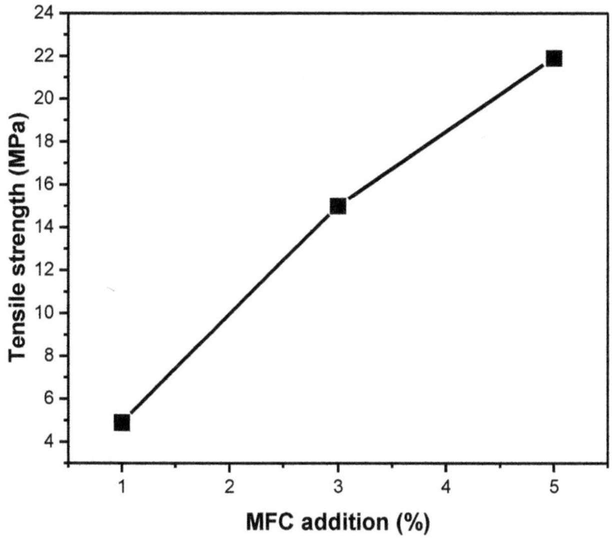

Figure 18.11: Effect of the MFC addition to the tensile strength of the fabricated thermoplastic films [95] (reprinted with permission from AIP Publishing).

polymers such as polyvinyl alcohol (PVA), polycaprolactone (PCLP), or polylactic acid (PLA) [97–101]. Low-density polyethylene (LDPE) and sago starch have been blended to form the composite film [98]. It has been found that the LDPE addition improved the mechanical strength by 800% compared to the thermoplastic films without LDPE. However, some issues are still found in the LLDPE/starch composite film, including low adhesion between starch and LDPE, the semi nonbiodegradable, and relatively higher composition of LDPE compared to the starch in the composite making [101]. If the starch composition is increased, the mechanical strength of the composite tends to decrease [99]. This finding was due to the unwell dispersion of LDPE and starch, which resulted in the nonhomogeneous distribution of constituents in the product [102]. It has been investigated that the conventional blending between LDPE and sago starch was not well mixed [98, 101]. There are some agglomerations on the fabricated composite. The agglomeration structure is reported to cause alteration in the mechanical properties of the composite [103]. An approach to making better blending between starch and LDPE has been found by using polyethylene grafted malic acid (PE-g-MA) that enhances the miscibility of polymer and starch, thus, resulting in better mechanical strength [104]. Some findings have been reported that the improved mechanical strength of the thermoplastic film from starch with LDPE addition can be attributed to the strong adhesion between starch and PE-g-MA [102–104].

Another approach to improve the applicability of sago starch has been reported by using PVA and different plasticizers, i.e., sorbitol and epoxidized soybean oil (ESBO) prepared using a twin-screw extruder machine [105]. It has been reported that the thermoplastic composite plasticized with ESBO gave a better mechanical property than sorbitol and the pristine sago starch film. It is due to the composite that plasticized with ESBO had a better dispersion than that of sorbitol, and it also had better compatibility with PVA/ sago starch composite. Another investigation revealed that starch/silica/PVA thermoplastic composite film cross-linked with sago starch improved the mechanical strength of the fabricated films better than that of pristine sago starch [106, 107]. The cross-linking structure can enhance the tensile strength and elongation at the break of the fabricated composites [14, 108]. Silviana and coworkers have reported that the addition of glycerol (20%-wt) as a plasticizer, chicken feather keratin (20%-wt), and lime juice's citric acid (5%-wt) as cross-linking agent resulted in an improvement of tensile strength of the sago starch thermoplastic composite films with a value of 6.32 MPa [14]. This tensile strength is higher than the pristine sago starch with glycerol as a plasticizer with a tensile strength of 4.2 MPa. Therefore, it can be concluded that the addition of polymers, compatibilizers, and cross-linking agents can significantly improve the mechanical properties of thermoplastic-based sago starch composite.

18.5.6 Challenges and future recommendations

Even though there have been numerous potential accomplishments on pilot and laboratory scale, there are still many challenges to bring out this modified sago starch-based thermoplastic films on the industrial scale. The main challenge that should be overcome to achieve a successful commercialization of the modified sago starch-based thermoplastic films is the brittleness of the fabricated films. An optimization is required to gain a composition of starch, additives, plasticizer, and cross-linker that resulted in the optimum properties of the sago starch-based thermoplastic films. It should be noted that these thermoplastic films are produced as a substitute for the already well-known petroleum-derived plastics. Therefore, general public may demand this particular novel material possesses comparable physical specifications, or even better if sago starch films could surpass commercially available materials.

From another point of view, however, commercialization of sago starch-based thermoplastic films means that some of annual sago production should be allocated for thermoplastic films production. As demand grows, sago starch industry should also be developed to support the already undersupplied food sector together with thermoplastic films manufacture. In this case, study of efficiency and process integration between one sector with another is required in order to fulfill the market demand while providing the novel product with competitive price. One thing that should be noted that most companies focus more on profit and return of investment. Thus, sago starch thermoplastic films production should be proposed as an interesting investment while also environmentally friendly. In these circumstances, some chemical and industrial engineering aspects, management studies, supply and demand, together with insights from economists may also play a role in facing this challenge.

As sustainability is the core of sustainable development goals, sago starch thermoplastic films, in a sense, is one solution in reaching this purpose. Therefore, this sustainability potential is an incomparable advantage to commercial plastics. Looking at the fact that these petroleum-derived materials have undergone major developments and efforts in cutting their production costs for tens of years, comparing laboratory-scale biodegradable materials to commercial plastics may not come up as a wise decision. Future studies on biodegradable materials are still wide open while fossil fuels and its derivatives are destined to disappear in the next ten to twenty years. It will be even better if one could assess available resources in an area and come up with the methods in creating similar sustainable materials and environment.

18.6 Conclusions

The thermoplastic product from starchy crops provides a strong opportunity for the massive production of biodegradable plastic products. Recently, sago crops have become a sustainable starch resource. Therefore, the study and use of sago starch have gained much attention due to its extensive availability for thermoplastic products. It is also important to note that the poor mechanical and water/moisture resistance properties remain the significant problems for thermoplastic products from sago and other starchy products. To overcome these problems, several types of plasticizers, nanoparticles, nanocellulose, fiber, crosslinkers, and polymers, were incorporated into sago starch with different methods of production. These enhancements of mechanical and water resistance properties of the thermoplastic products should meet the practical standard. Therefore, it can be concluded that sago starch has promising sustainability and availability as a starch source with a great opportunity in various applications, specifically in thermoplastic products.

Acknowledgment: The authors are grateful to the community of Advanced Material Laboratory (AMaL), Department of Chemical Engineering, Diponegoro University for their countless support and assistance in writing this study.

References

1. Karim AA, Tie AP-L, Manan DMA, Zaidul ISM. Starch from the sago (*Metroxylon sagu*) palm tree – properties, prospects, and challenges as a new industrial source for food and other uses. Compr Rev Food Sci Food Saf 2008;7:215–28.
2. Singhal RS, Kennedy JF, Gopalakrishnan SM, Kaczmarek A, Knill CJ, Akmar PF. Industrial production, processing, and utilization of sago palm-derived products. Carbohydr Polym 2008;72:1–20.
3. Darma. Improvement of sago processing machinery. In: Sago palm. Singapore: Springer; 2018:231–45 pp.
4. Lim TK. Cycas revoluta. In: Edible medicinal and non-medicinal plants. Netherlands: Springer; 2012: 732–8 pp.
5. Bujang K. Production, purification, and health benefits of sago sugar. In: Sago palm. Singapore: Springer; 2018:299–307 pp.

6. Adawiyah DR, Akuzawa S, Sasaki T, Kohyama K. A comparison of the effects of heat moisture treatment (HMT) on rheological properties and amylopectin structure in Sago (Metroxylon sago) and Arenga (Arenga pinnata) starches. J Food Sci Technol 2017;54:3404–10.
7. Zhu F. Relationships between amylopectin internal molecular structure and physicochemical properties of starch. Trends Food Sci Technol 2018;78:234–42.
8. Martinez MM, Li C, Okoniewska M, Mukherjee I, Vellucci D, Hamaker B. Slowly digestible starch in fully gelatinized material is structurally driven by molecular size and A and B1 chain lengths. Carbohydr Polym 2018;197:531–9.
9. Pinyo J, Luangpituksa P, Suphantharika M, Hansawasdi C, Wongsagonsup R. Effect of enzymatic pretreatment on the extraction yield and physicochemical properties of sago starch. Starch - Stärke 2016; 68:47–56.
10. Bhat R, Abdullah N, Din RH, Tay G-S. Producing novel sago starch based food packaging films by incorporating lignin isolated from oil palm black liquor waste. J Food Eng 2013;119:707–13.
11. Rindlav-Westling A, Stading M, Hermansson A-M, Gatenholm P. Structure, mechanical and barrier properties of amylose and amylopectin films. Carbohydr Polym 1998;36:217–24.
12. López Ov., García MA. Starch films from a novel (Pachyrhizus ahipa) and conventional sources: development and characterization. Mater Sci Eng C 2012;32:1931–40.
13. Phan TD, Debeaufort F, Luu D, Voilley A. Functional properties of edible agar-based and starch-based films for food quality preservation. J Agric Food Chem 2005;53:973–81.
14. Silviana, Zuhri LA, Muzakki MM, Rahmawati AK, Marthasari AA. Utilization of lime juice to enhance mechanical performance of sago-starch based biofilm reinforced chicken feather keratin. Adv Sci Lett 2017;23:5739–41.
15. Muscat D, Adhikari B, Adhikari R, Chaudhary DS. Comparative study of film forming behaviour of low and high amylose starches using glycerol and xylitol as plasticizers. J Food Eng 2012;109:189–201.
16. Myllärinen P, Partanen R, Seppälä J, Forssell P. Effect of glycerol on behaviour of amylose and amylopectin films. Carbohydr Polym 2002;50:355–61.
17. Abdorreza MN, Cheng LH, Karim AA. Effects of plasticizers on thermal properties and heat sealability of sago starch films. Food Hydrocolloids 2011;25:56–60.
18. Hizukuri S, Kaneko T, Takeda Y. Measurement of the chain length of amylopectin and its relevance to the origin of crystalline polymorphism of starch granules. Biochim Biophys Acta Gen Subj 1983;760:188–91.
19. Uthumporn U, Wahidah N, Karim AA. Physicochemical properties of starch from sago (*Metroxylon sagu*) palm grown in mineral soil at different growth stages. IOP Conf Ser Mater Sci Eng 2014;62:012026.
20. Takeda Y, Takeda C, Suzuki A, Hizukuri S. Structures and properties of sago starches with low and high viscosities on amylography. J Food Sci 1989;54:177–82.
21. Othman Z, Hassan O, Hashim K. Physicochemical and thermal properties of gamma-irradiated sago (Metroxylon sagu) starch. Radiat Phys Chem 2015;109:48–53.
22. Koganti N, Mitchell J, MacNaughtan W, Hill S, Foster T. Effect of granule organisation on the behaviour of starches in the NMMO (N-methyl morpholine N-oxide) solvent system. Carbohydr Polym 2015;116:103–10.
23. Ng J, Siew CK, Mamat H, Matanjun P, Lee J. Effect of acid methanol treatment and heat moisture treatment on in vitro digestibility and estimated glycemic index of raw and Gelatinized (*Metroxylon sagu*) starch. Starch - Stärke 2018;70:1700198.
24. Fouladi E, Mohammadi Nafchi A. Effects of acid-hydrolysis and hydroxypropylation on functional properties of sago starch. Int J Biol Macromol 2014;68:251–7.
25. Bertoft E, Piyachomkwan K, Chatakanonda P, Sriroth K. Internal unit chain composition in amylopectins. Carbohydr Polym 2008;74:527–43.
26. Schirmer M, Jekle M, Becker T. Starch gelatinization and its complexity for analysis. Starch - Stärke 2015;67: 30–41.
27. Vamadevan V, Bertoft E, Seetharaman K. On the importance of organization of glucan chains on thermal properties of starch. Carbohydr Polym 2013;92:1653–9.

28. Vamadevan V, Bertoft E. Impact of different structural types of amylopectin on retrogradation. Food Hydrocolloids 2018;80:88–96.
29. Zhang Y, Rempel C, McLaren D. Thermoplastic starch. In: Innovations in food packaging. Cambridge: Elsevier; 2014:391–412 pp.
30. Koch K, Gillgren T, Stading M, Andersson R. Mechanical and structural properties of solution-cast high-amylose maize starch films. Int J Biol Macromol 2010;46:13–9.
31. Mościcki L, Mitrus M, Wójtowicz A, Oniszczuk T, Rejak A, Janssen L. Application of extrusion-cooking for processing of thermoplastic starch (TPS). Food Res Int 2012;47:291–9.
32. Li M, Liu P, Zou W, Yu L, Xie F, Pu H, et al. Extrusion processing and characterization of edible starch films with different amylose contents. J Food Eng 2011;106:95–101.
33. Melo CPB, Grossmann MVE, Yamashita F, Youssef EY, Dall'Antônia LH, Mali S. Effect of manufacturing process and xanthan gum addition on the properties of cassava starch films. J Polym Environ 2011;19: 739–49.
34. Katekhong W, Wongphan P, Klinmalai P, Harnkarnsujarit N. Thermoplastic starch blown films functionalized by plasticized nitrite blended with PBAT for superior oxygen barrier and active biodegradable meat packaging. Food Chem 2022;374:131709.
35. Fourati Y, Tarrés Q, Mutjé P, Boufi S. PBAT/thermoplastic starch blends: effect of compatibilizers on the rheological, mechanical and morphological properties. Carbohydr Polym 2018;199:51–7.
36. Montoille L, Morales Vicencio C, Fontalba D, Ortiz JA, Moreno-Serna V, Peponi L, et al. Study of the effect of the addition of plasticizers on the physical properties of biodegradable films based on kefiran for potential application as food packaging. Food Chem 2021;360:129966.
37. Sa'adah AN, Milyawan GNA, Nadya T, Silviana S. A mini-review of bio-scrubber derived from bacterial cellulose impregnated by flavonoid of moringa leaves. IOP Conf Ser Earth Environ Sci 2022;963:012022.
38. Silviana S, Khusna EI, Susanto GAH, Sanyoto GJ, Hadiyanto H. Biocomposite of bacterial cellulose based from yam bean (Pachyrhizus erosus L. Urban) reinforced by bamboo microfibrillated cellulose through in situ method. AIP Conf Proc 2020;2197:050011.
39. Thunwall M, Kuthanova V, Boldizar A, Rigdahl M. Film blowing of thermoplastic starch. Carbohydr Polym 2008;71:583–90.
40. Silviana S, Sa'adah AN, Deastuti RP, Ramadhani NC, Simarmata N, Arianto LE, et al. Synthesis of silica-cellulose aerogel derived from bagasse through impregnation and ambient pressure drying methods as thermal insulator. IOP Conf Ser Earth Environ Sci 2022;963:012027.
41. Galdeano MC, Mali S, Grossmann MVE, Yamashita F, García MA. Effects of plasticizers on the properties of oat starch films. Mater Sci Eng C 2009;29:532–8.
42. Silviana S, Subagio A. Biocomposite characterization of bagasse starch derived from cassava reinforced by acetylated bamboo cellulose and plasticized by epoxidized waste cooking oil. Rasayan J Chem 2019;12: 1470–7.
43. Smits ALM, Kruiskamp PH, van Soest JJG, Vliegenthart JFG. The influence of various small plasticisers and malto-oligosaccharides on the retrogradation of (partly) gelatinised starch. Carbohydr Polym 2003;51: 417–24.
44. Zhang Y, Han JH. Mechanical and thermal characteristics of pea starch films plasticized with monosaccharides and polyols. J Food Sci 2006;71:E109–18.
45. Hazrati KZ, Sapuan SM, Zuhri MYM, Jumaidin R. Effect of plasticizers on physical, thermal, and tensile properties of thermoplastic films based on Dioscorea hispida starch. Int J Biol Macromol 2021;185:219–28.
46. Wang N, Yu J, Chang PR, Ma X. Influence of formamide and water on the properties of thermoplastic starch/poly(lactic acid) blends. Carbohydr Polym 2008;71:109–18.
47. Yu J, Chang PR, Ma X. The preparation and properties of dialdehyde starch and thermoplastic dialdehyde starch. Carbohydr Polym 2010;79:296–300.

48. Zullo R, Iannace S. The effects of different starch sources and plasticizers on film blowing of thermoplastic starch: correlation among process, elongational properties and macromolecular structure. Carbohydr Polym 2009;77:376–83.
49. Ma X. The plastcizers containing amide groups for thermoplastic starch. Carbohydr Polym 2004;57: 197–203.
50. Silviana S, Dzulkarom MC. Synthesis of cassava bagasse starch-based biocomposite reinforced woven bamboo fibre with Lime juice as crosslinker and Epoxidized Waste Cooking Oil (EWCO) as bioplasticizer. J Phys Conf 2019;1295:012076.
51. Silviana ., Anggoro DD. Use of epoxidized waste cooking oil as bioplasticizer of sago starch-based biocomposite reinforced microfibrillated cellulose of bamboo. Adv Sci Lett 2017;23:2591–4.
52. Chung Y-L, Ansari S, Estevez L, Hayrapetyan S, Giannelis EP, Lai H-M. Preparation and properties of biodegradable starch–clay nanocomposites. Carbohydr Polym 2010;79:391–6.
53. Cyras VP, Manfredi LB, Ton-That M-T, Vázquez A. Physical and mechanical properties of thermoplastic starch/montmorillonite nanocomposite films. Carbohydr Polym 2008;73:55–63.
54. de Carvalho AJF, Curvelo AAS, Agnelli JAM. A first insight on composites of thermoplastic starch and kaolin. Carbohydr Polym 2001;45:189–94.
55. Wu H, Liu C, Chen J, Chang PR, Chen Y, Anderson DP. Structure and properties of starch/α-zirconium phosphate nanocomposite films. Carbohydr Polym 2009;77:358–64.
56. Hou J, Yan X. Preparation of chitosan-SiO_2 nanoparticles by ultrasonic treatment and its effect on the properties of starch film. Int J Biol Macromol 2021;189:271–8.
57. Nazrin A, Sapuan SM, Zuhri MYM, Ilyas RA, Syafiq R, Sherwani SFK. Nanocellulose reinforced thermoplastic starch (TPS), polylactic acid (PLA), and polybutylene succinate (PBS) for food packaging applications. Front Chem 2020;8:213.
58. Ilyas RA, Sapuan SM, Ishak MR, Zainudin ES. Development and characterization of sugar palm nanocrystalline cellulose reinforced sugar palm starch bionanocomposites. Carbohydr Polym 2018;202: 186–202.
59. Mishra RK, Ha SK, Verma K, Tiwari SK. Recent progress in selected bio-nanomaterials and their engineering applications: an overview. J Sci Adv Mater Dev 2018;3:263–88.
60. Mondal S. Preparation, properties and applications of nanocellulosic materials. Carbohydr Polym 2017; 163:301–16.
61. Embuscado ME, Marks JS, BeMiller JN. Bacterial cellulose. II. Optimization of cellulose production by Acetobacter xylinum through response surface methodology. Food Hydrocoll 1994;8:419–30.
62. Liebner F, Haimer E, Wendland M, Neouze M-A, Schlufter K, Miethe P, et al. Aerogels from unaltered bacterial cellulose: application of $scCO_2$ drying for the preparation of shaped, ultra-lightweight cellulosic aerogels. Macromol Biosci 2010;10:349–52.
63. Picheth GF, Pirich CL, Sierakowski MR, Woehl MA, Sakakibara CN, de Souza CF, et al. Bacterial cellulose in biomedical applications: a review. Int J Biol Macromol 2017;104:97–106.
64. Silviana S, Giraldi VF, Purnomo LA, Saadah AN, Dalanta F. Bio-microbeads from bacterial cellulose incorporated with antimicrobial of chitosan and moringa leaves flavonoid. RASAYAN J Chem 2022;15: 1160–5.
65. Ashjaran A, Yazdanshenas ME, Rashidi A, Khajavi R, Rezaee A. Overview of bio nanofabric from bacterial cellulose. J Textil Inst 2013;104:121–31.
66. Thi T, Sugiyama J, Bulone V. Bacterial cellulose-based biomimetic composites. In: Biopolymers. London: Sciyo; 2010.
67. Silviana S, Hermawan F, Indracahya J, Kusumawardhani DAL, Dalanta F. Optimizing the environmentally friendly silica-cellulose aerogel composite for acoustic insulation material derived from newspaper and geothermal solid waste using a central composite design. J Sol Gel Sci Technol 2022;103:226–43.
68. Oksman K, Aitomäki Y, Mathew AP, Siqueira G, Zhou Q, Butylina S, et al. Review of the recent developments in cellulose nanocomposite processing. Compos Appl Sci Manuf 2016;83:2–18.

69. Ng H-M, Sin LT, Bee S-T, Tee T-T, Rahmat AR. Review of nanocellulose polymer composite characteristics and challenges. Polym-Plast Technol Eng 2017;56:687–731.
70. Klemm D, Cranston ED, Fischer D, Gama M, Kedzior SA, Kralisch D, et al. Nanocellulose as a natural source for groundbreaking applications in materials science: today's state. Mater Today 2018;21:720–48.
71. Dufresne A. Cellulose nanomaterial reinforced polymer nanocomposites. Curr Opin Colloid Interface Sci 2017;29:1–8.
72. Kargarzadeh H, Huang J, Lin N, Ahmad I, Mariano M, Dufresne A, et al. Recent developments in nanocellulose-based biodegradable polymers, thermoplastic polymers, and porous nanocomposites. Prog Polym Sci 2018;87:197–227.
73. Trache D, Hussin MH, Haafiz MKM, Thakur VK. Recent progress in cellulose nanocrystals: sources and production. Nanoscale 2017;9:1763–86.
74. Yang S, Tang Y, Wang J, Kong F, Zhang J. Surface treatment of cellulosic paper with starch-based composites reinforced with nanocrystalline cellulose. Ind Eng Chem Res 2014;53:13980–8.
75. Chen Y, Liu C, Chang PR, Cao X, Anderson DP. Bionanocomposites based on pea starch and cellulose nanowhiskers hydrolyzed from pea hull fibre: effect of hydrolysis time. Carbohydr Polym 2009;76:607–15.
76. Lu Y, Weng L, Cao X. Biocomposites of plasticized starch reinforced with cellulose crystallites from cottonseed linter. Macromol Biosci 2005;5:1101–7.
77. Moreira FKV, Marconcini JM, Mattoso LHC. Solid state ball milling as a green strategy to improve the dispersion of cellulose nanowhiskers in starch-based thermoplastic matrices. Cellulose 2012;19:2049–56.
78. Zainuddin SYZ, Ahmad I, Kargarzadeh H, Abdullah I, Dufresne A. Potential of using multiscale kenaf fibers as reinforcing filler in cassava starch-kenaf biocomposites. Carbohydr Polym 2013;92:2299–305.
79. Nasseri R, Mohammadi N. Starch-based nanocomposites: a comparative performance study of cellulose whiskers and starch nanoparticles. Carbohydr Polym 2014;106:432–9.
80. Teixeira Ede M, Pasquini D, Curvelo AAS, Corradini E, Belgacem MN, Dufresne A. Cassava bagasse cellulose nanofibrils reinforced thermoplastic cassava starch. Carbohydr Polym 2009;78:422–31.
81. Kaushik A, Singh M, Verma G. Green nanocomposites based on thermoplastic starch and steam exploded cellulose nanofibrils from wheat straw. Carbohydr Polym 2010;82:337–45.
82. Savadekar NR, Mhaske ST. Synthesis of nano cellulose fibers and effect on thermoplastics starch based films. Carbohydr Polym 2012;89:146–51.
83. Babaee M, Jonoobi M, Hamzeh Y, Ashori A. Biodegradability and mechanical properties of reinforced starch nanocomposites using cellulose nanofibers. Carbohydr Polym 2015;132:1–8.
84. Johansson E, Wågberg L. Tailoring the mechanical properties of starch-containing layer-by-layer films. Colloids Surf A Physicochem Eng Asp 2012;394:14–22.
85. Li M, Li D, Wang L, Adhikari B. Creep behavior of starch-based nanocomposite films with cellulose nanofibrils. Carbohydr Polym 2015;117:957–63.
86. Woehl MA, Canestraro CD, Mikowski A, Sierakowski MR, Ramos LP, Wypych F. Bionanocomposites of thermoplastic starch reinforced with bacterial cellulose nanofibres: effect of enzymatic treatment on mechanical properties. Carbohydr Polym 2010;80:866–73.
87. Wan YZ, Luo H, He F, Liang H, Huang Y, Li XL. Mechanical, moisture absorption, and biodegradation behaviours of bacterial cellulose fibre-reinforced starch biocomposites. Compos Sci Technol 2009;69:1212–7.
88. Martins IMG, Magina SP, Oliveira L, Freire CSR, Silvestre AJD, Neto CP, et al. New biocomposites based on thermoplastic starch and bacterial cellulose. Compos Sci Technol 2009;69:2163–8.
89. Prakobna K, Galland S, Berglund LA. High-performance and moisture-stable cellulose–starch nanocomposites based on bioinspired core–shell nanofibers. Biomacromolecules 2015;16:904–12.
90. Gilfillan WN, Nguyen DMT, Sopade PA, Doherty WOS. Preparation and characterisation of composites from starch and sugar cane fibre. Ind Crop Prod 2012;40:45–54.
91. Lu P, Zhang M, Liu Y, Li J, Xin M. Characteristics of vermiculite-reinforced thermoplastic starch composite films. J Appl Polym Sci 2012;126:E116–22.
92. Prachayawarakorn J, Hommanee L, Phosee D, Chairapaksatien P. Property improvement of thermoplastic mung bean starch using cotton fiber and low-density polyethylene. Starch - Stärke 2010;62:435–43.

93. Silviana S, Susanti S, Subagio A. Preliminary study for acetylation of cassava bagasse starch and microfibrillated cellulose of bamboo. MATEC Web Conf 2018;156:01019.
94. Silviana S, Rahayu P. Central composite design for optimization of starch-based bioplastic with bamboo microfibrillated cellulose as reinforcement assisted by potassium chloride. J Phys Conf 2019;1295:012073.
95. Silviana S, Hadiyanto H. Preparation of sago starch-based biocomposite reinforced microfibrillated cellulose of bamboo assisted by mechanical treatment. In: AIP conference proceedings on green process, material, and energy: a sustainable solution for climate change; 2017:0300241–5 pp. https://doi.org/10.1063/1.4985494.
96. Müller CMO, Laurindo JB, Yamashita F. Effect of cellulose fibers addition on the mechanical properties and water vapor barrier of starch-based films. Food Hydrocoll 2009;23:1328–33.
97. Soykeabkaew N, Supaphol P, Rujiravanit R. Preparation and characterization of jute- and flax-reinforced starch-based composite foams. Carbohydr Polym 2004;58:53–63.
98. Girija BG, Sailaja RRN. Low-density polyethylene/plasticized tapioca starch blends with the low-density polyethylene functionalized with maleate ester: mechanical and thermal properties. J Appl Polym Sci 2006; 101:1109–20.
99. Huneault MA, Li H. Preparation and properties of extruded thermoplastic starch/polymer blends. J Appl Polym Sci 2012;126:E96–108.
100. Luo X, Li J, Lin X. Effect of gelatinization and additives on morphology and thermal behavior of corn starch/PVA blend films. Carbohydr Polym 2012;90:1595–600.
101. Yu F, Prashantha K, Soulestin J, Lacrampe M-F, Krawczak P. Plasticized-starch/poly(ethylene oxide) blends prepared by extrusion. Carbohydr Polym 2013;91:253–61.
102. Stagner JA, Alves VD, Narayan R. Application and performance of maleated thermoplastic starch-poly(butylene adipate-*co*-terephthalate) blends for films. J Appl Polym Sci 2012;126:E135–42.
103. Majid RA, Ismail H, Taib RM. Effects of PE-g-MA on tensile properties, morphology and water absorption of LDPE/thermoplastic sago starch blends. Polym-Plast Technol Eng 2009;48:919–24.
104. Sabetzadeh M, Bagheri R, Masoomi M. Effect of corn starch content in thermoplastic starch/low-density polyethylene blends on their mechanical and flow properties. J Appl Polym Sci 2012;126:1–7.
105. Mohamed R, Mutalib NWA, Norizan MN, Hirzin RSN, Isa SAM. Effect of different plasticizers on tensile properties of PVA/sago starch system before and after weathering exposure; 2018;030005. https://doi.org/10.1063/1.5047163.
106. Beg MDH, Kormin S, Bijarimi M, Zaman HU. Preparation and characterization of low-density polyethylene/thermoplastic starch composites. Adv Polym Technol 2016;35. https://doi.org/10.1002/adv.21521.
107. Ismail H, Zaaba NF. Effect of unmodified and modified sago starch on properties of (sago starch)/silica/PVA plastic films. J Vinyl Addit Technol 2014;20:185–92.
108. Silviana S, Sa'adah AF. Mechanical and thermal properties of bacterial cellulose reinforced with bamboo microfibrillated cellulose and plasticized with epoxidized waste cooking oil. Cellul Chem Technol 2022;56: 331–9.

Supplementary Material: This article contains supplementary material (https://doi.org/10.1515/psr-2022-0017).

Dzun Noraini Jimat*, Yusilawati Ahmad Nor and
Noor Illi Mohamad Puad

19 Review on sago thermoplastic starch composite films reinforced with nanocellulose

Abstract: Recently, many studies have been carried out on developing green composites that are eco-friendly and more sustainable compared to fuel-derived composites. Thermoplastic starch (TPS) is one of the potential biopolymer materials that can be used as food packaging materials, pharmaceuticals and many other applications. Sago (*Metroxylon sagu*) is unlike other starch-based composite such as cassava and corn, which are highly utilized and popular option sources for variety applications in industries. This renewable starch can be a competitive starch source like other starches if rigorous scientific research study is conducted to explore other findings, which might be important for its production and usage in various industries. Research studies related to sago TPS are still lacked, and it is difficult to compare its performance with TPS made from other starches as its preparation differs based on the starch source. The TPS composite generally exhibits low mechanical strength and poor barrier properties. Reinforcing sago TPS nanocomposites with nanocellulose (NC) is expected to enhance its mechanical and barrier properties. This book chapter covers the overview of characteristics and extraction of sago from its pith as well as its usage to make thermoplastic starch composite. The characteristics of sago TPS nanocomposite reinforced with nanocellulose (NC) is also discussed based on a few research studies due to its published research findings are still limited.

Keywords: barrier properties; biopolymer; mechanical strength; nanocellulose; sago *Metroxylon sagu*.

19.1 Introduction

Products made from eco-friendly polymers have been significantly created with innovative technologies to deal with the crisis of products of non-degradable polymers, which are known to be one of the main pollutants of the environment. Thermoplastic starch (TPS) can be one of potential biopolymers that can be used in various applications

***Corresponding author: Dzun Noraini Jimat**, Department of Biotechnology Engineering, Kulliyyah of Engineering, International Islamic University of Malaysia, 53100, Selangor, Malaysia,
E-mail: jnoraini@iium.edu.my
Yusilawati Ahmad Nor and Noor Illi Mohamad Puad, Department of Biotechnology Engineering, Kulliyyah of Engineering, International Islamic University of Malaysia, 53100, Selangor, Malaysia

As per De Gruyter's policy this article has previously been published in the journal Physical Sciences Reviews. Please cite as: D. N. Jimat, Y. Ahmad Nor and N. I. Mohamad Puad "Review on sago thermoplastic starch composite films reinforced with nanocellulose" *Physical Sciences Reviews* [Online] 2023. DOI: 10.1515/psr-2022-0016 | https://doi.org/10.1515/9783110773606-019

specifically as a food packaging material due to their cheap and abundant sources, biodegradability and non-toxic properties. TPS is developed from the most abundant biopolymer, starch, where its coarse structure is transformed using a plasticizer such as water or glycerol. Starch is not only inexpensive but also non-toxic that can be obtained from renewable sources.

However, the drawbacks of thermoplastic starch (TPS) are less hydrophobic feature and weak mechanical strength that hindered its broad applications [1]. However, its structure and functional stability can be enhanced through several approaches such as blending, cross-linking, micro and nano-reinforcement.

Starch is among the demanding raw materials for food and many other industry applications due to its low cost, biodegradable, renewable and abundant availability material. It is a mixture of amylose and amylopectin where its typical ratio is 20:80 [2]. Starch naturally existed in numerous plant sources such as potato, wheat, tapioca, corn and sago. It is different in whole composition over granules size distribution, structure, amylose and lipid content, distribution of chain length in amylopectin and crystalline structure [2, 3]. Sago (*Metroxylon sagu*) is an unpopular source of starch compared to cassava and corn, and its selling price is higher compared to these both starch sources [4, 5]. Previous study reported that its yield per unit area could be superior to other starch sources (corn, wheat or cassava) [5]. This palm tree is native to Southeast Asia regions specifically Malaysia, Indonesia and Papua New Guinea.

This book chapter covers the overview of characteristics and extraction of sago from its pith as well as its usage to make thermoplastic starch composite. The characteristics of sago TPS nanocomposite reinforced with nanocellulose (NC) is also discussed based on a few research studies due to its published research findings are still limited as well as its development challenges.

19.2 Overview of sago starch processing

Sago (*M. sagu*) is unfavourable option for starch sources in industry due to its limited production, which is only produced for domestic demand in main countries producers namely Malaysia, Indonesia and Papua New Guinea [4]. This is due to lack of advanced processing technology and allocated budget to improve the upstream and downstream processes to obtain high yield and quality of sago starch. Figure 19.1 shows a summary of the challenges related with the sago starch production.

This palm can grow well in harsh swampy peat environment and tolerant to extreme climate condition (Figure 19.2). This versatile plant has a quite long juvenile phase in which it is ready to be harvest after 10 years if grown on good soil condition [4]. Some palm tree may take 15 years to reach maturity if grown on poor soil condition [4]. It is competitive to other starch crop sources as its gel clarity property give an advantage to food manufacturers who produce noodle and vermicelli [4]. Sago starch processing technologies adopted the ones from cassava processing with some alterations due to their physical differences [4]. Sago starch can be only extracted after the debarking the sago logs followed by milling its pith. Typically, the trunk of sago palm is 500–1800 cm in length

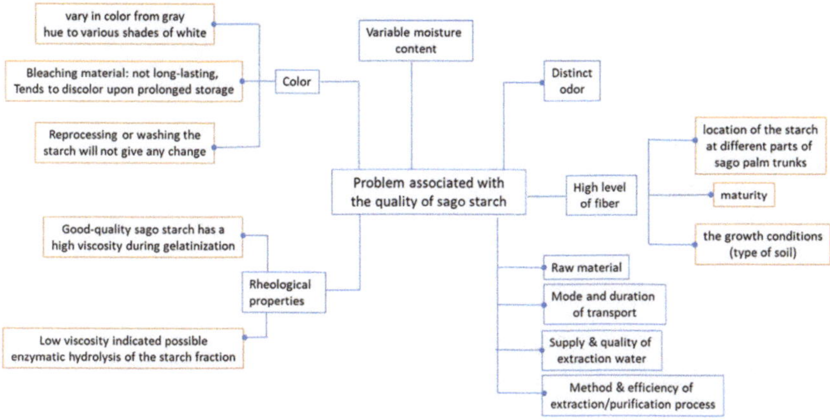

Figure 19.1: Summary of challenges related with the quality of sago starch [5].

Figure 19.2: Sago (*Metroxylon sagu*) palms planted in main countries producers named Indonesia, Papua New Guinea and Malaysia (source: https://craunresearch.com.my/) [6].

Figure 19.3: The sago palm trunks that have been cleaned and ready for the grinding process before extracting the starch (source: https://craunresearch.com.my/) [6].

while and 25–70 cm in diameter [4] (Figure 19.3). The sago starch is deposited in the pith tissue of the sago palm trunk. The trunk needs to be cut into 0.9–1.2 m long and the dense lignified bark is cut off manually or mechanically before extracting the starch-rich pith [4].

Typically, a saw-blade rasper is used to mill the pith of sago. The sago pith contents of high fibre and with its hardness and shape, it is good to chip it before the milling process to ensure the effectiveness of the rasping process [4]. Moreover, chipping the pith prior to feeding into saw-blade raspers may minimize the wear and tear of saw blades and prevent the sieve blocking. However, the whole pith section can be easily to rasp when the steel drums with a quick-fit nail-studded case are being introduced [4]. The overall processes to obtain sago starch are illustrated in Figure 19.4.

Finally, a few units of extractors are used to separate the starch from unwanted debris such as fibre and cell masses. Conventionally, sieving and sedimentation processes

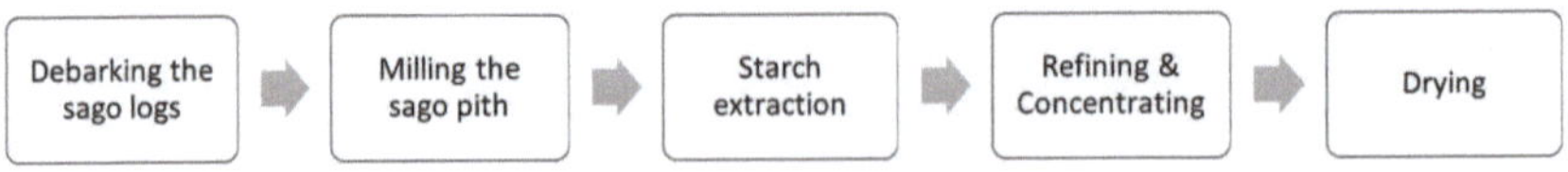

Figure 19.4: Downstream processes of sago starch.

are applied to remove coarse debris and finer debris, respectively [4]. These separation processes are cheap and efficient. However, they require huge space for installation. A proper maintenance of extractor is required to minimize malfunctioning of extractor sieve due to frequent clogged, overloaded, vibrated severely and tore [4]. Furthermore, frequent cleaning is required to prevent any fungal spore contamination to the extracted starch. Finally, the extracted starch slurry goes through several processes such as refining and concentrating, dewatering and drying using disc separators and decanters or hydro-cyclones, respectively, to produce sago starch [4].

The summary of characteristics and functional properties of sago starch is shown in Figure 19.5.

19.3 Sago thermoplastic starch (TPS) composite film

In Malaysia, sago (*M. sagu*) palm is planted mainly in Mukah districts, Sarawak, which is managed by small-scale sago farmers. Historically, it is staple food for Melanau communities in Sarawak [11]. It seems that sago starch is the least preferred material to be used in the industry due to lacking aggressive strategy promotion compared to other starch sources such as cassava, which is well-known to be used widely in many industries. Research and development (R&D) studies related to sago starch are also limited in contrast to cassava, corn and potato.

Previous study reported that pure starch could not be counted as a thermoplastic polymer due to its tough intermolecular and intramolecular hydrogen bonding in the amylose and amylopectin macromolecular chains [12]. However, it can be easily to melt and flow and ready for next process such as extruding, injection moulding or blowing

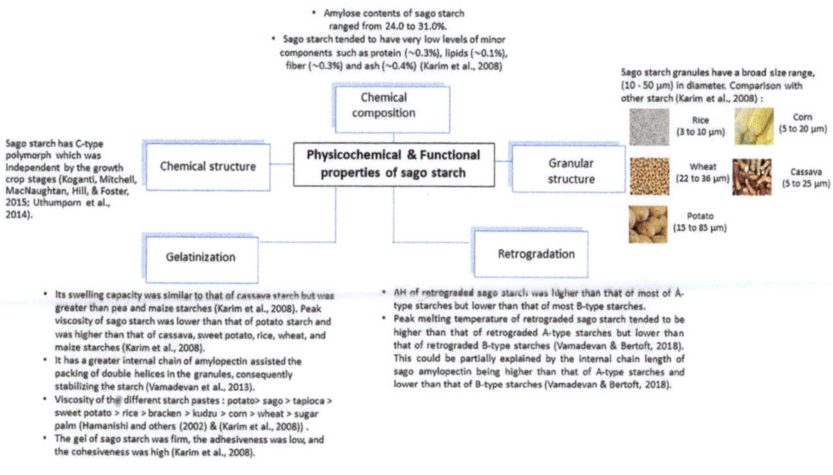

Figure 19.5: Physicochemical and functional properties of sago starch [7–10].

when blend with the plasticizers such as water, glycerine or sorbitol at high temperature (90 °C–180 °C) and under shear [12]. Plasticizers would destruct granular starch by disrupting the inter/intra-molecular hydrogen bonds, associated with partial depolymerization of the starch backbone, and dropping the melting temperature of starch below its decomposition temperature (230 °C) [1, 13]. It shows that the role of plasticisers is critical in producing good TPS with desired characteristics. Figure 19.6 below illustrates the role of plasticizer in synthesizing the TPS.

Moisture sorption behaviour, barrier properties, thermal, mechanical, surface tension and retrogradation of starch are the critical attributes of TPS that should be competitive with other composites [14]. The tensile properties of TPS depend on the original source of starch, additives, glass-transition temperature (*Tg*), crystallinity and ratio of amylose to amylopectin [1]. Tensile properties of TPS can be improved by inclusion of additives such as plasticizers, cross-linking agents, antimicrobial agents, antioxidants and texturizing agents, and many studies found that plasticizers have significant role in the structure and performance of TPS [1]. It is reported that the inclusion of plasticizer reduces the tensile strength (TS) and modulus of elasticity (EM) and enhances the elongation at breakage (E) to different degrees [1]. However, excessive plasticizer may cause the E value to decline [1].

Halimatul et al. [15] and co-workers prepared two sets of thermoplastic sago films via compression moulding using: (1) different sago starch concentrations (3–9 w/v%) and constant of plasticiser loadings (30 w/w%) and (2) different plasticizer loadings (10–50 w/w%) and constant of sago starch concentration (5 w/v%) [15]. It is found that the mechanical behaviour of starch films is considerably affected by the amount of both sago starch and plasticizer. They found that the tensile strength increased

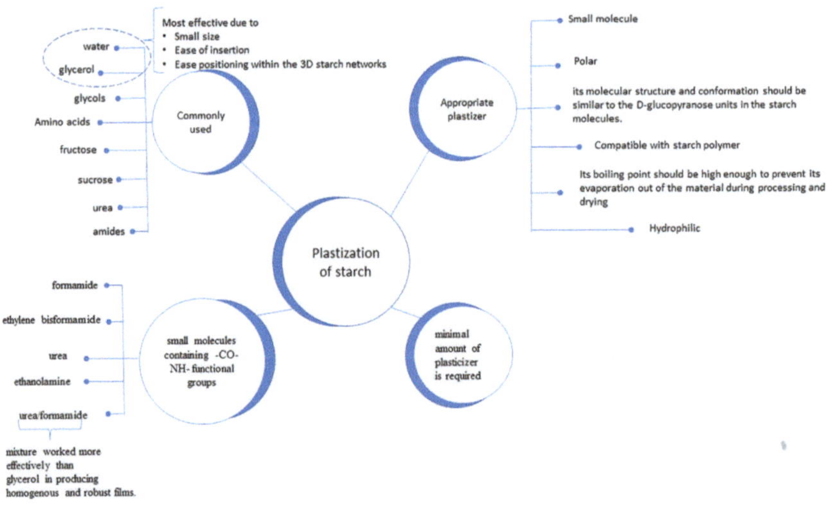

Figure 19.6: Role of plasticizer in synthesizing the TPS [1].

(associated by radially reduced tensile strain and enhanced tearing strength) when the starch increased. However, tensile and tearing reduced but tensile strain increased when the plasticizer concentration increased. After repeating soaking and drying tests, the sago films made with more plasticizer and low starch amount had lower mass loss contrasted to those films with lower plasticizer and higher starch concentration. This result showed that few cycles of soaking and drying might cause reduction of starch dissolution and was almost certainly associated with the retrogradation process inside the film.

Ismail and co-workers et al. 2012 prepared sago thermoplastics starch (TPS) by using blending and casting technique with fixed amount of starch and different various additives (glycerol fixed amount 3 g), polyethylene glycol (0.1–0.3 g) and glutaraldehyde (0.1–0.3 g) [16]. The tensile strength of the TPS improved up to 2.6 MPa and decreased when the amount of glutaraldehyde exceeded 0.2 g. Conversely, the increment of PEG loading caused the elongation at break of the films increased. It showed that the presence of plasticizer may defeat film brittleness caused by strong intermolecular forces. The result also indicates that the film is more flexible and extendable due to the addition of plasticizers to decrease the intermolecular forces and enhance the mobility of polymer chains. The study also found that the films exhibited improved biodegradability, water absorption and water vapour transmission when glycerol and polyethylene glycol were added.

19.4 Sago thermoplastic starch (TPS) nanocomposite film

In creating a high advanced polymer nanocomposite, several modifications need to be conducted on TPS to overcome the hygroscopic nature of starch that limits its application. It is reported that TPS has low mechanical, barrier, rheological and thermal-resistance properties.

Nanocellulose (NC) is categorized based on its dimensions, composition and properties, which mainly depend on the employment of its extraction and fibrillation methods. Typically, cellulose nanocrystals (CNCs) are slightly shorter in length, narrower in width and greater in crystallinity compared to the cellulose nanofibers (CNFs) [17]. The former NC exhibited better mechanical attributes compared to the latter NC [17].

Many studies reported that TPS reinforced with nanocellulose exhibits better performance in the aspect of its mechanical strength [18], gas permeability and water barrier resistance properties [19]. Cellulose nano sized possess a high aspect of ratio and specific surface area with a rough surface. These features could enhance the adhesion of the nanofiller/matrix consequently provide improved mechanical performance of the nanocomposites [17]. However, the biggest challenge of reinforcing nanocellulose into the thermoplastic starch film is its well dispersion within the polymer matrix as high NC loading could cause agglomeration, which lead to poor stress distribution in films [20]. Nasution et al. [21] and co-authors prepared sago biocomposite via solution casting with various amounts of cellulose nanocrystal rattan and co-plasticizer (citric acid), (1–4 wt%) and 10–40 wt%, respectively, and a fixed amount of glycerol (30 wt%) as plasticizer [21]. The density of sago biocomposite was increased when 3 wt% CNC was added and slightly decreased when the CNC loading increased to 4%. The result could be due to the high loadings of CNC that led to the formation of aggregates in the sago biocomposite.

Lisdayana and co-workers in 2018 studied the effect of different starch sources (corn, cassava and sago) to the thermoplastic starch–polyvinyl alcohol (PVA) nanocomposite film reinforced with nanocellulose, which was prepared via evaporation casting method [22]. The morphology, water vapour transmission rate (WVTR), mechanical properties and transparency of the prepared TPS nanocomposite film were investigated. The study observed that the morphology of the TPS sago–PVA–NC was rough compared to other types of TPS nanocomposite film. High amylose content of starch possesses a smooth surface with better rheological properties than the starch with low amylose [14]. However, the morphology TPS sago–PVA–NC nanocomposite was not smooth, even though the sago starch had slightly greater amylose content than the corn starch and the cassava starch, respectively [14]. The study reported that different starch sources require different operating conditions to produce TPS–PVA nanocomposite films. The optimum thermoplastic condition is required to produce TPS sago–PVA nanocomposite films with sufficient water and optimal gelatinization temperature. This study also found that inclusion of NC could decrease the WVTR nanocomposite films. It is reported that the nanometre-sized cellulose could enhance the hydrogen bonds in the nanocomposite films and consequently improved the density of the film [22]. TPS sago–PVA–NC has the smallest WVTR value compared to other types of TPS starch [22]. This result showed that the amylose content in the starch seems acting a key role in the ability to keep the moisture in the food packaging [22]. It is found that TPS sago–PVA nanocomposite films showed that the highest tensile strength probably due to its amylose content is the greatest compared to other starches [22]. Previous study reported that the starch with a high amylose content generally shows a higher stiffness and hardness than the starch with a high amylopectin because of the relatively less branching structure [14]. It is also observed that the addition of NC influenced the transparency of the TPS nanocomposite where the transparency value of TPS sago–PVA–NC was significantly decreased [22]. In some cases, the NC improves the transparency of the TPS nanocomposite in the condition of well dispersion and bonding of NC with the TPS matrix [20]. These results showed that the differences in water content and ratio of amylose to amylopectin of various native starches are reflected on the physical and barrier properties of thermoplastic starch as reported by previous study [23].

19.5 Development challenges

Development of green composite materials has increased tremendously and attracted many industrial investors since these materials exhibit biodegradability, renewability and are economical compared to petrochemical derived composites. Thus, this is a great opportunity to develop sago thermoplastic starch composite films reinforced with nanocellulose with well blending and compositing methods. The challenge of starch-based materials is that their susceptibility to moisture uptake, consequently, shows unsatisfactory performance, particularly water vapour permeability and tensile strength. Thus, comprehensive studies with broader experimental conditions need to be conducted to allow broader data in understanding the mechanism and behaviour of the produced nanocomposites. Additionally, knowledge development in harvest technology and sago processing are highly vital to ensure the sustainability of its supply in the global market and expedite the development of sago industries. The developed composites then potentially to be used as packaging materials such as food as they are safe and biodegradable.

19.6 Conclusions

1. Sago can be one of the sustainable starchy sources to produce degradable TPS nanocomposite with desired characteristics. Previous studies showed that NC could facilitate mechanical and barrier properties of other starch-based TPS and, thus, sago thermoplastic nanocomposite reinforced with nanocelluloses (NC) is possible too.
2. However, information is restricted. Intensive study with a wide range of experimental conditions needs to be conducted to allow wider structural and physicochemical characteristics data.
3. It is very important due to the produced TPS differs based on its starch source and processing methods.
4. It is expected that sago TPS nanocomposite can be one of the competitive nanocomposites similar like conventional nanocomposite to allow its usage in various industry applications. It can be a favourable choice for new advanced and green materials for the packaging industries.

References

1. Zhang Y, Rempel C, Liu Q. Thermoplastic starch processing and characteristics-A review. Crit Rev Food Sci Nutr 2014;54:1353–70.
2. Ahmad FB, Williams PA, Doublier JL, Durand S, Buleon A. Physico-chemical characterisation of sago starch. Carbohydr Polym 1999;38:361–70.

3. Jiang T, Duan Q, Zhu J, Liu H, Yu L. Starch-based biodegradable materials: challenges and opportunities. Adv Ind Eng Polym Res 2020;3:8–18.
4. Jong FS. An overview of sago industry development, 1980s–2015. In: Ehara H, Toyoda Y, Johnson D, editors. Sago palm. Singapore: Springer; 2018:75–89 pp.
5. Karim AA, Tie APL, Manan DMA, Zaidul ISM. Starch from the sago (Metroxylon sagu) palm tree – properties, prospects, and challenges as a new industrial source for food and other uses. Compr Rev Food Sci Food Saf 2008;7:215–28.
6. Craun Research Sdn Bhd; n.d. Available from: https://craunresearch.com.my/.
7. Zhu F. Recent advances in modifications and applications of sago starch. Food Hydrocolloids 2019;96: 412–23.
8. Koganti N, Mitchell J, Macnaughtan W, Hill S, Foster T. Effect of granule organisation on the behaviour of starches in the NMMO (N-methyl morpholine N-oxide) solvent system. Carbohydr Polym 2015;116:103–10.
9. Uthumporn U, Wahidah N, Karim AA. Physicochemical properties of starch from sago (Metroxylon Sagu) palm grown in mineral soil at different growth stages. IOP Conf Ser Mater Sci Eng 2014;62:1–12.
10. Vamadevan V, Bertoft E. Impact of different structural types of amylopectin on retrogradation. Food Hydrocolloids 2018;80:88–96.
11. Mohamad Naim H, Yaakub AN, Awang Hamdan DA. Commercialization of sago through estate plantation scheme in Sarawak: the way forward. Int J Agron 2016;8319542. https://doi.org/10.1155/2016/8319542.
12. Ma XF, Yu JG, Wang N. Fly ash-reinforced thermoplastic starch composites. Carbohydr Polym 2007;67:32–9.
13. Souza RCR, Andrade CT. Investigation of the gelatinization and extrusion processes of corn starch. Adv Polym Technol 2002;21:17–24.
14. Yossathorn T, Jutarat P. Effect of starch types on properties of biodegradable polymer based on thermoplastic starch process by injection molding technique. Songklanakarin J Sci Technol 2015;37:193–9.
15. Halimatul MJ, Sapuan SM, Jawaid M, Ishak MR, Ilyas RA. Effect of sago starch and plasticizer content on the properties of thermoplastic films: mechanical testing and cyclic soaking-drying. Polimery/Polymers 2019; 64:422–31.
16. Ismail H, Zaaba NF. Tensile properties, degradation behavior, and water absorption of sago starch plastic films. J Vinyl Addit Technol 2012;18:235–40.
17. Bangar SP, Whiteside WS. Nano-cellulose reinforced starch bio composite films-A review on green composites. Int J Biol Macromol 2021;185:849–60.
18. Balakrishnan P, Gopi S, Geethamma VG, Kalarikkal N, Thomas S. Cellulose nanofiber vs nanocrystals from pineapple leaf fiber: a comparative studies on reinforcing efficiency on starch nanocomposites. Macromol Symp 2018;380:1–7.
19. Pérez-Pacheco E, Canto-Pinto JC, Moo-Huchin VM, Estrada-Mota IA, Estrada-León RJ, Chel-Guerrero L. Thermoplastic starch (TPS)-cellulosic fibers composites: mechanical properties and water vapor barrier: a review. In: Composites from renewable and sustainable materials. London: Intech; 2016. https://doi.org/10.5772/65397.
20. Rivadeneira-Velasco KE, Utreras-Silva CA, Díaz-Barrios A, Sommer-Márquez AE, Tafur JP, Michell RM. Green nanocomposites based on thermoplastic starch: a review. Polymers 2021;13. https://doi.org/10.3390/polym13193227.
21. Nasution H, Harahap H, Al Fath MT, Afandy Y. Physical properties of sago starch biocomposite filled with Nanocrystalline Cellulose (NCC) from rattan biomass: the effect of filler loading and co-plasticizer addition. IOP Conf Ser Mater Sci Eng 2018;309. https://doi.org/10.1088/1757-899X/309/1/012033.
22. Lisdayana N, Fahma F, Sunarti TC, Iriani ES. Thermoplastic starch–PVA nanocomposite films reinforced with nanocellulose from oil palm empty fruit bunches (OPEFBs): effect of starch type. J Nat Fibers 2020;17: 1069–80.
23. Zhou Y, Wang D, Zhang L, Du X, Zhou X. Effect of polysaccharides on gelatinization and retrogradation of wheat starch. Food Hydrocolloids 2008;22:505–12.

Eduardo Valério de Barros Vilas Boas*, Rafael Carvalho do Lago and Ana Lázara Matos de Oliveira

20 Rice thermoplastic starch nanocomposite films reinforced with nanocellulose

Abstract: Rice starch is a raw material with proven potential in obtaining biodegradable polymers. Plasticization improves the flexibility characteristics of starch-based films, although they still tend to have low tensile strength and high hydrophilicity. The addition of nanocellulosic materials is an alternative to improve these characteristics. In this chapter, the effects of adding different sources and concentrations of nanocellulose (NC) on the properties of thermoplastic rice starch films (TRSF) are addressed. One can highlight as main effects the increase of tensile strength and transparency of the films, the reduction of water vapor permeability and water solubility. The type of NC used, the way it is obtained, as well as its interaction with starch molecules are factors that influence the effects. Further studies involving the interaction of TRSF and NC should be conducted in order to overcome the lack of information.

Keywords: biodegradable polymers, nanotechnology, Oryza sativa, renewable resources

20.1 Introduction

In the context of the elaboration of biodegradable polymeric materials, the use of starch is recurrent, since it is a low cost and not scarce raw material, with large occurrence in nature, and can be obtained from the most diverse sources, such as potato, cassava, wheat, rice, among others [1–3]. Considering the amount produced and the global importance – it is the second most produced cereal in the world and a crop staple that feeds more than half of the world's population [4, 5] – rice (*Oryza sativa* L.) stands out as a potential source of starch for applications in biodegradable polymers.

Rice starchrepresents one of the main agricultural products in the world and is used in numerous applications, food and non-food [6]. It has unique characteristics, such as hypoallergenicity, freeze and thaw stability, mild flavor, white color, digestibility, and higher acid resistance, besides the reduced size of the granules, which range from 2 to 7 μm [7, 8]. However, rice starch presents low solubility, lower resistant starch content

***Corresponding author: Eduardo Valério de Barros Vilas Boas**, Food Science Department, Federal University of Lavras, 37200-900, Lavras, Minas Gerais, Brazil, E-mail: evbvboas@ufla.br
Rafael Carvalho do Lago and Ana Lázara Matos de Oliveira, Food Science Department, Federal University of Lavras, 37200-900, Lavras, Minas Gerais, Brazil

As per De Gruyter's policy this article has previously been published in the journal Physical Sciences Reviews. Please cite as: E. V. de Barros Vilas Boas, R. C. do Lago and A. L. M. de Oliveira "Rice thermoplastic starch nanocomposite films reinforced with nanocellulose" *Physical Sciences Reviews* [Online] 2023. DOI: 10.1515/psr-2022-0022 | https://doi.org/10.1515/9783110773606-020

with reduced retrogradation and poor functional properties, which limits its application as a base material in the elaboration of biopolymers [9].

Plasticization of rice starch is therefore an alternative for improving the mechanical properties of these biopolymers mainly by increasing malleability [10]. Plasticizing agents can change the melting temperature and glass transition temperature of starch, as well as the mechanical properties of a polymer, because they reduce or even neutralize the intermolecular interaction forces, making it possible to obtain more flexible materials [11, 12]. Although they correct problems related to mechanical properties, the application of plasticizing agents tends to increase the hydrophilicity of starch-based polymers, increasing their water vapor permeability and solubility, which may limit their application for some purposes [13, 14].

In this context, the use of nanocellulose technology is one way to complement effects of plasticizing agents on the properties of starch-based polymers. Nanocellulosic materials are obtained from cellulose-rich matrices using chemical, physical, enzymatic, and other treatments, resulting in a material with nanometric dimensions (1–100 nm), high aspect ratio (length to width ratio), and high specific surface area (>100 m^2/g) [15, 16]; include cellulose nanofibers and cellulose nanocrystals, as well as bacterial cellulose [17–19]. Besides the natural strength, a consequence of the almost perfect crystalline arrangement, nanocellulosic materials are able to establish strong hydrogen bonds with the starch molecules, which increase the stiffness of the biopolymers and impose tortuous paths to the passage of gases and water vapor, improving the barrier properties [20–27].

Although the incorporation of nanocellulose and plasticizing agents is known to improve the mechanical and barrier properties of starch-based polymers, the study of the interaction of raw materials from different sources and in different concentrations is important to try to arrive at a biodegradable material with physical properties on par with traditional polymers. The lack of comprehensive information about the interaction between rice starch and endo/exogenous constituents for enhancing the physicochemical properties of biopolymers is a major challenge for researchers and its holistic understanding should be considered a prerequisite in the design of industrial products with enhanced functional attributes [9].

In this chapter, the effects of nanocellulose incorporation on the properties of rice thermoplastic starch-based polymers are discussed, based on studies previously conducted by several authors.

20.2 Rice thermoplastic starch-based biopolymers reinforced with nanocellulose

20.2.1 Materials composition

First, it is important to mention that there is a limited amount of studies involving the interaction of plasticized rice starch and nanocellulose. Such scarcity of data may be attributed to the intrinsic properties of native rice starch, which are considered inferior compared to other starch sources such as cassava, potato, and corn, for example [9, 27]. The main studies found in the literature are shown in Table 20.1.

The concentrations of rice starch added range from 2 to 5%, and both rice starch alone and rice flour have been used. In studies such as those by Almendárez-Camarillo et al. [29], Careli-Gondim [30] and Oliveira et al. [28], rice starch was combined with other polysaccharides such as chitosan and pectin. According to Tiozon et al. [9], the association of rice starch with other polysaccharides is necessary to improve its properties, given the limitations presented, previously discussed. Regarding plasticizing agents, the addition values range from 1% to 70%.

The most used plasticizer agents were sorbitol and glycerol. These substances belong to a class of low molecular weight non-volatile compounds, whose plasticizing effect has already been proven for rice starch-based films [13, 14, 31]. In addition,

Table 20.1: Rice thermoplastic starch nanocomposite films reinforced with nanocellulose.

Material	Nanocellulose sources	Concentration	Obtaining conditions	Reference
Films based on rice flour (5% w/v), pectin (5% w/v), and sorbitol (70% m/v)	Corn straw *Pine* Eucalyptus	0%, 33.3%, 50%, and 100% (replacing to rice flour)	Casting method (oven drying at 30 ± 2 °C to constant weight)	Oliveira et al. [28]
Films based on chitosan (2% w/v), rice starch (2% w/v), and sorbitol (1% v/v)	Pineapple leaves	0.3%, 0.7%, and 1.0%	Casting method (room temperature, 144 h)	Almendárez-Camarillo et al. [29]
Films based on rice starch (5% m/v) and glycerol (30% m/v)	Banana pseudostem	0%, 2%, 4%, 6%, 8%, and 10%	Casting method (drying at room temperature for 3 days and storage in a chamber at 25 ± 1 °C and 50 ± 3% RH for 2 days)	Jeevahan and Chandrasekaran [7]
Films based on rice starch (5% w/v), and glycerol (30% w/v)	Banana pseudostem	0%, 2%, 5%, 10%, and 15%	Casting method (drying at room temperature for 3 days and storage in a chamber at 25 ± 1 °C and 50 ± 3% RH for 2 days)	Jeevahan et al. [27]
Films based on rice flour (5% w/v), pectin (5% w/v), and sorbitol (70% m/v)	Rice husk	3%, 6%, and 9%	Casting method (oven drying at 30 ± 2 °C to constant weight)	Careli-Gondim [30]

they are biodegradable and safe for food applications. Regarding the sources for obtaining nanocellulose, there seems to be a trend to seek alternative sources, using under-utilized agro-industrial wastes, such as leaves and peels, which contributes to sustainability, meeting the concept of environmentally friendly packaging. The nanocellulose addition values range from 0.3% to 15%. In the study by Oliveira et al. [28], the addition of nanocellulose targeted the total or partial replacement of rice flour in the production of biocomposites, with replacement values of 0%, 33.3%, 50%, and 100%. All the materials reported were prepared by the casting method. This method is characterized by the use of water as a carrier for starch gelatinization and has been one of the most used in the elaboration of biocomposites, due to its simplicity and good performance in laboratory scale [32].

Although they do not involve the use of nanocellulose, it is worth noting as similar works the studies of Suwanprateep et al. [33], who prepared films based on rice starch (4% w/v) plasticized with glycerol (2.5% w/v) reinforced with palm mesocarp cellulose fibers at concentrations of 2%, 4%, 6%, and 8% (in relation to the mass of the starch); Prachayawarakorn et al. [34], who developed films of rice starch and glycerol (1:1) reinforced with cotton fibers (5%, 10%, and 15% by weight of starch and glycerol); and Piyada et al. [35], who obtained nanocrystals from rice starch to reinforce rice starch-based films (3% w/v) plasticized with sorbitol (40% by weight of starch) at concentrations of 5%, 10%, 15%, 20%, 25%, and 30%.

20.2.2 Interaction between the constituents

When making nanocomposite films, it is important to investigate the interaction between the different constituents in order to ensure the formation of well-structured films with a continuous and uniform surface; such characteristics will have a direct effect on properties such as mechanical strength and diffusion barrier. The interaction between the components is evaluated structurally using Fourier Transform Infrared (FTIR) analysis and visually using electron microscopy.

In general, FTIR spectra of thermoplastic rice starch-based films with and without the addition of nanocellulose do not show major differences, indicating structural similarity between the different components (starch, plasticizer, and nanocellulose). Decreased intensity of peaks assigned to the OH group with the addition of nanocellulose, often reported, indicates formation of hydrogen bonds between plasticized rice starch and nanocellulose [29, 34]. The intensity of this interaction, through the formation of hydrogen bonds, is crucial to the performance of the films in mechanical and barrier tests, as will be discussed later.

The observation, under an electron microscope, of the morphology of the films formed is fundamental to evaluate the formation of possible agglomerates of nanocellulose, which also compromises its effect on improving the properties of biocomposites. Through scanning electron microscopy (SEM) analysis, Careli-Gondim [30], for example, observed the incidence of rice husk nanofiber agglomerations in rice flour and pectin-based films. Oliveira et al. [28], meanwhile, found that films formed with rice flour and different concentrations and types of cellulosic nanofibers had a rough and homogeneous structure, with a low rate of isolated agglomerations (Figure 20.1). These results call attention to the fact that, besides the surface of the films formed, the morphological evaluation of the nanocellulose is also important, in order to ascertain aspect ratio, thickness, and incidence of agglomeration, which interferes in its ability to interact with the starch and, consequently, in the reinforcement property [36, 37].

20.3 Effects of nanocellulose addition in thermoplastic rice starch-based biopolymers

20.3.1 Optical properties

The evaluation of optical properties, such as transparency and opacity, is of great importance when it comes to polymeric materials for use in packaging. It is desirable that the films formed are light colored and transparent, so as not to interfere in the characteristics of the product to be packaged/coated [38, 39].

It is reported that nanocellulose addition to plasticized rice starch-based films tends to increase transparency and, consequently, decrease opacity. This behavior is due to the fact that micro-scale particles in a polymer solution, such as starch particles, disperse more light, reducing the transmittance; on the other hand, when nanoscale particles that well interact with the starch are added to the matrix, the effect of dispersion can be eliminated [40, 41].

In fact, Jeevahan et al. [27] and Jeevahan and Chandrasekaran [7] observed increased L* values in thermoplastic rice starch-based films proportional to the addition of nanocellulose; since the L* variable indicates the degree of clarity of the sample (0 = totally black – 100 = totally white), it is inversely proportional to opacity.

In contrast, Suwanprateep et al. [33] observed decreased transparency and consequent increased opacity in rice starch-based films reinforced with palm mesocarp cellulose fibers and attribute this to possible fiber clusters, which may lead to light scattering and subsequently decreased light transmittance of the film. This result may be attributed to the dimensions of the cellulose fibers, which in this case were not in the nanoscale. In fact, it has been proven that nano-scale cellulosic materials perform better than microcellulosic materials [42]. Nano-scale materials exhibit good aspect ratio and tend to establish strong bonds with starch molecules, and there are fewer incidences of isolated agglomerations and consequently less opacity. Therefore, the good performance in increasing the transparency of the film is dependent on the characteristics of the nanocellulose material as well as its degree of interaction with the starch.

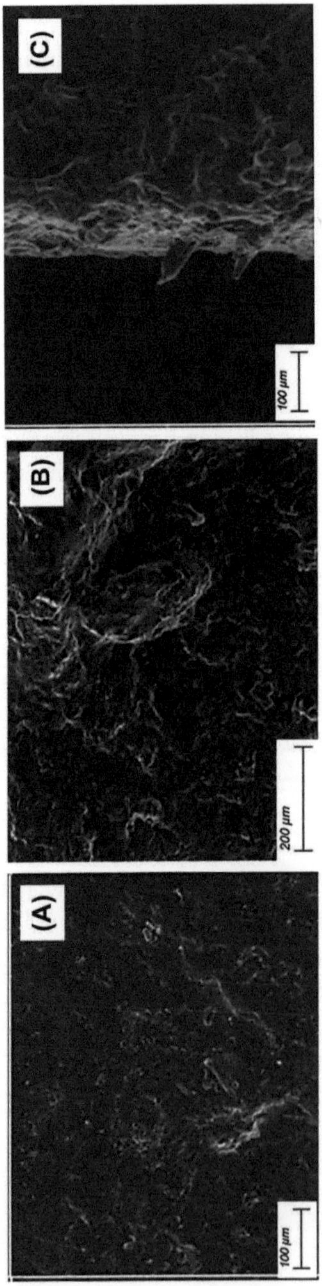

Figure 20.1: SEM images obtained for films based on rice flour reinforced with: pine nanofibers (A); corn straw + eucalyptus nanofibers (B) and; pine + eucalyptus nanofibers (C). Adapted from Oliveira et al. [28].

Oliveira et al. [28] and Careli-Gondim [30], meanwhile, found that the opacity of rice flour-based films increased with the addition of cellulose nanofibers from different sources. In addition to the aspect ratio of the nanocellulose material and its degree of interaction with the starch, the inherent coloration may be another point of influence on opacity; the darker the nanocellulose, the less transparency it will impart to the film. In this case, the application of pre-treatments, such as bleaching of the fibers, can have an influence on the coloring of the composites, as well as on the transparency and opacity. The influence of the coloration of the CNF used on the final coloration and transparency/opacity of the films is exemplified in Figure 20.2.

20.3.2 Mechanical properties

With regard to rice starch based films, there seems to be a consensus among the several authors: the addition of plasticizer promotes increased malleability, with reduced stiffness, while the addition of nanocellulose increases the stiffness of the films, decreasing, however, the elongation at break.

Laohakunjit and Noomhorm [14], for example, proved that the addition of plasticizers such as sorbitol and glycerol increased the elongation at break values of rice starch based films, while the tensile strength values decreased proportionally to the concentration of plasticizers. On the other hand, Almendárez-Camarillo et al. [29] observed that a reinforcement of only 0.7% cellulose nanofibers from pineapple leaves waste was sufficient to increase the storage modulus of chitosan and rice starch sorbitol-plasticized nanocomposites by 50%. Whereas in the study by Jeevahan and Chandrasekaran [7], the tensile strength and Young's Modulus values of plasticized rice starch-based films increased from 2.1 MPa and 21 MPa to 5.3 MPa and 187 MPa, respectively, with the addition of up to 10% nanocellulose; in contrast, elongation was reduced from 59% to 29%. This reverse trend was also observed by Jeevahan et al. [27] for addition values of up to 15% nanocellulose.

This behavior is explained by the fact that plasticizers have the ability to reduce or even neutralize the intermolecular interaction forces, increasing the mobility of the chains and making it possible to obtain more flexible polymers [11, 43]. In addition, the hygroscopic nature of plasticizing agents tends to provide additional water to the polymer matrix, softening its structure [44]. Nanocellulose, in turn, tends to establish strong hydrogen bonds with the starch molecules, which reinforces the network and limits chain movements, increasing rigidity and therefore reducing flexibility, and reducing the hydrophilicity of the films [45].

The reinforcing effect of nanocellulose in thermoplastic rice starch-based biopolymers seems to be dependent on the added concentration as well. While Jeevahan and Chandrasekaran [7] and Jeevahan et al. [27] observed increased tensile strength and Young's Modulus values proportional to increasing nanocellulose concentration, Almendárez-Camarillo et al. [29] noted that the storage modulus of chitosan and

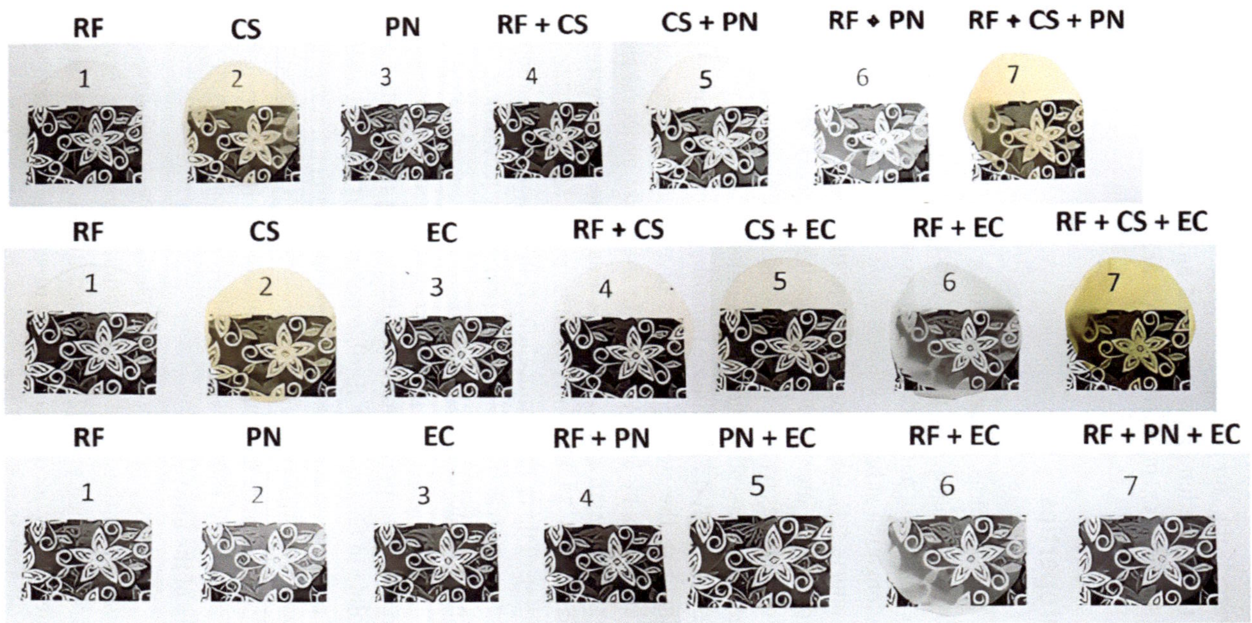

Figure 20.2: Rice flour-based bionanocomposites reinforced with different types of celullose nanofibers. RF, rice flour; CS, corn straw CNF; PN, pine CNF; EC, eucalyptus CNF.

plasticized rice starch based biocomposites reached a maximum value of 4709 MPa with nanocellulose addition values of 0.7%, and a reduction around 33% (3,140 MPa) was observed when the addition value increased to 1%. Similarly, Prachayawarakorn et al. [34] observed reduced tensile strength and Young's Modulus of glycerol-plasticized rice starch-based films with cotton fiber addition percentages above 10%. Authors suggest that large amounts of nanocellulose may induce the formation of isolated clusters on the films' surface, which leads to poorer performance. On another hand, cellulosic nanofibrils can form strong hydrogen bonding networks with each other and with the starch molecules, increasing stiffness. These factors seem to be conditioned on the type of nanocellulose, its aspect ratio, and the degree of interaction with the starch molecules.

In fact, Oliveira et al. [28] observed distinct behaviors in relation to different sources of nanocellulose as reinforcement of rice flour-based films; while nanofibers from eucalyptus and pine increased the values of tensile strength of the films concomitantly with increasing concentrations, nanofibers obtained from corn straw determined lower averages, with a drop in values at high concentrations. These results prove that the effects on the films' characteristics are conditioned, also, to the characteristics of the nanocellulose material, which highlights the importance of further studies on the morphology of the materials elaborated in electron microscopy, which are limited in the literature involving the thermoplastic rice starch and nanocellulose association.

Stiffness in polymeric materials is desirable, so that they can withstand minor mechanical stresses that may occur. However, it is important that they have certain malleability, so that they can be molded and manipulated. Therefore, the study of the ratio nanocellulose:plasticizer is a way to optimize the properties of composites, achieving a balance between stiffness and malleability.

20.3.3 Thermal properties

The thermal properties of polymers are important to evaluate their stability and behavior in relation to processes involving high temperatures, such as extrusion and injection, as well as to understand the physical characteristics and the modifications in the composites by the addition of other compounds. Thermogravimetric analysis (TGA), differential scanning calorimetry (DSC), and thermomechanical analysis are usually performed.

Of the studies available in the literature, few have explored the thermal properties of rice starch-based films reinforced with nanocellulose. In general, little difference is noted in the thermal behavior of the reinforced films, which is attributed to the structural similarity between the materials, confirmed by FTIR analyses [29]. The main differences observed correspond to thermal stability and physical properties.

Almendárez-Camarillo et al. [29], for example, observed in TGA analysis that the addition of nanocellulose determined increased mass loss of thermoplastic rice starch and chitosan-based polymers, suggesting lower thermal stability with nanocellulosic reinforcement; mass loss increased the higher the nanocellulose concentration was. The authors attribute this thermal instability to the easier degradation of the amorphous regions of the nanocellulose fibers, which underwent alkaline pretreatment, in the present study. Suwanprateep et al. [33], in contrast, observed that increasing the concentration of palm mesocarp cellulosic fibers improved the thermal stability of glycerol-plasticized rice starch-based polymers by increasing the degradation temperature and determining lower mass loss. Similar results were reported by Piyada et al. [35], who used rice starch nanocrystals as reinforcement for filling plasticized rice starch-based films and Prachayawarakorn et al. [34], who used cotton fibers as reinforcement. The authors attribute the increased thermal stability to the strong interactions formed between the reinforcing materials and the starch molecules, as well as the high crystallinity of cellulose, which imposes a greater apolar character on the films, increasing their carbon yield. The degradation temperatures of the films were in the ranges of 225 °C–260 °C [33], 220 °C–380 °C [29], 250 °C–340 °C [35] and 178 °C–408 °C [34].

Regarding thermo-mechanical analysis and DSC, Almendárez-Camarillo et al. [29] observed that the addition of nanocellulose promoted an increase in the values of storage modulus (*E'*) and glass transition temperature (*Tg*) of thermoplastic rice starch and chitosan-based films. A maximum increase of 90% over the control film in *E'* values was reported by adding 0.7% (w/w) of nanocellulose fibers. The *Tg* increased from 66 °C in the control film to 75 °C in the samples with 1% added nanocellulose. Similar results have been reported for plasticized rice starch-based films reinforced with starch nanocrystals [35]. The film structure is glassy, hard, and brittle below the *Tg*, and rubbery, soft, and flexible above it [43]. Therefore, the fact that the addition of nanocellulose increases the *Tg* is indicative of lower mobility of the molecules and, consequently, higher stiffness of the polymers, which corroborates the increase of *E'*. Such results confirm the effect of reinforcement on the mechanical properties of rice starch-based films by the addition of nanocellulose and are indicative of good interaction between nanocellulose and rice starch.

20.3.4 Barrier properties

One of the functions of packaging, especially food packaging, is to prevent or even reduce moisture transfer between the product and the surrounding atmosphere; therefore, water vapor permeability should be as low as possible [46]. Likewise, the film must be storage stable, so that it does not completely solubilize in water, especially when applied to matrices with high water activity.

To evaluate the barrier properties and hydrophilicity of starch-based polymeric materials, water vapor permeability (WVP) tests, water solubility, and contact angle evaluation are usually performed.

Rice starch-based films are naturally hydrophilic and it is known that the addition of plasticizers increases their polarity, due to the determination of greater spacing between the starch molecules, which optimizes the water ingress, as well as their hygroscopic properties [9, 14]. On the other hand, the addition of nanocellulose seems to promote the reduction of polarity and, consequently, reduce the water solubility and WVP of the biocomposites.

This behavior is observed in the different studies involving rice starch and nanocellulose. Jeevahan and Chandrasekaran [7], for example, obtained reduction of WVP values of rice starch-based films plasticized with glycerol around 64% with the insertion of 10% banana pseudostem nanocellulose; the average reduction was 36%. Water solubility decreased from 23.2% for the control film to 16.5% for the formulation with 10% nanocellulose. Similarly, Prachayawarakorn et al. [34] and Suwanprateep et al. [33] found reduced water absorption in thermoplastic rice starch biopolymers with the addition of cotton fibers and palm mesocarp cellulose fibers, respectively. Almendárez-Camarillo et al. [29], in turn, observed that the addition of nanocellulose in films based on chitosan and sorbitol-plasticized rice starch caused a slight temperature increase in an endothermic peak in DSC thermograms, attributed to water evaporation, which indicates hydrophobic behavior.

According to Jeevahan and Chandrasekaran [7], the interaction between rice starch and plasticizers, such as glycerol and sorbitol, can promote moisture adsorption. This can be attributed to the dimensional dissimilarities of rice starch and plasticizers, which are weakly bonded, facilitating moisture adsorption. This can be attributed to the dimensional dissimilarities of rice starch and plasticizers, which are poorly bonded, easing the adsorption of moisture. In contrast, nanocellulose and starch sources, such as rice starch, have similar chemical structure and dimensional similarity, which confers strong interaction between both, promoting a stiff and uniform structure that imposes a barrier for water interaction [24, 27, 47]. In addition, the high degree of molecular orientation and higher crystallinity of nanocellulosic materials gives them higher hydrophobicity compared to starch, which contributes to the reduction of film polarity.

Although most of available data in the literature point to improved barrier properties of thermoplastic rice starch films with the addition of nanocellulose, Careli-Gondim [30] observed reduced WVP in sorbitol-plasticized rice flour and pectin-based films with the insertion of rice husk nanofibers; the films were 100% water soluble in this case. The presence of isolated clusters of nanofibrils was the author's explanation for this behavior. Oliveira et al. [28], on the other hand, although he found a decrease in WVP of the films with the insertion of nanocellulose, observed a reduction in contact angle values with Pine and eucalyptus nanocellulose, and an increase with corn straw nanocellulose, which indicates increased and reduced hydrophilicity, respectively.

These results suggest that the type, concentration and morphology of the nanocellulose materials, as well as their degree of interaction with starch, have an influence on the barrier properties of the biocomposites, as they do on the optical and mechanical properties, and should be studied.

20.3.5 Biodegradability

Since the main focus in the elaboration of biopolymers is to obtain environmentally friendly materials that are easily degradable in nature, the study of the real biodegradability of biocomposites is of extreme importance. On a laboratory scale, biodegradability testing by burial in composite soil is most commonly used for starch-based films and consists of burying samples of the materials in soil and monitoring their mass over a period of time.

With regard to films based on thermoplastic rice starch and nanocellulose, there is a lack of specific studies regarding the evaluation of biodegradability. An example of a similar study can be cited as the work of Prachayawarakorn et al. [34], who used cotton fibers as reinforcement for plasticized rice starch biocomposites. The authors found that, although they showed lower mass than the control, the films with the addition of 10% cotton fiber degraded later – at 18 days, while the control sample showed complete degradation at day 6. The delay in degradation by the addition of the cellulose fibers can be attributed to the higher hydrophobicity of these materials, compared to pure rice starch, which makes it more stable over the test time. Note that the fibers used in the present study were not at the nanometer scale, but considering the inherent properties of nanocellulose and its effect on the hydrophobicity of the films, already discussed here, one can predict similar behavior.

Although there is an lack of specific studies with nanocellulose, the results obtained by Almendárez-Camarillo et al. [29] for the thermal behavior of plasticized rice starch and chitosan films reinforced with nanocellulose from pineapple leaves indicate an increase in biodegradability with the addition of nanocellulose; according to the authors, the greater thermal degradation observed in nanocellulose-reinforced films is related to the greater ease of degradation of the amorphous regions of nanocellulose, which is attributed to the alkaline treatment used to obtain it. In this case, the behavior is in contrast to that exposed by Prachayawarakorn et al. [34] and indicates the influence of the treatment applied to obtain the nanocellulose, as well as its aspect ratio, on the biodegradability of the films.

Given the diversity of nanocellulosic materials used as reinforcement, further detailed studies involving the biodegradability of rice thermoplastic starch-based films reinforced with nanocellulose are needed in order to seek a more holistic understanding and to address the lack of information in this regard.

It is emphasized that, despite the type and concentration of materials used, the biodegradability of starch and/or nanocellulose-based composites compared to conventional plastics is unquestionable.

20.4 Applications

In addition to investigating the properties of the films elaborated, it is important that *in vivo* application studies are carried out in order to evaluate their real performance and stability throughout storage, as well as their interaction with different matrices, making it possible to define the best applicability.

Regarding on rice thermoplastic starch-based films reinforced with nanocellulose, not a lot of data is available regarding applications. In this context, one can mention the study of Careli-Gondim et al. [48], who applied sorbitol-plasticized rice flour-based films reinforced with cellulosic nanofibers obtained from the husk of rice itself as edible coatings on avocados, evaluating their post-harvest quality in refrigerated storage. It was found that the coating contributed to the maintenance of the original coloration and firmness of the fruits, as well as reducing the respiration rate, thus extending their shelf life. Such results can be attributed to the barrier properties of the films, which were discussed earlier and which are enhanced by the addition of nanocellulose. On the other hand, the same film solution was applied to strawberries and, although there was an extension of the fruit shelf life, the application of the coating increased the respiration rate and the mass loss of the fruits [49] (Table 20.2). This behavior can be attributed to the high solubility exhibited by the films in combination with strawberries, which have high water activity.

These results reinforce the importance of studying the application of the films in different matrices, since each one has peculiar characteristics and behaves in a different way. Therefore, further studies involving the application of nanocellulose-reinforced thermoplastic rice starch-based biocomposites should be conducted, in different matrices, whether food-based or not.

20.5 Future studies

The amount of studies involving the association of rice thermoplastic starch and nanocellulose in the elaboration of bionanocomposites is limited, compared to other starch sources, such as corn and cassava, for example.

Studies exploring other sources of nanocellulose and different concentrations as well as different ratios of starch: plasticizer may optimize the properties of biopolymers and bring them closer to conventional plastics. In addition, the morphological study of the nanocellulose materials to be applied to the films should be conducted in order to aid in the understanding of their effects and their interaction with the other

Table 20.2: Examples for application of rice thermoplastic starch nanocomposite films reinforced with nanocellulose.

Material	Product applied	Main results	Reference
Sorbitol-plasticized rice flour-based films reinforced with cellulosic nanofibers obtained from the husk of rice	Avocado	Maintenance of color and firmness; reduction of respiration rate; shelf life extension	Careli-Gondim et al. [48]
Sorbitol-plasticized rice flour-based films reinforced with cellulosic nanofibers obtained from the husk of rice	Strawberry	Increase in the respiration rate and the mass loss of the fruits; Double time extension of the fruits shelf life	Careli-Gondim et al. [49]

constituents. The exploration of other methods of obtaining the polymers, such as extrusion, for example, is another point to consider for future investigations.

The absence of data regarding the real biodegradability of the films produced is also noteworthy. These investigations should be conducted in order to ensure the biodegradability of the materials and understand what effect the addition of nanocellulose has on it.

Similarly, there is a lack of data about the application of nanocellulose-reinforced rice thermoplastic starch-based bionanocomposites, which may instigate future studies. Investigating the application of the elaborated materials is important to evaluate their real performance, *in vivo*. In this sense, new bionanocomposites, as well as those already developed and that have shown good performance in mechanical and barrier tests, can be studied as edible coatings or even packaging in food matrices of both animal and vegetal origin, as well as non-food matrices, in order to confirm their potential use as packaging material and to direct their best applicability. Different methods of applying the films to the matrices can also be explored.

20.6 Conclusions

The addition of nanocellulose complements the effects of plasticizing agents on the mechanical and barrier properties of rice starch-based nanocomposite films, making their use as packaging and/or coating material feasible. The main effects are to increase the tensile strength and transparency of the films, as well as to reduce water vapor permeability and water solubility.

The type of nanocellulose used, the method of obtaining it, its aspect ratio, as well as its interaction with the starch molecules are factors that influence the effects on the biocomposites, as well as them *in vivo* performance as packaging/coating.

Further studies involving the interaction of rice thermoplastic starch and nanocellulose should be conducted in order to address the lack of information, especially regarding morphology, biodegradability, and application. New sources of nanocellulose, different concentrations and the study of the ratio between nanocellulose and plasticizer can be explored.

Acknowledgements: Authors thank to Fundação de Amparo à Pesquisa do Estado de Minas Gerais (FAPEMIG), Coordenacão de Aperfeiçooamento de Pessoal de Nível Superior (CAPES) and Conselho Nacional de Desenvolvimento Científico e Tecnológico (CNPq).

References

1. Fahma F, Sugiarto, Sunarti TC, Indriyani SM, Lisdayana N. Thermoplastic cassava starch-PVA composite films with cellulose nanofibers from oil palm empty fruit bunches as reinforcement agent. Int J Polym Sci 2017;2017:1–6.
2. López-Córdoba A, Medina-Jaramillo C, Piñeros-Hernandez D, Goyanes S. Cassava starch films containing rosemary nanoparticles produced by solvent displacement method. Food Hydrocolloids 2017;71:26–34.
3. Pelissari FM, Ferreira DC, Louzada LB, Dos Santos F, Corrêa AC, Moreira FKV, et al. Starch-based edible films and coatings: an eco-friendly alternative for food packaging. In: Starches for food application: chemical, technological and health properties. Cambridge, Massachusetts, United States: Elsevier; 2018.
4. FAO. FAOSTAT; 2021. Available from: http://www.fao.org/faostat/en/#data/QC.
5. Muthayya S, Sugimoto JD, Montgomery S, Maberly GF. An overview of global rice production, supply, trade, and consumption. Ann N Y Acad Sci 2014;1324:7–14.
6. Amagliani L, O'Regan J, Kelly AL, O'Mahony JA. Chemistry, structure, functionality and applications of rice starch. J Cereal Sci 2016;70:291–300.
7. Jeevahan J, Chandrasekaran M. Influence of nanocellulose additive on the film properties of native rice starch-based edible films for food packaging. Recent Pat Nanotechnol 2019;13:222–33.
8. Wani AA, Singh P, Shah M, Schweiggert-Weisz U, Gul K, Wani I. Rice starch diversity: effects on structural, morphological, thermal, and physicochemical properties-a review. Compr Rev Food Sci Food Saf 2012;11: 417–36.
9. Tiozon RJN, Bonto AP, Sreenivasulu N. Enhancing the functional properties of rice starch through biopolymer blending for industrial applications: a review. Int J Biol Macromol 2021;192:100–17.
10. Diyana ZN, Jumaidin R, Selamat MZ, Ghazali I, Julmohammad N, Huda N, et al. Physical properties of thermoplastic starch derived from natural resources and its blends: a review. Polymers 2021;13:5–20.
11. Abdollahzadeh E, Nematollahi A, Hosseini H. Composition of antimicrobial edible films and methods for assessing their antimicrobial activity: a review. Trends Food Sci Technol 2021;110:291–303.
12. Sari Y, Putri S, Intan N, Bahtiar A, Kurniati M. The effect of sorbitol and sweet sorghum to carrageenan ratio on the physicochemical properties of sweet sorghum/carrageenan bioplastics. Biomass Convers Biorefinery 2021;11. https://doi.org/10.1007/s13399-020-01254-3.
13. Alrimawi BH, Chan MY, Ooi XY, Chan SY, Goh CF. The interplay between drug and sorbitol contents determines the mechanical and swelling properties of potential rice starch films for buccal drug delivery. Polymers 2021;13:1–15.
14. Laohakunjit N, Noomhorm A. Effect of plasticizers on mechanical and barrier properties of rice starch film. Starch/Staerke 2004;56:348–56.
15. Gardner DJ, Oporto GS, Mills R, Samir MASA. Adhesion and surface issues in cellulose and nanocellulose. J Adhes Sci Technol 2008;22:545–67.

16. Ilyas RA, Sapuan S, Ishak M, Zainudin ES, Mahamud A. Nanocellulose reinforced starch polymer composites: a review of preparation, properties and application. In: 5th International Conference on Applied Sciences and Engineering Application (ICASEA 2018).
17. Bangar SP, Harussani MM, Ilyas RA, Ashogbon AO, Singh A, Trif M, et al. Surface modifications of cellulose nanocrystals: processes, properties, and applications. Food Hydrocolloids 2022;130:107689.
18. Bangar SP, Whiteside WS. Nano-cellulose reinforced starch bio composite films-a review on green composites. Int J Biol Macromol 2021;185:849–60.
19. Norfarhana AS, Ilyas RA, Ngadi N. A review of nanocellulose adsorptive membrane as multifunctional wastewater treatment. Carbohydr Polym 2022;291:119563.
20. Ilyas RA, Zuhri MYM, Norrrahim MNF, Misenan MSM, Jenol MA, Samsudin SA, et al. Natural fiber-reinforced polycaprolactone green and hybrid biocomposites for various advanced applications. Polymers 2022a;14:1–28.
21. Ilyas RA, Zuhri MYM, Aisyah HA, Asyraf MRM, Hassan SA, Zainudin ES, et al. Natural fiber-reinforced polylactic acid, polylactic acid. Polymers 2022b;14. https://doi.org/10.3390/polym14010202.
22. Ilyas RA, Aisyah HA, Nordin AH, Ngadi N, Yusoff M, et al. Nanocomposites for various advanced applications. Polymers 2022c;14:1–36.
23. Ferrer A, Pal L, Hubbe M. Nanocellulose in packaging: advances in barrier layer technologies. Ind Crop Prod 2017;95:574–82.
24. Noshirvani N, Ghanbarzadeh B, Fasihi H, Almasi H. Starch-PVA nanocomposite film incorporated with cellulose nanocrystals and MMT: a comparative study. Int J Food Eng 2016;12:37–48.
25. Omran AAB, Mohammed AABA, Sapuan SM, Ilyas RA, Asyraf MRM, Koloor SSR, et al. Micro-and nanocellulose in polymer composite materials: a review. Polymers 2021;13:1–30.
26. Nazrin A, Sapuan SM, Zuhri MYM, Ilyas RA, Syafiq R, Sherwani SFK. Nanocellulose reinforced thermoplastic starch (TPS), polylactic acid (PLA), and polybutylene succinate (PBS) for food packaging applications. Front Chem 2020;8:213.
27. Jeevahan J, Chandrasekaran M, Sethu A. Influence of nanocellulose addition on the film properties of the bionanocomposite edible films prepared from maize, rice, wheat, and potato starches. AIP Conf Proc 2019; 2162:020073.
28. Oliveira ALM, Bento JAC, Fidelis MC, Dias MC, de Barros HEA, Natarelli CVL, et al. Effect of pine, eucalyptus, and corn straw nanofibers on the structural properties of rice flour-based biodegradable films. Ind Crop Prod 2023;191:115929.
29. Almendárez-Camarillo A, Flores-Hernandez CG, Balcázar-Enríquez V, Aguirre-García MS, Martínez-Hernandez AL, Velasco-Santos C. Nanocellulose extraction of pineapple leaves for chitosan-starch nanocomposites. J Nat Fibers 2020;00:1–14.
30. Careli-Gondim Í. Uso de revestimentos comestíveis ativos com nanofibras celulosídicas de casca de arroz na pós-colheita de frutas. Goiânia, Goiás, Brazil: Universidade Federal de Goiás; 2017.
31. Vieira MGA, da Silva MA, dos Santos LO, Beppu MM. Natural-based plasticizers and biopolymer films: a review. Eur Polym J 2011;47:254–63.
32. Gao W, Zhu J, Kang X, Wang B, Liu P, Cui B, et al. Development and characterization of starch films prepared by extrusion blowing: the synergistic plasticizing effect of water and glycerol. LWT 2021;148:111820.
33. Suwanprateep S, Kumsapaya C, Sayan P. Structure and thermal properties of rice starch-based film blended with mesocarp cellulose fiber. Mater Today Proc 2019;17:2039–47.
34. Prachayawarakorn J, Sangnitidej P, Boonpasith P. Properties of thermoplastic rice starch composites reinforced by cotton fiber or low-density polyethylene. Carbohydr Polym 2010;81:425–33.
35. Piyada K, Waranyou S, Thawien W. Mechanical, thermal and structural properties of rice starch films reinforced with rice starch nanocrystals. Int Food Res J 2013;20:439–49.
36. Azeredo HMC, Mattoso LHC, Wood D, Williams TG, Avena-Bustillos RJ, McHugh TH. Nanocomposite edible films from mango puree reinforced with cellulose nanofibers. J Food Sci 2009;74:N31–5.

37. Silvério H, Flauzino Neto W, Dantas N, Pasquini D. Extraction and characterization of cellulose nanocrystals from corncob for application as reinforcing agent in nanocomposites. Ind Crop Prod 2013;44:427–36.
38. Alves JS, Dos Reis KC, Menezes EGT, Pereira FV, Pereira J. Effect of cellulose nanocrystals and gelatin in corn starch plasticized films. Carbohydr Polym 2015;115:215–22.
39. Martins MP, Dagostin JLA, Franco TS, de Muñiz GIB, Masson ML. Application of cellulose nanofibrils isolated from an agroindustrial residue of peach palm in cassava starch films. Food Biophys 2020;15:323–34.
40. Nobbmann U, Morfesis A. Light scattering and nanoparticles. Mater Today 2009;12:52–4.
41. Quinten M. The color of finely dispersed nanoparticles. Appl Phys B 2001;73:317–26.
42. Zhong F, Nsor-Atindana J. Chapter 16 – microcrystalline cellulose and nanocrystalline cellulose. In: Phillips GO, Williams PA, editors. Handbook of hydrocolloids, 3rd ed. Cambridge, Massachusetts, United States: Woodhead Publishing; 2021:509–36 pp.
43. Ashwar BA, Shah A, Gani A, Shah U, Gani A, Wani IA, et al. Physico-chemical, structural, pasting and thermal properties of starches of fourteen Himalayan rice cultivars. Int J Biol Macromol 2017;95:1101–7.
44. da Matta MD, Jr., Sarmento SBS, de Oliveira LM, Zocchi SS. Mechanical properties of pea starch films associated with xanthan gum and glycerol. Starch – Stärke 2011;63:274–82.
45. Nasri-Nasrabadi B, Behzad T, Bagheri R. Preparation and characterization of cellulose nanofiber reinforced thermoplastic starch composites. Fibers Polym 2014;15:347–54.
46. Gontard N, Duchez C, Cuq JL, Guilbert S. Edible composite films of wheat gluten and lipids: water vapour permeability and other physical properties. Int J Food Sci Technol 1994;29:39–50.
47. Jiang L, Walczyk D, Mcintyre G, Bucinell R. A new approach to manufacturing biocomposite sandwich structures: mycelium-based cores. In: Proceedings of the ASME 2016 11th international manufacturing science and engineering conference; 2016:1–11 pp.
48. Careli-Gondim Í, Mesquita TC, Vilas Boas EV de B, Caliari M, Soares Júnior MS. The effect of active coating and refrigerated storage on the quality of avocado cultivar, Quintal. J Food Sci Technol 2020;57:143–51.
49. Careli-Gondim Í, Lago RC do, Vilas Boas EV de B, Caliari M, Soares Júnior MS. Revestimento ativo de nanofibras celulosídicas e sorbato de potássio na pós-colheita de morango (Fragaria x ananassa Duch.). Anais Do V Congresso Mineiro de Engenharia de Alimentos 2019;1:696.

Mohd Nor Faiz Norrrahim*, Nurjahirah Janudin,
Mohd Saiful Asmal Rani, Mohd Azwan Jenol, Nur Sharmila Sharip,
Norizan Mohd Nurazzi, Muhammad Rizal Muhammad Asyraf and
Rushdan Ahmad Ilyas

21 Wheat thermoplastic starch composite films reinforced with nanocellulose

Abstract: The rising costs of non-renewable plastic and environmental concerns with their industrial usage have encouraged the study and development of renewable products. As an alternative, biological-based materials create a huge opportunity for a healthy and safe environment by replacing non-renewable plastic in a variety of applications. Wheat is one of the world's most widely cultivated crops. Due to its mechanical and physical properties, wheat starch is vital in the biopolymer industry. Wheat thermoplastic starch exhibits useable properties when plasticizers, elevated temperatures and shear are present. Thus, make it very suitable to be used as packaging material. However, this material suffers from low mechanical properties, which limit its applications. Several studies looked at the feasibility of using plant components which is nanocellulose as a reinforcing agent in wheat starch thermoplastic composites.

***Corresponding author: Mohd Nor Faiz Norrrahim**, Research Centre for Chemical Defence, Universiti Pertahanan Nasional Malaysia, Kem Perdana Sungai Besi, 57000 Kuala Lumpur, Malaysia, E-mail: faiznorrrahim@gmail.com, faiz@upnm.edu.my

Nurjahirah Janudin, Research Centre for Chemical Defence, Universiti Pertahanan Nasional Malaysia, Kem Perdana Sungai Besi, 57000 Kuala Lumpur, Malaysia

Mohd Saiful Asmal Rani, School of Materials and Mineral Resources Engineering, Universiti Sains Malaysia, 14300 Nibong Tebal, Pulau Pinang, Malaysia

Mohd Azwan Jenol, Department of Bioprocess Technology, Faculty of Biotechnology and Biomolecular Sciences, Universiti Putra Malaysia, UPM, 43400 Serdang, Selangor, Malaysia

Nur Sharmila Sharip, Research and Development Department, Nextgreen Pulp & Paper Sdn Bhd, Taman Tun Dr Ismail, 60000 Kuala Lumpur, Malaysia

Norizan Mohd Nurazzi, Bioresource Technology Division, School of Industrial Technology, Universiti Sains Malaysia, 11800 Pulau Pinang, Malaysia

Muhammad Rizal Muhammad Asyraf, Engineering Design Research Group (EDRG), School of Mechanical Engineering, Faculty of Engineering, Universiti Teknologi Malaysia, 81310 Johor Bahru, Johor, Malaysia; and Centre for Advanced Composite Materials (CACM), Universiti Teknologi Malaysia, 81310 UTM Johor Bahru, Johor, Malaysia

Rushdan Ahmad Ilyas, Department of Chemical Engineering, Faculty of Chemical and Energy Engineering, Universiti Teknologi Malaysia, 81310 Johor Bahru, Johor, Malaysia; Centre for Advanced Composite Materials (CACM), Universiti Teknologi Malaysia, 81310 UTM Johor Bahru, Johor, Malaysia; Institute of Tropical Forest and Forest Products, Universiti Putra Malaysia, 43400 UPM Serdang, Selangor, Malaysia; and Centre of Excellence for Biomass Utilization, Universiti Malaysia Perlis, 02600 Arau, Perlis, Malaysia. https://orcid.org/0000-0001-6622-2632

As per De Gruyter's policy this article has previously been published in the journal Physical Sciences Reviews. Please cite as: M. N. F. Norrrahim, N. Janudin, M. S. Asmal Rani, M. A. Jenol, N. S. Sharip, N. M. Nurazzi, M. R. M. Asyraf and R. A. Ilyas "Wheat thermoplastic starch composite films reinforced with nanocellulose" *Physical Sciences Reviews* [Online] 2023. DOI: 10.1515/psr-2022-0012 | https://doi.org/10.1515/9783110773606-021

Overall, the addition of nanocellulose can improve the performance of wheat thermoplastic starch, especially for its mechanical properties. It can potentially be used in several areas of packaging and biomedical. The objective of this review is to discuss several achievements regarding wheat starch/nanocellulose-based composites. Several important aspects of the mechanical performance and the thermal properties of the composites were evaluated. The discussion on wheat starch and nanocellulose was also tackled in this review.

Keywords: composite; films; nanocellulose; thermoplastic starch; Wheat.

21.1 Introduction

Plastic is a polymeric substance that can be moulded or formed, typically using heat and pressure. Plasticity, when combined with other specific features such as low density, low electrical conductivity, transparency and toughness, enables plastics to be produced into various products.

The high demand on investigation of plastics is highly due to this type of material can also be used to mimic a variety of things especially in human activities (for example, electronics, medical supplies, construction, and packaging) [1–5]. These include polyethylene terephthalate (PET) drinking bottles that are durable and lightweight, polyvinyl chloride (PVC) garden hoses that are flexible, foamed polystyrene (PS) food containers and polymethyl methacrylate (PMMA) shatterproof windows [6]. Plastic can be categorized into two groups based on their chemical composition. First category of plastics is that made of polymers with only aliphatic (linear) carbon atoms in their backbone chains. Heterochain polymers make up another type of plastic. In addition to carbon, the backbone chains of these compounds contain elements such as oxygen, nitrogen, or sulphur. Most of the engineering plastics are made up of heterochain polymers [7].

As previously stated, thermoplastics and thermosets are the two major types of polymers categorized as plastics. Thermoplastics like polyethylene (PE), polypropylene (PP), polyvinyl chloride (PVC), nylon and polystyrene (PS) can be formed and remoulded multiple times. A foamed-polystyrene cup, for example, can be heated and moulded into a new shape, such as a dish. Individual molecules are separated from one another and flow past one another in the polymer structure associated with thermoplastics. The molecules may have a low or extremely high molecular weight, and their shape may be branching or linear, but separability and mobility are the most important characteristics [8]. Thermosets such as epoxy, silicone, polyurethane and phenolic on the other hand cannot be reprocessed after being reheated. During the early stages of thermosetting resin manufacturing, a chemical reaction occurs, resulting in an infusible and permanent network. Consequently, the final heated product is essentially one huge compound. For instance, when the epoxy polymer used to manufacture a fibre-reinforced laminate for a golf club is moulded at high temperatures, it experiences a cross-linking process. Additional heating does not soften the material enough to allow it to be reshaped, but it may even break it down.

Despite of their remarkable properties, plastic can be the cause for major pollutant in our environment [9]. Poor recycling systems and limited storage capacity leads to accumulation of plastic in dumping grounds. Another major drawback with plastics is there are numerous additives applied in manufacturing level and not biodegradable as well as highly toxic. Therefore, the primary motivation is to discover a way to replace oil-based products with bio-based products. Numerous studies have been conducted to discover sustainable and environmentally beneficial alternatives to the excessive production of plastic derived from fossil fuels. The major goal is to develop low-cost, biodegradable materials with great mechanical properties and quality. Biopolymers derived from biomass or organic waste, which are abundant and inexpensive in natural sources, are one of the alternatives [10, 11]. Consequently, they are more cost-effective than synthetic oil-based products. Thermoplastic starch (TPS) is a promising candidate to achieve this goal. Essentially, starch is not naturally thermoplastic. At high temperatures and shear, it reacts with plasticizers to produce thin films. Corn (82%), wheat (8%), potato (5%), rice, barley, oats, avocado, tapioca, pomegranate, pea, banana, sugar palm, cassava and soy are the most common sources of starch. It's a polysaccharide made up of amylose linked by α (1–4) bonds and amylopectin linked by α (1–4) chains and α (1–6) branches. Because of amylopectin, the degree of crystallinity ranges from 15 to 45%, impacting biodegradability, solubility and mechanical strength. When the amylose level is higher, the film is stronger and more flexible, whereas amylopectin causes poor mechanical qualities such as brittleness. TPS possess low gas permeability (O_2 and CO_2), great hygroscopicity, viscoelastic behaviours, poor mechanical strength, density and thermal properties. Figure 21.1 depicts application of TPS in our daily uses.

21.2 Wheat

21.2.1 Wheat starch

Starch has grown in popularity as a natural polymer due to its intrinsic biodegradability, abundance in nature and high renewability. It can be extracted from different parts of plant such as seeds, fruits, roots and tubers. Starch is found as a mixture of linear

Figure 21.1: Application of thermoplastic starch. Reproduced from ref. [12].

amylose and branched amylopectin organised as highly ordered and tightly packed glucan chains within massive, insoluble granules. Because of their low material cost and flexibility to be produced using existing plastic processing tools, it provides a low-cost platform for new biodegradable polymers. As previously said, there are several sources of starch, one of them is from wheat grains.

Wheat starch is a by-product of essential gluten production, but it can also be extracted from wheat flour via wet milling. The primary application of vital gluten is in fermentation bread products, particularly variety breads where the load of non-functional components is rather considerable and flour protein is diminished. Vital gluten facilitates in the preservation of product quantity and interior structure. The residual protein in the wheat starch gives it a flour-like odour, flavour and texture. Essentially, wheat starch granules have a bimodal size distribution, with enormous lenticular granules measuring ~25 µm in diameter and small spherical granules are at less than 10 µm in diameter [13].

Wheat starch represents for around 63–65% of the weight of the wheat kernel, with percentages greater for soft wheat than hard wheat. Wheat starch has less thickening capacity than corn starch, but the paste texture, clarity and strength are similar. Wheat starch has a lower gelatinization temperature than corn starch, which benefit in folded adhesives application. Wheat starch is preferred for laundry shrinking since it yields a firmer finish at a lower temperature than corn starch. Wheat starch is also favoured for baked products since it is produced without the use of hazardous chemicals [14]. Table 21.1 depicts the conversion method from starch to TPS.

Table 21.1: Conversion method from starch to thermoplastic starch [15].

Solvent casting method	Industrial method
– Starch and water-soluble plasticizers are used as reactant	Addition of starch, plasticizer and additives
– The process started with gelatinization, which the starch heated with water, then cast to evaporate until dry and produce thin films	– The process started with compounding using extrusion – Extrusion is conducted to destroy crystalline nature of starch at 100–150 °C
– Fragmentation of film due to degradation of starch occurs when high amylose starch is heated at 130–150 °C	– Thermoforming is performed using compression moulding, blow moulding and injection moulding
Advantages:	Advantages:
– Can be used in laboratory scale	– Can be used for large scale production
– Simple preparation method	– Produce high quality of starch
Disadvantages:	Disadvantages:
– Need huge amount of water	– Extrusion difficult for high amylose starch due to high melt viscosity
– Need long time of drying	
– Energy consuming	– Low amylopectin and 25 wt% of glycerol lead to shrinkage around 18%

However, the shortage of oil and the rising interest in lowering the environmental impact of petrochemically derived polymers have spurred the development of biodegradable starch-based materials for a wide range of applications such as packaging, biodegradable films, polymer electrolytes, textile, pharmaceuticals and health [16]. Furthermore, starch-based materials' mechanical qualities, particularly tensile strength, had limited their performance [17]. Several attempts have been made to improve their mechanical qualities, such as blending with other biodegradable polyesters or reinforcing with mineral or natural fillers, as well as the addition of additives to strengthen cross-linking interaction between starch and other compounds. Natural fibres and polysaccharide-based cellulose, particularly at the nanoscale, have piqued the interest of researchers working to improve the mechanical properties of starch-based biodegradable composites. It was discovered that nanocellulose and starch can both enhance tensile strength and Young's modulus [18]. Figure 21.2 illustrated the crystalline structure starch which composed of three polymorphisms that are classified as types A, B and C (as seen in Figure 21.2). Type A is made up of orthogonally packed double helices with a firm inclusion of structural water, type B has a larger orthogonal packing with 36 water molecules per unit cell (some of which are inserted in the hexagonal packing's channels) and type C contains both variants, A and B.

21.3 Nanocellulose

In recent years, natural and biodegradable materials such as cellulose, chitosan, protein, starch, lignin and natural gums emerged as an attractive renewable feedstock for unique material applications. One of the most significant green materials has been the creation of novel nanomaterials such as nanocellulose and cellulose with nanoscale structural dimensions.

Researchers are interested in these materials because of their excellent properties, such as being the most abundant polymer on the planet, having a high aspect ratio, great mechanical capabilities, biodegradability, renewability and nontoxicity [20, 21]. Furthermore, because nanocellulose has an abundance of hydroxyl groups, it is easily chemically modified. Cellulose can be found in numerous polymorphs namely, cellulose I, II, III_I, III_{II}, IV_I and IV_{II}, based on their van der Waals interaction, intra- and intermolecular interaction, molecular orientations isolation and treatment procedure, which it can be transform from one to another via chemical or thermal method [22–25]. Essentially, nanocellulose is divided into three types: cellulose nanocrystals, bacterial nanocellulose and cellulose nanofibers (seen in Figure 21.3) [26–30].

Nanocellulose can be prepared using either an up-down or a down-up approach. A down-up process is commonly used to create bacterial nanocellulose, in which cellulose producing bacteria, such as *Agrobacterium, Aerobacter, Achromobacter, Azotobacter, Rhizobium, Sarcina, Acetobacter xylinum, Salmonella* and *Escherichia* are utilised [31]. It contains two critical process, which are fermentation and purification. The unique

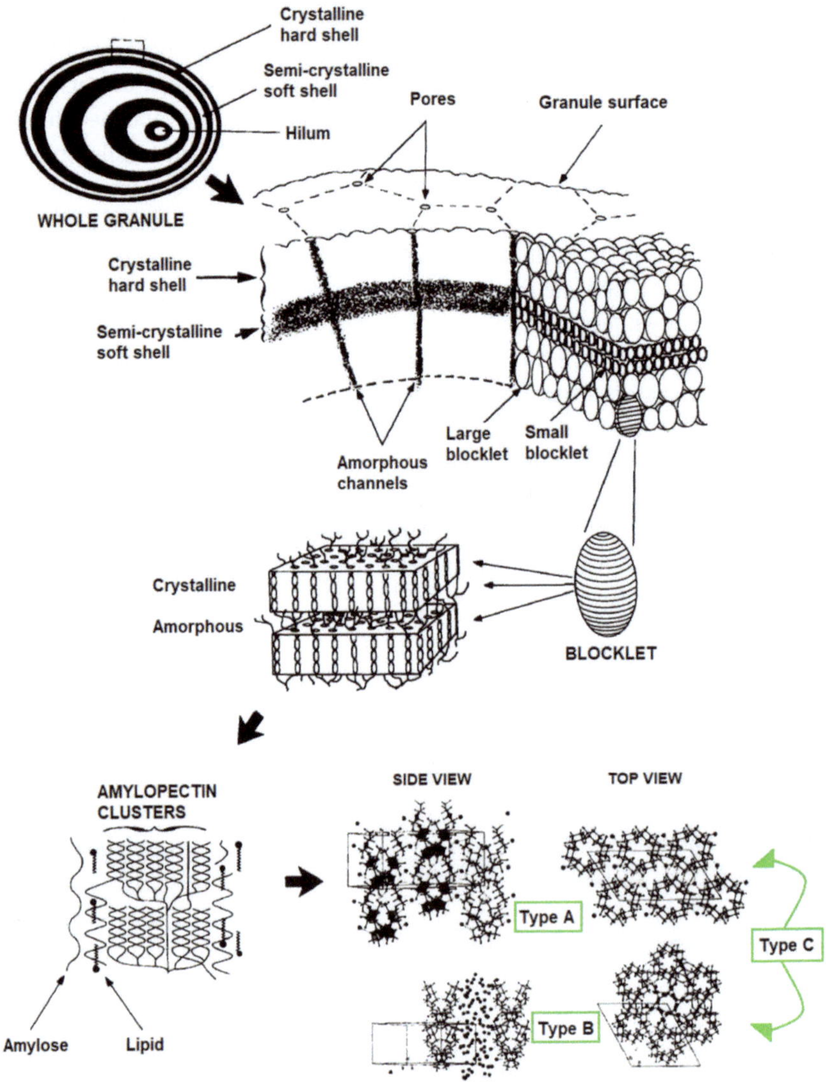

Figure 21.2: Crystalline structure of starch. Reproduced from ref. [19].

properties of bacterial nanocellulose are excellent chemical purity (lack of impurities such as hemicellulose, lignin or pectin), excellent mouldability and lesser cross section of fibres that contributes to high porosity of this material [32].

The up-down approach, on the other hand, is used to create cellulose nanocrystals and cellulose nanofibre. Inorganic acid (e.g., sulphuric acid, phosphoric, hydrochloric and hydrobromic acid) is used in the acid hydrolysis of pure cellulose to produce cellulose nanocrystals. In order to minimise power usage, mechanical disintegration using high

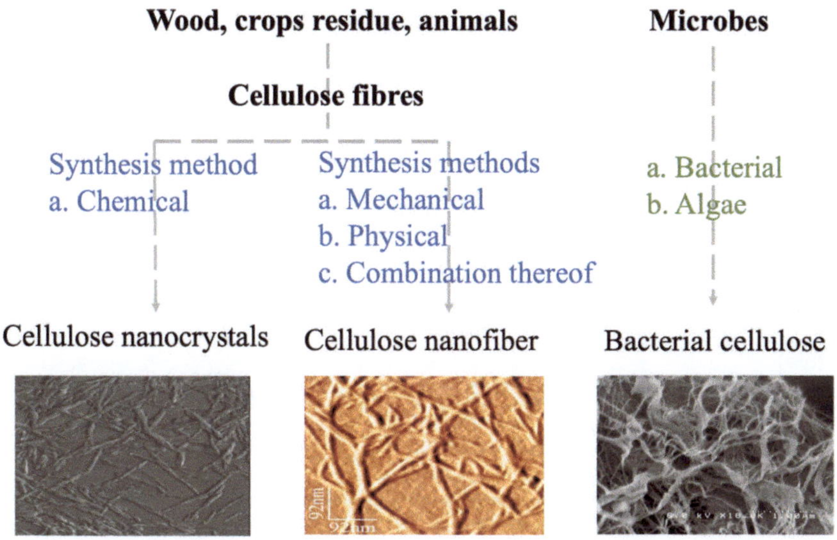

Figure 21.3: From cellulose to nanocellulose. Various procedure of nanocellulose in different form. Reproduced from ref. [26].

shear pressures, in conjunction with chemical or enzymatic pre-treatment, used to produce cellulose nanofibres [26]. In up-down process, cellulose disintegrated from lignocellulosic sources to their nano size. There are enormous application using nanocellulose such as water treatment [33, 34], 3D ink printing [35], automotive [20, 36], polymer composites [20, 37–41], tissue engineering [42, 43], drug delivery [44], food packaging [5, 45–48], sensor technology [28, 49, 50] and many others. Table 21.2 represents properties of cellulose nanofibres, cellulose nanocrystal and bacterial nanocellulose.

21.4 Performance evaluation on nanocellulose-reinforced wheat thermoplastic starch composite

21.4.1 Mechanical performance of nanocellulose-reinforced wheat thermoplastic starch composite

Incorporating nanocellulose to a wheat TPS for desirable mechanical properties of composite is very important for several applications. The performance of the wheat TPS composite reinforced with nanocellulose can be observed in several ways, including the mechanical properties. Several studies had shown that the reinforcement of nanocellulose in TPS managed to improve the mechanical performance as compared to the

Table 21.2: Properties of nanocellulose [22, 51].

Properties	Cellulose nanofibres	Cellulose nanocrystal	Bacterial nanocellulose
Morphology	Entangled network structure with flexible, long and wide nanofibres	Whisker shape or a short-rod-like shape	Dense and well-organised three-dimensional network of fibrils that forms porous sheets
Length (nm)	500–2000	100–500	0.001
Diameter (nm)	1–100	2–20	100
Young's modulus (GPa)	11.45	110–180	0.72
Crystallinity index (%)	49–72.5	>70	84–89
Tensile strength	42.3	7500 MPa	0.235

neat TPS [52]. For an example, Savadekar and Mhaske [53] reported at very low concentration (0.4 wt%) of nanocellulose, the TPS composite showed improvement in the mechanical properties. Where, the tensile strength was improved 46.1% as compared to the neat TPS. Table 21.3 summarized the mechanical performance of wheat TPS incorporated with several sources of nanocellulose, in comparison to several other sources of TPS.

Based on Table 21.3, generally the performances of wheat TPS can be seen as potential applicable to produce composite reinforcing with nanocellulose. Overall, the tensile strength, Young's modulus as well as percent of elongation are comparable to the existed starch in reviews. However, further improvement is needed for the wheat TPS to be potentially utilized as enhanced nanocellulose composite. Salehudin et al. [55] revealed that 2% nanocellulose incorporated with wheat TPS has higher tensile strength as compared to starch-chitosan hybrid film, and further increment of nanocellulose ratio to TPS has no significant effect on its performance. The authors also revealed the unique properties of this composite, in which significantly inhibit the Gram-negative bacterial growth.

On the other hand, Bruni et al. [54] have demonstrated the utilization of native and phosphorylated wheat TPS in the production of nanocellulose composite incorporated with several sources of nanocellulose. The results showed that native wheat TPS exhibited slightly superior mechanical performance in comparison to phosphorylated wheat TPS. The authors suggested that the high resistance and flexible nanocellulose composites obtained were suited for the food industry application.

Table 21.3: The mechanical performance of starch thermoplastic nanocellulose composite.

Source of TPS	Source of nanocellulose	Tensile strength (MPa)	Young's modulus (MPa)	Elongation (%)	Reference
Wheat	Eucalyptus nanocellulose crystal	4.32	70.81	34.90	[54]
Wheat	Rice nanocellulose crystal	3.64	34.86	61.60	[54]
Wheat	Oil palm empty fruit bunch cellulose nanofibre	4.30–5.25	-	25.04–27.74	[55]
Phosphorylated wheat	Eucalyptus nanocellulose crystal	3.08	30.12	57.05	[54]
Phosphorylated wheat	Rice nanocellulose crystal	3.78	31.94	62.85	[54]
Corn	Eucalyptus cellulose nanocrystals	10.8–12.9	479–491	26.9–29.5	[54]
Sugar palm	Sugar palm nano-fibrillated cellulose	6.80–10.68	59.02–121.26	25.38–37.70	[56]
Potato	Coconut fibres cellulose nanocrystals	4.09–8.20		23.71–30.40	[57]

21.4.2 Thermal analysis nanocellulose-reinforced wheat thermoplastic starch composite

Thermogravimetry analysis (TGA) is an excellent tool to evaluate the stability of all the composites because of its simplicity and effective information. Many studies used the technique to analyse starch in the past. A few researchers used the TGA method to investigate the thermal stability of the main TPS components and discovered a difference in thermal resistance between amylose and amylopectin in wheat starch. The TGA technique can be employed to estimate the contents of amylose and amylopectin in starch due to the difference in thermal resistance. Furthermore, the TGA can be used to study the thermal degradation and stability of starch and starch-based materials under various processing conditions [58].

Montero et al. [59] reported the decomposition profiles of wheat starch granules, pure glycerol (Gly) and nanocellulose. The wheat TPS encountered an initial weight loss of approximately 6.3%, owing primarily to water evaporation. A single degradation step occurred with an onset temperature of 309 °C and a peak temperature of 329 °C at the maximum degradation rate. The nanocellulose profile exhibits a similar pattern, which is expected given the structural similarities between starch and cellulose molecules. The water evaporation step results in a mass loss of about 4.1%. The degradation process begins at a temperature of 334 °C. The char yield obtained from wheat starch at 800 °C was 13%, while nanocellulose yielded 5%. For nanocellulose, a second step at higher temperatures is performed, leaving a minor starch residue. This second step is caused by the breakdown of a more thermally resistant component found in cellulose materials [60]. All films exhibit three major stages of decomposition; the step at low temperatures (50–120 °C) is attributed to water evaporation [61].

The second stage occurs at temperatures ranging from 120 to 280 °C and is caused by the decomposition of the TPS's glycerol-rich phase. The mass loss in this stage decreases as the nanocellulose content of the film increases, indicating a more complete trapping of glycerol molecules in the plasticized starch matrix. The third stage occurs as temperatures exceeds 280 °C. This step corresponds to the partially decomposed starch-rich phase oxidation and the cellulose nanoparticles degradation. The char residue obtained at 800 °C for neat wheat TPS films is clearly lower than that obtained at the same temperature for wheat TPS/nanocellulose films, as can be expected given the filler's higher degradation resistance. In terms of the filler effect on the temperature values observed in differential thermal gravimetry (DTG) curves for the maximum rate of degradation, wheat TPS nanocomposite reinforced with nanocellulose exhibits a slight shift towards higher values when compared to the neat wheat TPS, possibly due to a good interaction between nanocellulose and the wheat TPS matrix and the higher stability of the crystalline structure of nanocellulose [59].

21.5 Conclusions

Conventional petroleum-based plastics cause significant environmental damage and are one of humanity's greatest issues. Using wheat TPS reinforcing with nanocellulose to produce nanocomposite materials is a promising alternative for petroleum-based plastics. It has been proven that the presence of nanocellulose in the wheat TPS composite has better performance than the neat wheat TPS. Moreover, the mechanical performance of nanocellulose-reinforced wheat TPS as discussed here have comparable properties as compared to other available TPS. On the whole, the wheat TPS composite reinforced with nanocellulose is suitable to be used for packaging application.

References

1. Alyousef R, Ahmad W, Ahmad A, Aslam F, Joyklad P, Alabduljabbar H. Potential use of recycled plastic and rubber aggregate in cementitious materials for sustainable construction: a review. J Clean Prod 2021;329: 129736.
2. de Jongh FW, Sanches EE, Pouwels S, ten Harkel, Timen C, Ingels KJAO. E-health and telemedicine applications in plastic surgery and the treatment of facial palsy. Health Sci Rev 2022;2:100009.
3. Hermawan A, Abu Bakar MB, Mohd SH. Plastic in home building and construction applications. Amsterdam, The Netherlands: Elsevier; 2021.

4. Norrrahim MNF, Ariffin H, Hassan MA, Ibrahim NA, Nishida H. Performance evaluation and chemical recyclability of a polyethylene/poly(3-hydroxybutyrate-co-3-hydroxyvalerate) blend for sustainable packaging. RSC Adv 2013;3:24378–88.
5. Norrrahim MNF, Sapuan SM, Yasim-Anuar TAT, Padzil FNM, Sharip NS, Ng LYF, et al. Antimicrobial studies on food packaging materials. In: Food packaging. Boca Raton: CRC Press; 2020:141–70 pp.
6. Trivedi DN, Rachchh NV. Graphene and its application in thermoplastic polymers as nano-filler – a review. Polymer 2022;240:124486.
7. Greene JP. 8 – engineering plastics. In: Greene JPBT-AP, editor. Plastics design library. Norwich, New York, USA: William Andrew Publishing; 2021:107–25 pp.
8. Asim M, Jawaid M, Saba N, Ramengmawii, Nasir M, Sultan MTH. Processing of hybrid polymer composites—a review. In: Thakur VK, Thakur MK, Gupta RK, editors. Hybrid polymer composite materials. Sawston, UK: Woodhead Publishing; 2017:1–22 pp.
9. Ilyas RA, Sapuan SM, Ibrahim R, Atikah MSN, Asyraf MRM, Norrrahim MNF, et al. Environmental advantages and challenges of bio-based packaging materials. In: Bio-based packaging: material, environmental and economic aspects. Hoboken, New Jersey, USA: John Wiley & Sons; 2021.
10. Faiz Norrrahim MN, Ahmad Farid MA, Lawal AA, Tengku Yasim-Anuar TA, Samsudin MH, Zulkifli AA. Emerging technologies for value-added use of oil palm biomass. Environ Sci Adv 2022;1:259–75.
11. Thiruganasambanthan T, Ilyas RA, Norrrahim MNF, Kumar TSM, Siengchin S, Misenan MSM, et al. Emerging developments on nanocellulose as liquid crystals: a biomimetic approach. Polymers 2022;14: 1–28.
12. Jiang T, Duan Q, Zhu J, Liu H, Yu, L. Starch-based biodegradable materials: challenges and opportunities. Adv Ind Eng Poly Res 2020;3:8–18.
13. Cornell HJ. Chapter 2 – the chemistry and biochemistry of wheat. In: Cauvain SPBT-B, editor. Woodhead publishing series in food science, technology and nutrition, 3rd ed. Sawston, UK: Woodhead Publishing; 2012:33–80 pp.
14. Knight JW, Olson RM. Chapter XV – wheat starch: production, modification, and uses. In: Whistler RL, Bemiller JN, Paschall EF, editors. Food science and technology, 2nd ed. San Diego: Academic Press; 1984: 491–506 pp.
15. Kaushik A, Kaur R. Thermoplastic starch nanocomposites reinforced with cellulose nanocrystals: effect of plasticizer on properties. Compos Interfaces 2016;23:701–17.
16. Shaffie AH, Misenan MSM, Isa MIN, Khiar ASA. Effect of ionic liquid bmimno3 to chitosan-starch blend biopolymer electrolyte system. Solid State Phenom 2019;290:177–82.
17. Ribba L, Garcia NL, D'Accorso N, Goyanes S. Chapter 3 – disadvantages of starch-based materials, feasible alternatives in order to overcome these limitations. In: Villar MA, Barbosa SE, García MA, Castillo LA, López OV, editors. Starch-based materials in food packaging. Cambridge, Massachusetts: Academic Press; 2017:37–76 pp.
18. Ali A, Xie F, Yu L, Liu H, Meng L, Khalid S, et al. Preparation and characterization of starch-based composite films reinfoced by polysaccharide-based crystals. Compos B Eng 2018;133:122–8.
19. Rivadeneira-Velasco KE, Utreras-Silva CA, Díaz-Barrios A, Sommer-Márquez AE, Tafur JP, Michell RM. Green nanocomposites based on thermoplastic starch: a review. Polymers 2021;13:1–36.
20. Norrrahim MNF, Yasim-Anuar TAT, Jenol MA, Mohd Nurazzi N, Ilyas RA, Sapuan SM. Performance evaluation of cellulose nanofiber reinforced polypropylene biocomposites for automotive applications. In: Biocomposite and synthetic composites for automotive applications. Amsterdam, Netherland: Woodhead Publishing Series; 2020:119–215 pp.
21. Nor M, Norrrahim F, Azilah N, Kasim M, Knight VF, Janudin N, et al. Mini review on nanofibrillation techniques to obtain cellulose nanofiber from lignocellulosic biomass article info abstract. Zulfaqar J Defence Manag Soc Sci Humanit 2021;4:134–45.
22. Trache D, Tarchoun AF, Derradji M, Hamidon TS, Masruchin N, Brosse N, et al. Nanocellulose: from fundamentals to advanced applications. Front Chem 2020;8:1–33.

23. Norrrahim MNF, Ilyas RA, Nurazzi NM, Rani MSA, Atikah MSN, Shazleen SS. Chemical pretreatment of lignocellulosic biomass for the production of bioproducts: an overview. Appl Sci Eng Prog 2021;14:588–605.
24. Norrrahim MNF, Huzaifah MRM, Farid MAA, Shazleen SS, Misenan MSM, Yasim-Anuar TAT, et al. Greener pretreatment approaches for the valorisation of natural fibre biomass into bioproducts. Polymers 2021;13: 1–23.
25. Norrrahim MNF, Ariffin H, Yasim-Anuar TAT, Ghaemi F, Hassan MA, Ibrahim NA, et al. Superheated steam pretreatment of cellulose affects its electrospinnability for microfibrillated cellulose production. Cellulose 2018;25:3853–9.
26. Norrrahim MNF, Kasim NAM, Knight VF, Misenan MSM, Janudin N, Shah NAA, et al. Nanocellulose: a bioadsorbent for chemical contaminant remediation. RSC Adv 2021;11:7347–68.
27. Norrrahim MNF, Norizan MN, Jenol MA, Farid MAA, Janudin N, Ujang FA, et al. Emerging development on nanocellulose as antimicrobial material: an overview. Mater Adv 2021;2:3538–51.
28. Norrrahim MNF, Kasim NAM, Knight VF, Ujang FA, Janudin N, Razak MAIA, et al. Nanocellulose: the next super versatile material for the military. Mater Adv 2021;2:1485–506.
29. Norrrahim MNF, Mohd Kasim NA, Knight VF, Ong KK, Mohd Noor SA, Abdul Halim N, et al. Emerging developments regarding nanocellulose-based membrane filtration material against microbes. Polymers 2021;13:3249.
30. Norrrahim MNF, Kasim NAM, Knight VF, Ong KK, Noor SAM, Jamal SH, et al. Cationic nanocellulose as promising candidate for filtration material of COVID-19: a perspective. Applied Science and Engineering Progress 2021;14:580–7.
31. Budaeva VV, Gismatulina YA, Mironova GF, Skiba EA, Gladysheva EK, Kashcheyeva EI, et al. Bacterial nanocellulose nitrates. Nanomaterials 2019;9:1–12.
32. Noremylia MB, Hassan MZ, Ismail Z. Recent advancement in isolation, processing, characterization and applications of emerging nanocellulose: a review. Int J Biol Macromol 2022;206:954–76.
33. Li Y, Guo C, Shi R, Zhang H, Gong L, Dai L. Chitosan/nanofibrillated cellulose aerogel with highly oriented microchannel structure for rapid removal of Pb (II) ions from aqueous solution. Carbohydr Polym 2019;223: 115048.
34. Dai Z, Ottesen V, Deng J, Helberg RML, Deng L. A brief review of nanocellulose based hybrid membranes for CO_2 separation. Fibers 2019;7:1–18.
35. Fourmann O, Hausmann MK, Neels A, Schubert M, Nyström G, Zimmermann T, et al. 3D printing of shape-morphing and antibacterial anisotropic nanocellulose hydrogels. Carbohydr Polym 2021;259:1–11.
36. Ilyas RA, Sapuan SM, Mohd Nurazzi N, Norrrahim MNF, Ibrahim R, Atikah MSN, et al. Macro to nanocscale natural fibre composites for automotive components: research, development, and application. In: Biocomposite and synthetic composites for automotive applications. Amsterdam, Netherland: Woodhead Publishing Series; 2020:51–105 pp.
37. Wang H, Qian X, An X. Tb-coordination polymer-anchored nanocellulose composite film for selective and sensitive detection of ciprofloxacin. Carbohydr Polym 2022;287:119337.
38. Agumba DO, Kumar B, Latif M, Panicker PS, Pham HD, Kim HC, et al. High-performance esterified-poly (vinyl alcohol)-citric acid-lignin resin and its application to wet-spun nanocellulose filament-reinforced polymer composite. Composites Part A 2022;153:1–7.
39. Norrrahim MNF, Ariffin H, Hassan MA, Ibrahim NA, Yunus WMZW, Nishida H. Utilisation of superheated steam in oil palm biomass pretreatment process for reduced chemical use and enhanced cellulose nanofibre production. Int J Nanotechnol 2019;16:668–79.
40. Norrrahim MNF, Kasim NAM, Knight VF, Halim NA, Shah NAA, Noor SAM, et al. Performance evaluation of cellulose nanofiber reinforced polymer composites. Funct Compos Struct 2021;3. https://doi.org/10.1088/2631-6331/abeef6.
41. Norrrahim MNF, Ariffin H, Yasim-Anuar TAT, Hassan MA, Ibrahim NA, Yunus WMZW, et al. Performance evaluation of cellulose nanofiber with residual hemicellulose as a nanofiller in polypropylene-based nanocomposite. Polymers 2021;13:1064.

42. Luo H, Cha R, Li J, Hao W, Zhang Y, Zhou F. Advances in tissue engineering of nanocellulose-based scaffolds: a review. Carbohydr Polym 2019;224:115144.
43. Patil TV, Patel DK, Dutta SD, Ganguly K, Santra TS, Lim KT. Nanocellulose, a versatile platform: from the delivery of active molecules to tissue engineering applications. Bioact Mater 2022;9:566–89.
44. Das S, Ghosh B, Sarkar K. Nanocellulose as sustainable biomaterials for drug delivery. Sensors Int 2022;3: 100135.
45. Zhang W, Zhang Y, Cao J, Jiang W. Improving the performance of edible food packaging films by using nanocellulose as an additive. Int J Biol Macromol 2021;166:288–96.
46. Yasim-Anuar TAT, Norrrahim MNF, Sapuan SM, Ilyas RA, Jenol MA, Mamat NA, et al. Polyhydroxyalkanoates for packaging application. In: Bio-based packaging: material, environmental and economic aspects. Hoboken, New Jersey, USA: John Wiley & Sons; 2021.
47. Norrrahim MNF, Yasim-Anuar TAT, Sapuan SM, Ilyas RA, Hakimi MI, Najmuddin SUFS, Jenol MA. Nanocellulose reinforced polypropylene and polyethylene composite for packaging application. In: Bio-based packaging: material, environmental and economic aspects. Hoboken, New Jersey: Wiley Online Library; 2021.
48. Ariffin H, Norrrahim MNF, Yasim-Anuar TAT, Nishida H, Hassan MA, Ibrahim NA, Yunus WMZW. Oil palm biomass cellulose-fabricated polylactic acid composites for packaging applications. In: Bionanocomposites for packaging applications. Midtown Manhattan, New York City: Springer; 2018:95–105 pp.
49. Misenan MSM, Janudin N, Idayu MA, Norrrahim MNF, Jamal SH, Wan Yusoff WY, et al. Cellulose nanofiber as potential absorbent material for chloride ion. Solid State Phenom 2021;317:263–9.
50. Janudin N, Kasim NAM, Feizal Knight V, Norrrahim MNF, Razak MAIA, Abdul Halim N, et al. Fabrication of a nickel ferrite/nanocellulose-based nanocomposite as an active sensing material for the detection of chlorine gas. Polymers 2022;14:1–16.
51. Pradhan D, Jaiswal AK, Jaiswal S. Emerging technologies for the production of nanocellulose from lignocellulosic biomass. Carbohydr Polym 2022;285:119258.
52. Jeevahan J, Chandrasekaran M. Influence of nanocellulose additive on the film properties of native rice starch-based edible films for food packaging. Recent Pat Nanotechnol 2019;13:222–33.
53. Savadekar NR, Mhaske ST. Synthesis of nano cellulose fibers and effect on thermoplastics starch based films. Carbohydr Polym 2012;89:146–51.
54. Bruni GP, de Oliveira JP, Fonseca LM, da Silva FT, Dias ARG, da Rosa Zavareze E. Biocomposite films based on phosphorylated wheat starch and cellulose nanocrystals from rice, oat, and eucalyptus husks. Starch-Stärke 2020;72:1900051.
55. Salehudin MH, Salleh E, Mamat SNH, Muhamad II. Starch based active packaging film reinforced with empty fruit bunch (EFB) cellulose nanofiber. Procedia Chem 2014;9:23–33.
56. Ilyas RA, Sapuan SM, Ibrahim R, Abral H, Ishak MR, Zainudin ES, et al. Effect of sugar palm nanofibrillated cellulose concentrations on morphological, mechanical and physical properties of biodegradable films based on agro-waste sugar palm (Arenga pinnata (Wurmb.) Merr) starch. J Mater Res Technol 2019;8: 4819–30.
57. Cerqueira JC, Penha JDS, Oliveira RS, Guarieiro LLN, Melo PDS, Viana JD, et al. Production of biodegradable starch nanocomposites using cellulose nanocrystals extracted from coconut fibers. Polímeros 2017;27: 320–9.
58. Nurazzi NM, Asyraf MRM, Rayung M, Norrrahim MNF, Shazleen SS, Rani MSA, et al. Thermogravimetric analysis properties of cellulosic natural fiber polymer composites: a review on influence of chemical treatments. Polymers 2021;13:1–33.
59. Montero B, Rico M, Barral L, Bouza R, López J, Schmidt A, et al. Preparation and characterization of bionanocomposite films based on wheat starch and reinforced with cellulose nanocrystals. Cellulose 2021; 28:7781–93.

60. Jia R, Katano T, Yoshimoto Y, Gao Y, Nakazawa N, Osako K, et al. Effect of small granules in potato starch and wheat starch on quality changes of direct heated surimi gels after freezing. Food Hydrocolloids 2020;104: 105732.
61. Sainorudin MH, Abdullah NA, Rani MS, Mohammad M, Abd Kadir NH, Razali H, et al. Investigation of the structural, thermal and morphological properties of nanocellulose synthesised from pineapple leaves and sugarcane bagasse. Curr Nanosci 2022;18:68–77.

Nur Amalina Amirullah, Mohd Hafif Samsudin,
Mohd Nor Faiz Norrrahim*, Rushdan Ahmad Ilyas*,
Norizan Mohd Nurazzi, Mohd Azwan Jenol, Husain Siti Nor Hawanis and
A. A. N. Gunny

22 Regulations for food packaging materials

Abstract: Food safety and quality are essential concerns for consumers worldwide. One way to ensure that the food we consume meets these standards is through proper food regulation. While there are various aspects of food regulation, one critical aspect is the regulations for food packaging materials. As such, this book chapter has been written to provide an in-depth overview of the regulations for food packaging materials in different continents, namely Asia, Europe, and America. The chapter delves into several countries in each continent, such as Malaysia, Japan, Australia, China, the United States, and South Africa, and discusses their specific food packaging regulations. This chapter also emphasizes the importance of considering the composition, migration properties, and potential for contamination of food packaging materials. It highlights the need for regulations to ensure that food packaging materials do not adversely affect the safety and quality of food products. In light of the evolving food packaging industry, the chapter further underscores the need for continuous evaluation and improvement of food packaging regulations. This is crucial to keep up with emerging risks and new technologies that may arise and impact the safety and quality of food.

***Corresponding authors: Rushdan Ahmad Ilyas**, Department of Chemical Engineering, Faculty of Chemical and Energy Engineering, Universiti Teknologi Malaysia, 81310, Johor Bahru, Johor, Malaysia; Centre for Advanced Composite Materials (CACM), Universiti Teknologi Malaysia, 81310, Johor Bahru, Johor, Malaysia; Institute of Tropical Forest and Forest Products (INTROP), Universiti Putra Malaysia, 43400 UPM Serdang, Selangor, Malaysia; and Centre of Excellence for Biomass Utilization, Universiti Malaysia Perlis, 02600, Arau, Perlis, Malaysia, E-mail: ahmadilyas@utm.my. https://orcid.org/0000-0001-6622-2632 (R.A. Ilyas); and **Mohd Nor Faiz Norrrahim**, Research Center for Chemical Defence, Universiti Pertahanan Nasional Malaysia, Kem Sungai Besi, 57000 Kuala Lumpur, Malaysia, E-mail: faiz@upnm.edu.my. https://orcid.org/0000-0003-2101-5642 (M.N.F. Norrrahim)

Nur Amalina Amirullah, Faculty of Science, University of Malaya, Institute of Biological Sciences, 50603 Kuala Lumpur, Malaysia

Mohd Hafif Samsudin and Mohd Azwan Jenol, Department of Bioprocess Technology, Faculty of Biotechnology and Biomolecular Sciences, Universiti Putra Malaysia, 43400 UPM Serdang, Selangor, Malaysia

Norizan Mohd Nurazzi, Bioresource Technology Division, School of Industrial Technology, Universiti Sains Malaysia, 11800, Pulau Pinang, Malaysia; and Green Polymer, Coatings and Packaging Cluster, School of Industrial Technology, Universiti Sains Malaysia, 11800, Pulau Pinang, Malaysia

Husain Siti Nor Hawanis, Department of Chemical Engineering, Faculty of Chemical and Energy Engineering, Universiti Teknologi Malaysia, 81310, Johor Bahru, Johor, Malaysia

A. A. N. Gunny, Centre of Excellence for Biomass Utilization, Universiti Malaysia Perlis, 02600, Arau, Perlis, Malaysia; and Faculty of Chemical Engineering Technology, Universiti Malaysia Perlis, Kompleks Pusat Pengajian Jejawi 3, Kawasan Perindustrian Jejawi, 02600, Arau, Perlis, Malaysia

As per De Gruyter's policy this article has previously been published in the journal Physical Sciences Reviews. Please cite as: N. A. Amirullah, M. H. Samsudin, M. N. F. Norrrahim, R. A. Ilyas, N. M. Nurazzi, M. A. Jenol, H. S. Nor Hawanis and A. A. N. Gunny "Regulations for food packaging materials" *Physical Sciences Reviews* [Online] 2023. DOI: 10.1515/psr-2022-0033 | https://doi.org/10.1515/9783110773606-022

Keywords: continents; food packaging; regulations

22.1 Introduction

The food industry relies heavily on food packaging as it serves a multitude of essential functions. Among its primary roles is safeguarding food products from potential contamination during transportation and storage, thereby ensuring their safety for consumption [1, 2]. In addition to protect food products from being exposed to various types of harm, including physical, chemical, and biological, packaging also plays a crucial role in extending their shelf life. This, in turn, reduces food waste and increases the availability of food products to consumers. Moreover, packaging materials are utilized as a means of promoting and distinguishing products in the market. They offer a platform for branding and communication with consumers, enabling companies to effectively convey the unique attributes and benefits of their products [3, 4].

The history of food packaging begins in ancient times when humans first used natural materials such as leaves, gourds, and animal skins to wrap and store food. For instance, the use of leaves to wrap food is still common in many cultures, including Southeast Asia and Africa. However, as human civilization advanced, so did the packaging industry. Technological advancements led to the development of innovative packaging materials such as composites, plastics, metals, and laminates [5–9].

In modern times, food packaging has become an essential component of the food industry, with manufacturers relying on packaging to transport and store their products. The use of different types of packaging materials has allowed food manufacturers to expand their product offerings and improve their supply chain efficiency [10]. Moreover, the progress in packaging technologies has resulted in the innovation of modified atmosphere packaging that can prolong the shelf life of perishable food items such as fresh fruits, vegetables, and meat.

In recent times, the use of various packaging materials for food products has raised concerns about their safety and potential health risks [11]. This is because some of these materials may contain harmful substances that can migrate into the food products, causing contamination and posing risks to consumers' health. For example, plastic containers made of bisphenol A (BPA) have been found to leach BPA into food, leading to concerns about its potential adverse health effects, such as hormonal disruptions.

To address these concerns and safeguard consumers' health, the regulation of food packaging materials has become necessary. Regulations for food packaging materials are introduced to set standards for their composition, migration properties, labeling requirements, and testing requirements. These regulations help to ensure that food

Figure 22.1: Characteristic properties of food packaging material.

packaging materials are safe and free from harmful substances that can contaminate food products.

The composition of food packaging materials is an essential aspect that needs to be regulated. The materials used for packaging food products should be safe for human consumption, and their chemical properties should not pose any risks to human health. Figure 22.1 shows the characteristic properties of food packaging materials. Migration properties of food packaging materials are also critical since they determine whether harmful substances can leach from the packaging materials into the food products. The regulation of migration properties ensures that food packaging materials do not pose any risks of contamination to food products.

Labeling requirements for food packaging materials are also essential. These requirements help to inform consumers about the composition of the packaging materials and any potential risks associated with them. For instance, labeling requirements can inform consumers of the presence of allergens or any substances that may cause adverse health effects. Besides that, testing requirements for food packaging materials ensure that the packaging materials meet the required safety standards. These tests can include assessments of the chemical composition, migration properties, and overall safety of the packaging materials.

Therefore, this chapter discusses the regulations for food packaging materials in different continents, including Asia, Europe, and North America. The chapter evaluates the regulations in several countries in each continent, such as Malaysia, Japan, Australia, China, and the United States, and discusses their regulatory frameworks, testing requirements, and labeling requirements. The chapter also highlights the importance of considering the composition, migration properties, and potential for contamination of food packaging materials to ensure the safety of food products.

22.2 General knowledge of safety and regulations for food packaging

Basic knowledge of safety regulations for food packaging involves a set of guidelines and regulations to ensure that food packaging materials are safe for consumers to use. The primary goal of these regulations is to prevent any potential contamination of food by packaging materials [12, 13].

The safety of food packaging materials is evaluated based on several parameters, including the chemical composition of the packaging material, its interaction with the food product, and the potential migration of any harmful substances from the packaging material to the food product. The migration of any harmful substances from the packaging material to the food product is evaluated based on various factors, including the type of food product, the storage conditions, and the duration of storage.

Several regulatory agencies, such as the United States Food and Drug Administration (FDA), the European Food Safety Authority (EFSA), and the Japanese Ministry of Health, Labor, and Welfare (MHLW) have set limits on the number of certain substances that can migrate from packaging materials to food products. These limits are based on the potential risk to human health and are regularly updated to reflect any new scientific findings.

To ensure compliance with safety regulations, food packaging manufacturers must conduct extensive testing on their packaging materials to assess their safety for use with various food products. They must also maintain strict quality control procedures to ensure that their products meet all relevant safety standards.

While most global regulatory systems for food packaging are grounded in similar scientific principles, the methods employed by different governing bodies can vary significantly. The regulation of food packaging varies across different jurisdictions, and this can have an impact on trade between countries in the global economy. Therefore, some countries are re-evaluating their laws and regulations regarding food packaging to align with other jurisdictions, particularly the EU, to facilitate international trade.

22.3 Asia

Over the last seven decades, several Asian countries have implemented various regulations and guidelines about packaging and container materials, with a particular emphasis on food packaging. These regulations were created in response to growing concerns related to the environmental and health impacts of harmful and toxic materials, as well as the accumulation of plastic packaging waste.

The formulation of these regulations involved the input of expert groups, such as nutritionists, dieticians, physicians, and food scientists. Their expertise was instrumental in identifying the risks and hazards associated with different packaging materials and guiding the development of regulations that would ensure the safety of food products and protect the environment.

In this section, we will delve into the regulations governing food packaging and container materials in several Asian countries, including Malaysia, China, Japan, and India. We will examine the specific regulations implemented by these countries and the rationale behind them, as well as their effectiveness in addressing the identified risks and hazards. Through this exploration, we hope to gain a comprehensive understanding of the various approaches taken by these countries to regulate food packaging materials and to identify the best practices that can be adopted by other countries to promote the safe and sustainable use of packaging materials in the food industry.

22.3.1 Malaysia

The Ministry of Health Malaysia has recently expressed concern regarding the use of single-use plastics in consumer products, including food packaging. As a result, state governments in Malaysia have implemented regulations aimed at providing policies for all stakeholders, including local industries and communities. These regulations governing food containers and packaging materials in Malaysia are in line with the relevant provisions of the Food Act of 1983 and the Food Regulations of 1985. However, assessing the risk of chronic contaminant ingestion from food packaging can be challenging, as chemicals present in packaging and containers may migrate into food items. To address this issue, the Food Regulations of 1985 provide resolutions for food packaging, as outlined in Section VI, No. 27–36 A, according to Claudio [14], these regulations stipulate the requirements for food packaging materials used in Malaysia, including permissible levels of heavy metals and toxic substances. Additionally, they outline the necessary labeling and identification of food packaging materials to ensure transparency and enable consumers to make informed decisions. Here are the regulations for food packaging/container materials in Malaysia:

Regulation 27: Prohibits the use of food packaging or containers that may cause food contamination with harmful substances.

Regulation 28: Food containers must not contain harmful substances unless they pass the tests outlined in Schedule 13.

Regulation 29: Prohibits the use of food packaging or containers made from polyvinyl chloride that contains more than 1 mg/kg of vinyl chloride monomer.

Regulation 30: Containers made of polyvinyl chloride that contain more than 0.05 mg/kg of vinyl chloride monomer cannot be used for the sale of food products.

Regulation 31: Prohibits the use of food packaging/containers that are not designated for non-food products.

Regulation 32: Plays a vital role in maintaining the safety and quality of food products by prohibiting the reuse of certain packaging materials that may pose a risk to human health.

Regulation 33: Restricts reusing packaging that was previously used for another product.

Regulation 33A: Certain packaging for certain food products are permitted to be recycled such as glass bottles for alcoholic beverages and shandy or box/crate for vegetables and fruits.

Regulation 34: Assumes that a specific food container carrying a mark/label for another food has been used for the food as shown by the mark/label.

Regulation 35: Prohibits the use of a package/container with damages.

Regulation 36: Restricts the sale of non-food items in food or packaging materials.

Regulation 36A: Specifies the permissible substances for sachets containing reduced iron powder and mandates the inclusion of specific labeling requirements on the packaging.

The food packaging regulations in Malaysia prioritize consumer health and safety over addressing the challenges of waste management. Despite this, the country faces significant environmental problems due to the mismanagement of plastic waste. In response to this issue, the Ministry of Energy, Science, Technology, Environment, and Climate Change (MESTECC) released a report that aims to promote the use of eco-friendly materials in the plastic industry as a substitute for single-use plastics.

To achieve this goal, the Malaysia Roadmap Towards Zero Single-Use Plastics 2018–2030 was launched in October 2018. The roadmap has three phases, with Phase 1 (2018–2021) focusing on the implementation of a minimum charge for plastic bags and a "no straw by default" policy. Additionally, the use of biodegradable and compostable materials for food packaging was encouraged.

According to a survey as shown in Figure 22.2, plastic bags were the most commonly used item, with only 37–47 % awareness of their reusability. In Phase 2 (2022–2025), the use of biodegradable products will be expanded, and R&D funding and incentives will be provided. The final phase, Phase 3 (2026–2030), aims to increase the use of sustainable packaging and promote health and food safety. By implementing the Malaysia Roadmap Towards Zero Single-Use Plastics 2018–2030, Malaysia hopes to reduce plastic waste and promote the use of eco-friendly materials in the plastic industry, resulting in a cleaner and healthier environment for its citizens.

22.3.2 Japan

Japan is renowned for its advanced technology, which has greatly influenced the productivity of its food packaging industry. Although there is a high demand for Japanese

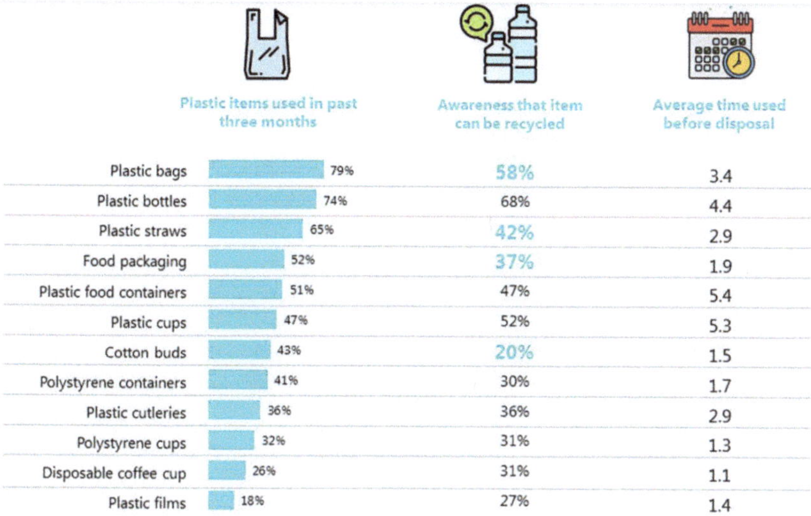

Figure 22.2: Malaysia's awareness regarding recyclable plastic items (from ref. [18]).

food products globally, the country prioritizes the quality of their manufactured goods. As a marketing strategy to gain consumer trust, the Japanese government is exploring alternative options, especially in food packaging.

Japan has two national laws to ensure food safety, namely the Food Sanitation Law, which was introduced in 1947, and the Food Safety Basic Law. These laws are rooted in The Constitution of Japan Article 25 which guarantees a "minimum standard of healthy and cultured living" for all citizens. The Food Sanitation Law is responsible for ensuring the safety of all food products, appliances, food additives, and non-drug packages/containers that are regulated under the Pharmaceutical Affairs Law. This law serves as a regulation for food safety management. In 1959, the Ministry of Health and Welfare established Notification No. 370, which specifies the standards for food, food additives, and other materials.

22.3.3 China

In China, there are various materials used for food packaging and containers, such as plastic, porcelain, bamboo, paper, metal, rubber, and glass products. The China Food Hygiene (CFHL) Act, which was implemented on 30 October 1995, is the primary legislation governing food packaging and container materials in China. The Ministry of Health is responsible for formulating and updating hygiene standards for food packaging materials and food seasoning following the CFHL. The CFHL establishes rules and regulations on food hygiene, including 40 compulsory hygiene standards for commonly used plastics. Approval from the Minister of Health is necessary for new resin-finished plastic

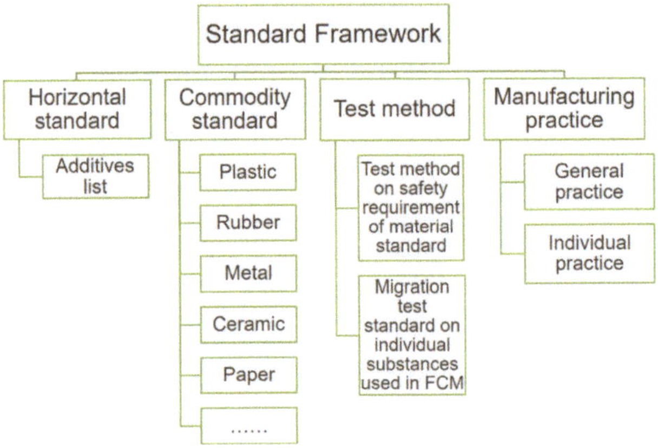

Figure 22.3: Current framework of food packaging material standard in China.

articles. Moreover, hygiene control standards for plastics include GB 9691-88 for PE resin and GB 9687-88 for polyethylene finished products, which must be followed in compliance with the Food Hygiene Act and other relevant regulations. Figure 22.3 illustrates the current framework for food packaging materials in China.

The food safety law contains significant regulations on packaging materials, including:

- Article 20 sets a restriction on the usage of harmful substances in food, such as pathogens, heavy metals, pesticide residues, and other hazardous materials.
- Requiring food producers to review the manufacturing licenses and certificates of raw material suppliers in Article 36. Products without certificates must be inspected before use to ensure compliance with safety standards.
- Mandating quality checks on all goods before they are released to the market in Article 38.
- Article 44 mandates the submission of safety assessments for any new raw materials used in food products to the health administrative agency of the State Council.
- Article 62 specifies that all food products released for the international market must comply with national food safety standards.
- Updating Article 44 and specifying compliance requirements for food products exported to other countries in Article 63.

22.3.4 India

India's Food Safety and Standards Authority (FSSAI) has updated its regulations on food packaging to improve food safety and prevent contamination. The FSSAI's packaging regulations were implemented in 2011 and are regularly reviewed and updated to keep up with changing trends and concerns. The regulations cover various aspects of food packaging, including the use of materials, labeling requirements, and safety measures. The new guidelines prohibit the use of recycled plastics, newspaper materials, and similar materials due to their carcinogenic effects. The primary goal of the updated regulations is to prevent food contamination and maintain the food's properties during packaging. Additionally, India has also reviewed ink standards used in food packaging materials [16]. The country has been making significant efforts to protect the environment.

22.4 Europe

In Europe, food contact materials are regulated at both national and societal levels. The European Union (EU) Framework Regulation oversees food contact materials such as packaging, machinery, and kitchenware at the EU level. Meanwhile, various directives and laws have been established at the community level. The Framework Regulation allows for the use of 17 different food contact materials, each with specific requirements that must be met.

The use of recycled paper in food contact products is governed by two different approaches, depending on the method of recycling. For depolymerized plastics, the same criteria as virgin plastics outlined in EC 10/2011 must be adhered to. Mechanically recycled plastics must meet the requirements specified in the European Food and Safety Authority (EFSA) regulation. Regenerated cellulose films are regulated under Directive 2007/42/EC, which contains a list of authorized substances. In cases where cellulose meets food during marketing, a written declaration is necessary.

Active materials and articles, which are intended to enhance the condition and shelf-life of food products, are classified under Regulations 1935/2004/EC and 450/2009/EC. These materials contain components that either release or absorb substances from the food or its environment [17]. On the other hand, intelligent food contact materials are designed to monitor the conditions of packaged food or the surrounding environment.

Food packaging regulations in Europe have evolved (Figure 22.4). Originally, laws were implemented at the individual member state level. However, with the creation of the EU, laws were harmonized to create a single industry and eliminate trade obstacles. EU laws aim to protect consumer safety by ensuring that no substances in contact with food will alter its characteristics [18].

Active packaging extends food shelf-life, while intelligent packaging monitors food conditions [19]. Different types of active and intelligent products and technologies are available [20]. The active material should be incorporated into packaging material safely and regulated to simplify the process and increase consumer safety. Active and intelligent materials may consist of different parts from different sources.

Active packaging refers to the incorporation of additives or freshness enhancers that interact with packaging materials, thereby enhancing the preservation functions

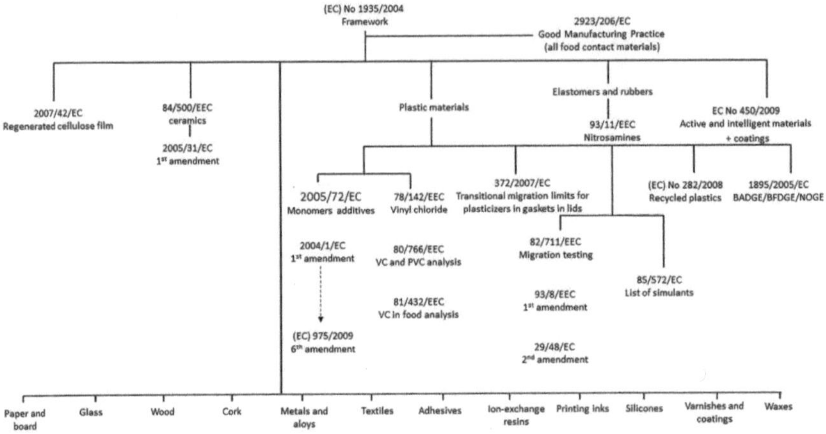

Figure 22.4: EU food-contact materials legislations (ref: [18]).

of the primary packaging system. Additives used in active packaging include those that can scavenge or transport oxygen, carbon dioxide, ethylene, humidity, and odor or flavor. Active packaging systems may also involve the release of preservatives, antimicrobials, sorbates, antioxidants, and various gases such as oxygen, carbonic acid, and alcohol. Additionally, some active packaging systems are designed for humidity control (refer to Table 22.1). Active packaging has numerous applications, and it can be particularly useful in extending the shelf-life of specific food products such as meat, whose spoilage processes are well-known and can be regulated. There have been several studies on active packaging implementation (refer to Table 22.2) [21–23] and techniques (refer to Table 22.3) [20, 24].

The EU Framework Regulation aims to ensure the protection of human health and consumer interests by authorizing the use of food contact materials, including active and intelligent packaging, under Regulation 1935/2004/EC. This regulation permits the

Table 22.1: Examples of active packaging applications for use within the food industry.

Absorbing/scavenging properties	**Oxygen, carbon dioxide, moisture, ethylene, flavors, taints, UV light**
Releasing/emitting properties	Ethanol, carbon dioxide, antioxidants, preservatives, sulfur dioxide, flavors, pesticides
Removing properties	Catalysing food component removal: Lactose, cholesterol
Temperature control	Insulating materials, self-heating and self-cooling packaging, microwave susceptors and modifiers, temperature-sensitive packaging
Microbial and quality control	UV and surface-treated packaging materials

Table 22.2: Applications of active packaging technologies.

Type of Application	Foods
Oxygen scavengers	Ground coffee, tea, roasted nuts, potato chips, chocolate, fat powdered milk, powdered drinks, bread, tortillas, pizza crust, refrigerated fresh pasta, fruit tortes, cakes, cookies, beer, deli meats, smoked and cured meats, fish, cheese
Carbon dioxide absorbers	Ground coffee
Carbon dioxide emitters	Meat, fish
Moisture absorbers	Dry and dehydrated products, meat, poultry, fish
Ethylene scavengers	Kiwifruit, banana, avocados, persimmons
Ethanol emitters	Bread, cakes, fish
Antimicrobial releasing films	Dry apricots
Antioxidant releasing films	Cereals
Flavor releasing films	Ground coffee
Color containing films	Surimi
Anti-fogging films	Some fresh fruit and vegetable packages
Anti-sticking films	Soft candies, cheese slices
Light absorbers	Pizza, milk
Time-temperature indicators	Microwavable pancake syrup, refrigerated pasta, deli items
Gas permeable/breathable films	Ready-to-eat salads
Microwave susceptors	Ready-to-eat meals

use of active and intelligent materials provided they improve the safety, quality, and shelf-life of packaged foods [25]. The Regulation applies to all materials and articles designed to be in contact with food, including active and intelligent packaging (Figure 22.5). Only food additives or flavorings can be released into the food environment, and additional safety criteria are set out in EC 450/2009 directive. Materials used for active and intelligent packaging must be traceable and have appropriate identification systems.

Regulation 450/2009/EC ensures the safe use of active and intelligent packaging by encompassing the general prerequisites of Regulation 1935/2004/EC. This regulation also lays out the marketing requisites for packaging materials and articles designed for food contact, whether they are separate containers or integrated into the packaging material.

To summarize, the primary aim of food packaging is to ensure the safety, freshness, and quality of food products. Therefore, it is crucial to carefully select materials, comply with the guidelines of the Environmental Protection Agency (EPA), and assess the environmental impact of packaging materials to address the environmental issues caused by packaging waste.

Table 22.3: Examples of some currently known active packaging systems.

Type of active packaging system	Substances used and mode of action
Oxygen scavengers	Enzymatic systems (glucose oxidase-glucose, alcohol oxidase-ethanol vapour), chemical systems (powdered iron oxide, catechol, ferrous carbonate, iron-sulfur, sulfite salt-copper sulfate, photosensitive dye oxidation, ascorbic acid oxidation, catalytic conversion of oxygen by platinum catalyst)
Carbon dioxide absorbing/emitting	Iron powder-calcium hydroxide, ferrous carbonate-metal halide
Moisture absorbing	Silica gel, propylene glycol, polyvinyl alcohol, diatomaceous earth
Ethylene absorbing	Activated charcoals, silica gel-potassium, permanganate, Kieselguhr, bentonite, Fuller's earth, silicon dioxide powder, powdered Oya stone, zeolite, ozone
Ethanol emitting	Encapsulated ethanol
Antimicrobial releasing	Sorbates, benzoates, propionates, ethanol, ozone, peroxide, sulfur dioxide, antibiotics, silver-zeolite, quaternary ammonium salts
Antioxidant releasing	BHA, BHT, TBHQ, ascorbic acid, tocopherol
Flavor absorbing	Baking soda, active charcoal
Flavor releasing	Many food flavors
Color containing	Many food colors
Anti-fogging and anti-sticking	Biaxially oriented vinyl on, compression rolled oriented HDPE
Light absorbing/regulating	UV blocking agents, hydroxy benzophenone
Monitoring	Time-temperature indicators
Temperature controlling	Non-woven microperforated plastic
Gas permeable/breathable	Surface-treated, perforated or microporous films
Microwave susceptors	Metallized thermoplastics
Insect repellent	Low toxicity fumigants (pyrethrins, permethrin)

22.5 America

The section discusses food packaging regulations in North and South America, with a focus on the United States. Food packaging serves various functions, including physical protection, product preservation, and food safety. Effective packaging can also convey information about the product. Advanced packaging examples include susceptors for browning foods in microwaves, organic or chemical preservatives/stabilizers, and monitoring capabilities. Regulations on food packaging are crucial for consumer safety and are recognized in the food industry.

Figure 22.6 depicts the food packaging materials most commonly used for both direct and indirect contact with food and beverages. Regulations governing food contact materials are stringent to ensure the safety of consumers. In the United States, 21CFR174 mandates that food contact materials do not release hazardous chemicals into the food. It is crucial to comprehend the chemical composition of food packaging

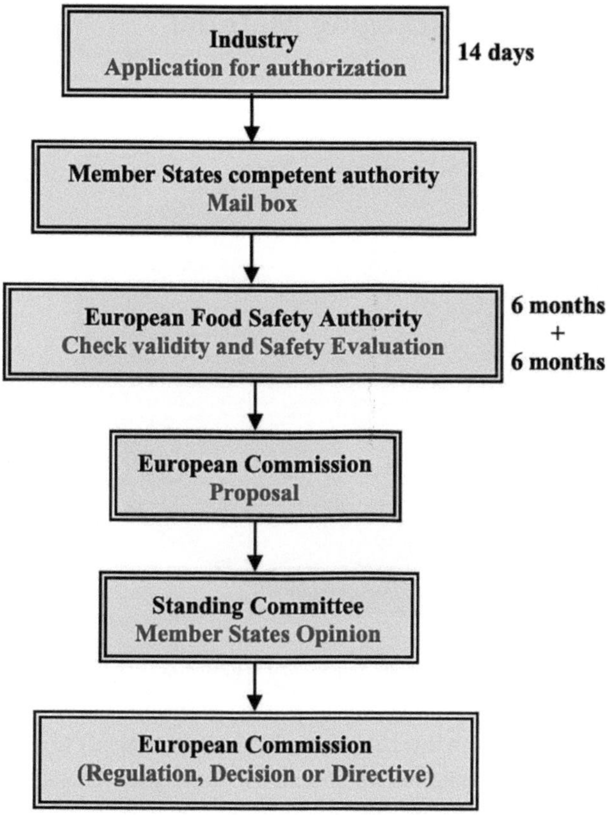

Figure 22.5: Authorization procedure as defined by reg. 1935/2004 EC (ref. [25]).

materials to assess the potential health risks from migration. Commonly used materials for food packaging include ceramics, glass, metal, paper and board, plastics, printing inks, wax, and wood, among others [26, 27].

22.5.1 History of formal food packaging regulations in the United States

Food packaging regulations were established in the US with the Food Additives Amendment in 1958. This amendment defines a "food additive," and certain substances can be exempted from this classification if they are considered "Generally Recognized as Safe" or listed as an exclusion in the Act. In the US, food contact substances are regulated as food additives and those that are not approved by the FDA for their intended use are considered to adulterate food under Section 409 of the Act.

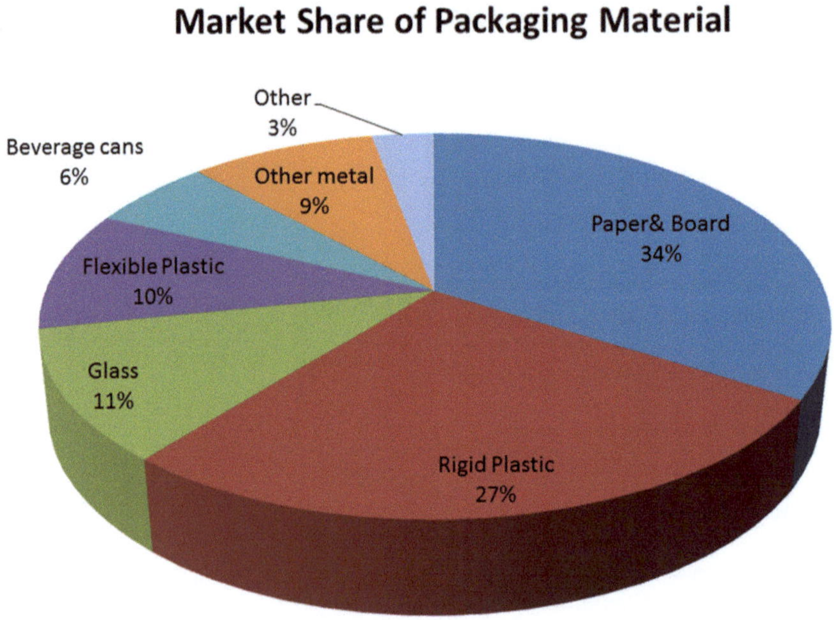

Figure 22.6: Percentages of food packaging materials that are widely used in the market (ref. [28]).

However, there are three exceptions to this rule, one of which is commonly known as the "prior sanctioned" substance exception. Before 1958, some food packaging manufacturers would consult with the FDA or the United States Department of Agriculture (USDA) to ensure that a particular packaging material would not be objected to if used. If no objection was made, these materials would be considered "prior sanctioned" substances and would be exempt from the food additive requirements for further clearance or authorization by the FDA.

In the 1950s, there were concerns that food additives could cause serious health issues, including cancer. This led to hearings on the topic, but no concrete evidence was found to support these claims. In 1956, legislation was proposed to require FDA pre-clearance for materials used in the production and packaging of food. Although there is no evidence that packaging materials can harm food or humans, they are still regulated as food additives.

While there is no all-encompassing federal law in the United States to regulate the packaging industry, there are several federal agencies and laws that oversee different aspects of packaging materials and their use in food, drug, and cosmetic products. These regulations and agencies aim to ensure consumer safety and maintain trust in the products they purchase.

22.5.2 The United States food packaging regulations

The FDA plays a crucial role in regulating the packaging and labeling of food to improve food safety and ensure that consumers are informed about the products they consume. Packaging that comes into direct contact with food is classified as a "food contact substance," which is regulated by the Office of Food Additive Safety and the Center for Food Safety and Applied Nutrition (CFSAN). This includes materials such as coatings, plastics, papers, food colorants, and adhesives used in food packaging, which must be deemed safe for use.

To prevent the growth of microbes, mold, mildew, parasites, discoloration, and unpleasant odors, the Environmental Protection Agency (EPA) requires that plastic and materials used in food packaging incorporate antimicrobial technology. The FDA also mandates that packaging labeling includes information such as "best if used by" dates, expiration dates, preparation and handling instructions, nutrition facts, and the packaging company's contact information. Since 2006, the Food Allergen Labeling and Consumer Act have required allergen labeling on food packaging, which means that food labels must list potential allergens such as peanuts, fish, milk, eggs, shellfish, soybeans, or wheat. If the item was manufactured or packaged in a facility that handles any of these allergens, the notification must also be included on the label.

22.5.3 The ecological consequences of materials employed in the packaging of food

Under the Federal Food, Drug, and Cosmetic Act, the FDA has the responsibility to ensure that food packaging materials do not contaminate food and to assess the safety of both new and previously used food contact materials. Apart from safety concerns, the National Environmental Policy Act (NEPA) requires the FDA to evaluate the environmental impact of these materials. An Environmental Assessment (EA) and a Finding of No Significant Impact (FONSI) are required for all food packaging clearances and FDA approval must be obtained. The FDA also requires information on the recycling effects of packaging materials when evaluating their environmental effects. Companies that wish to use new packaging materials must justify why commonly recycled materials, such as rigid single-use food containers or PET, cannot be used.

22.5.4 Containers made of inflexible plastic

The state of California introduced the Rigid Plastic Packaging Container (RPPC) Law as a measure to reduce the quantity of plastic waste in landfills and promote the use of recycled plastics. This law applies to plastic containers that can retain their shape and can be opened and closed, with a minimum volume of 8 fluid ounces and a maximum capacity of 5 gallons. According to this law, the name of the manufacturer must be displayed on the container. Similar legislation has also been enacted by other states, including Oregon and Wisconsin, which permit manufacturers to use recyclable or reusable materials or to incorporate previously used materials.

22.5.5 Regulations

To guarantee the safety of food products from food contact materials (FCM), the food industry must adopt an integrated approach. Table 22.4 summarizes the treatments implemented by various countries in North and South America to achieve this goal. Sharing information and knowledge across all levels of the supply chain is crucial to ensure the effective implementation of these measures.

22.5.6 The United States' exposure approach to FCM legislation

It is crucial to acknowledge the variations in the United States and the EU's approach to global FCM control systems when discussing the subject. Despite some similarities, such as toxicology assessments and analyses, the two systems differ fundamentally and are almost complete regulatory opposites.

In contrast to the United States, the EU takes a toxicological approach that involves evaluating each substance and adding it to a list of acceptable materials, whereas the US uses an exposure approach that evaluates the material's use and its impact on food and determines the degree of toxicological testing required based on the Paracelsian principle 'the dose makes the poison.' As a result, substances are exempted if projected exposure levels are low or minimal.

Another significant distinction between the US and EU FCM control systems is the allocation of responsibilities between packaging manufacturers and regulatory authorities. In the US, manufacturers have the autonomy to interpret regulations and determine whether their products require approval. The FDA believes that manufacturers should take responsibility for packaging matters and considers it unnecessary and impractical to be consulted.

22.5.7 The process of enforcing regulations in the United States regarding food packaging materials

The FDA is responsible for enforcing the regulatory framework for food contact materials in the United States under the Federal Food, Drug, and Cosmetic Act (FDCA). The Center for Food Safety and Applied Nutrition (CFSAN), through its Office for Food Additive Safety (OFAS), has been assigned the specific task of ensuring compliance with these guidelines. Furthermore, the Department of Agriculture's Food Safety and

Table 22.4: A summary of various country treatments.

Country	Authority responsible	Type of regulation for FCM
United States, US	US Food and Drug Administration USFDA Science Board, USFDA Food Advisory Committee Center for Food Safety and Applied Nutrition (CFSAN)	Main regulatory body. Modified positive list system with pre-marketing clearance. The Federal Food, Drug and Cosmetic Act (FFDCA, Title 21 United States Code 348). 21 CFR170.3 defines the term FCM. – FCM fall into – positive lists (indirect additives) (21CFR, 174–179), GRAS (21CFR, 182–186), Prior sanctioned substances (21CFR 181) or comply with the Threshold of regulation (21 CFR 170.39). Advisory scientific body. "Indirect Additive" database – consists of regulated indirect food additives
Countries having non-specific/few regulations for FCM		
North America – Canada South American countries E. g Argentina Brazil Mexico Other South American countries – Chile, Bolivia, Ecuador and Peru	Canada's food directorate Surveillance – Canadian food inspection agency and bureau of chemical safety in health Common market of the south – MERCOSUR (Argentina, Brazil, Paraguay, Uruguay, Venezuela)	Responsible for food safety from FCM – Positive list system – Letter of no objection (LONO) voluntarily – Controlled separately under division 23, part B of the food and drug act and regulations. Harmonized regulations. – Positive list approach based on EU's regulation 2011 and the USFDA. – But legal confirmation for a specific substance is to be taken from concerned authorities. – Final FCM/package needs to have premarket approval/registration. – If FCM is not present on the positive list, they must be petitioned for inclusion in MERCOSUR positive lists to national sanitary authorities. – GMC 26/03 – defines FCM. Ministry of Health – regulatory agency CONAL (National Food Commission) – advisory scientific body. Ministry of Health – through its regulatory agency – ANVISA – National Agency of Sanitary Surveillance. In Argentina and Brazil, FCMs are defined and regulated independently.

Table 22.4: (continued)

Country	Authority responsible	Type of regulation for FCM
		Ministry of Health (SSA) – through COFEPRIS. – Have no specific regulations on FCM and no specific regulations for their use. – Have regulations on packaging materials. – Have no specific regulations on FCM. – Only limited to a few food packaging standards in their legislation for F & B. E.g. Some standards on Ecuador, Bolivia and Peru food packaging material regulations.

Inspection Service (FSIS) is in charge of ensuring the safety of meat and poultry products, with food additive substance status playing a critical role in the evaluation and implementation of regulations [28]. Moreover, the Department of the Treasury's Bureau of Alcohol, Tobacco requires food additive status clearance for packaging materials [29].

22.5.8 A pragmatic strategy for navigating the regulatory framework of food packaging materials in the United States

Manufacturers need to follow a methodical approach when deciding on the preclearance of food contact materials to comply with regulations and minimize toxicological testing costs [29]. The first step is to verify if the substance and its use have been precleared under Food Additive Petition (FAP) or Food Contact Notification (FCN) regulations. Manufacturers must confirm that there is a possibility of migration, and if not, the substance is not regulated. Exemptions like GRAS, the threshold of the regulation (TOR), and polymer doctrine may exempt substances from being classified as food additives. Manufacturers have two protocols to follow for preclearance, submitting the data through FCN or applying for a Food Additive Petition.

Recent advancements in technology have led to further scrutiny of the safety of food packaging materials. Lobbyists have raised concerns about importing food packaging made with perfluorinated substances, which were banned by FDA in 2016 due to similar toxic substances and uncertain harm. Some states and local jurisdictions have also implemented bans on certain food packaging substances. As food packaging becomes increasingly scrutinized, it is essential to remain informed about the latest scientific findings and align with legal and regulatory requirements to ensure the safety of packaging materials and their use with food products.

22.6 Australia and Africa

22.6.1 Regulations governing food packaging materials in Australia

Packaged food products are globally traded, including those from Australia and New Zealand [30]. Food packaging plays a crucial role in preserving and transporting food, preventing contamination, extending shelf life, and providing consumer convenience and information. However, food packaging involves a variety of chemicals, and there is an increasing need for recycled packaging materials. The use of recycled materials can present the potential risk of chemical migration from a combination of reused packaging materials.

Compliance with regulations, CoPs, guidelines, and self-imposed safety requirements is crucial for ensuring the safety of food in Australia and New Zealand. Municipal waste in Australia is growing unsustainably due to several factors, such as demographics, regulations, market competition, and voluntary consumer protection measures [30].

The National Notification and Assessment Scheme for Industrial Chemicals (NIC-NAS) in Australia has defined industrial nanomaterials and provided guidelines for testing their impact on health [31]. Furthermore, Food Standards Australia New Zealand (FSANZ) has assessed the suitability of Australia and New Zealand's current regulatory framework for addressing potential human health risks associated with nanotechnology [31].

22.6.2 Minimizing environmental damage in the natural surroundings

Reducing environmental harm in natural environments is a complex issue that involves multiple challenges. Littering is often linked to consumer behavior in specific circumstances, while eco-friendly packaging materials may potentially compromise product or consumer safety [32]. The problem is further compounded when it comes to takeout meals since they involve multiple companies, products, locations, and customers. Despite the complexity of the issue, it is still a reality that needs to be addressed. The global environmental crisis is further highlighted by the fact that there are approximately 18,000 pieces of plastic floating in every square kilometer of the ocean, which significantly impacts marine life in Australia [30]. The Australian Packaging Covenant Strategic Plan 2017–2022 provides a roadmap for tackling this issue, as outlined in Table 22.5 [32].

Table 22.5: Australian Packaging Covenant Strategic Plan 2017–2022.

Pillars		Key objectives	Strategic priorities and initiatives	Key performance indicators
1. Resource efficiency	A Sustainable design: a life cycle approach	Increase the adoption of sustainable design of packaging by: – Embedding packaging sustainability in business strategy and process; – Enabling informed packaging design decisions through eco-design assessment; and – Ensuring packaging sustainability guidelines adapt and remain relevant to industry ad environmental needs.	Development of tools and resources to support and encourage industry performance: 1. Guide design, through the continuous evolution of packaging sustainability guidelines, eco-design tools, education and information, and 2. Encourage high performance, through awards and recognition schemes.	By 2020, 90 % of signatories have a long-term sustainability strategy in place that: – Incorporates the APC packaging sustainability guidelines; – Is integrated into business processes; and – Has clear and measurable targets By 2020, 60 % of signatories have achieved or exceeded sector-specific renewable and recycled content targets based on 2017 levels.
	B Circular economies: waste of resources	Find supply chain solutions to recover packaging and create sustainable closed-loop economies.	Bring industry together to identify the barriers to the recovery and reuse of waste packaging, and develop strategies to address and build innovation across industry sectors and the supply chain.	By 2020, APCO will have developed proven viable approaches to remove 50 % of current problem packaging types or materials from the waste stream. By 2020, 90 % of signatories will have actively participated in closed-loop collaboration.

22.6.3 Regulations governing food packaging materials in Africa

South Africa's food system plays a crucial role in the country's economic growth and food security. Despite this, the industry is confronted with numerous food safety concerns, including foodborne illnesses, food fraud, and inadequate regulation enforcement. The government's ineffective regulation of the food sector has increased the risk of food safety issues. A total of 978 laboratory listeriosis cases were reported in South Africa between January 2017 and March 2018, with 674 cases confirmed and a fatality rate of approximately 30 % [33], which is higher than that reported in the EU in 2015 [33].

Mycotoxins, particularly deoxynivalenol (DON), pose a significant threat to global agriculture and food industries. DON, a type B trichothecene mycotoxin produced in cereal commodities by *Fusarium crookwellense, Fusarium culmorum, and Fusarium graminearum* can cause oxidative damage, vomiting, digestive disorders, and reproductive toxicity if consumed in contaminated food [34].

22.6.4 Production of food based on cereals and wheat

The potential risk of exposure to DON among infants and young children was evaluated by examining the major mycotoxins found in maize grains sold in stores [34]. The results revealed that children in the savanna zone of northern Guinea faced a significantly high risk of DON exposure, with tolerable daily intake (TDI) levels exceeding the recommended limit of 1 µg/kg body weight (bw)/day by 63 and 25 times. Similarly, in Nigeria, children and young children who consumed contaminated maize were also susceptible to adverse health effects caused by mycotoxins [34]. The authors suggested promoting exclusive breastfeeding, particularly in the first six months of life, as well as incorporating other food ingredients, including alternative cereals such as millet and sorghum, which were less prone to mycotoxin contamination [34].

22.6.5 Beers

The degree of exposure to DON varies among countries and agroecological regions. Some African countries with more advanced industries have lower levels of DON contamination in raw materials compared to others. Nevertheless, the possibility of DON exposure for infants and children remains. Although many African countries have put into effect tactics to reduce mycotoxin incidence in food, some nations still do not have adequate prevention and control measures for mycotoxins in food [34].

22.6.6 Food packaging; reuse, reduce, and recycle

The amount of packaging waste ending up in South African landfills is on the rise due to a combination of increased consumption and a lack of recycling awareness [35]. This trend not only contributes to the depletion of valuable landfill space but also leads to the loss of recyclable resources like steel, glass, and plastics. These materials could be reprocessed into new products and reduce the demand for virgin raw materials [35].

Increasing recycling rates alone will not be enough to address the problem of packaging waste in South Africa. The country's growing demand for packaging, fueled by consumerism and increased purchasing power, means that unless efforts to reduce packaging quantities are also implemented, landfill sites will continue to receive increasing amounts of packaging waste. Therefore, a combination of regulations aimed at reducing packaging quantities and industry-led recycling initiatives will be necessary. However, achieving this will require political will, despite possible opposition to regulations, as well as awareness-raising efforts to encourage source separation among consumers [35].

22.7 Conclusions

In conclusion, food packaging materials play a critical role in protecting and preserving food products, but the use of certain materials can also pose risks to human health and the environment. Regulations for food packaging materials are necessary to ensure the safety and sustainability of the food industry. Different countries have developed their regulations and guidelines for food packaging materials, but challenges remain in terms of enforcement and monitoring of these regulations.

One of the challenges in regulating food packaging materials is balancing the need for safe and sustainable packaging with the demand for packaging in a growing consumer market. It is essential to reduce the amount of packaging material and increase recovery rates through recycling efforts, but this must be done in a way that does not compromise food safety or quality. Governments, industry stakeholders, and consumers all have a role to play in achieving this balance. Another challenge is the risk of food contamination from natural toxins. This highlights the importance of not only regulating food packaging materials but also addressing the wider issue of food safety and ensuring that measures are in place to prevent and detect contamination. In addition to legislation, industry actions and increasing consumer knowledge are critical in promoting sustainable packaging and minimizing the food industry's environmental effect. By working together, we can create a safer and more sustainable food system for future generations.

Looking to the future, regulations and industry initiatives will play a critical role in addressing the challenges of food packaging materials. Governments and regulatory bodies need to work together to develop effective and enforceable regulations that aim to reduce the number of packaging materials produced and improve recycling rates. Industry initiatives, including the development of more sustainable and eco-friendly packaging materials and increased efforts to promote source separation and recycling, will also be key to achieving these goals. Increased public awareness and education will be necessary in addition to these efforts. Consumers need to be informed about the impact of food packaging on the environment and the risks associated with certain materials, such as those containing harmful chemicals or toxins. Increased awareness-raising campaigns and educational programs can help to promote more sustainable consumption habits and encourage individuals to take action to reduce their environmental footprint. Overall, the issue of food packaging materials is a complex and multifaceted one, with many different factors at play. However, with a concerted effort from governments, industry, and consumers alike, it is possible to develop more sustainable and responsible approaches to food packaging that balance economic, environmental, and health considerations.

Acknowledgement: The authors are appreciative for the assistance, support and the facilities provided by Universiti Pertahanan Nasional Malaysia and Universiti Teknologi Malaysia throughout this study. The authors also would like to thank the editors R.A. Ilyas, S.M. Sapuan and Mohd Nor Faiz Norrrahim for their guidance and review of this article before its publication.

References

1. Marsh K, Bugusu B. Food packaging—roles, materials, and environmental issues. J Food Sci 2007;72: R39–55.
2. Kaiser K, Schmid M, Schlummer M. Recycling of polymer-based multilayer packaging: a review. Recycling 2018;3. https://doi.org/10.3390/recycling3010001.
3. Piergiovanni L, Limbo S. Plastic packaging materials. In: Food packaging materials. Basel, Switzerland: Springer; 2016:33–49 pp.
4. Yasim-Anuar TAT, Norrrahim MNF, Sapuan SM, Ilyas RA, Jenol MA, Mamat NA, et al. Polyhydroxyalkanoates for packaging application. In: Bio-based packaging: material, environmental and economic aspects. Hoboken, New Jersey: John Wiley; 2021:279–93 pp.
5. Silva FAGS, Dourado F, Gama M, Poças F. Nanocellulose bio-based composites for food packaging. Nanomaterials 2020;10:1–29.
6. Norrrahim MNF, Yasim-Anuar TAT, Sapuan SM, Ilyas RA, Hakimi MI, Najmuddin SUFS, et al. Nanocellulose reinforced polypropylene and polyethylene composite for packaging application. In: Bio-based packaging: material, environmental and economic aspects. Hoboken, New Jersey: Wiley Online Library; 2021.
7. Ilyas RA, Sapuan SM, Ibrahim R, Atikah MSN, Asyraf MRM, Norrrahim MNF, et al. Environmental advantages and challenges of bio-based packaging materials. In: Bio-based packaging: material, environmental and economic aspects. Hoboken, New Jersey: John Wiley; 2021:371–80 pp.
8. Ariffin H, Norrrahim MNF, Yasim-Anuar TAT, Nishida H, Hassan MA, Ibrahim NA, et al. Oil palm biomass cellulose-fabricated polylactic acid composites for packaging applications. In: Bionanocomposites for packaging applications. Switzerland: Springer; 2018:95–105 pp.
9. Norrrahim MNF, Ariffin H, Hassan MA, Ibrahim NA, Nishida H. Performance evaluation and chemical recyclability of a polyethylene/poly(3- hydroxybutyrate-co-3-hydroxyvalerate) blend for sustainable packaging. RSC Adv 2013;3:24378–88.
10. Coles R, McDowell D, Kirwan MJ. Food packaging technology. Boca Raton, Florida: CRC Press; 2003, 5.
11. Dıblan S, Kaya S. Antimicrobials used in active packaging films. Food Health 2018;4:63–79.
12. Akelah A. Polymers in food packaging and protection. In: Functionalized polymeric materials in agriculture and the food industry. US: Springer; 2013:293–347 pp.
13. Fang Z, Zhao Y, Warner RD, Johnson SK. Active and intelligent packaging in meat industry. Trends Food Sci Technol 2017;61:60–71.
14. Claudio L. Our food: packaging & public health. Environ Health Perspect 2012;120:A232–7.
15. Ipsos Sdn Bhd. The last straw: discouraging single use plastic [Internet]. Kuala Lumpur; 2019. Available from: www.ipsos.com/en-my.

16. Keller and Heckman. India announces new food packaging regulations | packaging. Law.com; 2019. https://www.packaginglaw.com/news/india-announces-new-food-packaging-regulations.
17. Yildirim S, Röcker B, Pettersen MK, Nilsen-Nygaard J, Ayhan Z, Rutkaite R, et al. Active packaging applications for food. Compr Rev Food Sci Food Saf 2018;17:165–99.
18. Restuccia D, Spizzirri UG, Parisi OI, Cirillo G, Curcio M, Iemma F, et al. New EU regulation aspects and global market of active and intelligent packaging for food industry applications. Food Control 2010;21:1425–35.
19. Han JH, Ho CHL, Rodrigues ET. Intelligent packaging. In: Innovations in food packaging. Amsterdam: Elsevier; 2005:138–55 pp.
20. Ozdemir M, Floros JD. Active food packaging technologies. Crit Rev Food Sci Nutr 2004;44:185–93.
21. Kerry JP, O'Grady MN, Hogan SA. Past, current and potential utilisation of active and intelligent packaging systems for meat and muscle-based products: a review. Meat Sci 2006;74:113–30.
22. Kruijf DD, Beest VV, Rijk R, Sipiläinen-Malm T, Losada PP, Meulenaer DD. Active and intelligent packaging: applications and regulatory aspects. Food Addit Contam 2002;19:144–62.
23. Vermeiren L, Devlieghere F, Van Beest M, De Kruijf N, Debevere J. Developments in the active packaging of foods. Trends Food Sci Technol 1999;10:77–86.
24. Curcio M, Puoci F, Iemma F, Parisi OI, Cirillo G, Spizzirri UG, et al. Covalent insertion of antioxidant molecules on chitosan by a free radical grafting procedure. J Agric Food Chem 2009;57:5933–8.
25. Restuccia D, Salomone R, Spizzirri UG, Saija G, Ioppolo G, Parisi OI, et al. Industrial applications. In: Barros-Velázquez J, editor. Antimicrobial food packaging. San Diego: Elsevier; 2016:215–27 pp.
26. Muncke J. Food packaging materials: Food Packaging Forum; 2012.
27. Department of Health and Human Services. Code of federal regulations title 21: U.S. Food & Drug Administration; 2019.
28. Food Packaging Forum. Food packaging materials | Food packaging materials [Internet]. https://www.foodpackagingforum.org/food-packaging-health/food-packaging-materials [Accessed 29 Mar 2023].
29. Heckman JH. Food packaging regulation in the United States and the European Union. Regul Toxicol Pharmacol 2005;42:96–122.
30. Australian Packaging Covenant Organisation. Australian Packaging Covenant Strategic Plan 2017–2022; 2019:1-16 pp.
31. Sothornvit R. Nanostructured materials for food packaging systems: new functional properties. Curr Opin Food Sci 2019;25:82–7.
32. Australian Packaging Covenant Organisation. Australian packaging covenant strategic plan 2017–2022-DCCEEW [Internet]. https://www.dcceew.gov.au/environment/protection/waste/publications/australian-packaging-covenant-strategic-plan-2017-2022 [Accessed 29 Mar 2023].
33. Boatemaa S, Barney M, Drimie S, Harper J, Korsten L, Pereira L. Awakening from the listeriosis crisis: food safety challenges, practices and governance in the food retail sector in South Africa. Food Control 2019;104: 333–42.
34. Wang L, Liao Y, Peng Z, Chen L, Zhang W, Nüssler AK, et al. Food raw materials and food production occurrences of deoxynivalenol in different regions. Trends Food Sci Technol 2019;83:41–52.
35. Mohanty CRC. Reduce, reuse and recycle and resource efficiency as the basis for sustainable waste management; 2011.

Isabella Cícera Dias Miranda, Jorge Alberto López and
Maria Lucila Hernández-Macedo*

23 Environmental advantages and challenges of nanocellulose reinforced starch-based packaging

Abstract: Sustainable packaging development with lower environmental impact is urgent due to the petroleum-based plastic accumulation. Thereby, nanocellulose arouses the interest of several industrial sectors as renewable sources material and its mechanical properties for packaging manufacture. Biopolymers such as starch have been an alternative in packaging preparation, although its low mechanical and water resistance represents a challenge for its application. An alternative to improve these starch characteristics is nanocellulose incorporation. Thus, this chapter proposes obtaining nanocellulose from renewable sources, and its environmental advantage to replace plastics. Furthermore, types of nanocellulose, methods for obtaining, and the main physicochemical characteristics of nanocellulose are discussed since this polymer versatility allows its application in several industrial sectors, especially in packaging development. Finally, this review presents the incorporation nanocellulose-based nanocomposites in matrices, aiming to obtain formulations containing active agents to improve mechanical strength and water resistance and packaged product quality.

Keywords: bionanocamposites; biopolymers; nanocellulose; packaging; starch.

23.1 Introduction

The high consumption of non-renewable source materials, such as petroleum-based packaging plastics, impact the environmental due to their low degradability, leading to implementing policies to replace and reduce this material consumption [1]. Thus, the search for sustainability, through the material development from renewable, biodegradable, and green

***Corresponding author: Maria Lucila Hernández-Macedo,** Industrial Biotechnology Graduation, Tiradentes University, Av. Murilo Dantas 300, 49032-490, Aracaju, SE, Brazil; and Molecular Biology Laboratory, Research and Technology Institute - ITP, Av. Murilo Dantas 300, Aracaju, SE, Brazil, E-mail: lucyherma@gmail.com. https://orcid.org/0000-0003-1050-9807
Isabella Cícera Dias Miranda, Industrial Biotechnology Graduation, Universidade Tiradentes, Av. Murilo Dantas 300, 49032-490, Aracaju, SE, Brazil; and Molecular Biology Laboratory, Research and Technology Institute - ITP, Av. Murilo Dantas 300, Aracaju, SE, Brazil
Jorge Alberto López, Multidisciplinary Research Laboratory, Health Sciences Center, Federal University of Rio Grande do Norte, R. Gen. G. Cordeiro de Farias, s/n, 59012-570, Natal RN, Brazil

As per De Gruyter's policy this article has previously been published in the journal Physical Sciences Reviews. Please cite as: I. C. D. Miranda, J. A. López and M. L. Hernández-Macedo "Environmental advantages and challenges of nanocellulose reinforced starch-based packaging" *Physical Sciences Reviews* [Online] 2023. DOI: 10.1515/psr-2022-0032 | https://doi.org/10.1515/9783110773606-023

sources to replace conventional plastics, has aroused a global scale interest, justifying studies to develop materials such as bioplastics mainly in the packaging industry [2].

At this point, several natural polymers or biopolymers are tested for bioplastic development. In this concern, cellulose from plant raw biomass displays several advantages such as renewability, biodegradability, compatibility, and physical and mechanical properties, making these materials potential substitutes for petroleum-derived inputs [3]. Cellulose is remarkably the most abundant renewable biological organic matter derived mainly from biomass, corresponding to >50 % of the planet's carbon content. Thus, cellulose as a sustainable natural polymer has been applied over millennia in the form of fiber to manufacture a wide range of materials and products (e.g., paper, textiles and packaging materials) [4].

Structurally, cellulose is constituted by repeating glucose units, covalently linked by acetal function, containing numerous repeating units of OH groups. These OH groups along the polymeric chains allow modifications upon interaction with other functional groups, resulting in a wide range of cellulose derivatives. Furthermore, the glucose unit reactivity is influenced by the steric effects of the reagents and by the supramolecular structure, whereby esterification, acylation, grafting, and etherification reactions have been applied to obtain cellulosic derivatives. Hence, this polysaccharide becomes a promising candidate for sustainable input production considering its availability, renewability, and biodegradability [5].

Natural cellulose possesses a unique hierarchical structural arrangement, consisting of nanofibrils and nanocrystallites, which can be extracted by mechanical, biological (microbial enzymes) and chemical processes, or by a combination of these techniques [6, 7]. Concerning isolated cellulose nanofibrils, these structures are long, thin, and flexible, containing alternating crystalline and amorphous domains. On the other hand, cellulose nanocrystals configure rod-shaped crystalline particles obtained after the division of amorphous domains. Still, in the context of structural arrangement, studies have reported other types of nanocellulose, such as amorphous nanocellulose and cellulose nanowires or electrospun nanofibers [4, 8].

Nanocellulose (NC) has attracted significant scientific attention, which focuses on different applications based on its availability, renewability, lightweight, nano-dimensionality, in addition to its unique morphology, particular physical and chemical properties, low thermal expansion coefficient, and low density [9, 10].

Renewable sources such as agro-industrial residues have shown potential for obtaining nanocellulose since this input is abundant and low cost. Thereby, studies have reported the extraction of NC from sugarcane bagasse and *Imperata brasiliensis* lignocellulosic fibers as suitable sources for obtaining nanocellulose. Also, its chemical, physical and thermal and eco-friendly properties indicated its potential applications in food packaging, plastic packaging, and container fabrication (Figure 23.1) [9, 11].

Biopolymers such as starch have been an alternative in packaging preparation due to their high availability and low cost. Notwithstanding, starch displays not suitable mechanical strength and barrier characteristics for its use as a bioplastic [12]. Thus,

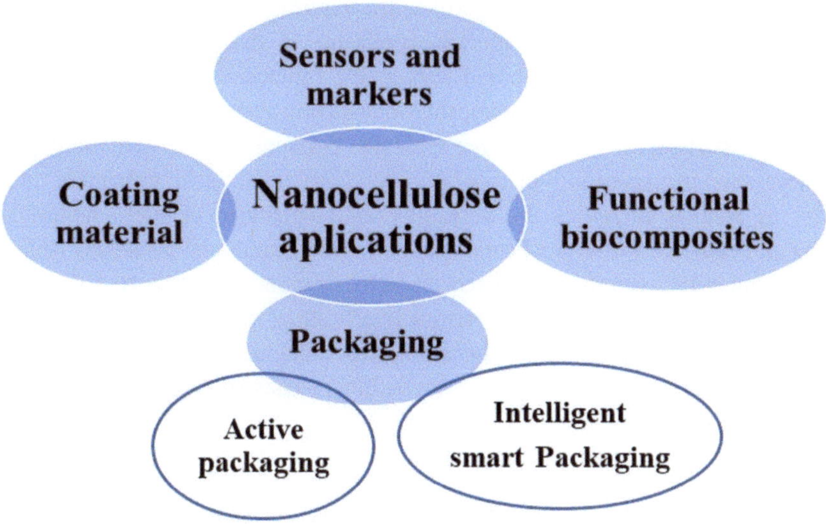

Figure 23.1: Applications of nanocellulose and its composites as reinforcement in polymer matrix composites to develop food packaging.

thermal treatments and the nanomaterial addition such as cellulose nanofibers (CNF) and nanocrystals (CNC) are proposed as a new approach to improve the polymeric film physicochemical properties used as bioplastics. The NC properties such as thermal resistance, mechanical performance, light material, high surface area, high tensile strength, low density, in addition to the good chemical interaction with biopolymers justify the potential use of these nanomaterials in the composition of polymeric matrices [13]. Furthermore, biopolymers exhibit a high potential for applications in several sectors, including food, electronics, agriculture, textiles, biomedical, pharmaceuticals, and cosmetics, representing an alternative to synthetic composites due to their sustainability [10]. Nowadays, knowledge and research about biodegradable polymers and their fiber-reinforced composites are a challenge concerning future opportunities and emerging applications of sustainable biopolymer composites as a suitable solution in the current environmental scenario [14].

Concerning biopolymer-based food packaging materials, the growing interest in nanocellulose is due to its many advantages, such as interface stability, high surface area, mechanical strength, surface chemical versatility, and easy modification to obtain derivatives, in addition to being a renewable raw material [15]. Thus, the nanocellulose mechanical and barrier properties allow its use as a reinforcing material to develop composites and other inputs for a wide range of applications. Hence, cellulose and its derivatives deserve attention in the production of reinforced nanocomposites and their incorporation into different materials, like natural matrices, forming the so-called hybrid composites [9]. This incorporation provides non-toxic materials with efficient tensile strength and water vapor permeability, which are important and distinctive features to developing flexible packaging materials applicable in the food sector. Overall, the implementation of biopolymer-based sustainable composites for food packaging applications represents an approach to improve food quality and safety by reducing food losses and spoilage, which is inserted in the context of environmental sustainability [14]. Furthermore, the incorporation of additives to these composites enhances their properties as well as increases the versatility and dynamism of these materials, enhancing and expanding their applications as active/intelligent food packaging [16].

In this chapter discusses the methods to obtain NC from renewable sources, its types, and the main physicochemical characteristics which allow its industrial application, mainly in packaging development. Furthermore, nanocellulose incorporation in starch formulations is discussed as a strategy to improve film mechanical strength and barrier properties.

23.2 Obtaining nanocellulose from renewable sources and its environmental advantage to replace plastics

The proper use of natural sources as input in product manufacture becomes an essential factor to establish sustainable industry and low environmental impact [17]. Nowadays, the world faces the challenge of excessive consumption of petroleum-based plastics, a light, resistant, and durable material widely used mainly in the food packaging industry, and electronic and medical devices. This consumption generates approximately 300 million tons of waste annually, of which only 7 % is recycled, and around 8 % is incinerated [18]. Thus, the conventional plastic low degradability associated with its non-renewable origin stimulate the search for renewable and biodegradable materials as an urgent alternative due to the environmental burden [19, 20].

In this scenario, an interesting resource is cellulose, the most abundant, biodegradable, and biocompatible biopolymer on earth, which can be extracted from different renewable sources, such as agricultural residues [21]. Notwithstanding, the lack of length and strength of its fibers for standard use can limit this biomass value. Hence, an attractive route to produce high value-added materials from this waste is to convert cellulose into nanocellulose (NC), which, due to size reduction <100 nm, displays new properties such as thermal resistance and mechanical performance [13], light material, high surface area, high tensile strength, and low density, giving it application as a renewable resource [22].

In the global context, the estimated growth of the NC market is estimated to be US$963 million up to 2026 due to its applicability and unique structural properties [7], driving the development of new research in the area nanotechnology and biodegradable and biocompatible materials, beneficial for the environment [23]. Pharmaceutical, food packaging, tissue engineering as well as composite technology industries including electronic devices are promising areas of NC application [23, 24]. Furthermore, the NC use as a reinforcing material incorporated in natural polymer matrices has been applied as a strategy to obtain nanocomposites with improved mechanical and barrier properties, compatible with those of synthetic polymers [25, 26].

23.2.1 Nanocellulose extraction from natural fibers

Nanocellulose is the most abundant cellulosic material extracted from plant fibers, which occurs as nanocrystal cellulose and nanofibrillated cellulose (Figure 23.2). Its efficient mechanical properties, low thermal expansion and high biodegradability have aroused biotechnological interest in technological and industrial applications [27]. However, its application depends on extraction methods and subsequent chemical treatments to improve the properties of NC and its composites [4].

Concerning nanocellulose extraction, several methods have been described, involving using residual lignocellulosic agro-industrial biomass due to the high cellulosic content and availability associated with the low cost of producing NC [9, 11, 28]. The NC extraction from lignocellulosic biomass requires a pre-treatment to eliminate non-cellulosic compounds (lignin, hemicellulose, pectin, and other compounds) before its extraction from cellulose fibrils by applying several procedures [4].

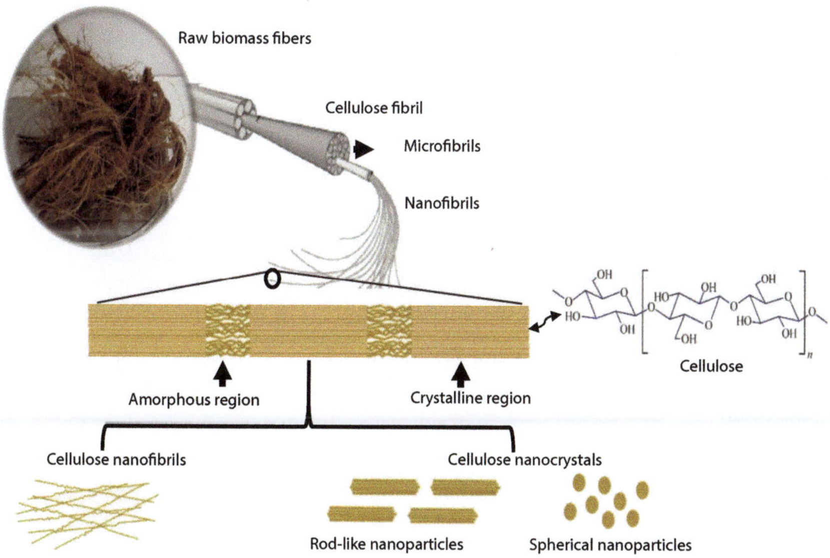

Figure 23.2: Scheme depicting the location and production of nanocellulose by enzymatic hydrolysis from raw plant biomass fibers.

The obtained nanocellulosic material characteristics depend on the extraction technique applied and its conditions with their consequent effects for an efficient yield and subsequent functionalization of the nanomaterial surface [4]. Regarding the selection of the extraction method for NC, some parameters must be considered such as porosity, crystallinity, surface area and chain length of the nanocellulose, as well as the lignocellulosic plant biomass [4]. Several biomasses, such as (*e.g.*, cereal and fruit husks, sisal, sugarcane bagasse, and cotton fibers) have been used to obtain NC due to their high quality and low content of non-cellulosic components. compared to wood, which, despite its abundance, requires a vigorous chemical and/or mechanical extraction due to its resistant and rigid structure [9, 11, 27]. Among the various procedures applied to isolate nanocellulose, the chemical, mechanical, and physical-chemical routes stand out, besides biological methods using enzymes [4].

The nanocellulose production depends on both extractive procedure and isolation steps, closely related to the non-cellulosic compound removal during pre-treatment, allowing intermolecular interactions and determining the final behavior and properties of the extracted material [4, 28]. According to Owonubi et al. [29], NC extraction can be achieved through the following steps (I) Pre-treatment (dewaxing, alkalinization, bleaching); (II). Depolymerization (chemical, mechanical, or biological methods); and (III). Isolation. Figure 23.3 depicts the flowchart of the steps involved in the NC extraction process. Furthermore, it is noteworthy that these extractive steps are closely associated with the type of lignocellulosic fibers used, which requires a previous compositional analysis of the fiber biomass before the extraction process [28].

Chemical (bases, acids, gases, oxidizing and reducing agents, and organic solvents), physical (grinding, high-intensity ultrasonication, and high-pressure homogenization), radiation (electron rays, γ rays, and microwaves), temperature (pyrolysis and steam

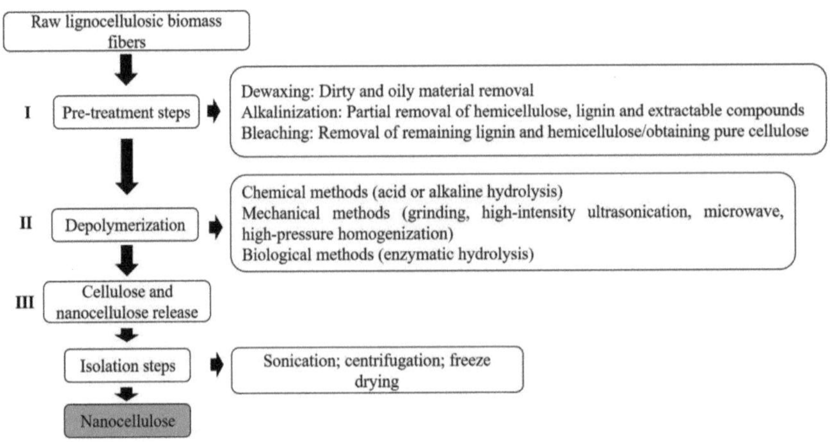

Figure 23.3: Flowchart showing the steps of the nanocellulose extraction process from untreated fibers. Adapted from [29].

explosion), and biological (microorganisms/enzymes) pre-treatments or combinations thereof have been applied to degrade lignin and hemicellulose, aiming to release the cellulose [4].

Towards physical treatment, this method alters the specific surface area of the biomass particle size, crystallinity index, or degree of biomass polymerization. Although they are eco-friendly methods for cleaving recalcitrant structures and improving cellulose digestibility without generating toxic compounds, these procedures are disadvantageous since they require the application of other methods to remove lignin [4].

A prominent physical method is microwave treatment, which applies non-ionizing electromagnetic radiation at the frequency of radio and infrared waves, whose low-energy vibration excites molecules without breaking their chemical bonds. The electric field of the microwave transfers the energy to the molecules, generating thermal energy. The advantages of microwave treatment are lower energy consumption and reaction times, in addition to cleaving recalcitrant structures and improving cellulose digestibility, which facilitates the application of other procedures to isolate nanocellulose [4].

On the other hand, chemical treatment increases the specific surface area of the substrate, moisturizing fibers, as well as modifying or removing hemicellulose and/or lignin, which improves the accessibility of cellulose. This pre-treatment uses diluted acids such as 0.1–0.7 % H_2SO_4 or HCl and/or alkaline solutions such as NaOH, KOH, and $Ca(OH)_2$ to de-structure the biomass. Nevertheless, the high chemical load, toxicity, environmental pollution, and high cost restrict the application of these methods on an industrial scale [4, 30].

The biological pre-treatment applying filamentous fungi producing ligninolytic enzymes or isolated enzymes is another procedure used to depolymerize lignin and obtain NC. This method requires long processing times and continuous monitoring of the process. Some of the advantages of the procedure are its recyclability, cleavage of lignin-carbohydrate linkers, relatively low cost, elimination of lignin and hemicellulose, low energy consumption, plus mild reaction conditions, and no release of hazardous and harmful compounds [30].

Despite the diversity of methods, the extraction efficiency of target compounds (cellulose, hemicellulose, and lignin), mainly nanocellulose, is improved by combining chemical and/or biological treatments associated with physical methods due to the rigidity of the plant wall [4].

23.3 Types of nanocellulose, methods for obtaining and the main physicochemical characteristics

Nanocellulose refers to cellulose particles with a size <100 nm, classified into three types called cellulose nanocrystals (CNC), cellulose nanofibers (CNF), and bacterial nanocellulose (NC) [31]. Although with similar in chemical composition, these structures differ in morphology, particle size, crystallinity, and extraction methods [32, 33].

Cellulose nanocrystals (CNC) are similar to rigid needles with dimensions 5–70 nm in width and 200–500 nm in length, exhibiting a sulfate group on their surface and variable morphology depending on the cellulose source [34]. CNCs stand out for their internal properties, such as nanodimension, high surface area and crystallinity, low density, high mechanical strength, unique morphology, availability, renewability, and biodegradability [25]. These characteristics allow its application as nanocomposite reinforcement, drug carrier, cosmetics, paints, coatings, and nanostructured adsorbents [31].

Cellulose nanofibers (NFC) are colloidal with a diameter of 5–50 nm and a length of a few micrometers containing crystalline and amorphous regions [34]. This nanomaterial has properties such as lightness, rigidity, non-toxicity, water insolubility, biodegradability, and renewability, making it attractive for different applications in the pharmaceutical, cosmetic, food, and polymer composite industries [35].

Bacterial nanocellulose (BNC) is composed of ultra-fine fibers from 20 to 100 nm in diameter, arranged in an ultra-fine network, with high purity and a higher polymerization degree than plant nanocellulose [36]. BNC is produced as exopolysaccharides by bacteria such as *Acetobacter*, *Achromobacter*, *Bacillus*, *Sarcina*, *Aerobacter*, *Agrobacterium*, *Escherichia*, *Azotobacter*, *Rhizobium*, *Enterobacter*, *Klebsiella*, *Salmonella*, and mainly by *Gluconacetobacter xylinus* [37]. Although it is a challenge due to the high capital investments, production cost, and lack of well-organized scale-up processes, in its low production, its properties have stimulated the development of products such as tissue engineering scaffolding, materials for dressings, dental implants, artificial blood vessels, surgical meshes [38], and biomedical applications [36].

Beside the NC types, the methods of obtaining it include chemical (bases, acids, and solvents), physical (high-pressure homogenization, mechanical grinding, ultrasonic treatment, and radiation), and biological or enzymatic (microorganisms) pre-treatments or a combination of them [6, 7]. Chemical methods are the most to obtain nanocellulose and it include alkaline treatment (NaOH), ionic liquids, eutectic liquids, bleaching treatment, and acid hydrolysis (sulfuric acid) [39]. Although, this method has long reaction time, low yield and high energy and corrosive chemicals consumption to obtain NC [40].

Physical methods such as ball mill, fibrilization, ultrasonication, and high-pressure homogenizer treatment are also commonly used for obtaining nanocellulose, however, this method would generally consume more energy in processing. In this sense, the combination of physical and chemical methods such as 2,2,6,6-tetramethylpiperidine-1-oxyl (TEMPO) -mediated oxidation to remove lignin and hemicellulose by sodium hydroxide, followed by Soxhlet extraction, acetic acid, sodium acetate, and sodium chlorite treatment has been proposed as a strategy to obtain NC [41].

Biological method by microbial enzymes has been proposed as an eco-friendly alternative to obtain NC. This method allows an accurately controlled of the structural characteristics and obtains a stable and high-quality nanocellulose [40]. However, biological hydrolysis requires a longer reaction, although a combination method can improve treatment time and yield [42].

With respect to BNC, no specific pre-treatments are required to obtain BNC since this type of cellulose is hemicellulose and lignin free compared to plant cellulose [43]. Table 23.1 presents different pre-treatment methods to obtain NC from natural fibers and microorganisms in the last five years.

23.4 Nanocelullose applied in the packaging development

The technology advancement has stimulated industrial interest in nanocellulose (NC) for develop innovative biomaterial. This interest is due to its versatility in exhibiting a unique combination of properties [13, 26, 56].

The NC structural characteristics allow its application to improve its mechanical and thermal properties in nanocomposites to obtain biodegradable products [26, 57]. These biomaterials have wide application in industrial areas, such as food packaging and coatings, pharmaceutical and cosmetic sectors, biomedical devices, capacitors, sensors, and environmental application [30, 58]. Thereby, studies reported the NC application to reinforce biomaterial, and also its chemical modification to improve mechanical and physical features and also provide antimicrobial and antioxidant nanocomposites [17]. Therefore, some perspectives of its incorporation into materials raise challenges,

Table 23.1: Extraction pre-treatment methods of cellulose nanofibers (NFC), cellulose nanocrystals (NCC), and bacterial cellulose.

Cellulosic substrate	Nanocellulose	Method	Ref.
Orange bagasse	NFC	Ionic liquids	[44]
Banana pseudo-stem	NFC	TEMPO-oxidacion/ultrasonic/acidolysis	[45]
Sugar beet	NFC	Chemical and enzymatic pre-treatment	[46]
Bagasse sugar cane	NFC	Enzymatic pre-treatment	[47]
Tea stalk	NCC	Chemical treatment (H_2O_2/acetic acid solution)	[48]
Pineapple leaf	NCC	Acid hydrolysis	[49]
Dietary fiber	NCC	Acid sulfuric	[50]
Oil palm empty fruit bunches	NCC	HCl (3 M/2 h/80 °C)	[51]
Komagataeibacter xylinus (BCC529) strain	BNC	Static culture (96 h/30 °C) Agitated culture (300 rpm/30 °C)	[52]
Komagataeibacter xylinus ATCC	BNC	Apple juice BOD (30 °C/10 days)	[53]
Komagataeibacter	BNC	Medium Hestrin–Schramm (HS) pectina cítrica/maça e 12,5 mM de $CaCl_2$	[54]
Gluconacetobacter hansenii UCP	BNC	Medium Hestrin–Schramm (HS)	[55]

targeting its potential industrial application as a functionalized raw material of added value [17, 30].

In the packaging context, nanocellulose use is the focus of studies, aiming at its technological applications [59]. Nowadays, NC emerges as a promising biomaterial due to its physicochemical properties to obtain high-added-value products and low environmental impact [60, 61]. Furthermore, the consumer's perception of the chemical product risks and the household waste accumulation influence the packaging innovation development [62].

Thereby, the innovative flow of technological knowledge reflects a trend in material search, such as nanocellulose, and their incorporation as scaffolding packing products [30, 63]. This is a strategy related to the packaging functionality and market requirements to preserve the product quality during transport, storage and shelf life [64]. Therefore, NC displays industrial potential, associated with the perception of natural and biodegradable raw materials to develop sustainable biomaterials [17, 54].

Despite its properties, nanocellulose is hydrophobic and non-thermostable, limiting its dissemination in food packaging manufacture [65]. Notwithstanding, structural chemical modifications can provide a hydrophobic character to obtain high value-added materials, which allows the NC application to develop packaging and coating inputs [66, 67].

Thus, research with modified nanocellulose have been conducted as a reinforcing agent by its incorporation into polymeric matrices such as starch to improve mechanical, barrier, and polythermal properties [12, 30, 68]. These NC-nanocomposite effects improve the performance of starch-based film packaging related to mechanical and barrier strength, required for food packaging [64, 69, 70].

With increasing consumer and industrial demands for eco-friendly materials, the development of sustainable or renewable polymeric materials has become mandatory [71]. Nevertheless, the mechanical performance of renewable polymer matrices is sometimes inferior compared to petroleum-based polymers since these polymer matrices are brittle and have low thermal distortion temperatures. In this concern, combining polymers with renewable reinforcements could improve the performance of the mechanical and physical-chemical properties of these matrices, generating composites of polymers reinforced with natural fibers [72]. Therefore, nanocellulosic material incorporation is one of the best methods used to increase the physical and chemical properties of different polymeric matrices and, consequently, their functionalization aiming at the material development and their application [73].

Studies have reported the use of nanocellulose as reinforcement in polymer matrix composites to improve their functionality. Thus, various bionanocomposites applied in food packaging materials have been produced using NC as nanofillers and natural polymer matrixes, such as starch, pullulan, poly (vinyl alcohol) (PVA), and gelatin used in food packaging materials. Although these matrices are widely used for their ability to form films, some exhibit low mechanical strength and water vapor permeability, limiting their application as a packaging material [72, 74, 75]. Hence, the nanocellulose incorporation leads to composites with better mechanical performance since the nanocellulose use as reinforcement in polymeric matrix composites is a strategy to develop polymeric composites with high performance. For this reason, nanocellulose is often considered the next-generation renewable reinforcement for the production of high-performance biocomposites [76].

Active packaging development is another approach as a barrier to improving packaged food conservation against environmental conditions [77–79]. Hence, the packaging industry is focused on developing packaging with the ability to absorb or release substances to extend the food shelf life compared to conventional packaging. Therefore, the research of multifunctional biodegradable polymeric materials as active agents is crucial to reducing environmental impact and improving product quality [59]. NC is a support to develop antioxidant and antimicrobial agents for active-based packaging, incorporated as functional resources to improve their performance concerning product quality and safety [16, 62].

Regarding antimicrobial capacity, nanocellulose-titanium dioxide nanocomposites offer physicochemical and functional properties [16]. Its incorporation into polymeric matrices such as starch confers antioxidant and antimicrobial effects, and hydrophobic barrier, expanding its use as a coating in the food field [80]. These features also indicate the excellent potential of this nanomaterial for industrial applications [81].

NC also displays significant attention as a material for biosensing applications. This approach allows specific functionalization by modifying its surface area to develop biocomposites used as sensing systems and specific interaction markers capable of detecting chemicals without influencing the NC particular properties [82]. These biosensing composites can be incorporated into the packaging to provide information about packaged foods conditions (*e.g.*, temperature, humidity, and gas concentrations) to ensure product quality and safety during storage and transport [82, 83].

Likewise, other studies have evaluated the NC application to develop nanocomposites with UV-barrier properties and oxygen transmission rate control. These nanocomposites exhibit potential use in both active packaging and smart food packaging as moisture sensors to preserve products susceptible to microbial and oxidative spoilage [83].

Concerning UV shielding, it is an important strategy in active packaging to reduce food photodegradation and prolong its shelf life. Although NC displays a low UV blocking ability, the cellulose-based nanocompound synthesis containing inorganic UV absorbers such as Ag, ZnO, TiO2, and ZnS is the subject of research by its incorporation into packaging polymeric matrices [84].

23.5 Biocomposites

Composites are engineering materials made from materials with different properties, resulting in a structure that incorporates the superior characteristics of each material. Nowadays, biocomposite has become a widely used term to designate nanomaterials based on biodegradable materials, encompassing naturally occurring polymers incorporated with an inorganic portion. Based on their size, these materials are defined on the nanometer scale as particles comprising dimensions ranging from 1–100 nm [85]. Furthermore, nanocomposites with an inorganic nanoportion in their structure belong to a class of materials commonly called bionanocomposites, although the terms green composites and biohybrids are also applied [86].

The nanostructure dimensionality is the pivotal factor for its different properties, such as thermal stability, water solubility, biocompatibility, and biodegradability, which determine its functionalities and areas of application. In biocomposites, materials with different properties are combined forming a composite, whose individual constituents remain interconnected through physical cooperation, which allows them to maintain their physical or composite properties. Thus, the resulting nanostructures display unique characteristics in many aspects compared to their individual components, such as physical and thermal properties [87].

Biocomposites, in addition to reducing synthetic fiber use, are biodegradable and recyclable and their mechanical properties (*e.g.*, strength, elastic modulus, and dimensional stability) and the hydrophilic nature of the reinforcement can be improved according to the preparation method [88]. These properties play an important role in the functionality, format, and field of nanomaterial application. Moreover, bionanocomposites differ from biocomposites since the latter can be constituted by biopolymers without having nanometric additives in their structure [87].

23.6 Use of starch in bionanocomposites formulations containing nanocellulose to improve mechanical strength and water resistance

The use of biopolymers to prepare different materials for packaging or medical and pharmaceutical applications is largely discussed in scientific publications. The growing interest in biopolymers in these application fields is due to their biocompatibility, biodegradability, and to reduce environmental burden resulting from the conventional plastic consumption [2].

Biopolymers are chain-like molecules composed of repeating chemical units produced or derived from living organisms, exhibiting renewable and environmentally degraded characteristics [89]. Biopolymers such as starch, cellulose, pectin, chitosan, alginate, casein, collagen, and gelatin display high application potential in film and composite preparation as characteristics “green” or environmentally friendly materials [90].

Cellulose and starch are the most abundant biopolymers in nature, starch is low cost and allows for its wide use in the film elaboration. Starch is composed of two polysaccharides, amylose and amylopectin, both used as energy reserves in plants and are accumulated mainly in roots, seeds, and tubers. The variation in the proportion of amylose and amylopectin depends on the starch plant origin that including maize (73 %), tapioca (11 %) wheat (9 %), and potatoes (6 %); used mainly as food [91]. Furthermore, this biopolymer is also explored in the paper, pharmaceutical, and packaging industries.

Industrially, the starch is classified into three main groups native starch (odorless and without any treatment after its extraction); modified starch (starch altered or modified with physical, chemical, enzymatic, or combination method treatment); and hydrolyzed starch, (whose polymeric chains are broken into simple sugars like maltose) [92].

The native starch use is challenging in film and packaging formulations due to its low mechanical strength and sensitivity to moisture. Currently, this problem is overcome by applying a heat treatment between 90 and 180 °C associated with a plasticizer, usually, glycerol forming the so-called thermoplastic starch (TPS) that improves the starch physical and mechanical properties [12]. Other polymers such as polylactic acid (PLA) and polybutylene succinate (PBS) have been used in association with starch to ameliorate its mechanical strength and sensitive to moisture [67]. Despite these improving, synthetic films and plastics display superior properties compared to bioplastic [12].

In this regard, nanotechnology studies have been carried out to solve the weak mechanical properties and the high permeability to water vapor of biopolymers through of the development nanocomposites preparation, or biocomposites with desirable physicochemical attributes. Hence, biomaterials are formulated into biocomposite materials containing cellulose nanofibers (CNF) as an alternative to reinforce the bioplastic's mechanical and barrier properties [93]. The CNF incorporation into starch-based biocomposites, for example, forms bionanocomposites are considered a new generation of composites combining biopolymers to improve the mechanical, thermal, and functional properties of the material by forming an environment of nanocross or nanofiller linker between its molecules. The CNF used as bionanocomposite preparation strategy is due to its nanometric size, high crystallinity, and chemical structure that act with starch, improving their film properties [94].

The NCF-starch chemical interaction occur through strong hydrogen bonds since starch is constituted by amylose, a linear chain stabilized by hydrogen bonds, and amylopectin, a branched polymeric chain of shorter length, forming inter-chain hydrogen bonds. This combination allows the obtention of promising bionanocomposite, which is biodegradable, safe, and sustainable [95].

The amylose content is important for the starch structural stability since the linear chains influence its tensile strength and elongation, and allow the formation of biodegradable films [96]. Overall, the mechanical properties requirements for packing-based-bioplastic are good tensile strength (TS), adequate elongation, flexibility, and resistance to drilling to provide a physical barrier against external agents such as gas and moisture [97]. The measure TS expresses the maximum deformation of the film to the breaking point, while the elongation percentage represents the increase in film length after applying opposing surface forces [98]. Regarding the drilling resistance, it is the result of the maximum film deformation values up to the breaking point, considering the maximum force exerted on a known area of the film surface [99].

The crystallinity of both cellulose nanocrystals and cellulose nanofibers allows them a modulus of elasticity of 85 and 65 J/g^{-1}, respectively, higher than steel with approximately 25 $J\ g^{-1}$ [100]. Thus, these physicochemical characteristics of nanocellulose can achieve higher properties for nanocomposites which can be improved by adjusting the proportions nanocellulose: polymer matrix used for the film manufacture [101].

23.7 Conclusions

Petroleum-based plastic packaging is responsible for an environmental burden, and its replacement with bioplastics from biopolymers such as starch is an alternative to solve this problem. Starch-based bioplastics have low cost and excellent biodegradability. Furthermore, starch is renewable and its chemical structure allows interactions with other molecules. Notwithstanding, its low mechanical strength and barrier properties are a challenge for its application to replace conventional plastics. Thus, nanocellulose as biological nanomaterial displays promising results for improving the starch-based film properties. Both cellulose nanofibers and cellulose nanocrystals interact with amylose by hydrogen bonds, improving bioplastic mechanical proprieties. Furthermore, nanocellulose is a low-cost input extracted from cellulose, a renewable source, including lignocellulosic residues. Despite improving bioplastic barrier, mechanical, and tensile properties, the accuracy and quantification of nanocellulose interaction with the polymer starch matrix become a challenge due to restrictions of the current analytical methods available.

References

1. Di Bartolo A, Infurna G, Dintcheva NT. A review of bioplastics and their adoption in the circular economy. Polymers 2021;13:1229.
2. Rai P, Mehrotra S, Priya S, Gnansounou E, Sharma SK. Recent advances in the sustainable design and applications of biodegradable polymers. Bioresour Technol 2021;325:124739.
3. Aziz T, Farid A, Haq F, Kiran M, Ullah A, Zhang K, et al. A Review on the modification of cellulose and its applications. Polymers 2022;14:3206.
4. Kargarzadeh H, Ioelovich M, Ahmad I, Thomas S, Dufresne A. Methods for extraction of nanocellulose from various sources. In: Kargarzadeh H, Ahmad I, Thomas S, Dufresne A, editors. Handbook of nanocellulose and cellulose nanocomposites. Germany: Wiley VCH; 2017. Chapter 1.
5. Etale A, Onyianta AJ, Turner SR, Eichhorn SJ. Cellulose: a review of water interactions, applications in composites, and water treatment. Chem Rev 2023;123:2016–48.
6. Teo HL, Wahab RA. Towards eco-friendly deconstruction of agro-industrial biomass and preparation of renewable celulose nanomaterials: a review. Int J Biol Macromol 2020;15:1414–30.
7. Dhali K, Ghasemlou M, Daver F, Cass P, Adhikari. A review of nanocellulose as new material toward environmental sustainability. Sci Total Environ 2021;25:775.

8. Nagarajan KJ, Ramanujam NR, Sanjay MR, Siengchin S, Rajan BS, Sathick Basha K, et al. A comprehensive review on cellulose nanocrystals and cellulose nanofibers: pretreatment, preparation, and characterization. Polym Compos 2021;42:1588–630.
9. Gond RK, Gupta MK, Jawaid M. Extraction of nanocellulose from sugarcane bagasse and its characterization for potential applications. Polym Compos 2021;42:5400–12.
10. Khalid MY, Arif ZU, Sheikh MF, Nasir MA. Mechanical characterization of glass and jute fiber-based hybrid composites fabricated through compression molding technique. Int J Material Form 2021;14:1085–95.
11. Benini KCCC, Voorwald HJC, Cioffi MOH, Milanese AC, Ornaghi AL Jr. Characterization of a new lignocellulosic fiber from Brazil: *Imperata brasiliensis* (Brazilian satintail) as an alternative source for nanocellulose extraction. J Nat Fibers 2017;14:112–25.
12. Nazrin A, Sapuan MS, Zuhri MYM, Ilyas RA, Syafiq R, Sherwani SFK. Nanocellulose reinforced thermoplastic starch (TPS), polylactic acid (PLA), and polybutylene succinate (PBS) for food packaging applications. Front Chem 2020;8:213.
13. Kaur P, Sharma N, Munagala M, Rajkhowa R, Aallardyce B, Shastri Y, et al. Nanocellulose: resources, physio-chemical properties, current uses and future applications. Front Nanotechnol 2021;3:747329.
14. Arif ZU, Khalid MY, Sheikh MF, Zolfagharian A, Bodaghi M. Biopolymeric sustainable materials and their emerging applications. J Environ Chem Eng 2022;10:108159.
15. Khalid MY, Arif ZU. Novel biopolymer-based sustainable composites for food packaging applications: a narrative review. Food Packag Shelf Life 2022;33:100892.
16. Abdalkarim SYH, Chen L-M, Yu H-Y, Li F, Chen X, Zhou Y, et al. Versatile nanocellulose-based nanohybrids: a promising-new class for active packaging applications. Int J Biol Macromol 2021;182:1915–30.
17. Khalid MY, Al Rashid A, Arif ZU, Ahmed W, Arsha H. Recent advances in nanocellulose-based different biomaterials: types, properties, and emerging applications. J Mater Res Technol 2021;14:2601–23.
18. Atiwesh G, Mikhael A, Parrish CC, Banoub J, Le TAT. Environmental impact of bioplastis use: a review. Heliyon 2021;7:e07918.
19. Yates J, Deeney M, Rolker HB, White H, Kalamatianou S, Kadiyala S. A systematic scoping review of environmental, food security and health impacts of food system plastics. Nat Food 2021;2:80–7.
20. Zhang Y, Duan C, Bokka SH, He Z, Ni Y. Molded fiber and pulp products as green and sustainable alternatives to plastics: a mini review. J Bioresour Bioprod 2022;7:14–25.
21. Freitas LC, Barbosa JR, Costa ALC, Bezerra FWF, Pinto RHH, Carvalho Junior RN. From waste to sustainable industry: how can agro-industrial wastes help in the development of new products? Resour Conserv Recycl 2021;169:105466.
22. Zubair M, Arshad M, Ullah A. Nanocellulose: a sustainable and renewable material for water and wasterwater treatment. In: Kalia S, editor. Natural polymers-based green adsorbents for water treatment. Chennai, India: Elsevier; 2021. Chapter 4.
23. Wang Q, Yao Q, Liu J, Sun J, Zhu Q, Chen H. Processing nanocelulose to bulk materials: a review. Cellulose 2019;26:7585–617.
24. Raghav N, Sharma MR, Kennedy JF. Nanocellulose: a mini-review on types and use in drug delivery systems. Carbohydr Polym Technol Appl 2021;2:100031.
25. Chen Q, Liu Y, Chen G. A comparative study on the starch-based biocomposite films reinforced by nanocellulose prepared from different non-wood fibers. Cellulose 2019;26:2425–35.
26. Thakur V, Guleria A, Kumar S, Sharma S, Singh K. Recent advances in nanocellulose processing, functionalization and applications: a review. Mater Adv 2021;2:1872–95.
27. Tahir D, Karim MRA, Hu H, Naseem S, Rehan M, Ahmad M, et al. Sources, chemical functionalization, and commercial applications of nanocellulose and nanocellulose-based composites: a review. Polymers 2022; 14:4468.
28. Norizan MN, Shazleen SS, Alias AH, Sabaruddin FA, Asyraf MRM, Zainudin ES, et al. Nanocellulose-based nanocomposites for sustainable applications: a review. Nanomaterials 2022;12:3483.

29. Owonubi SJ, Agwuncha SC, Malima NM, Shombe GB, Makhatha EM, Revaprasadu N. Non-woody biomass as sources of nanocellulose particles: a review of extraction procedures. Front Energy Res 2021;9:608825.
30. Trache D, Tarchoun AF, Derradji M, Hamidon TS, Masruchin N, Brosse N, et al. Nanocellulose: from fundamentals to advanced applications. Front Chem 2020;8:392.
31. Vincent S, Kandasubramanian B. Cellulose nanocrystals from agricultural resources: extraction and functionalization. Eur Polym J 2021;160:110789.
32. Blanco A, Monte MC, Campano C, Balea A, Merayo N, Negro C. Nanocellulose for industrial use: cellulose nanofibers (CNF), cellulose nanocrystals (CNC) and bacterial cellulose (BC). In: Hussain CM, editor. Handbook of Nanomaterials for Industrial Applications. Chennai, India: Elsevier; 2018. Chapter 5.
33. Pradhan D, Jaiswal AK, Jaiswal S. Emerging technologies for the production of nanocellulose from lignocellulosic biomass. Carbohydr Polym 2022;285:119258.
34. Joseph B, Sagarika VK, Sabu C, Kalarikkal N, Thomas S. Cellulose nanocomposites: fabrication and biomedical applications. J Bioresour Bioprod 2020;5:223–37.
35. Raza M, Abu-Jdayil B. Cellulose nanocrystals from lignocellulosic feedstock: a review of production technology and surface chemistry modification. Cellulose 2022;29:685–722.
36. Barja F. Bacterial nanocellulose production and biomedical applications. J Biomed Res 2021;35:310–7.
37. Sharma C, Bhardwaj NK. Bacterial nanocellulose: present status, biomedical applications and future perspectives. Mater Sci Eng C 2019;104:109963.
38. Choi SM, Shin EJ. The nanofication and functionalization of bacterial cellulose and its applications. Nanomaterials 2020;10:406.
39. Moriana R, Vilaplana F, Ek M. Cellulose nanocrystals from forest residues as reinforcing agents for composites: a study from macro- to nano-dimensions. Carbohydr Polym 2016;139:139–49.
40. Yu S, Sun J, Shi Y, Wang Q, Wu J, Liu J. Nanocellulose from various biomass wastes: its preparation and potential usages towards the high value-added products. Environ Sci Ecotech 2021;5:100077.
41. Pascoli DU, Dichiara A, Roumeli E, Gustafson R, Bura R. Lignocellulosic nanomaterials production from wheat straw via peracetic acid pretreatment and their application in plastic composites. Carbohydr Polym 2022;295:119857.
42. Gupta GK, Shukla. Lignocellulosic biomass for the synthesis of nanocellulose and its eco-friendly advanced applications. Front Chem 2020;8:601256.
43. Oun AA, Shankar S, Rhim JW. Multifunctional nanocellulose/metal and metal oxide nanoparticle hybrid nanomaterials. Crit Rev Food Sci Nutr 2020;60:435–60.
44. Menezes DB, Diz FM, Ferreira LFR, Corrales Y, Baudrit JRV, Costa LP, et al. Starch-based biocomposite membrane reinforced by orange bagasse cellulose nanofibers extracted from ionic liquid treatment. Cellulose 2021;28:4137–49.
45. Meng F, Wang G, Du X, Wang Z, Xu S, Zhang Y. Extraction and Characterization of cellulose nanofibers and nanocrystals from liquefied banana pseudo-stem residue. Compos B Eng 2019;160:341–7.
46. Tao P, Wu Z, Xing C, Zhang Q, Wei Z, Nie S. Effect of enzymatic treatment on the termal stability of cellulose nanofibrils. Cellulose 2019;26:7717–25.
47. Perzon A, Jorgensen B, Ulvskov P. Sustainable production of celullose nanofiber gels and paper from sugar beet waste using enzymatic pre-treatment. Carbohydr Polym 2020;230:115181.
48. Guo Y, Zhang Y, Zheng D, Li M, Yue J. Isolation and characterization of nanocellulose crystals via acid hydrolysis from agricultural waste-tea stalk. Int J Biol Macromol 2020;163:927–33.
49. Santos RM, Neto WPF, Silvério HA, Martins DF, Dantas NO, Pasquini D. Cellulose nanocrystals from pineapple leaf, a new approach for the reuse of this agro-waste. Ind Crops Prod 2013;50:707–14.
50. Ma Y, Chai X, Bao H, Huang Y, Dong W. Study on nanocellulose isolated from waste from chilli stems processing as dietary fiber in biscuits. PLoS One 2023;18:e0281142.
51. Hastuti N, Kanomata K, Kitaoka T. Hydrochloric acid hydrolysis of pulps from oil palm empty fruit bunches to produce cellulose nanocrystals. J Polym Environ 2018;26:3698–709.

52. Gao H, Sun Q, Han Z, Li J, Liao B, Hu L, et al. Comparison of bacterial nanocellulose produced by diferentt strains under static and agitated culture conditions. Carbohydr Polym 2020;227:115323.
53. Silva SMF, Ribeiro HL, Mattos ALA, Borges MF, Rosa MF, Azeredo HMC. Films from cashew byproducts: cashew gum and bacterial cellulose from cashew apple juice. J Food Sci Technol 2021;58:1979–86.
54. Kaczmarek M, Jedrzejczak M, Ludwicka. Comparative analysis of bacterial cellulose membranes synthesized by chosen *Komagataeibacter* strains and their application potential. Int J Mol Sci 2022;23:3391.
55. Albuquerque RMB, Meira HM, Silva IDL, Silva CJG, Almeida FCGA, Amorim JDP, et al. Production of a bacterial cellulose/poly(3-hydroxybutyrate) blend activated with clove essential oil for food packaging. Polym Polym Compos 2021;29:259–70.
56. Noremylia MB, Zaki MZ, Ismail Z. Recent advancement in isolation, processing, characterization and applications of emerging nanocellulose: a review. Int J Biol Macromol 2022;206:954–76.
57. Ghasemlou M, Daver F, Ivanova EP, Habibi Y, Adhikari B. Surface modifications of nanocellulose: from synthesis to high-performance nanocomposites. Prog Polym Sci 2021;119:101418.
58. Hosakun W, Tsalagkas D, Csóka L. Nanocellulose-based composites for environmental applications: a review. In: Deans M, editor. Innovation in nano-polysaccharides for eco-sustainability. Chennai, India: Elsevier; 2022. Chapter 15.
59. Ahankari SS, Subhedar AR, Bhadauria SS, Dufresne A. Nanocellulose in food packaging: a review. Carbohydr Polym 2021;255:117479.
60. Piccinno F, Hischier R, Seeger S, Som C. Predicting the environmental impact of a future nanocellulose production at industrial scale: application of the life cycle assessment scale-up framework. J Clean Prod 2018;174:283–95.
61. Mu R, Hong X, Ni Y, Li Y, Pang J, Wang Q, et al. Recent trends and applications of cellulose nanocrystals in food industry. Trends Food Sci Technol 2019;93:134–44.
62. Al-Obadi M, Ayad H, Pokharel S, Ayari MA. Perspectives on food waste management: prevention and social innovations. Sustain Prod Consum 2022;31:190–208.
63. Milanez DH, Amaral RM, Faria LIL, Gregolin JAR. Assessing nanocellulose developments using science and technology indicators. Mater Res 2013;16:635–41.
64. Salgado PR, Di Giorgio L, Musso YS, Mauri NA. Recent developments in smart food packaging focused on biobased and biodegradable polymers. Front Sustain Food Syst 2021;5:630393.
65. Azeredo HMC, Rosa MF, Mattoso LHC. Nanocellulose in bio-based food packaging applications. Ind Crop Prod 2017;97:664–71.
66. Fotie G, Limbo S, Piergiovanni L. Manufacturing of food packaging based on nanocellulose: current advances and challenges. Nanomaterials 2020;10:1726.
67. Shen R, Xue S, Xu Y, Liu Q, Feng Z, Ren H, et al. Research progress and development demand of nanocellulose reinforced polymer composites. Polymers 2020;12:2113.
68. Perumal AB, Nambiar RB, Moses JA, Anandharamakrishnan C. Anandharamakrishnana. Nanocellulose: recent trends and applications in the food industry. Food Hydrocolloids 2022;127:107484.
69. Noorbakhsh-Soltani SM, Zerafat MM, Sabbaghi S. A comparative study of gelatin and starch-based nano-composite films modified by nano-cellulose and chitosan for food packaging applications. Carbohydr Polym 2018;189:48–55.
70. Trivedi R, Prajapati P. Nanocellulose in packaging industry. In: Oraon R, Rawtani D, Singh P, Hussain CM, editors. Nanocellulose materials. New Delhi, India: Elsevier; 2022. Chapter 3.
71. Chakrabarty A, Teramoto Y. Recent advances in nanocellulose composites with polymers: a guide for choosing partners and how to incorporate them. Polymers 2018;10:517.
72. Peng N, Huang D, Gong C, Wang Y, Zhou J, Chang C. Controlled arrangement of nanocellulose in polymeric matrix: from reinforcement to functionality. ACS Nano 2020;14:16169–79.
73. Martínez-Barrera G, Garduño-Jaimes IZ, Vigueras-Santiago E, Cruz-Olivares J, González-Rivas N, Gencel O. Green composites from sustainable cellulose nanofibrils. In: Thomas S, Balakrishnan P, editors. Green composites. Singapore: Springer; 2021.

74. Lee K-Y, Aitomäki Y, Berglund LC, Oksman K, Bismarck A. On the use of nanocellulose as reinforcement in polymer matrix composites. Compos Sci Technol 2014;105:15–27.
75. Ding Z, Chang X, Fu X, Kong H, Yu Y, Xu H, et al. Fabrication and characterization of pullulan-based composite films incorporated with bacterial cellulose and ferulic acid. Int J Biol Macromol 2022;219: 121–37.
76. De S, James B, Ji J, Wasti S, Zhang S, Kore S, et al. Biomass-derived composites for various applications. Adv Bioenergy 2023. https://doi.org/10.1016/bs.aibe.2023.01.001. In press.
77. Anjos HA, Luna S, Hernández-Macedo ML, López JA. Antimicrobial and antioxidant active food packaging: technological and scientific prospection. Recent Pat Biotechnol 2020;14:99–111.
78. Firouz MS, Mohi-Alden K, Omid M. A critical review on intelligent and active packaging in the food industry: research and development. Food Res Int 2021;141:110113.
79. Atta OM, Manan S, Shahzad A, Ul-Islam M, Ullah MW, Yang G. Biobased materials for active food packaging: a review. Food Hydrocolloids 2022;125:107419.
80. Cherian RM, Tharayil A, Varghese RT, Antony T, Kargarzadeh H, Chirayil CJ, et al. A review on the emerging applications of nano-cellulose as advanced coatings. Carbohydr Polym 2022;288:119123.
81. Darwish MSA, Mostafa MH, Al-Harbi LM. Polymeric nanocomposites for environmental and industrial applications. Int J Mol Sci 2022;23:1023.
82. Zinge C, Kandasubramanian B. Nanocellulose based biodegradable polymers. Eur Polym J 2020;133: 109758.
83. Pirsa S, Sani IK, Mirtalebi SS. Nano-biocomposite based color sensors: investigation of structure, function, and applications in intelligent food packaging. Food Packag Shelf Life 2022;32:100789.
84. Farooq A, Patoary MK, Zhang M, Mussana H, Li M, Naeem MA, et al. Cellulose from sources to nanocellulose and an overview of synthesis and properties of nanocellulose/zinc oxide nanocomposite materials. Int J Biol Macromol 2020;154:1050–73.
85. Lee CH, Lee SH, Padzil FNM, Ainun ZMA, Norrrahim MNF, Chin KL. Biocomposites and nanocomposites. In: Sachdeva A, Singh PK, Rhee HW, editors. composite materials. Boca Raton: CRC Press; 2021.
86. Pande VV, Sanklecha VM. Bionanocomposite: a review. Austin J Nanomed Nanotechnol 2017;5:1045.
87. Arora B, Bhatia R, Attri P. Bionanocomposites: green materials for a sustainable future. In: Hussain CM, Mishra AK, editors. New polymer nanocomposites for environmental remediation. India: Elsevier; 2018.
88. Khalid MY, Arif ZU, Al Rashid A. Investigation of tensile and flexural behavior of green composites along with their impact response at different energies. Int J Precis Eng Manuf.-Green Tech 2022;9:1399–410.
89. Mohiuddin M, Kumar B, Haque S. Biopolymer composites in photovoltaics and photodetectors. In: Sadasivuni KK, Ponnamma D, Kim J, Cabibihan JJ, AlMaadeed MA, editors. Biopolymer composites in electronics. Amsterdam: Elsevier; 2017. Chapter 17.
90. Lisitsyn A, Semenova A, Nasonova V, Polishchuk E, Revutskaya N, Kozyrev I, et al. Approaches in animal proteins and natural polysaccharides application for food packaging: edible film production and quality estimation. Polymers 2021;13:1592.
91. Bashir K, Aggarwal M. Physicochemical, structural and functional properties of native and irradiated starch: a review. J Food Sci Technol 2019;56:513–23.
92. Chung HJ, Lee SY, Kim JH, Lee JW, Byun MW, Lim ST. Pasting characteristics and in vitro digestibility of γ-irradiated RS_4 waxy maize starches. J Cereal Sci 2010;52:53–8.
93. Ilyas RA, Sapuan SM, Ibrahim R, Abral H, Ishak MR, Zainudin ES, et al. Effect of sugar palm nanofibrillated cellulose concentrations on morphological, mechanical and physical properties of biodegradable films based on agro-waste sugar palm (*Arenga pinnata* (Wurmb.) Merr) starch. J Mater Res Technol 2019;8: 4819–30.
94. Pattanashetti NA, Heggannavar GB, Kariduraganavar MY. Smart biopolymers and their biomedical applications. Procedia Manuf 2017;12:263–79.

95. Gopi S, Amalraj A, Jude S, Thomas S, Qipeng G. Bionanocomposite films based on potato, tapioca starch and chitosan reinforced with cellulose nanofiber isolated from turmeric spent. J Taiwan Inst Chem Eng 2019;96:664–71.
96. Wang HJ, An DS, Rhim JW, Lee DS. A multi-funcional biofilm used as an active insert in modified atmosphere packaging for fresh produce. Packag Technol Sci 2015;28:999–1010.
97. Martins VG, Romani VP, Martins PC, Filipini GS. Innovative packaging that saves food. In: Galanakis CM, editor. Saving food. Chennai, India: Academic Press; 2019. Chapter 6.
98. Ghasemlou M, Aliheidari N, Fahmi R, Aliabadi SS, Keshavarz B, Cran MJ, et al. Physical, mechanical and barrier properties of corn starch films incorporated with plant essential oils. Carbohydr Polym 2013;98: 1117–26.
99. Song Y, Zhang KF, Wang ZR, Diao FX, Yan YN, Zhang RJ. Coupled thermo-mechanical analysis for plastic thermoforming. Polym Eng Sci 2000;40:1736–46.
100. Dufresne A. Cellulose nanomaterial reinforced polymer nanocomposites. Curr Opin Colloid Interface Sci 2017;29:1–8.
101. Chen W, Yu H, Liu Y, Chen P, Zhang M, Hai Y. Individualization of cellulose nanofibers from wood using high-intensity ultrasonication combined with chemical pretreatments. Carbohydr Polym 2011;83:1804–11.

Index

https://doi.org/10.1515/9783110773606-024

Bei Fragen zur Produktsicherheit wenden Sie sich bitte an:
If you have any questions regarding product safety,
please contact:

Walter de Gruyter GmbH
Genthiner Straße 13
10785 Berlin
productsafety@degruyterbrill.com